Lecture Notes in Computer Science

T0230045

Vol. 352: J. Díaz, F. Orejas (Eds.), TAPSOFT 89. Volume 2. Proceedings, 1989. X, 389 pages. 1989.

Vol. 353: S. Hölldobler, Foundations of Equational Logic Programming. X, 250 pages. 1989. (Subseries LNAI).

Vol. 354: J.W. de Bakker, W.-P. de Roever, G. Rozenberg (Eds.), Linear Time, Branching Time and Partial Order in Logics and Models for Concurrency. VIII, 713 pages. 1989.

Vol. 355: N. Dershowitz (Ed.), Rewriting Techniques and Applications. Proceedings, 1989. VII, 579 pages. 1989.

Vol. 356: L. Huguet, A. Poli (Eds.), Applied Algebra, Algebraic Algorithms and Error-Correcting Codes. Proceedings, 1987. VI, 417 pages. 1989.

Vol. 357: T. Mora (Ed.), Applied Algebra, Algebraic Algorithms and Error-Correcting Codes. Proceedings, 1988. IX, 481 pages. 1989.

Vol. 358: P. Gianni (Ed.), Symbolic and Algebraic Computation. Proceedings, 1988. XI, 545 pages. 1989.

Vol. 359: D. Gawlick, M. Haynie, A. Reuter (Eds.), High Performance Transaction Systems. Proceedings, 1987. XII, 329 pages. 1989.

Vol. 360: H. Maurer (Ed.), Computer Assisted Learning – ICCAL '89. Proceedings, 1989. VII, 642 pages. 1989.

Vol. 361: S. Abiteboul, P.C. Fischer, H.-J. Schek (Eds.), Nested Relations and Complex Objects in Databases. VI, 323 pages. 1989.

Vol. 362: B. Lisper, Synthesizing Synchronous Systems by Static Scheduling in Space-Time. VI, 263 pages. 1989.

Vol. 363: A.R. Meyer, M.A. Taitslin (Eds.), Logic at Botik '89. Proceedings, 1989. X, 289 pages. 1989.

Vol. 364: J. Demetrovics, B. Thalheim (Eds.), MFDBS 89. Proceedings, 1989. VI, 428 pages. 1989.

Vol. 365: E. Odijk, M. Rem, J.-C. Syre (Eds.), PARLE '89. Parallel Architectures and Languages Europe. Volume I. Proceedings, 1989. XIII, 478 pages. 1989.

Vol. 366: E. Odijk, M. Rem, J.-C. Syre (Eds.), PARLE '89. Parallel Architectures and Languages Europe. Volume II. Proceedings, 1989. XIII, 442 pages. 1989.

Vol. 367: W. Litwin, H.-J. Schek (Eds.), Foundations of Data Organization and Algorithms. Proceedings, 1989. VIII, 531 pages. 1989.

Vol. 368: H. Boral, P. Faudemay (Eds.), IWDM '89, Database Machines. Proceedings, 1989. VI, 387 pages. 1989.

Vol. 369: D. Taubner, Finite Representations of CCS and TCSP Programs by Automata and Petri Nets. X. 168 pages. 1989.

Vol. 370: Ch. Meinel, Modified Branching Programs and Their Computational Power. VI, 132 pages. 1989.

Vol. 371: D. Hammer (Ed.), Compiler Compilers and High Speed Compilation. Proceedings, 1988. VI, 242 pages. 1989.

Vol. 372: G. Ausiello, M. Dezani-Ciancaglini, S. Ronchi Della Rocca (Eds.), Automata, Languages and Programming. Proceedings, 1989. XI, 788 pages. 1989.

Vol. 373: T. Theoharis, Algorithms for Parallel Polygon Rendering. VIII, 147 pages. 1989.

Vol. 374: K.A. Robbins, S. Robbins, The Cray X-MP/Model 24. VI, 165 pages. 1989.

Vol. 375: J.L.A. van de Snepscheut (Ed.), Mathematics of Program Construction. Proceedings, 1989. VI, 421 pages. 1989.

Vol. 376: N.E. Gibbs (Ed.), Software Engineering Education. Proceedings, 1989. VII, 312 pages. 1989.

Vol. 377: M. Gross, D. Perrin (Eds.), Electronic Dictionaries and Automata in Computational Linguistics. Proceedings, 1987. V, 110 pages. 1989.

Vol. 378: J.H. Davenport (Ed.), EUROCAL '87. Proceedings, 1987. VIII, 499 pages. 1989.

Vol. 379: [...]dations of Compu[...] 1989.

Vol. 380: [...]ntals of Computat[...]

Vol. 381: J. Dassow, J. Kelemen (Eds.), Machines, Languages, and Complexity. Proceedings, 1988. VI, 244 pages. 1989.

Vol. 382: F. Dehne, J.-R. Sack, N. Santoro (Eds.), Algorithms and Data Structures. WADS '89. Proceedings, 1989. IX, 592 pages. 1989.

Vol. 383: K. Furukawa, H. Tanaka, T. Fujisaki (Eds.), Logic Programming '88. Proceedings, 1988. VII, 251 pages. 1989 (Subseries LNAI).

Vol. 384: G.A. van Zee, J.G.G. van de Vorst (Eds.), Parallel Computing 1988. Proceedings, 1988. V, 135 pages. 1989.

Vol. 385: E. Börger, H. Kleine Büning, M.M. Richter (Eds.), CSL '88. Proceedings, 1988. VI, 399 pages. 1989.

Vol. 386: J.E. Pin (Ed.), Formal Properties of Finite Automata and Applications. Proceedings, 1988. VIII, 260 pages. 1989.

Vol. 387: C. Ghezzi, J.A. McDermid (Eds.), ESEC '89. 2nd European Software Engineering Conference. Proceedings, 1989. VI, 496 pages. 1989.

Vol. 388: G. Cohen, J. Wolfmann (Eds.), Coding Theory and Applications. Proceedings, 1988. IX, 329 pages. 1989.

Vol. 389: D.H. Pitt, D.E. Rydeheard, P. Dybjer, A.M. Pitts, A. Poigné (Eds.), Category Theory and Computer Science. Proceedings, 1989. VI, 365 pages. 1989.

Vol. 390: J.P. Martins, E.M. Morgado (Eds.), EPIA 89. Proceedings, 1989. XII, 400 pages. 1989 (Subseries LNAI).

Vol. 391: J.-D. Boissonnat, J.-P. Laumond (Eds.), Geometry and Robotics. Proceedings, 1988. VI, 413 pages. 1989.

Vol. 392: J.-C. Bermond, M. Raynal (Eds.), Distributed Algorithms. Proceedings, 1989. VI, 315 pages. 1989.

Vol. 393: H. Ehrig, H. Herrlich, H.-J. Kreowski, G. Preuß (Eds.), Categorical Methods in Computer Science. VI, 350 pages. 1989.

Vol. 394: M. Wirsing, J.A. Bergstra (Eds.), Algebraic Methods: Theory, Tools and Applications. VI, 558 pages. 1989.

Vol. 395: M. Schmidt-Schauß, Computational Aspects of an Order-Sorted Logic with Term Declarations. VIII, 171 pages. 1989 (Subseries LNAI).

Vol. 396: T.A. Berson, T. Beth (Eds.), Local Area Network Security. Proceedings, 1989. IX, 152 pages. 1989.

Vol. 397: K.P. Jantke (Ed.), Analogical and Inductive Inference. Proceedings, 1989. IX, 338 pages. 1989 (Subseries LNAI).

Vol. 398: B. Banieqbal, H. Barringer, A. Pnueli (Eds.), Temporal Logic in Specification. Proceedings, 1987. VI, 448 pages. 1989.

Vol. 399: V. Cantoni, R. Creutzburg, S. Levialdi, G. Wolf (Eds.), Recent Issues in Pattern Analysis and Recognition. VII, 400 pages. 1989.

Vol. 400: R. Klein, Concrete and Abstract Voronoi Diagrams. IV, 167 pages. 1989.

Vol. 401: H. Djidjev (Ed.), Optimal Algorithms. Proceedings, 1989. VI, 308 pages. 1989.

Vol. 402: T.P. Bagchi, V.K. Chaudhri, Interactive Relational Database Design. XI, 186 pages. 1989.

Vol. 403: S. Goldwasser (Ed.), Advances in Cryptology – CRYPTO '88. Proceedings, 1988. XI, 591 pages. 1990.

Vol. 404: J. Beer, Concepts, Design, and Performance Analysis of a Parallel Prolog Machine. VI, 128 pages. 1989.

Vol. 405: C.E. Veni Madhavan (Ed.), Foundations of Software Technology and Theoretical Computer Science. Proceedings, 1989. VIII, 339 pages. 1989.

Vol. 406: C.J. Barter, M.J. Brooks (Eds.), AI '88. Proceedings, 1988. VIII, 463 pages. 1990 (Subseries LNAI).

Vol. 407: J. Sifakis (Ed.), Automatic Verification Methods for Finite State Systems. Proceedings, 1989. VII, 382 pages. 1990.

Lecture Notes in Computer Science

Edited by G. Goos and J. Hartmanis

447

J.R. Gilbert   R. Karlsson   (Eds.)

SWAT 90

2nd Scandinavian Workshop on Algorithm Theory
Bergen, Sweden, July 11–14, 1990
Proceedings

Springer-Verlag

Berlin Heidelberg New York London Paris Tokyo Hong Kong

Editors

John R. Gilbert
Xerox Palo Alto Research Center and
University of Bergen
Xerox PARC, 3333 Coyote Hill Road
Palo Alto, California 94304, USA

Rolf Karlsson
Department of Computer Science
Lund University
Box 118, S-221 00 Lund, Sweden

CR Subject Classification (1987): E.1–2, F.1–2, G.2–3

ISBN 3-540-52846-6 Springer-Verlag Berlin Heidelberg New York
ISBN 0-387-52846-6 Springer-Verlag New York Berlin Heidelberg

Printing and binding: Druckhaus Beltz, Hemsbach/Bergstr.
2145/3140-543210 – Printed on acid-free paper

FOREWORD

The papers in this volume were presented at the Second Scandinavian Workshop on Algorithm Theory held on July 11–14, 1990, in Bergen, Norway. The workshop, which continues the tradition of SWAT 88 and WADS 89, is intended as a forum for researchers in the area of design and analysis of algorithms. The call for papers sought contributions on original research on algorithms and data structures, in all areas, including combinatorics, computational geometry, parallel computing, and graph theory. There were 61 papers submitted, of which the program committee selected 34 for presentation. In addition, invited lectures were presented by Juris Hartmanis (Structural complexity theory: recent surprises), Robert Tarjan (New themes in algorithm design), and David Johnson (Data structures for traveling salesmen). This proceedings includes the first invited paper and an abstract of the third.

The organizing committee consisted of Bengt Aspvall, chairman (University of Bergen), Hjálmtýr Hafsteinsson (University of Iceland), Rolf Karlsson (Lund University), Erik M. Schmidt (University of Aarhus), and Esko Ukkonen (University of Helsinki).

The program committee consisted of Svante Carlsson (Lund University), John Gilbert, chairman (Xerox PARC and University of Bergen), Johan Håstad (Royal Institute of Technology, Stockholm), Thomas Lengauer (University of Paderborn), Andrzej Lingas (Lund University), Olli Nevalainen (University of Turku), Andrzej Proskurowski (University of Oregon), Jörg Sack (Carleton University), Raimund Seidel (University of California, Berkeley), and Jeffrey Vitter (Brown University).

The program committee wishes to thank the following subreferees who aided in evaluating the papers: Arne Andersson, Marshall Bern, Jingsen Chen, Juergen Doenhardt, Anders Edenbrandt, Joerg Heistermann, Franz Hoefting, Jyrki Katajainen, Christos Levcopoulos, Christer Mattsson, Bengt Nilsson, Ola Petersson, Rolf Mueller, Egon Wanke, Charlotte Wieners-Lummer, and everybody else who helped with this process. Thanks are also due Katie Hendrickson for administrative service beyond the call of duty. We are very grateful to the Swedish Natural Science Research Council, the Norwegian Research Council for Science and Humanities, the Meltzer Foundation, University of Bergen, Alliant Computer Systems AB, the Bergen High Technology Center Limited, for sponsoring the workshop, and to the Xerox Palo Alto Research Center for financial support of the program committee's work.

Palo Alto and Lund, April 1990   John Gilbert
                                                                            Rolf Karlsson

TABLE OF CONTENTS

Structural complexity theory: Recent surprises (Invited)
J. Hartmanis, R. Chang, D. Ranjan, and P. Rohatgi 1

Approximating maximum independent sets by excluding subgraphs
R. Boppana and M.M. Halldórsson 13

Generating sparse spanners for weighted graphs
I. Althöfer, G. Das, D. Dobkin, and D. Joseph 26

Finding the *k* smallest spanning trees
D. Eppstein 38

The file distribution problem for processor networks
G. Kant and J. van Leeuwen 48

Translating polygons with applications to hidden surface removal
M. de Berg 60

Output-sensitive generation of the perspective view of isothetic parallelepipeds
(extended abstract)
F.P. Preparata, J.S. Vitter, and M. Yvinec 71

Graphics in Flatland revisited
M. Pocchiola 85

The visibility diagram: A data structure for visibility problems and motion planning
G. Vegter 97

Fast updating of well-balanced trees
A. Andersson and T.W. Lai 111

How to update a balanced binary tree with a constant number of rotations
T. Ottmann and D. Wood 122

Ranking trees generated by rotations
S.W. Bent 132

Expected behaviour analysis of AVL trees
R. Baeza-Yates, G.H. Gonnet, and N. Ziviani 143

Analysis of the expected search cost in skip lists
T. Papadakis, J.I. Munro, and P.V. Poblete 160

Lower bounds for monotonic list labeling
P.F. Dietz and J. Zhang 173

Sorting shuffled monotone sequences
C. Levcopoulos and O. Petersson 181

A faster parallel algorithm for a matrix searching problem
M.J. Atallah 192

A rectilinear Steiner minimal tree algorithm for convex point sets
D.S. Richards and J.S. Salowe 201

Finding shortest paths in the presence of orthogonal obstacles using a combined L_1 and link metric
M.T. de Berg, M.J. van Kreveld, B.J. Nilsson, and M.H. Overmars 213

Input-sensitive compliant motion in the plane
J. Friedman, J. Hershberger, and J. Snoeyink 225

Fast algorithms for greedy triangulation
C. Levcopoulos and A. Lingas 238

Star unfolding of a polytope with applications (extended abstract)
P.K. Agarwal, B. Aronov, J. O'Rourke, and C.A. Schevon 251

Space-sweep algorithms for parametric optimization (extended abstract)
D. Fernández-Baca 264

Approximating finite weighted point sets by hyperplanes
N.M. Korneenko and H. Martini 276

Data structures for traveling salesmen (Invited, abstract)
D.S. Johnson 287

Efficient parallel algorithms for shortest paths in planar graphs
G.E. Pantziou, P.G. Spirakis, and C.D. Zaroliagis 288

The pathwidth and treewidth of cographs
H.L. Bodlaender and R.H. Möhring 301

Canonical representations of partial 2- and 3-trees
S. Arnborg and A. Proskurowski 310

On matroids and hierarchical graphs
D. Fernández-Baca and M.A. Williams 320

Fast algorithms for two-dimensional and multiple pattern matching (preliminary version)
R. Baeza-Yates and M. Régnier 332

Boyer-Moore approach to approximate string matching (extended abstract)
J. Tarhio and E. Ukkonen 348

Complete problems with L-samplable distributions
P. Grape 360

Upper envelope onion peeling
J. Hershberger 368

Applications of a semi-dynamic convex hull algorithm
J. Hershberger and S. Suri 380

Intersection queries in sets of disks
M. van Kreveld, M. Overmars, and P.K. Agarwal 393

Dynamic partition trees
H. Schipper and M.H. Overmars 404

Structural Complexity Theory: Recent Surprises[†]

Juris Hartmanis   Richard Chang[‡]   Desh Ranjan   Pankaj Rohatgi

Department of Computer Science
Cornell University
Ithaca, New York 14853

Abstract

This paper reviews the impact of some recent results on the research paradigms in structural complexity theory.

> *On a paper submitted by a physicist colleague:*
> *"This isn't right. This isn't even wrong."*
> —*Wolfgang Pauli*

1   Introduction

Computational complexity theory studies the quantitative laws which govern computing. It seeks a comprehensive classification of problems by their intrinsic difficulty and an understanding of what makes these problems hard to compute. The key concept in classifying the computational complexity of problems is the complexity class which consists of all the problems solvable on a given computer model within a given resource bound.

Structural complexity theory is primarily concerned with the relations among various complexity classes and the internal structure of these classes. Figure 1 shows some major complexity classes. Although much is known about the structure of these classes, there have not been any results which separate any of the classes between P and PSPACE. We believe that all these classes are different and regard the problem of proving the exact relationship between these classes as the Grand Challenge of complexity theory.

The awareness of the importance of P, NP, PSPACE, etc, has lead to a broad investigation of these classes and to the use of relativization. Almost all of the major results in recursive function theory also hold in relativized worlds. Quite the contrary happens in complexity theory. It was shown in 1975 [BGS75] that there exist oracles A and B such that

$$P^A = NP^A \text{ and } P^B \neq NP^B.$$

[†]This research was supported in part by NSF Research Grant CCR 88-23053.
[‡]Supported in part by an IBM Graduate Fellowship.

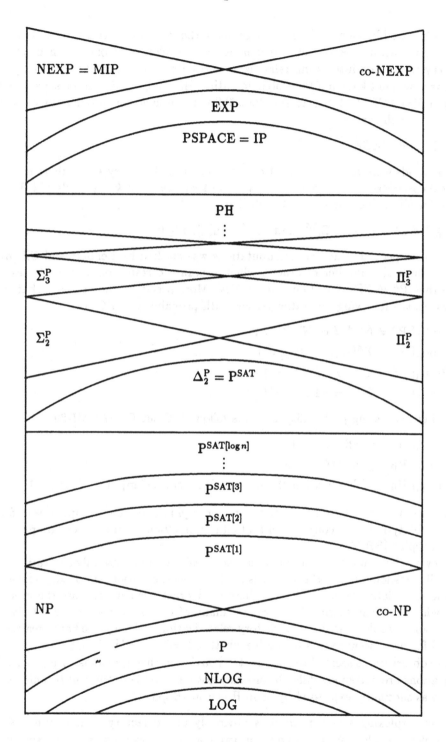

Figure 1: Some standard complexity classes

This was followed by an extensive investigation of the structure of complexity classes under relativization. An impressive set of techniques was developed for oracle constructions and some very subtle and interesting relativization results were obtained. For example, for a long time it was not known if the Polynomial-time Hierarchy (PH) can be separated by oracles from PSPACE. In 1985, A. Yao [Yao85] finally resolved this problem by constructing an oracle A, such that

$$PH^A \neq PSPACE^A.$$

These methods were refined by Ko [Ko88] to show that for every $k \geq 0$ there is an oracle which collapses PH to exactly the kth level and keeps the first $k-1$ levels of PH distinct. That is, for all k, there exists an A such that

$$\Sigma_0^{P,A} \neq \Sigma_1^{P,A} \neq \cdots \neq \Sigma_k^{P,A} \text{ and } \Sigma_k^{P,A} = \Sigma_{k+i}^{P,A}, \; i \geq 0.$$

Another aspect of relativized computations was studied by Bennett and Gill who decided to count the number of oracles which separate certain complexity classes. They showed that $P^A \neq NP^A$ for almost all oracles. More precisely, they showed that for random oracles the following separations occur with probability 1 [BG81]:

$$Prob_A[\; P^A \neq NP^A \neq \text{co-}NP^A \;] = 1$$
$$Prob_A[\; SPACE^A[\log n] \neq P^A \;] = 1$$
$$Prob_A[\; PSPACE^A \neq EXP^A \;] = 1$$
$$Prob_A[\; P^A = RP^A = BPP^A \;] = 1.$$

Many other interesting probability 1 results followed [Cai86,Cai87,KMR89]:

$$Prob_A[\; BH^A \text{ is infinite} \;] = 1$$
$$Prob_A[\; PH^A \subsetneq PSPACE^A \;] = 1$$
$$Prob_A[\; \text{Under relativization the Berman-Hartmanis Conjecture fails} \;] = 1.$$

The last result asserts that there exist non-isomorphic many-one complete sets for NP^A with probability 1. It was conjectured that all NP many-one complete sets are polynomial-time isomorphic [BH77].

Surveying the rich set of relativization results, we can make several observations. First, almost all questions about the relationship between the major complexity classes have contradictory relativizations. That is, there exist oracles which separate the classes and oracles which collapse them. Furthermore, many of our proof techniques relativize and cannot resolve problems with contradictory relativizations. Finally, we have unsuccessfully struggled for over twenty years to resolve whether P =? NP =? PSPACE.

These observations seemed to support the conviction that problems with contradictory relativizations are extremely difficult and may not be solvable by current techniques. This opinion was succinctly expressed by John Hopcroft [Hop84]:

> This perplexing state of affairs is obviously unsatisfactory as it stands. No problem that has been relativized in two conflicting ways has yet been solved, and this fact is generally taken as evidence that the solutions of such problems are beyond the current state of mathematics.

How should complexity theorists remedy "this perplexing state of affairs"? In one approach, we assume as a working hypothesis that PH has infinitely many levels. Thus, any assumption which would imply that PH is finite is deemed incorrect. For example, Karp, Lipton and Sipser [KL80] showed that if NP $\subseteq$ P/poly, then PH collapses to Σ_2^P. So, we believe that SAT does not have polynomial sized circuits. Similarly, we believe that the Turing-complete and many-one complete sets for NP are not sparse, because Mahaney [Mah82] showed that these conditions would collapse PH. One can even show that for any $k \geq 0$, $P^{SAT[k]} = P^{SAT[k+1]}$ implies that PH is finite [Kad88]. Hence, we believe that $P^{SAT[k]} \neq P^{SAT[k+1]}$ for all $k \geq 0$. Thus, if the Polynomial Hierarchy is indeed infinite, we can describe many aspects of the computational complexity of NP.

A second approach uses random oracles. Since the probability 1 relativization results agree with what complexity theorists believe to be true in the base case and since random oracles have no particular structure of their own, it seemed that the probability 1 behavior of complexity classes should be the same as the base case behavior. This lead Bennett and Gill to postulate the Random Oracle Hypothesis [BG81] which essentially states that whatever holds with probability 1 for relativized complexity classes holds in the unrelativized case—i.e., the real world.

In the following, we will discuss a set of results about interactive proofs which provide dramatic counterexamples to the belief that problems with contradictory relativizations cannot be resolved with known techniques. They also add a striking new counterexample against the Random Oracle Hypothesis. There have been other counterexamples in the literature [Kur82,Har85], but none as compelling. Thus, contradictory relativizations should no longer be viewed as strong evidence that a problem is beyond our grasp. We hope that these results will encourage complexity theorists to renew the attack on problems with contradictory relativizations.

2 A Review of IP

The class IP is the set of languages that have interactive proofs or protocols. IP was first defined as way to generalize NP. NP can be characterized as being precisely those languages for which one can present a polynomially long proof to certify that the input string is in the language. Moreover, the proof can be checked in polynomial time. It is this idea of presenting and checking the proof that the definition of IP generalizes.

Is there a way of giving convincing evidence that the input string is in a language without showing the whole proof to a verifier? Clearly, if we do not give a complete proof to a verifier which does not have the power or the time to generate and check a proof, then we cannot expect the verifier to be completely convinced. This leads us to a very fascinating problem: *how can the verifier be convinced with high probability that there is a proof? and how rapidly can this be done?*

This problem has been formulated and extensively studied in terms of *interactive protocols* [Gol89]. Informally, an interactive protocol consists of a *Prover* and a *Verifier*. The Prover is an all powerful Turing Machine (TM) and the Verifier is a TM which operates in time polynomial in the length of the input. In addition, the Verifier has a random source (e.g., a fair coin) not visible to the Prover. In the beginning of the interactive protocol the Prover and the Verifier receive the same input string. Then, the Prover tries to convince the Verifier, through a series of queries and answers, that the input string belongs to a given language. The Prover succeeds if the Verifier accepts with probability greater than 2/3. The probability is computed over all possible coin tosses made by the Verifier. However, the Verifier must guard against imposters masquerading as the real Prover. The Verifier must not be convinced to accept a string not in the language with probability greater than 1/3—even if the Prover lies.

Definition IP: Let V be a probabilistic polynomial time TM and let P be an arbitrary TM. P and V share the same input tape and communicate via a communication tape. P and V forms an interactive protocol for a language L if

1. $x \in L \Longrightarrow$ Prob[P-V accepts x] $> \frac{2}{3}$.

2. $x \notin L \Longrightarrow \forall P^*$, Prob[P^*-V accepts x] $< \frac{1}{3}$.

A language L is in IP if there exist P and V which form an interactive protocol for L.

Clearly, IP contains all NP languages, because in polynomial time the Prover can give the Verifier the entire proof. In such a protocol, the Verifier cannot be fooled and never accepts a string not in the language. To illustrate how randomness can generalize the concept of a proof, we look at an interactive protocol for a language not known to be in NP. Consider GNI, the set of pairs of graphs that are not isomorphic. GNI is known to be in co-NP and it is believed not to be in NP. However, GNI does have an interactive protocol [GMW86]. For small graphs, the Verifier can easily determine if the two graphs are not isomorphic. For sufficiently large graphs, the Verifier solicits help from the Prover to show that G_i and G_j are not isomorphic, as follows:

1. The Verifier randomly selects G_i or G_j and a random permutation of the selected graph. This process is independently repeated n times, where n is the number of vertices in G_j. If the graphs do not have the same number of vertices, they are clearly not isomorphic. This sequence of n randomly chosen, randomly permuted graphs is sent to the Prover. Recall that the Prover has not seen the Verifier's random bits. This assumption is not necessary, but simplifies the exposition.

2. The Verifier asks the Prover to determine, for each graph in the sequence, which graph, G_i or G_j, was the one selected. If the Prover answers correctly, then the Verifier accepts.

Suppose the two original graphs are not isomorphic. Then, only one of the original graphs is isomorphic to the permuted graph. The Prover simply answers by picking that graph. If the graphs are isomorphic, then the Prover has at best a 2^{-n} chance of answering all n questions correctly. Thus, the Verifier cannot be fooled often. Therefore, GNI $\in$ IP.

Note that GNI is not known to be complete for co-NP. So, the preceding discussion does not show that co-NP $\subseteq$ IP. For a while, it was believed that co-NP is not contained in IP, because there are oracle worlds where co-NP $\not\subseteq$ IP [FS88]. In fact, the computational power of interactive protocols was not fully appreciated until Lund, Fortnow, Karloff and Nisan [LFKN89] showed that IP actually contains the entire Polynomial Hierarchy. This result then led Shamir [Sha89] to completely characterize IP by showing that

IP = PSPACE.

Very recently, Babai, Fortnow and Lund [BFL90] characterized the computational power of multi-prover interactive protocols

MIP = NEXP.

In both cases, it is interesting to see that interactive proof systems provide alternative definitions of classic complexity classes. Thus, they fit very nicely with the overall classification of feasible computations. Furthermore, both of these problems have contradictory relativizations [FS88]. That is, there exist oracles A and B such that

$$IP^A = PSPACE^A \text{ and } IP^B \neq PSPACE^B,$$

and similarly for the multi-prover case. Thus, these results provide the first *natural* counterexamples to the belief that problems with contradictory relativizations are beyond our proof techniques.

3 The Random Oracle Hypothesis ...

In this section we observe that the proof of IP = PSPACE does not relativize and show that for almost all oracles the two relativized classes differ:

$$\text{Prob}_A[\ IP^A \neq PSPACE^A\ ] = 1.$$

It is easily seen that

$$IP^{PSPACE} = PSPACE^{PSPACE}$$

and using standard methods [BGS75] one can construct an A such that

$$IP^A \neq PSPACE^A.$$

This shows that the IP =? PSPACE problem has contradictory relativizations and that the IP = PSPACE proof does not relativize. Similarly, we can see that the MIP =? NEXP problem has contradictory relativizations. In the following we show that these theorems also supply counterexamples to the Random Oracle Hypothesis.

3.1 ...it isn't right ...

Theorem 1 $\text{Prob}_A[\ PSPACE^A \not\subseteq IP^A\ ] = 1.$

Proof: For each oracle A, define the language:

$$\mathcal{L}(A) = \{1^n\ :\ |A^{=n}|\ \text{is odd}\}.$$

Clearly, $\mathcal{L}(A) \in PSPACE^A$ for all A. We will show that for any verifier V, the set

$$C = \{A : V^A \text{ is an IP verifier for } \mathcal{L}(A)\}$$

has measure zero. (C stands for "correct".) Let $p(n)$ be an upper bound on the running time on V. Fix n to be large enough so that $p(n) < \frac{1}{300}2^n$ and let $\mathcal{P}$ be the power set of $\Sigma^{\leq p(n)}$. Since $\mu(C) \leq |C \cap \mathcal{P}|/|\mathcal{P}|$, we can concentrate on counting the finite sets in $\mathcal{P}$.

Define the following classes of oracles:

$$\text{ODD} = \{A \in \mathcal{P} \;:\; |A^{=n}| \text{ is odd}\},$$
$$\text{EVEN} = \{A \in \mathcal{P} \;:\; |A^{=n}| \text{ is even}\},$$
$$S_0 = \text{ODD} \cap C,$$
$$S_1 = \text{EVEN} \cap \overline{C}.$$

ODD is the set of oracles where V should accept 1^n; and EVEN the set where V should reject. S_0 is the set of oracles where V should accept and does; S_1 the set oracles where V should reject but does not.

The idea of the proof is simple. We show that $0.99 \cdot |S_0|$ is a lower bound on the size of S_1. This means that the number of oracles where V messes up is bounded below by roughly one-half of all the oracles in $\mathcal{P}$, because

$$
\begin{aligned}
|\mathcal{P} \cap \overline{C}| &= |\text{ODD} \cap \overline{C}| + |S_1| \\
&\geq |\text{ODD} \cap \overline{C}| + 0.99 \cdot |S_0| \\
&\geq 0.99 \cdot (|\text{ODD} \cap \overline{C}| + |\text{ODD} \cap C|) \\
&= 0.495 \cdot |\mathcal{P}|.
\end{aligned}
$$

Thus, $|\mathcal{P} \cap C| < 0.51|\mathcal{P}|$ which tells us that the probability (over A) that V^A is part of an IP protocol for $\mathcal{L}(A)$ is less than 0.51. By repeating this proof for larger and larger n, we can drive the probability down to zero.

So, all we need to show now is that $0.99 \cdot |S_0|$ is a lower bound on the size of S_1. For every oracle $A \in S_0$, there is some prover P, which convinces V^A to accept in two thirds of the computation paths. From P and A we can construct many oracles A' such that $A' \in \text{EVEN}$ but P still convinces $V^{A'}$ to accept in at least one third of the computation paths. (This implies $A' \in S_1$.) To construct such an A' we simply add or remove a single string z from A. If P convinces $V^{A'}$ to accept 1^n in less than one third of the computation paths, then z must have appeared in more than one third of the computation paths of the $V^A(1^n)$ computation. However, the number of strings queried in at least one third of the computation paths is bounded by $3p(n)$. So, we can construct $2^n - 3p(n) \geq 0.99 \cdot 2^n$ oracles in S_1.

Formally, define the function $\eta : \text{ODD} \to \text{EVEN}$ by

$$
\eta(A, z) = \begin{cases} A - \{z\} & \text{if } z \in A \\ A \cup \{z\} & \text{if } z \notin A. \end{cases}
$$

For each $A \in S_0$ define the set R_A to be

$$R_A = \{z \;:\; |z| = n, \; \eta(A, z) \in S_1\}.$$

We claim that $|R_A| \geq 0.99 \cdot 2^n$. To see this, let $r = 2^n - |R_A|$. As mentioned before, if $|z| = n$ and $z \notin R_A$, then z must be queried in at least one third of the computation paths of $V^{\eta(A,z)}$. So, the total number of queries to strings not in R_A is at least $\frac{1}{3}r2^{p(n)}$.

However, the total number of queries in the entire computation tree is at most $p(n)2^{p(n)}$. Thus, $r \leq 3p(n)$. Since we assumed that $p(n) \leq \frac{1}{300}2^n$, $|R_A| \geq 0.99 \cdot 2^n$.

Now, define

$$D = \bigcup_{A \in S_0} \{(A, z) \mid z \in R_A\}.$$

Clearly, $|D| \geq 0.99 \cdot 2^n |S_0|$. For each $(A, z) \in D$, $\eta(A, z) \in S_1$. However, η is not a one to one function, so $|D|$ is not a lower bound on $|S_1|$. In any case, notice that $|\eta^{-1}(A')| \leq 2^n$, because $\eta(A, z) = A'$ implies that A and A' differ at only one string of length n and there are only 2^n strings on length n. Now, let's consider the total number of distinct oracles in S_1 constructed.

$$D' = \{A' : \exists (A, z) \in D, \eta(A, z) = A'\}.$$

An easy lower bound on $|D'|$ is $|D|/2^n$. So, we get $|D'| \geq 0.99 \cdot |S_0|$. Since $D' \subseteq S_1$, $|S_1| \geq 0.99 \cdot |S_0|$. □

Using standard techniques [BGS75,BG81,FS88], the proof of Theorem 1 can be modified to yield the following theorems. For the sake of brevity, we omit the proofs.

Theorem 2 $\text{Prob}_A[\text{ co-NP}^A \not\subseteq \text{IP}^A] = 1$.

Theorem 3 $\text{Prob}_A[\text{ NEXP}^A \not\subseteq \text{MIP}^A] = 1$.

3.2 ...it isn't even wrong.

The IP = PSPACE and MIP = NEXP results provided natural examples against the Random Oracle Hypothesis. To give a more complete understanding of the behavior of these classes with random oracles, we define a less restrictive acceptance criterion for interactive protocols and denote the class of such languages by IPP. (See also [Pap83]). We show that

$$\forall A, \text{IPP}^A = \text{PSPACE}^A.$$

Using the theorem in the previous section, we can provide both an example and a counterexample to the Random Oracle Hypothesis.

$$\text{Prob}_A[\text{ IP}^A \neq \text{PSPACE}^A] = 1 \quad \text{and} \quad \text{Prob}_A[\text{ IPP}^A = \text{PSPACE}^A] = 1.$$

This severely damages the already battered hypothesis because it shows that the Random Oracle Hypothesis is sensitive to small changes in the definition of complexity classes. Thus, it cannot be used to predict what happens in the real world.

Definition IPP: Let V be a probabilistic polynomial time machine and let P be an arbitrary TM. P and V share the same input tape and they communicate via a communication tape. V forms an unbounded interactive protocol for a language L if

1. $x \in L \implies \text{Prob}[\text{ P-V on } x \text{ accept }] > \frac{1}{2}$.

2. $x \notin L \implies \forall P^*, \text{Prob}[\text{ P}^*\text{-V on } x \text{ accept }] < \frac{1}{2}$.

A language L is said to be in the class IPP if it has an unbounded interactive protocol.

procedure CHECKCOMP(C_1, C_2, s) ;

{ This procedure tries to detect if M^A can reach configuration C_2 from configuration C_1 in s steps. }

begin
    if $s = 1$ **then**
        { This may involve querying the oracle. }
        if $C_1 \rightarrow C_2$ in one step **then**
            accept
        else
            reject
    else
        Ask the prover for the middle configuration C_3 between C_1 and C_2.
        Toss a coin.
        if the coin toss is heads **then**
            CHECKCOMP$(C_1, C_3, s/2)$
        else
            CHECKCOMP$(C_3, C_2, s/2)$
end { procedure }

Figure 2: Pseudo-code for procedure CHECKCOMP

Theorem 4 *For all oracles A,* IPP^A = PSPACE^A.

Proof:
IPP^A $\subseteq$ PSPACE^A: Let L be a language in IPP^A. Using standard techniques [GS86, Con87], it can be shown that IPP^A with private coins is the same as IPP^A with *public* coins. Protocols with public coins are easy to simulate because the responses from the prover can be guessed. In fact, a PSPACE^A machine can traverse the entire probabilistic computation tree and compute the acceptance probability. Thus, a PSPACE^A machine can determine if there is a prover which makes the verifier accept in more than half of the computation paths. So, IPP^A $\subseteq$ PSPACE^A.

PSPACE^A $\subseteq$ IPP^A: This proof is similar to the proof that NP $\subseteq$ PP [Gil77]. Let L be a language in PSPACE^A. Then there is a machine M^A accepting L which runs in space $p(n)$ and halts in exactly $2^{q(n)}$ steps for some polynomials p and q. We claim that the following verifier V forms an unbounded interactive protocol for L. Given input x where $|x| = n$ do the following:

1. Toss $q(n) + 1$ coins. If all of them are "heads", then goto step 3.

2. Toss another coin. If "heads" then reject; otherwise, goto step 3.

3. Let I and F be the unique initial and final configurations of $M^A(x)$.
   Run CHECKCOMP$(I, F, 2^{q(n)})$.

In order to prove the correctness of the protocol, we need the following lemma.

Lemma: Let $\text{WRONG}(C_1, C_2, s)$ be the proposition that configuration C_2 does not follow from configuration C_1 in exactly s steps. Then,

$$\text{WRONG}(C_1, C_2, s) \Longrightarrow$$
$$\forall u, \ 0 \le u \le s, \ C_3, \ \text{WRONG}(C_1, C_3, u) \lor \text{WRONG}(C_3, C_2, s-u).$$

Now, if $x \in L$, then with the prover that always tells the truth, the probability of acceptance is the same as the probability that the verifier reaches step 3, which is greater than 1/2. On the other hand, if $x \notin L$, then $\text{Prob}[\, V \text{ rejects } x \mid V \text{ reaches step 3}\,] \ge 2^{-q(n)}$. (This follows from the lemma.) In this case,

$$
\begin{aligned}
\text{Prob}\,[\, V \text{ rejects } x\,] \ &= \ \text{Prob}[\, V \text{ rejects } x \text{ at step 2}\,] \\
&\quad + \text{Prob}[\, V \text{ reaches step 3}\,] \\
&\quad\ \cdot \text{Prob}[\, V \text{ rejects } x \mid V \text{ reaches step 3}\,] \\
&\ge \ (\tfrac{1}{2} - 2^{-q(n)-2}) + (\tfrac{1}{2} + 2^{-q(n)-2}) \cdot 2^{-q(n)} \\
&> \ \tfrac{1}{2}.
\end{aligned}
$$

Therefore, $\text{Prob}[\, V \text{ accepts } x\,] = 1 - \text{Prob}[\, V \text{ rejects } x\,] < \tfrac{1}{2}$. $\qquad\square$

4 Conclusion

We have shown that probability 1 results do not reliably predict the base case behavior of complexity classes. On the other hand, the meaning of probability 1 results for random oracles needs to be clarified and remains an interesting problem. It would be very interesting to know if there are identifiable problem classes for which the probability 1 results do point in the right direction.

In addition, we would like to note that the IP = PSPACE and MIP = NEXP results demonstrated *equality* in the base case. In many other problems with contradictory relativizations, we expect the unrelativized complexity classes to be different (e.g., we expect that P $\ne$ NP $\ne$ PSPACE, etc). The next big challenge for complexity theorists is to resolve one of these problems and *separate*—if not P and NP—any two classes with contradictory relativizations.

References

[BFL90] L. Babai, L. Fortnow, and C. Lund. Non-deterministic exponential time has two-prover interactive protocols. Technical Report 90-03, Department of Computer Science, University of Chicago, March 1990.

[BG81]   C. Bennett and J. Gill. Relative to a random oracle A, $P^A \ne NP^A \ne$ co-NP^A with probability 1. *SIAM Journal on Computing*, 10(1):96–113, February 1981.

[BGS75]   T. Baker, J. Gill, and R. Solovay. Relativizations of the P =? NP question. *SIAM Journal on Computing*, 4(4):431–442, December 1975.

[BH77]   L. Berman and J. Hartmanis. On isomorphism and density of NP and other complete sets. *SIAM Journal on Computing*, 6:305–322, June 1977.

[Cai86]   J. Cai. With probability one, a random oracle separates PSPACE from the polynomial-time hierarchy. In *ACM Symposium on Theory of Computing*, pages 21–29, 1986.

[Cai87]   J. Cai. Probability one separation of the Boolean Hierarchy. In *4th Annual Symposium on Theoretical Aspects of Computer Science*, Springer-Verlag *Lecture Notes in Computer Science # 247*, pages 148–158, 1987.

[Con87]   A. Condon. Computational models of games. Technical Report 87-04-04, Department of Computer Science, University of Washington, 1987.

[FS88]   L. Fortnow and M. Sipser. Are there interactive protocols for co-NP languages? *Information Processing Letters*, 28(5):249–251, August 1988.

[Gil77]   J. Gill. Computational complexity of probabilistic Turing machines. *SIAM Journal on Computing*, 6(4):675–695, December 1977.

[GMW86]   O. Goldreich, S. Micali, and A. Widgerson. Proofs that yield nothing but their validity and a methodology of cryptographic protocol design. In *Proceedings IEEE Symposium on Foundations of Computer Science*, pages 174–187, 1986.

[Gol89]   S. Goldwasser. Interactive proof systems. In *Computational Complexity Theory*, Proceedings of Symposia in Applied Mathematics, Volume 38, pages 108–128. American Mathematical Society, 1989.

[GS86]   S. Goldwasser and M. Sipser. Private coins versus public coins in interactive proof systems. In *ACM Symposium on Theory of Computing*, pages 59–68, 1986.

[Har85]   J. Hartmanis. Solvable problems with conflicting relativizations. *Bulletin of the European Association for Theoretical Computer Science*, 27:40–49, Oct 1985.

[Hop84]   J. E. Hopcroft. Turing machines. *Scientific American*, pages 86–98, May 1984.

[Kad88]   J. Kadin. The polynomial time hierarchy collapses if the Boolean hierarchy collapses. *SIAM Journal on Computing*, 17(6):1263–1282, December 1988.

[KL80]   R. Karp and R. Lipton. Some connections between nonuniform and uniform complexity classes. In *ACM Symposium on Theory of Computing*, pages 302–309, 1980.

[KMR89]   S. A. Kurtz, S. R. Mahaney, and J. S. Royer. The isomorphism conjecture fails relative to a random oracle. In *ACM Symposium on Theory of Computing*, pages 157–166, 1989.

[Ko88]   K. Ko. Relativized polynomial time hierarchies having exactly k levels. In *ACM Symposium on Theory of Computing*, pages 245–253, 1988.

[Kur82]   S. A. Kurtz. On the random oracle hypothesis. In *ACM Symposium on Theory of Computing*, pages 224–230, 1982.

[LFKN89] C. Lund, L. Fortnow, H. Karloff, and N. Nisan. The polynomial-time hierarchy has interactive proofs. Unpublished manuscript, 1989.

[Mah82]   S. Mahaney. Sparse complete sets for NP: Solution of a conjecture of Berman and Hartmanis. *Journal of Computer and System Sciences*, 25(2):130–143, 1982.

[Pap83]   C. H. Papadimitriou. Games against nature. In *Proceedings IEEE Symposium on Foundations of Computer Science*, pages 446–450, 1983.

[Sha89]   A. Shamir. IP = PSPACE. Unpublished manuscript, 1989.

[Yao85]   A. C. Yao. Separating the polynomial-time hierarchy by oracles. In *Proceedings IEEE Symposium on Foundations of Computer Science*, pages 1–10, 1985.

Approximating Maximum Independent Sets by Excluding Subgraphs

Ravi Boppana[1]   and   Magnús M. Halldórsson[2]
Department of Computer Science   Department of Computer Science
New York University   Rutgers University
New York, NY 10012   New Brunswick, NJ 08903
and
Rutgers University

Abstract

An approximation algorithm for the maximum independent set problem is given, improving the best performance guarantee known to $\mathcal{O}(n/(\log n)^2)$. We also obtain the same performance guarantee for graph coloring. The results can be combined into a surprisingly strong *simultaneous* performance guarantee for the clique and coloring problems.

The framework of *subgraph excluding* algorithms is presented. We survey the known approximation algorithms for the independent set (clique), coloring, and vertex cover problems and show how almost all fit into that framework. It is shown that among subgraph excluding algorithms, the ones presented achieve the optimal asymptotic performance guarantees.

1 Introduction

An *independent set* in a graph is a set of vertices with no edges connecting them. The problem of finding an independent set of maximum size is one of the classical $\mathcal{NP}$-hard problems. We consider polynomial time algorithms that find an approximation of guaranteed size. The quality of the approximation is given by the ratio of the size of the maximum independent set to the size of the approximation found, and the largest such ratio over all inputs gives the *performance guarantee* of the algorithm.

A few other problems are closely related to the independent set problem. A *clique* is a set of mutually connected vertices. Since finding the maximum size clique in a graph is equivalent to finding the maximum independent set in the complement of the graph, the clique problem is for our purposes the identical problem.

A *vertex cover* is a set of vertices with the property that every edge in the graph is incident to some vertex in the set. Note that vertices not in a given vertex cover must be independent, hence finding a maximum independent set is equivalent to finding a minimum vertex cover. Approximations of the two problems, however, are widely different.

The third related problem is *graph coloring*, namely finding an assignment of as few colors as possible to the vertices so that no adjacent vertices share the same color. Because the colors induce a partition of the graph into independent sets, the problems of approximating independent set and coloring are closely related.

---

[1]Supported in part by National Science Foundation Grant number CCR-8902522
[2]Supported in part by Center for Discrete Mathematics and Theoretical Computer Science fellowship

The analysis of approximation algorithms for these problems started with Johnson who showed that the greedy algorithm colors k-colorable graphs with $\mathcal{O}(n/\log_k n)$ colors, obtaining a performance guarantee of $\mathcal{O}(n/\log n)$. Several years later, Wigderson [20] introduced an elegant algorithm that colors k-colorable graphs with $\mathcal{O}(kn^{1/(k-1)})$ colors, which, when combined with Johnson's result, yields an $\mathcal{O}(n(\log\log n/\log n)^2)$ performance guarantee. Very recently, Berger and Rompel [3] gave an algorithm that improves on Johnson's idea to obtain an $\mathcal{O}(n/(\log_k n)^2)$ coloring. When combined with Wigderson's method, they obtain an $\mathcal{O}(n(\log\log n/\log n)^3)$ performance guarantee. Finally, Blum has improved the best ratio for small values of k, in particular for 3-coloring from the $\mathcal{O}(\sqrt{n})$ of Wigderson and the $\mathcal{O}(\sqrt{n}/\log n)$ of Berger and Rompel, to $n^{0.4+o(1)}$ [6] and later to $n^{0.375+o(1)}$ [5].

We shall present an efficient graph coloring algorithm that colors k-colorable graphs with $\mathcal{O}(n^{1/(k-1)}/k)$ colors when $k \leq 2\log n$, and $\mathcal{O}(\log n/\log \frac{k}{\log n})$ when $k \geq 2\log n$. The algorithm strictly improves on both Johnson's and Wigderson's method, and can be combined with the method of Berger and Rompel in place of Wigderson's method to improve the performance guarantee slightly (although not asymptotically).

Folklore (see [11, pp. 134] attributed to Gavril) tells us that any maximal matching approximates the minimum vertex cover by a factor of two. This was slightly improved independently by Bar-Yehuda and Even [2], and Monien and Speckenmeyer [17] to a factor of $2 - \Omega(\frac{\log\log n}{\log n})$, but no further improvements have yet been found.

Approximating the independent set problem has seen less success. No approximation algorithm yielding a non-trivial performance guarantee has been found in the literature. One of the main results of this paper is an algorithm that obtains an $\mathcal{O}(n/(\log n)^2)$ performance guarantee for the independent set problem on general graphs, as well as several results on graphs with a high independence number.

For all these problems the optimal approximation ratios are unknown, and the gaps between the upper and lower bounds are large. The vertex cover could possibly have a polynomial time approximation schema, i.e. being approximable within any fixed constant greater than one. The independent set problem might also have a polynomial time approximation schema, while otherwise it cannot be approximated within any fixed constant [11, pp. 146]. Recent results by Berman and Schnitger [4] give evidence that it may not be approximable within anything less than some fixed power of n. Finally, graph coloring cannot be approximated within less than a factor of two (assuming $\mathcal{P} \neq \mathcal{NP}$) [11, pp. 144], and the results of Linial and Vazirani [15] also suggest that some fixed power of n may be the best approximation we can hope for.

We will present lower bounds of a different kind, namely for a fixed class of algorithms, similar to the work of Chvátal [7]. We show how the best known approximation algorithms for these problems revolve around the concept of *excluding subgraphs* and how no algorithm within that framework can do significantly better than the algorithms presented here. The techniques used have a striking connection with graph Ramsey theory and the Ramsey-theoretic results may be of independent interest.

Some graph notation: For an undirected graph $G = (V, E)$, $|G|$ is the *order* of G or the number of vertices, $\alpha(G)$ is the *independence number* of the graph or the size of the largest independent set, $i(G)$ is the *independence ratio* or the independence number divided by the order of the graph, $cl(G)$ is the *clique number*, and $\chi(G)$ is the *chromatic number* (the number of colors needed to vertex color G). For a vertex v, $N(v)$ refers to the subgraph induced by the neighbors of v and $\overline{N}(v)$ similarly the subgraph induced by vertices non-adjacent to v. Unless otherwise stated, G is the input graph, n is the order of G, and H is a fixed forbidden (not necessarily induced) subgraph.

2 The Ramsey Algorithm

When one measure of a graph is fixed or restrained, others often become restricted as well. In particular, we are interested in constructive lower bounds on the independence number in terms of the clique number.

It was first proven by the English mathematician Frank Ramsey that for any integers s and t there is an integer n for which all graphs of order n contain either a clique of size s or an independent set of size t. Denote the minimal such value n by the function $R(s, t)$. A good upper bound for $R(s, t)$ was found by Erdös and Szekeres [9] in 1935. We shall see how their elegant proof will lead us to an efficient algorithm.

Theorem 1

$$R(s, t) \leq \binom{s + t - 2}{t - 1}$$

Proof. Trivially, $R(s, 1) = R(1, s) = 1$. Also, since a graph with no edge must be an independent set, and, vice versa, a graph without non-adjacent nodes must be a clique, $R(s, 2) = R(2, s) = s$.

Now consider a graph G of order $\binom{(s-1)+(t-1)}{t-1}$. Pick any vertex v in the graph and look at the subgraphs $N(v)$ and $\overline{N}(v)$. Since $\binom{x-1}{y} + \binom{x-1}{y-1} = \binom{x}{y}$, either the order of $N(v)$ is at least $\binom{(s-2)+(t-1)}{t-1}$ or the order of $\overline{N}(v)$ is at least $\binom{(s-1)+(t-2)}{t-2}$.

Hence by the inductive hypothesis, either $N(v)$ contains an $(s - 1)$-clique or t-independent set, or $\overline{N}(v)$ contains an s-clique or $(t - 1)$-independent set. In any case we are done because if $N(v)$ has a clique of size $s - 1$ then adding v produces a clique of size s, and similarly if $\overline{N}(v)$ has an independent set of size $t - 1$ then adding v will give us one of size t. ∎

For convenience, let us define $r(s, t)$ to be the above upper bound $\binom{s+t-2}{t-1}$ for $R(s, t)$ for all positive integers, and 1 if either s or t are non-positive. Also let $t_k(n)$ be the minimal t for which $r(k, t) \geq n$. We approximate $t_k(n)$ by $kn^{1/(k-1)}$ for $k \leq 2\log n$, and $\log n / \log \frac{k}{\log n}$ for $k \geq 2\log n$.

To produce an algorithm to find either the independent set or the clique guaranteed, consider a graph of order at least $r(s, t)$. Pick any vertex and look at the subgraphs induced by its neighbors and non-neighbors. At least one of those will satisfy our order constraints. If the order of the neighborhood subgraph is at least $r(s-1,t)$ we add the vertex to our clique, otherwise the order of the non-neighborhood subgraph is at least $r(s,t-1)$, in which case we add the vertex to the independent set. In either case, we restrict our attention to the subgraph chosen and continue this process until the graph has been exhausted. The last vertex, however, can be placed in both the clique and the independent set.

```
Ramsey (G, s)
begin
  I ← ∅; C ← ∅;
  while |G| > 1 do
    choose some v ∈ V(G)
    t ← t_s(|G|)
    if |N(v)| ≥ r(s-1,t)
      then C ← C ∪ {v}; G ← N(v); s ← s - 1
      else I ← I ∪ {v}; G ← N̄(v)
  od
  C ← C ∪ V(G); I ← I ∪ V(G)
  output (C, I)
end
```

Algorithm 2.1: The Ramsey Algorithm

Algorithm 2.1, Ramsey, will find both an independent set I and a clique C, obeying the inequality $r(|C|,|I|) \geq n$. It follows that $|I||C| \geq c(\log n)^2$, for some constant c. It can be shown to hold for $c = \frac{1}{4}$.

When carefully implemented, algorithm 2.1 involves $\mathcal{O}(n + \sum_{v \in I \cup C} d(v)) \leq \mathcal{O}(n + m)$ work. When selecting a pivot node v, search sequentially starting from the last pivot node. When restricting the graph, mark the neighbors with some new "color" if the non-adjacent nodes are to be deleted, and otherwise mark the neighbors as deleted.

Clique removal

Suppose we are trying to find a large independent set. The Ramsey algorithm provides us with the means of obtaining either a sizable clique or a sizable independent set — the problem is that we have no direct control over which one we will get. One way of influencing the result is by eliminating choice: if the input graph contains no clique of size k, the independent set returned must be of size at least $t_k(n)$.

The naive bound of $t_{cl(G)+1}(n)$ is poor, but if we can remove all cliques of size k while retaining a good portion of the graph, the above bound holds on the derived graph. A key observation is that a clique and an independent set can share no more than a single vertex. If the independence number of the graph is at least $(1/k + \epsilon)n$ for some constant $\epsilon > 0$, then at least a fraction $\epsilon/(1 - \frac{1}{k})$ of the vertices remain.

Algorithm 2.2 will find an independent set of size $t_k(N)$ for graphs with independence ratio

```
CliqueRemoval (G, k)
begin
  (C, I) ← Ramsey (G,k)
  while |C| ≥ k do
    G ← G - C
    (C, I) ← Ramsey (G,k)
  od
  output I
end
```

Algorithm 2.2: Algorithm for finding independent set in graphs with high independence ratio

strictly more than $1/k + \epsilon$, where $N \geq \epsilon n$ is the number of vertices remaining after the last clique has been removed. The algorithm repeats calling Ramsey and removing cliques, until a sizable independent set is found. Since no more than n/k cliques of size k can be removed from the graph, the complexity of the algorithm is $\mathcal{O}(nm/k)$.

It may appear that because ϵ can be arbitrarily small the approximation must be weak. We can however always make do with removing only $(k+1)$-cliques, in which case $\epsilon \geq (\frac{1}{k+1} - \frac{1}{k})$ and $N \geq (\frac{1}{k+1} - \frac{1}{k})\frac{1}{1-\frac{1}{k}}n \geq \frac{n}{k^2}$. Since the approximation is $N^{1/(k-1)} \geq (\frac{n}{k^2})^{1/(k-1)} = 2^{(\log n - 2\log k)/(k-1)}$, the factor of k^2 does not affect the asymptotic bound.

A technique by Ajtai, Komlós, and Szemerédi [1] can be used in addition to our method to improve the performance by some logarithmic factor. As treated by Shearer [18], the technique can be thought of as a *randomized* greedy algorithm, which we can be make deterministic. For a fixed k, it will find an independent set in k-clique-free graphs of size $\Omega(n^{1/(k-1)}(\log n)^{(k-2)/(k-1)})$ in polynomial time.

In general we don't know the k for which CliqueRemoval will yield the best possible approximation guarantee. Fortunately, a conservative estimate on the inverse independence ratio of the graph suffices. We can guess a value for k, and if CliqueRemoval(G,k) returns too small an approximation (less than $(\frac{n}{k^2})^{1/k}$), then the estimate was too low. If we find a k' such that CliqueRemoval fails on $k'-1$ but succeeds on k', then the estimate is conservative. To find such a k' we employ a form of binary search [20], thereby adding a $\log n$ factor to the time complexity. The performance guarantee of this algorithm is no worse than $\min_k \frac{n/k}{t_{k+1}(n)} = \mathcal{O}(n/(\log n)^2)$ with minima at $k \approx \log n$.

Graph coloring

Any approximation algorithm A_{IS} for independent set can be turned into an approximation algorithm for graph coloring using the ColorByExcavating heuristic [13]. This method of coloring is used (directly or indirectly) by all known approximate coloring methods.

Since a k-colorable graph contains no $(k+1)$-cliques, we can use Ramsey($G, k+1$) as the independent set approximation algorithm. When the chromatic number is not known, we simply use the size of the largest clique found so far plus one as a parameter. The complexity of the

```
ColorByExcavating (G)
begin
  p ← 0
  while G ≠ ∅ do
  I ← A_IS(G)
  Color I with color p
  G ← G − I
  p ← p + 1
  od
end
```

Algorithm 2.3: Coloring graphs by "excavating" independent sets

algorithm is clearly bounded by the the number of colors found times the number of edges and vertices in the graph.

The number of colors needed will be the inverse of the independent set approximation, namely $n/t_{k+1}(n) \approx n^{(k-1)/k}/k$, times some overhead factor. Johnson [14] showed that when using an optimal independent set algorithm the overhead is $\theta(\log n)$. In our case, however, the excavation is so piecemeal that the overhead is only a small constant. In fact, $\lim_{n \to \infty} \text{COLORS} = \frac{n}{|I|}$.

The performance guarantee is $\min_k \frac{n^{(k-1)/k}/k}{k} = \mathcal{O}(n/(\log n)^2)$. This can also be obtained using the previously noted fact that $|I| \geq \frac{(\log n)^2}{4|C|}$, conservatively obtaining $\text{COLORS} \leq \frac{5n}{(\log n)^2}|C|$. This was, in fact, noticed by Erdös in a rather obscure paper [8] from 1967.

Now consider the product of the performance guarantees for the clique problem R_{cl} and the coloring problem R_χ:

$$R_{cl} \cdot R_\chi \leq \frac{cl(G)}{|C|} \frac{\text{COLORS}}{\chi(G)} \leq \frac{5n}{(\log n)^2} \frac{cl(G)}{\chi(G)}$$

Since $cl(G)$ is never greater than $\chi(G)$, this bound immediately yields the individual bounds. Clearly, if one approximation is as far off as the individual bound indicates, then the other one must be within a constant factor of optimal. Also, for classes of instances for which the measures are apart, the performance guarantee is even stronger. In particular, random graphs almost always have a clique number asymptotically $2 \log n$ and chromatic number $n/(2 \log n)$, and for graphs with these parameters $R_{cl} \cdot R_\chi = \mathcal{O}(1)$.

3 Subgraph-Excluding Algorithms

The algorithms presented in the previous section all fall into the category of subgraph-excluding (alias Ramsey-type) algorithms.

Definition 1 A_H *is a* Ramsey-type *algorithm if, given arbitrary graph G, it is of the form:*

1. *Ensure G contains no copy of the subgraph H, and*
2. *Find an independent set in G, using only the property that G contains no copy of H.*

There are a few ways in which such an algorithm can exclude a subgraph H:

Remove: All copies of the subgraph, or parts of it, can be pulled out of the graph sequentially. A necessary and sufficient precondition for the removal process to retain at least a constant fraction of the vertices is that $i(H) < i(G) + \epsilon$, for some constant $\epsilon > 0$.

Forbid: The exclusion of the subgraph can be built into the statement of the problem. This applies particularly to the graph coloring problem. For instance, the clique on $k + 1$ vertices is forbidden in k-colorable graphs.

Merge: In certain cases, vertices can be fused together, causing a certain type of subgraphs to become non-existent.

The issue that remains is finding graphs H that force graphs free of H to contain large independent sets, as well as coming up with algorithms to actually finding those independent sets in H-free graphs. The previous section described algorithms that use cliques. Other subgraphs discussed in this section include odd cycles, wheels, and color-critical subgraphs. The following section will then illustrate that these subgraphs are in some sense the best of their kind.

Wheels

A *wheel*, denoted by $W_{p,m}$, is a graph that consists of an odd cycle of $m \geq 3$ nodes, and $p \geq 0$ *spokes* which are nodes that connect to all other nodes in the graph. A wheel with p spokes is referred to as a p-wheel. The clique number of an p-wheel is $p+2$ (except when $m = 3$), whereas the chromatic number is $p + 3$.

Define $R(W_p, K_t)$ to be the minimal n such that all graphs of order n contain some p-wheel or an independent set of size t. Note that if a graph does not contain a p-wheel, then its neighborhood graph cannot contain a $(p-1)$-wheel nor can its non-neighborhood graph contain a p wheel. Hence we obtain the same characterization as for the regular Ramsey numbers: $R(W_p, K_t) \leq R(W_{p-1}, K_t) + R(W_p, K_{t-1})$.

Observe that $R(W_0, K_t) = 2t - 1$ and $R(W_p, K_2) = p + 3$. An inductive argument shows that $R(W_p, K_t) \leq 2r(p+2,t)$, only a factor of two from the upper bound of the regular Ramsey function. We use these properties to design a variation of the Ramsey algorithm with bipartite graphs as a base case.

Given a graph with no $(k - 2)$-wheels, WheelFreeRamsey finds an independent set of size at least $\Omega(kn^{1/(k-1)})$. Applying algorithm 2.3, we can color a graph without $(k - 2)$-wheels using $\mathcal{O}(n^{(k-2)/(k-1)}/k)$ colors.

Algorithm 3.1 is strongly related to Wigderson's coloring algorithm. By considering the whole uncolored portion of the graph in each iteration, instead of fully coloring the pivot nodes' neighborhoods before coloring their non-neighbors, WheelFreeRamsey improves the approximation by a factor of k. Also, by focusing alternately on neighborhoods and non-neighborhoods, another factor of k is gained. Wigderson's method, however, has the advantage of $\mathcal{O}(\chi(G) (n + m))$ time complexity, compared to $\mathcal{O}(\text{COLORS} (n + m))$.

```
WheelFreeRamsey (G, s)
begin
  I ← ∅; C ← ∅;
  while s > 2 and G ≠ ∅ do
    choose some v ∈ V(G)
    t ← t_s(n/2)
    if |N(v)| ≥ 2 r(s - 1, t)
      then C ← C ∪ v; G ← N(v); s ← s - 1
      else I ← I ∪ v; G ← N̄(v)
  od
  if s ≤ 2   { G should be bipartite }
    then I ← I ∪ IndepSetInBipartiteGraph(G)
      C ← C ∪ {some edge in G, if it exists}
  output (C, I)
end
```

Algorithm 3.1: Ramsey Algorithm for Wheels

In comparison with the graph coloring algorithm deduced from the Ramsey algorithm for clique-free graphs, this algorithm improves the exponent from $(k - 1)/k$ to $(k - 2)/(k - 1)$.

Short odd cycles

For graphs with independence ratio in the range of $(\frac{1}{3} + \epsilon, \frac{1}{2})$, the Ramsey algorithm obtains an independent set approximation of $\Omega(\sqrt{n})$ by removing triangles. Families of odd cycles as excluded subgraphs allow us to refine the approximations in this range.

The method starts by removing all odd cycles of size up to $2k + 1$. Note that a cycle of length $2k + 1$ has an independence ratio $\frac{k}{2k+1}$. So if $i(G) > \frac{k}{2k+1}$, we can remove these cycles and then apply algorithm 3.2.

```
OddCycleFreeApproximation (G, k)
  { Graph G contains no odd cycles of length 2k + 1 or shorter}
begin
  while G ≠ ∅ do
    choose any vertex v in V(G).
    V_i ← vertices of distance i from v.
    S_i ← V_i ∪ V_{i-2} ∪ ...
    Determine i such that |S_{i+1}| ≤ n^{1/(k+1)} |S_i|.
    I ← I ∪ S_i
    G ← G - S_i - S_{i+1}
  od
  return I
end
```

Algorithm 3.2: Algorithm for independent sets on graphs with no short odd cycles

Since each independent set S_i selected causes only $n^{1/(k+1)}$ times as many other nodes to be removed from the graph, the graph is not exhausted until an independent set of at least

$n^{k/(k+1)}$ has been collected. Assume there was no i satisfying $|S_{i+1}| \leq n^{1/(k+1)}|S_i|$. Then $|S_k| > n^{1/(k+1)}|S_{k-1}| > n^{2/(k+1)}|S_{k-2}| > \cdots > n^{k/(k+1)}|S_0| = n^{k/(k+1)}$, and the problem is solved.

Since each vertex and each edge are looked at only once, the algorithm runs in linear time. On the other hand, when applied to general graphs the algorithm must be run for many different values of k, in which case it may be useful to combine the cycle removal process (see [17]).

The technique of Ajtai, Komlós, and Szemerédi can also be applied here. When k is fixed, we can find an independent set of size $\Omega(n^{k/(k+1)}(\log n)^{1/(k+1)})$ in polynomial time for graphs with no odd cycles of length $2k + 1$ or less.

Color-critical graphs

A. Blum [6, 5] recently introduced improved approximation algorithms for k-colorable graphs, where k is fixed, in particular improving approximate 3-coloring to $\tilde{O}(n^{3/8})$ colors. His complicated method can be summarized in the following three steps:

1. Destroy all copies of the subgraphs $K_4 - e$ and 1-2-3 graphs by collapsing certain pairs of nodes.

2. Classify vertices according to degree, producing a polynomial number of subgraphs, one of which has an independence ratio close to half.

3. Apply algorithm 3.2 on each of these subgraphs.

The graph $K_4 - e$ is the clique on 4 vertices with one edge removed. "1-2-3 graphs" is our terminology for the graphs with three specific parts: A, consisting of two disconnected nodes; B, an independent set of at least 3 nodes; and C, an odd cycle, where parts A and C are completely disconnected, A and B are completely connected, and the connections between B and C are such that each node in C is connected to some node in B. Since C requires three colors, B needs two, and thus the two nodes in A must have the same color under any legal 3-coloring of the subgraph, hence the name. Similarly, the two disjoint nodes in $K_4 - e$ must share the same color.

The first and the third steps are strictly Ramsey, whereas the second does use the size of the independent sets promised by the k-colorability property, hence, the algorithm appears to lack the "forgetfulness" property of Ramsey-style algorithms.

4 Limitation results

The main result of this section is that excluding subgraphs other than cliques and series of odd cycles does not help much in forcing a graph to contain a large independent set. This implies that no subgraph removal algorithms, even super-polynomial ones, can yield asymptotically better performance guarantees for the maximum independent set, graph coloring, and vertex cover problem than the algorithms given.

Extend the Ramsey function from cliques to arbitrary graphs. Let $R(H, K_t)$ denote the minimal n such that every graph on n vertices either contains a copy of the graph H or has an independent set of size t. Note that H does not need to be isomorphic to a *vertex induced* subgraph of G, only that all the edges of H be contained in such a subgraph. It immediately follows that $R(H, K_t) \geq R(H', K_t)$ whenever H' is an edge-subset of H. Obtaining an upper bound on $R(H, K_t)$ shows that not all H-free graphs contain all that large independent sets, showing a limitation on the power of excluding H.

A few definitions are in order. For a graph H, let $e(H)$ be the number of edges, and $\rho(H)$ denote the maximum of $e(H')/|H'|$ over all subgraphs H' of H. Extend these definitions to a collection $\mathcal{H}$ of graphs. Define $i(\mathcal{H})$ to be the maximum of $i(H)$ over all H in $\mathcal{H}$. Define $\rho(\mathcal{H})$ and $\chi(\mathcal{H})$ to be the minimum of $\rho(H)$ and $\chi(H)$, respectively, over all H in $\mathcal{H}$. Also, $R(\mathcal{H}, K_t)$ is the minimal n such that every graph on n vertices either contains a copy of some H in $\mathcal{H}$ or has an independent set of size t.

There are some well-known relations between these quantities. One relation is $\chi(H)i(H) \geq 1$, which holds because a coloring is just a partition into independent sets. Another relation is $\chi(H) \leq 2\rho(H) + 1$, which holds because H has a vertex of degree $2\rho(H)$ or less. Both relations generalize to a collection of graphs.

We will give the central theorem for a function slightly stronger than ρ. Define $\rho'(H) = \min \frac{e(H')-1}{|H'|-2}$ where H' is a subgraph of H on at least 3 vertices. Similarly extend it to a collection of graphs $\mathcal{H}$. The value of ρ' is always at least as large as ρ, and for small graphs the improvement makes a difference.

Theorem 2 $R(\mathcal{H}, K_t) = \Omega((\frac{t}{\log t})^{\rho'(\mathcal{H})})$

The omitted proof follows the probabilistic method using the Lovász local lemma (see [19]). We will first apply theorem 2 to obtain a limitation for forbidden subgraphs of a minimum chromatic value.

Recall that Blum's algorithm made use of subgraphs that contain two nodes that must be of the same color under any legal 3-coloring. A graph is *k-avoidable* iff it has a pair of vertices that get assigned the same color for every k-coloring of the graph. Note that this is vacuously true for non-k-colorable graphs. Alternatively, k-avoidable graphs can be characterized as being no more than one edge away from being $(k + 1)$-chromatic. A collection $\mathcal{H}$ is k-avoidable iff every H in $\mathcal{H}$ is.

Corollary 1 *For every positive integer k, if $\mathcal{H}$ is k-avoidable, then $R(\mathcal{H}, K_t) = \Omega((t/\log t)^{k/2})$.*

Proof. If $H \in \mathcal{H}$ is k-avoidable, then $H + e$ is $(k + 1)$-chromatic for some edge e. Hence $\rho(H + e) \geq \frac{k}{2}$. But $\rho'(H) \geq \min_{H' \in H+e} \frac{e(H')-2}{|H'|-2} \geq \rho(H + e)$, when $\rho(H + e) \geq 1$. This holds for all H in $\mathcal{H}$, hence the conclusion follows from theorem 2 ∎

This result implies that a Ramsey-type algorithm on a k-colorable graph that relies solely on the lack of some set of k-avoidable subgraphs, cannot guarantee finding an independent set of size more than $\mathcal{O}(n^{2/k}\log n)$, and hence cannot guarantee a coloring with less than $\Omega(n^{1-2/k}/\log n)$ colors. As an example, no such algorithm can guarantee coloring a 3-colorable graph with less than $\Omega(n^{1/3}/\log n)$ colors.

We can make a stronger statement regarding the 3-coloring problem.

Theorem 3 *If $\mathcal{H}$ is 3-avoidable, then $\rho'(\mathcal{H}) \geq \frac{3}{2} + \frac{1}{26}$.*

Proof. Let H be a 3-avoidable graph in $\mathcal{H}$, and $H + e$ be 4-chromatic. A *4-critical* graph is a 4-chromatic graph with the property that removing any node will make it 3-colorable. Gallai [10] showed that 4-critical graphs, with the exception of K_4, have an edge-to-vertex ratio of at least $3/2 + 1/26$. If $H + e$ contains a K_4, then $\rho'(H) \geq \rho'(K_4 - e) = \frac{5-1}{4-2} = 2$. Otherwise, $\rho(H + e) \geq \frac{e(H^*)}{|H^*|} \geq \frac{3}{2} + \frac{1}{26}$, by Gallai's result. In either case, $\rho'(H) \geq \frac{3}{2} + \frac{1}{26}$ for any H in $\mathcal{H}$. ∎

As a result, Ramsey-type algorithms require at least $\Omega(n^{1-1/(\frac{3}{2}+\frac{1}{26})}/\log n) \approx \Omega(n^{.35})$ colors on 3-colorable graphs. Notice that Blum's technique also breaks down in the region of $n^{1/3}$ [5], even though it is not known to be of a subgraph excluding type.

Let us now derive a limitation for general graphs. It can be shown, in the spirit of the bounds on the diagonal Ramsey function $R(s,s)$, that if $\mathcal{H}$ is a t-avoidable collection then $R(\mathcal{H}, K_t) = 2^{\Omega(t)}$. Hence if all graphs of order n contain either a subgraph H in the t-chromatic collection $\mathcal{H}$ or an independent set of size t, then t must be no more than $\mathcal{O}(\log n)$. Hence no Ramsey-type algorithm that relies solely on the lack of avoidable subgraphs can obtain a better than $\Omega(n/(\log n)^2)$ performance guarantee for graph coloring.

Our emphasis so far on graph coloring is for a good reason, namely because the lower bounds for graph coloring are also lower bounds for the independent set problem. Since $i(H) < 1/k$ implies that $\chi(H) \geq k + 1$, corollary 1 holds as well for graphs with large independence ratio. Similarly, the limitation result on performance guarantees for the general coloring problem carries immediately over to the maximum independent set problem.

Limitation results for odd cycles

Our next goal is to show that our cycle-based algorithm is close to optimal for graphs with independence ratio near $\frac{1}{2}$. Our results will also imply that the algorithm of Monien and Speckenmeyer for approximately solving the vertex cover problem is essentially optimal among subgraph-removal algorithms.

We need the following structural result on graphs without short odd cycles.

Theorem 4 *For every positive integer k, if $\rho(H) \leq 1 + \frac{1}{4k+2}$, and H contains no odd cycles of length $2k - 1$ or less, then $i(H) \geq \frac{k}{2k+1}$.*

Proof. By induction on the number of vertices in H. If there is a vertex v of degree 0 or 1, then remove v and its neighbor from the graph. By induction, the remaining graph has independence ratio at least $\frac{k}{2k+1}$. But adding v to the largest independent set of the remaining graph shows that H itself has independence ratio greater than $\frac{k}{2k+1}$.

Thus, assume that every vertex has degree 2 or more. Suppose there is a cycle passing through only vertices of degree exactly 2. Since H has no odd cycles of length $2k-1$ or less, the independence ratio of this cycle is at least $\frac{k}{2k+1}$. Then we could apply induction to the remainder of the graph and be finished.

Thus, assume there are no such cycles. Let H' be the subgraph induced by the vertices of degree exactly 2. The subgraph H' must be the disjoint union of paths, so $i(H') \geq \frac{1}{2}$. Since $\rho(H) \leq 1 + \frac{1}{4k+2}$, the subgraph H' contains at least a fraction $1 - \frac{1}{2k+1}$ of all the vertices. Therefore $i(H)$ is at least $\frac{1}{2}(1 - \frac{1}{2k+1}) = \frac{k}{2k+1}$, which completes the proof. ∎

Finally, we can prove the following limitation result.

Corollary 2 *For every positive integer k, if $i(\mathcal{H}) < \frac{k}{2k+1}$, then $R(\mathcal{H}, K_t) = \Omega((t/\log t)^{1+1/(4k+2)})$.*

Proof. By theorem 4 above, every H in $\mathcal{H}$ either has an odd cycle of length $2k-1$ or less, or satisfies $\rho(H) \geq 1 + 1/(4k+2)$. The first case implies that $\rho'(H) \geq (2k-2)/(2k-3) = 1 + 1/(2k-3)$, thus in either case $\rho'(H) \geq 1 + 1/(4k+2)$. ∎

This result implies that for a graph with independence ratio $\frac{k}{2k+1}$, no subgraph-removal algorithm can guarantee an independent set larger than $\mathcal{O}(n^{1-1/(4k+3)})$. Recall that our cycle-based algorithm will find an independent set of size $\Omega(n^{1-1/k})$ and thus the cycle-based algorithm is close to optimal. The above result also implies that for approximately solving the vertex cover problem, no subgraph-removal algorithm can achieve a performance guarantee better than $2 - \theta(\log \log n / \log n)$, the performance guarantee obtained by the algorithm of Monien and Speckenmeyer.

5 Discussion

The central open question is what the best possible approximation guarantees for the independent set and graph coloring problems are. The results of Berger and Rompel, and Blum show that there do exist more intelligent methods than mere subgraph exclusion. Their results, however, rely on special properties of k-colorable graphs, thus it is still not clear if better methodologies exist for the independent set and vertex cover problems.

One direction may be in approximating the number $\alpha(G)$ (or $cl(G)$) instead of actually constructing an independent set (clique). One such candidate is the ϑ function introduced in a seminal paper by Lovász [16], which has been shown to be polynomial time computable [12]. Its value always lies between the clique and the chromatic number of the graph, but it is open as to how far it can stray from each measure. Is it perhaps the case that ϑ approximates both the clique and the chromatic number within a factor of $\sqrt{n}$?

References

[1] M. Ajtai, J. Komlós, and E. Szemerédi. A note on Ramsey numbers. *J. Combin. Theory A*, 29:354–360, 1980.

[2] R. Bar-Yehuda and S. Even. A $2 - \frac{\log \log n}{2 \log n}$ performance ratio for the weighted vertex cover problem. Technical Report #260, Technion, Haifa, January 1983.

[3] B. Berger and J. Rompel. A better performance guarantee for approximate graph coloring. *Algorithmica*, 1990. to appear.

[4] P. Berman and G. Schnitger. On the complexity of approximating the independent set problem. In *Proc. Symp. Theoret. Aspects of Comp. Sci.*, pages 256–268. Springer-Verlag Lecture Notes in Comp. Sci. # 349, 1989.

[5] A. Blum. Some tools for approximate 3-coloring. Unpublished manuscript, 1989.

[6] A. Blum. An $\tilde{O}(n^{\cdot 4})$ approximation algorithm for 3-coloring. In *Proc. ACM Symp. Theory of Comp.*, pages 535–542, 1989.

[7] V. Chvátal. Determining the stability number of a graph. *SIAM J. Comput.*, 6(4), Dec. 1977.

[8] P. Erdös. Some remarks on chromatic graphs. *Colloq. Math.*, 16:253–256, 1967.

[9] P. Erdös and G. Szekeres. A combinatorial problem in geometry. *Compositio Mathematica*, 2:463–470, 1935.

[10] T. Gallai. Kritische graphen I. *Publ. Math. Inst. Hungar. Acad. Sci.*, 8:165–192, 1963. (See Bollóbas, B. *Extremal Graph Theory*. Academic Press, 1978).

[11] M. R. Garey and D. S. Johnson. *Computers and Intractibility: A Guide to the Theory of NP-completeness*. Freeman, 1979.

[12] M. Grötschel, L. Lovász, and A. Schrijver. *Geometric Algorithms and Combinatorial Optimization*. Springer-Verlag, 1988.

[13] D. S. Johnson. Approximation algorithms for combinatorial problems. *J. Comput. Syst. Sci.*, 9:256–278, 1974.

[14] D. S. Johnson. Worst case behaviour of graph coloring algorithms. In *Proc. 5th Southeastern Conf. on Combinatorics, Graph Theory, and Computing*, pages 513–527. Utilitas Mathematicum, 1974.

[15] N. Linial and V. Vazirani. Graph products and chromatic numbers. In *Proc. IEEE Found. of Comp. Sci.*, pages 124–128, 1989.

[16] L. Lovász. On the Shannon capacity of a graph. *IEEE Trans Info. Theory*, IT-25(1):1–7, Jan. 1979.

[17] B. Monien and E. Speckenmeyer. Ramsey numbers and an approximation algorithm for the vertex cover problem. *Acta Inf.*, 22:115–123, 1985.

[18] J. B. Shearer. A note on the independence number of triangle-free graphs. *Preprint*, 1982. (See Bollóbas, B. *Random Graphs*. Academic Press, 1985).

[19] J. Spencer. *Ten Lectures on the Probabilistic Method*, volume 52. SIAM, 1987.

[20] A. Wigderson. Improving the performance guarantee for approximate graph coloring. *J. ACM*, 30(4):729–735, 1983.

GENERATING SPARSE SPANNERS FOR WEIGHTED GRAPHS [1]

INGO ALTHÖFER – Fakultät fur Mathematik, Universität Bielefeld

GAUTAM DAS – Department of Computer Sciences, University of Wisconsin

DAVID DOBKIN – Department of Computer Science, Princeton University

DEBORAH JOSEPH – Department of Computer Sciences, University of Wisconsin

---

ABSTRACT. Given a graph G, a subgraph G' is a t-spanner of G, if for every u, $v \in V$, the distance from u to v in G' is at most t times longer than the distance in G. In this paper we give a very simple algorithm for constructing sparse spanners for arbitrary weighted graphs. We then apply this algorithm to obtain specific results for planar graphs and Euclidean graphs. We discuss the optimality of our results and present several nearly matching lower bounds.

---

1. INTRODUCTION

Let $G = (V, E)$ be a connected n-vertex graph with arbitrary positive edge weights. A subgraph $G' = (V, E')$ is a t-*spanner* if, between any pair of vertices the distance in G' is at most t times longer than the distance in G. The value of t is the *stretch factor* associated with G'. We consider the problem of determining t-spanners for graphs where the spanners are sparse and t is a constant independent of the size of the graph. Sparsity will be measured according to two criteria. Let $Weight(G)$ denote the sum of all edge weights of graph G, and $Size(G)$ denote the number of edges. A graph is sparse in *size* if it has few edges. Similarly, a graph is sparse in *weight* if its total edge weight is small. Our results separate graphs into classes where spanners with linearly many edges achieve constant stretch factors, and classes where a non-linear number of edges are necessary.

Problems of this type appear in numerous applications. Spanners appear to be the underlying graph structure in various constructions in distributed systems and communication networks [Aw, PUl, PU]. They also appear in biology in the process of reconstructing phylogenetic trees from matrices, whose entries represent genetic distances among contemporary living species [BD]. Robotics researchers have studied spanners under the constraints of *Euclidean* geometry, where vertices of the graph are points in space, and edges are line segments joining pairs of points [C, DFS, DJ, K, KG, LL].

---

[1] The work of the second and fourth authors was supported by NSF PYI grant DCR-8402375. The work of the third author was supported by NSF grant CCR-8700917.

In the above applications, previous research has focussed on graphs with specific constraints. In distributed computation, the design of synchronizers [Aw, PUl] and the design of succinct routing tables [PU] implicitly generated spanners for graphs with *unit edge weights*. For example, in designing routing tables [PU], the routes follow the edges of a sparse spanner. For any stretch factor $O(t)$, the size of these spanners are $O(n^{1 + 1/t})$. These designs are based upon a *clustering algorithm*, which is more complex than the algorithm in this paper. Moreover, it is not easy to generalize the clustering algorithm to graphs with arbitrary edge weights. Recently the problem of designing succinct routing tables has been considered for graphs with arbitrary edge weights [ABLP, AP]. For any t, the scheme in [ABLP] routes along paths which are at most $O(t^2 9^t)$ longer than the shortest paths, while the total memory required for the tables is $O(tn^{1 + 1/t}logn)$. In [AP], the routes are $O(t^2)$ longer, while the memory required is $O(n^{1 + 1/t}log^2 n log D)$, where D is the *diameter* of the graph. Spanners have been considered for special classes of graphs in [PS], however these graphs have unit edge weights.

In all the above research, sparseness has been achieved in the size of spanners, but not in the weight. In robotics, graphs with varying edge weights have been examined, but the weights are Euclidean distances and not arbitrary. Because of this restriction, it has been possible to construct linearly sized spanners, unlike the above examples. For instance, the *Delaunay* and other triangulations approximate complete straight-edged graphs on the plane [C, DFS, DJ, LL, KG]. A few papers have considered weight sparseness of spanners for these graphs [DJ, LL]. For any t, there exist spanners for the complete graph on the plane with stretch factor $O(t)$, and weight within a $O(1 + 1/t)$ multiple of the weight of the minimum spanning tree. Weight sparseness has also been considered in [A, D, S] for the special case of 1-spanners under a general model where spanners may have auxiliary vertices.

In this paper we approach the problem from a very broad perspective. Our graphs have no special embeddings, and we allow arbitrary positive edge weights. For any t, we show that every such graph has a spanner with $O(t)$ stretch factor, and $O(n^{1 + 1/t})$ size. We also provide weight bounds for our spanners. The contributions of this paper are: a very simple polynomial time algorithm for constructing sparse spanners (in *both* size and weight) of arbitrary weighted graphs, using these ideas for constructing spanners of *planar* graphs, some lower bound results, and some results on spanners in Euclidean spaces of arbitrary dimensions and norms. Since any spanner with appropriate sparseness and stretch factor can be used for constructing synchronizers [PUl], our algorithm provides a simple alternative to the clustering algorithm previously used for constructing synchronizers. Similarly, we hope that our results will simplify the construction of succinct routing tables for arbitrary weighted graphs.

The next section describes how to construct sparse spanners for general as well as planar graphs with arbitrary edge weights. Section 3 deals with several lower bound results. Section 4 discusses spanners for Euclidean graphs. We conclude with some open problems. Due to lack of space, we omit details of some of the proofs, which may be found in the full paper [ADDJ].

2. CONSTRUCTION OF SPARSE SPANNERS

Let G be a n-vertex, connected, weighted graph. The minimum spanning tree of G, denoted $MST(G)$, is obviously the sparsest spanner, however, it is not hard to see that its stretch factor can be as bad as $\Omega(n)$. Instead, we would like to look for spanners whose stretch factors are constants, independent of n. Our results are encouraging, because we show that any graph has spanners with constant stretch factors, whose sparseness can be made arbitrarily close to that of $MST(G)$. They are summarized by the following two theorems.

Theorem 1: Given a n-vertex graph G and a $t \geq 1$, there is a polynomially constructible $(2t + 1)$-spanner G' such that
1) $Size(G') < n \cdot \lceil n^{1/t} \rceil$,
2) $Weight(G') < Weight(MST(G)) \cdot (1 + n/2t)$.

Note that the stretch factor is independent of the number of vertices in the graph, and of the edge weights. Thus, even for a dense graph with $\Omega(n^2)$ edges, arbitrarily sparse spanners exist which have good stretch factors. Our lower bound results will show that the size bound is almost tight, though the weight bound is loose. Previous research [Aw, PS] has produced spanners with the same size and stretch factors, but the graphs considered there were unweighted, and sparseness was measured by size only. Also, our algorithm is simpler than the clustering algorithm employed. Recently the problem of designing succinct routing tables has been considered for graphs with arbitrary edge weights [ABLP, AP], again based on the clustering algorithm. For any t, the scheme in [ABLP] routes along paths which are at most $O(t^2 9^t)$ longer than the shortest paths, while the total memory required for the tables is $O(tn^{1 + 1/t}logn)$. In [AP], the routes are $O(t^2)$ longer, while the memory required is $O(n^{1 + 1/t}log^2 nlogD)$, where D is the *diameter* of the graph.

Theorem 2: Given a n-vertex planar graph G and a $t \geq 1/2$, there is a polynomially constructible $(2t + 1)$-spanner G' such that
1) $Size(G') \leq (n - 2) \cdot (1 + 2/\lfloor 2t \rfloor)$,
2) $Weight(G') < Weight(MST(G)) \cdot (1 + 1/t)$.

Here we again observe that the stretch factor is independent of the number of vertices in the planar graph, or the edge weights. Thus, arbitrarily sparse spanners exist which have good stretch factors. (Note that the maximum size of a planar graph can be $3n - 6$, so in effect we demonstrate that even the multiplicative constant in the size can be reduced). Theorem 2 is stronger in flavor than our first theorem, because our lower bound results will show that both size and weight bounds are tight. We later give an interesting application of this in connection with Euclidean graphs.

Before we prove our results, we introduce an algorithm for constructing spanners. It constructs a sparse subset of edges so that a required stretch factor is achieved. The algorithm, called $SPANNER(G, r)$, takes as input a weighted graph G, and a positive parameter r. The weights need not be unique. It produces as output a subgraph G'.[2]

Algorithm $SPANNER(G = (V, E), r)$;

begin

    Sort E by nondecreasing weight;
    $G' := \phi$;

    for every edge $e = [u, v]$ in E **do**
        begin
        Compute $P(u, v)$, the shortest path from u to v in G';
        if $(r \cdot Weight(e) < Weight(P(u, v)))$ **then** add e to G';
        end;

    Output G';

end;

While our algorithm is very simple, the subgraph it generates has many interesting properties. The following lemmas describe these properties.

Lemma 1: G' is a r-spanner of G.

Proof. Consider any edge $[u, v]$ in $G - G'$. At the instant it is examined by the algorithm, there is a path $P(u, v)$ from u to v in the current graph of length $\leq r \cdot Weight([u, v])$. Thus, in the final output each deleted edge $[u, v]$ is associated with a short path $P(u, v)$. Now consider any shortest path in G of length l. For every deleted $[u, v]$ edge along this path we replace by $P(u, v)$, and thus obtain an alternate path in G' of length $\leq l \cdot r$, which proves the lemma. •

Lemma 2: Let C be any simple cycle in G'. Then $Size(C) > r + 1$.

Proof. Assume that a cycle C remained with size $\leq r + 1$. Let $[u, v]$ be the last edge in C to be examined by the algorithm. Clearly, it is one of the (possibly many) edges in the cycle with the largest weight. When $[u, v]$ is being considered by the algorithm, there is a clearly path from u to v (the remaining portion of the cycle) whose length is $\leq r \cdot Weight([u, v])$. Thus, $[u, v]$ should not have been added, which is a contradiction. •

Lemma 3: Let C be any simple cycle in G', and let e be any edge in C. Then $Weight(C - \{e\}) > r \cdot Weight(e)$.

Proof. Assume that a cycle C remained which violates the above weight condition for some edge e in C. Clearly, it violates this condition for the last edge in C to be

---

[2] We understand that this algorithm has been independently discovered by Bern, [Be].

examined by the algorithm, say $[u, v]$. As in the above lemma, it is one of the (possibly many) edges in the cycle with the largest weight. Thus, the remaining portion of the cycle has length $\leq r \cdot Weight([u, v])$. Thus, $[u, v]$ should not have been added, which is a contradiction. •

Lemma 4: $MST(G)$ is contained in G'.

Proof. Before we prove this lemma, we observe that our algorithm is essentially a generalized minimum spanning tree algorithm, because for an infinite r, it outputs $MST(G)$. In fact, for an infinite r its behavior is exactly like Kruskal's minimum spanning tree algorithm [T].

Our proof (sketch) is by induction on the order in which edges are examined. Let the sequence $\phi = G'_0, G'_1, \cdots, G'_{|E|} = G'$ represent the growth of G', where G'_i represents the partially constructed subgraph after the ith edge has been examined. Now let us consider Kruskal's algorithm, which also examines edges by increasing weight. In this case, let the sequence $\phi = M_0, M_1, \cdots, M_{|E|} = MST(G)$ represent the growth of the minimum spanning tree, where M_i represents the partially constructed tree after the ith edge has been examined. It is not hard to prove by induction that, for all i, M_i is contained in G'_i, which in turn proves the lemma. •

The algorithm is clearly polynomial and easy to implement. In proving the two theorems, Lemma 2 will be used in proving size sparseness, while Lemma 3 and Lemma 4 will be used in proving weight sparseness of the resulting subgraphs.

We still need two more lemmas. Let the *size* of a face of a planar graph be the number of edges encountered while traversing the face (with repetitions allowed). Lemma 5 bounds the size of a planar graph, given a minimum face size.

Lemma 5: If all the faces of an n-vertex planar graph G have sizes $\geq r$, then $Size(G) \leq (n - 2) \cdot (1 + 2/(r - 2))$.

Proof. Let m be the size of the graph. Euler's formula for planar graphs states that $n - m + f = 2$, where f is the number of faces in the graph. Thus, $f = m + 2 - n$. If we traverse each face and mark the edges encountered, every edge in the graph will eventually be marked twice. We thus have $f \cdot r \leq 2m$. Substituting for f from above and simplifying, we have $m \leq (n - 2) \cdot (1 + 2/(r - 2))$, which proves the lemma. •

Our next lemma is from extremal graph theory (which is derivable from Theorem 3.7, Chapter III, [B]). Let the *girth* of a graph be the size (number of edges) of its smallest simple cycle. This lemma provides an upper bound on the size of a graph with a given girth.

Lemma 6: Let G be an n-vertex graph with girth $> r$. Then $Size(G) < n \cdot \lceil n^{2/(r - 1)} \rceil$.

We first prove Theorem 2, then use some of the ideas in proving Theorem 1.

Proof of Theorem 2. We run the *SPANNER* algorithm on a given n-vertex planar graph G and a $t \geq 1/2$, after setting $r = 2t + 1$. By Lemma 1, the resulting graph has stretch factor $2t + 1$. Also, by Lemma 2, the girth of the output is $> 2t + 1$, that is, $\geq \lfloor 2t + 2 \rfloor$. This clearly implies that the minimum face size is $\geq \lfloor 2t + 2 \rfloor$. Thus, by Lemma 5, the size of the output is $\leq n \cdot (1 + 2/\lfloor 2t \rfloor)$. In the next section we will show that this size bound is tight.

We now prove the weight bound, which requires a different approach (the outline is similar to the method in [LL]). By Lemma 4, the output subgraph has to contain $MST(G)$. Consider a planar drawing of the subgraph, with $MST(G)$ embedded in the subgraph. If we walk around the tree (counting each edge twice), our path will resemble a "skinny" polygon with perimeter being $2 \cdot Weight(MST(G))$. Our accounting strategy will be to "grow" this polygon outwards by absorbing neighboring edges of the subgraph, until it becomes the outer face of the graph. At any stage, an edge is selected which, along with a portion of the polygon's current boundary circumscribes a face adjacent to the polygon. Consider Stage i. Let the length of the polygon be W_i (with $W_0 = 2 \cdot Weight(MST(G))$). Let T_i be the total length of all non minimum spanning tree edges encountered so far. Thus $T_0 = 0$. Let the edge selected to be added be $[u, v]$, and the portion of the polygon's boundary from u to v be $P(u, v)$. By Lemma 3, $Weight([u, v]) < (1/(2t + 1)) \cdot Weight(P(u, v))$. Thus,

1) $W_{i + 1} < W_i - Weight([u, v]) \cdot (1/(2t + 1) - 1)$,
2) $T_{i + 1} = T_i + Weight([u, v])$.

We state without proof that T converges to at most $Weight(MST(G)) \cdot (1/t)$. Thus $Weight(G') < Weight(MST(G)) \cdot (1 + 1/t)$. We later show that this weight bound is tight because there are graphs for which one can do no better. ●

Proof of Theorem 1. We are given an n-vertex graph G and a $t \geq 1$ as input. We set $r = 2t + 1$ and run the *SPANNER* algorithm. By Lemma 1, the resulting graph has stretch factor $\leq 2t + 1$. Also, by Lemma 2, the girth of the output is $> 2t + 1$. Thus, by Lemma 6, the size of the output is $< n \cdot \lceil n^{1/t} \rceil$. In the next section we will show that this size bound is quite tight because there exist graphs for which one cannot do much better.

We now prove the weight bound. By Lemma 4, $MST(G)$ is contained in the subgraph. For each vertex v, consider the graph G_v, composed of $MST(G)$ and the edges in G' incident to v and not in $MST(G)$. Let the set of latter edges be denoted as E_v. It is easy to see that this graph is planar. Hence by methods similar to the proof of Theorem 2, we can show that $Weight(E_v) < Weight(MST(G)) \cdot 1/t$. Thus $\Sigma_{v \in V} Weight(E_v) < Weight(MST(G)) \cdot n/t$. In the above summation each edge in the E_v's have been counted twice, thus $Weight(G') < Weight(MST(G)) \cdot (1 + n/2t)$, which proves the weight bound. In the next section we shall see that this bound is not as tight as our other results. ●

3. LOWER BOUNDS

In this section we will show various lower bounds for spanners of graphs.

Our first result concerns general graphs with arbitrary positive weights. In [PS], using results from extremal graph theory [B], it has already been shown that, for every $t \geq 1$, there exist infinitely many n-vertex graphs with unit edge weights such that every $(2t + 1)$-spanner requires $\Omega(n^{1 + 1/(2t + 3)})$ edges. Thus the size result of Theorem 1 is tight up to a constant factor in the exponent of n. The above lower bound gives a corresponding lower bound for weights of spanners, that is, for every $t \geq 1$, there exist infinitely many n-vertex weighted graphs G such that every $(2t + 1)$-spanner has weight $\Omega(Weight(MST(G)) \cdot n^{1/(2t + 3)})$. Thus the weight result of Theorem 1 is not as tight as the size result.

Our next result concerns planar graphs with arbitrary positive weights.

Theorem 3: For infinitely many n and t, there exists a n-vertex planar graph G with unit edge weights such that every $(2t + 1)$-spanner requires $\Omega(n \cdot (1 + 1/t))$ edges, and has weight $\Omega(Weight(MST(G)) \cdot (1 + 1/t))$.

Proof. For infinitely many n and t, we state without proof that it is possible to construct a planar n-vertex graph with unit edge weights, such that each face is a regular $(2t + 4)$-gon, and such that the girth is also $\geq 2t + 4$. From Euler's formula it can be derived that the size of this graph is $\Omega(n \cdot (1 + 1/t))$. Clearly any proper subgraph will have stretch factor $\geq 2t + 3$, and the result follows. •

This shows that the size and weight results of Theorem 2 are tight.

In what follows we shall state lower bound results for various generalizations of spanners. Consider graphs with arbitrary positive weights. Let $Dist(u, v, G)$ be the distance from vertex u to vertex v in graph G. Let $G_1 = (V_1, E_1)$ and $G_2 = (V_2, E_2)$ be two graphs with V_1 a subset of V_2. G_2 is called a *generalized t-spanner* of G_1 if, for all $u, v \in V_1$, $Dist(u, v, G_1) \leq Dist(u, v, G_2) \leq t \cdot Dist(u, v, G_1)$. Thus, the spanner is not simply a subgraph of G_1, rather it may contain auxiliary vertices and edges. An important point is, the spanner is not allowed to "cheat", that is, paths in G_2 are never shorter than those in the original graph G_1, though they may use auxiliary vertices and edges.

For special graphs generalized spanners can be substantially smaller than simple spanners. For example, consider the complete n-vertex graph with unit edge weights. Clearly every simple 1-spanner requires all edges. But the *star graph* with one auxiliary vertex attached to all n original vertices via n additional edges of weight $1/2$ is a generalized 1-spanner with only n edges. However, the following lower bound result shows that such constructions are not always possible for all graphs. In fact, there exist graphs such that generalized spanners cannot be much smaller than simple spanners.

Theorem 4: For infinitely many n and t, there exists a n-vertex graph G with unit edge weights such that every generalized t-spanner requires $\Omega(\frac{1}{\log n} \cdot n^{1 + 1/(t + 2)})$ edges.

Before we prove the theorem, we need some definitions and lemmas. Let $t \geq 1$, and m be a positive integer. Let g be a function that maps each unweighted n-vertex graph to a string of m bits. We say g is an (n, t, m)-*compressor* if the following is true. For all pairs of graphs G_1, G_2, if $g(G_1) = g(G_2)$, then for all pairs of vertices u, v, $Dist(u, v, G_1) \leq t \cdot Dist(u, v, G_2)$. Informally, the compressor is just like a hash function. It partitions all graphs into groups, such that in each group the graphs have approximately the same distances between any given pair of vertices. Clearly, as t increases, it should be possible to construct compressors with fewer groups. The following lemma provides a lower bound on the number of groups, given any t.

Lemma 7: For an (n, t, m)-compressor to exist, $m \geq 1/4 \cdot n^{1 + 1/(t + 2)}$.

Proof. For infinitely many n and t, there exist n-vertex graphs with girth $\geq t + 2$ and at least $1/4 \cdot n^{1 + 1/(t + 2)}$ edges [B, pp 104]. Let $G = (V, E)$ be such a graph, and let G_1, G_2 be two different subgraphs of G. By the girth condition of G, both subgraphs have to belong to different groups of the compressor. Thus the total number of groups is $\geq 2^{|E|}$, thus $m \geq |E|$. •

Two further observations will be useful in proving the theorem. First, existence of spanners (even generalized spanners) for all graphs implies the existence of a compressor. For example, consider a collection of t-spanners, one for each n-vertex graph. Let the set of spanners (without repetitions) be $\{G'_1, \cdots, G'_k\}$. Then, setting $g(G) = binary(i)$ if G'_i is a spanner of G, yields an $(n, t, \lceil log k \rceil)$-compressor.

Second, we show how to encode an unweighted graph as a bit string. Every n-vertex graph with m edges can be encoded by $2 \cdot \lceil log n \rceil \cdot m$ bits, by representing every edge as a pair of $\lceil log n \rceil$ bit strings, each in turn representing a vertex.

Proof of Theorem 4. We shall first prove the theorem for unweighted spanners, then sketch the proof for weighted spanners. To prove the first part, consider any collection of t-spanners, one for each n-vertex graph. Since we are interested in sparse generalized spanners of n-vertex graphs, even with auxiliary vertices these spanners should have no more than $n + 2 \cdot n(n - 1)/2 = n^2$ vertices. Let m be the size of the largest spanner in this set. Each spanner can be encoded in $\leq 4 \cdot \lceil log n \rceil \cdot m$ bits. Thus by our previous observation, these spanners imply the existence of an $(n, t, 4 \cdot \lceil log n \rceil \cdot m)$-compressor. But by Lemma 7, $4 \cdot \lceil log n \rceil \cdot m \geq 1/4 \cdot n^{1 + 1/(t + 2)}$. Solving for m proves the theorem.

To prove the second part, the idea is to encode the edge weights of the spanners as bit strings. The problem arises because weights could be arbitrarily large, or require many precision bits. However, it can be shown that by ignoring very large weights, and rounding off the remaining in a certain way, we can get suitably encodable spanners with slightly larger stretch factors. •

Note that we have no lower bounds on the weight sparseness of generalized spanners. We now generalize spanners in another direction. Let $t \geq 1$. $G' = (V, E')$ is a t-*approximator* of $G = (V, E)$ if, for all u, $v \in V$, $1/t \cdot Dist(u, v, G) \leq Dist(u, v, G') \leq t \cdot Dist(u, v, G)$. The following theorem (without proof) provides a lower bound on the size of approximators.

Theorem 5: For infinitely many n and t, there exists a n-vertex graph G with unit edge weights such that every t-approximator has $\Omega(\frac{1}{\log n} \cdot n^{1 + 1/(t + 2)})$ edges.

We finish this section by stating a lower bound for *complete* graphs with weights that are *proper*. The latter mean that the weight of any edge is never more than the length of any path between its end vertices.

Theorem 6: For infinitely many n and t, there exists a n-vertex complete graph G with proper edge weights, such that every $(2t - 1)$-spanner requires $\Omega(1/t \cdot n^{1 + 1/(2t + 1)})$ edges.

All the above lower bound results emphasize robustness, that is, they hold even after allowing more general spanners, or more restrictive graphs. In the next section we discuss spanners of Euclidean graphs.

4. SPANNERS IN EUCLIDEAN GRAPHS

In Euclidean graphs, the weights assigned to the edges are not arbitrary, hence the lower bounds of Section 3 do not apply. It has been possible to construct spanners for such graphs with a linear number of edges.

Our first result is on Euclidean graphs on the plane. The vertices of such a graph are a set V of n points on the plane. The edges are line segments joining pairs of vertices and each edge weight is the Euclidean distance (measured in some norm, $|| \cdot ||$) between the vertices. These graphs may be either planar or nonplanar. We let $K(V)$ be the complete Euclidean graph (which is clearly nonplanar). We pose the question: are there sparse *planar* spanners of $K(V)$? This problem has been extensively studied in the past. In [C, DFS] it was shown that *Delaunay triangulations* in the $|| \cdot ||_1$ and $|| \cdot ||_2$ norms are spanners with constant stretch factors. Using a general framework, [DJ] showed that other planar graphs such as *greedy triangulations* and *minimum weight triangulations* also have constant stretch factors (with different constant values). In [KG] the stretch factor for Delaunay triangulations in the $|| \cdot ||_2$ norm was improved to 2.42 (current best). In [DJ, LL] it was shown that there exist extremely short Euclidean planar graphs (that is, almost as short as the minimum spanning tree) that have constant stretch factors. The algorithm in [LL] produces arbitrarily short graphs based upon a parameter, though it is not obvious how small are the *sizes* of these graphs. Using Theorem 2 we can produce planar spanners that are both short, as well as small, in size.

Let $t \geq 1/2$, and consider the Delaunay triangulation over V in the $|| \cdot ||_2$ norm. It is known that this triangulation contains $MST(K(V))$. We now apply the *SPANNER* algorithm on the triangulation with $r = 2t + 1$. Because spanners are transitive, it is easy to see that the output (denoted as G') satisfies the following property.

Theorem 7: G' is a $(2.42) \cdot (2t + 1)$-spanner of $K(V)$, such that
1) $Size(G') \leq (n - 2) \cdot (1 + 2/\lfloor 2t \rfloor)$,
2) $Weight(G) < Weight(MST(K(V))) \cdot (1 + 1/t)$.

In higher dimensions planarity of spanners is not an issue. Our next result is on constructing linear sized spanners for complete graphs of point sets V in $(R^d, ||\cdot||)$, for all dimensions $d \geq 2$, and all norms $||\cdot||$. Fix some angle $\delta > 0$. The key idea is to cover R^d by finitely many open cones C_1, ..., $C_{s(\delta)}$, all with the same focus at the origin O, such that for all points u, v in the same cone, $\angle u\, O\, v < \delta$. Such a covering exists for every d and every $||\cdot||$ by the theorem of Heine-Borel [CS]. Similar ideas for constructing spanners for fixed norms have been considered in [K].

We construct the spanner G' as follows. For every $v \in V$, consider the covering of R^d by the cones $C_1 + v$, ..., $C_{s(\delta)} + v$, where $C + v$ represents a shifting of the cone C to a new origin v, in the spirit of Minkowski. For every cone $C_i + v$, let u be the vertex in the cone such that $||u - v||$ is minimized. We add $[u, v]$ to G', and u is known as the ith neighbor of v.

Clearly, the size of G' is $\leq s(\delta) \cdot n$, and is therefore linear. It remains to show that its stretch factor is small. Consider u, $v \in V$. A short path between them is constructed in the following way. Let u be inside the cone $C_i + v$. Go from v to its ith neighbor, and proceed from there in the same way. We show that this path is not too long with respect to $||v - u||$.

Lemma 8: For every $\epsilon < 1/(2 + \sqrt{2})$, there is an angle $\delta(\epsilon) > 0$ such that $Distance(v, u, G') < ||v - u||/(1 - \epsilon(2 + \sqrt{2}))$.

The estimation of the angle $\delta(\epsilon)$ is rather technical to prove and we omit it from this version of the paper. The basic idea is to bound distances in $||\cdot||$ from above and below by constant multiples of corresponding distances in $||\cdot||_2$. The lemma leads to the following theorem.

Theorem 8: For every $t > 0$, dimension d, and norm $||\cdot||$ of R^d, there exists a constant $c(t, d, ||\cdot||)$ such that every finite set V has a t-spanner with at most $c \cdot n$ edges.

5. OPEN PROBLEMS

We conclude with some open problems.

1) In Theorem 1, the bound in the weight does not agree with the lower bound as nicely as in the other results. Can it be improved? We feel that our strategy of dividing the graph into planar components cannot be extended to yield the optimal answer.

2) Even in the other results, there are gaps between upper and lower bounds. For instance, in Theorem 1, the upper bound for the size is $O(n^{1 + 1/t})$, while the lower bound is $\Omega(n^{1 + 1/(2t + 3)})$. Only for $t = 1$, the known bounds, $\Theta(n^{3/2})$, are of the same order. The lower bound was proved in [L], and the upper bound is mentioned in [PU].

3) For dimensions higher than 2, Euclidean spanners with linear sizes exist. Do Euclidean spanners exist with weights within a constant multiple of the weight of the minimum spanning tree?

4) What spanners do random graphs have? The Euclidean random case has been examined in [SV].

5) Consider R^d, some fixed norm, and $t > 1$. What are the worst (or at least bad) point configurations V with respect to the number of edges in optimal t-spanners?

6) Do spanners have other applications?

6. ACKNOWLEDGEMENTS

Thanks are due to Torsten Sillke for his help in simplifying the proof of Lemma 8.

7. REFERENCES

[A] Althöfer: On Optimal Realizations of Finite Metric Spaces by Graphs: Discrete and Computational Geometry 3, 1988, 103-122.

[Aw] Awerbuch: Complexity of Network Synchronization: JACM, 1985, 804-823.

[ABLP] Awerbuch, Bar-Noy, Linial, Peleg: Compact Distributed Data Structures for Adaptive Routing: STOC, 1989, 479-489.

[ADDJ] Althöfer, Das, Dobkin, Joseph: Generating Sparse Spanners for Weighted Graphs: submitted to Discrete and Computational Geometry.

[AP] Awerbuch, Peleg: Routing with Polynomial Communication-Space Tradeoff: Manuscript, 1989.

[B] Bollobas: Extremal Graph Theory: Academic Press, 1978.

[BD] Bandelt, Dress: Reconstructing the Shape of a Tree from Observed Dissimilarity Data: Advances in Appl. Maths, 7, 1986, 309-343.

[Be] Bern: private communication to David Dobkin, 1989.

[C] Chew: There is a Planar Graph Almost as Good as the Complete Graph: ACM Symposium on Computational Geometry, 1986, 169-177.

[CS] Conway, Sloane: Sphere Packing, Lattices, and Groups: Springer, New York, 1988.

[D] Dress: Trees, Tight Extensions of Metric Spaces: Adv. in Math. 53, 1984, 321-402.

[DFS] Dobkin, Friedman, Supowit: Delaunay Graphs are Almost as Good as Complete Graphs: FOCS, 1987, 20-26.

[DJ] Das, Joseph: Which Triangulations Approximate the Complete Graph?: International Symposium on Optimal Algorithms, 1989 (LNCS, Springer-Verlag).

[K] Keil: Approximating the Complete Euclidean Graph: SWAT, 1988 (LNCS, Springer-Verlag).

[KG] Keil, Gutwin: The Delaunay Triangulation Closely Approximates the Complete Euclidean Graph: WADS, 1989 (LNCS, Springer-Verlag).

[L] Longyear: Regular d-valent Graphs of Girth 6 and $2(d * d - d + 1)$ Vertices: Journal of Combin. Theory 9, 1970, 420-422.

[LL] Levcopoulos, Lingas: There are Planar Graphs Almost as Good as the Complete Graphs and as Short as Minimum Spanning Trees: International Symposium on Optimal Algorithms, 1989 (LNCS, Springer-Verlag).

[PS] Peleg, Schäffer: Graph Spanners: Journal of Graph Theory, Vol 13 No 1, 1989, 99-116.

[PU] Peleg, Upfal: A Tradeoff Between Space and Efficiency for Routing Tables: STOC, 1988, 43-52.

[PUl] Peleg, Ullman: An Optimal Synchronizer for the Hypercube: SIAM Journal of Computing, Aug 1989, 740-747.

[S] Simoes-Pereira: A Note on the Tree Realizability of a Distance Matrix: Journal of Combin. Theory 6, 1969, 303-310.

[SV] Sedgewick, Vitter: Shortest Paths in Euclidean Graphs: Algorithmica 1, 1986, 31-48.

[T] Tarjan: Data Structures and Network Algorithms: Society for Industrial and Applied Mathematics, 1983.

Finding the k Smallest Spanning Trees

David Eppstein

Xerox Palo Alto Research Center

Palo Alto, CA 94304

Abstract

We give improved solutions for the problem of generating the k smallest spanning trees in a graph and in the plane. Our algorithm for general graphs takes time $O(m \log \beta(m, n) + k^2)$; for planar graphs this bound can be improved to $O(n + k^2)$. We also show that the k best spanning trees for a set of points in the plane can be computed in time $O(\min(k^2 n + n \log n, k^2 + kn \log(n/k)))$. The k best orthogonal spanning trees in the plane can be found in time $O(n \log n + kn \log \log(n/k) + k^2)$.

1 Introduction

One of the fundamental problems in graph theory is the computation of a *minimum spanning tree* (MST). Given an undirected graph G with weights on each edge, the MST of G is the tree spanning G having the minimum total edge weight among all possible spanning trees. This problem has been intensely studied, and good algorithms are known; currently the best known bound, by Gabow et al [13], is $O(m \log \beta(m, n))$ for a graph with n vertices and m edges. Here $\beta(m, n)$ is defined to be $\min \{i \mid \log^{(i)} n \le m/n\}$, and $\log^{(i)} x$ denotes the log function iterated i times. This is extremely close to linear time. We should also note that, for planar graphs, a MST can be found in linear time [7].

Minimum spanning trees have applications in many areas, including network design, VLSI, and geometric optimization. Yet in many cases what is desired is not necessarily the best spanning tree, but rather a "good" tree with some other properties that may be difficult to quantify. For instance, minimum spanning trees can be used to approximate a Euclidean travelling salesman tour, but it might be the case that some tree other than the minimum could yield a better tour. An approach to this difficulty is to generate a number of "good" trees, and then choose among them by whatever other criteria are desired. This leads to the problem of generating these trees.

A natural formulation of this problem is to find the k least weight spanning trees, for some input parameter k. This problem is not so well known as the usual MST problem, but it has been previously studied. It should be contrasted with the much harder problem of finding the k best possible spanning tree weights [15].

The main previous result, by Katoh et al [14], is that the k best spanning trees can be found in time $O(m \log \beta(m, n) + km)$. This result improved several prior results by Burns and Haff [5], Camerini et al [6], and Gabow [12]. Apparently, Dov Harel has further improved this for certain ranges of the input parameters, to $O(m \log \beta(m, n) + kn \log^2 n)$ [11].

Frederickson [11] considered the problem for small k, in particular $k = O(\sqrt{m})$. His algorithm uses a technique for maintaining a MST in a dynamically changing graph, and runs in time $O(m \log \beta(m, n) + k^2\sqrt{m})$. Frederickson also gave a version of his algorithm for planar graphs that runs in time $O(n + k^2 \log^2 n)$. Recently, Frederickson's technique for maintaining a MST in a dynamic planar graph has been improved by Eppstein et al [10], giving a bound for the k best spanning trees problem of $O(n + k^2 \log n)$; this improves the bound of [14] whenever $k = O(n/\log n)$.

Another well known MST problem is that of finding the MST of a set of points in the plane, with the weight of an edge connecting points x and y being the Euclidean distance between the two. It turns out that this can be solved in time $O(n \log n)$, using the fact that all MST edges must appear in the Delauney triangulation of the points [2]. Strangely, the problem of finding the k best minimum spanning trees of a set of points in the plane does not seem to have been studied; however it can clearly be solved in time $O(kn^2)$ using the technique of [14] on the complete graph of all pairs of points.

In this paper we again consider the problem of generating the k best spanning trees, both in a graph and in the plane, again for small k. We show that the k best spanning trees in any graph can be found in time $O(m \log \beta(m, n) + k^2)$; this is better than the previous $O(m \log \beta(m, n) + k^2\sqrt{m})$ bound. For planar graphs, our algorithm can be made to run in time $O(n + k^2)$, improving the previous $O(n + k^2 \log n)$ bound. Both algorithms improve the $O(m \log \beta(m, n) + km)$ bound of [14] whenever $k = O(m)$.

The general graph approach clearly leads to an $O(n^2 + k^2)$ bound for the Euclidean k best spanning trees problem, which improves the previous $O(kn^2)$ bound when $k = O(n^2)$. We describe two modifications to this technique. The first finds the k best spanning trees in time $O(k^2n + n \log n)$; it is best when $k = O(\log n)$, and is $O(n \log n)$ for $k = O(\sqrt{\log n})$. The second achieves time $O(k^2 + kn \log(n/k))$, and is better than $O(n^2 + k^2)$ when $k = O(n)$. We also consider alternate planar metrics; in particular we give an $O(k^2 + kn \log \log(n/k) + n \log n)$ algorithm to find the k best orthogonal spanning trees.

The algorithm for finding spanning trees in graphs consists of three stages. First, we show that many edges of the MST must be contained in every one of the k best spanning trees; therefore, we can contract the input graph G to a new graph G' having only $O(k)$ vertices. Then, we show that many edges not in the MST can also not be in any of the k best spanning trees; therefore, we can remove them from the graph, leaving only $O(k)$ edges. Finally, we apply the algorithm of [14], which takes time $O(k^2)$ on the reduced graph.

Throughout this paper we allow graphs to have multiple edges between the same pair of vertices. Therefore we do not denote an edge by its adjacent vertices, but rather treat it as a separate entity. The presence of multiple adjacencies between two vertices cannot affect the MST of a graph, but may affect the outcome of the k best spanning trees computation.

2 Contracting Required Edges

Given a graph G, and an edge e connecting vertices x and y in G, we define the *contraction* $G \cdot e$ to be the graph G' having as vertices $V(G) - y$, and edges $c_e(E(G))$, where the function c_e throws away edges connecting x and y, substitutes for each edge e' connecting y and any vertex $z \neq x$ a new edge e'' connecting x and z, and leaves unchanged all remaining edges. The contraction function c_e is defined to preserve the weight $w(e)$ of each edge not thrown away.

Lemma 1. *For any edge e in a spanning tree T of graph G, $c_e(T)$ is the MST of $G \cdot e$ if and only if T is the least weight spanning tree of G containing e.*

Proof: Obvious from the definitions. □

The following characterization of MSTs is well known:

Lemma 2. (folklore) *Let v be any vertex of G, and e be the least weight edge adjacent to v. Let T be the MST of $G \cdot e$. Then $T + e$ is the MST of G.*

Now for any edge e in the spanning tree, disconnecting the tree into two components T_1 and T_2, define the *replacement edge* $r_G(e)$ to be the least weight edge in G, other than e, between a vertex in T_1 and T_2.

Lemma 3. (folklore) *For any edge e in the MST T of a graph G, such that $G - e$ is connected, $T - e + r_G(e)$ is the MST of $G - e$.*

Proof: If G has only two vertices, the lemma is obvious. Otherwise, find some edge $f \neq e$ that is a leaf in the MST T of G, and let the leaf vertex adjacent to f be x. Then by lemma 2, f is the least weight edge adjacent to x, or it would not be in T, and by lemma 1 $c_f(T)$ is the MST of $G \cdot f$. Also, contracting f does not change the components T_1 and T_2 formed by removing e, and therefore $r_{G \cdot f}(e) = r_G(e)$. By induction $c_f(T) - e + r_G(e)$ is the MST of $(G \cdot f) - e = (G - e) \cdot f$, and using lemma 1 again gives that $T - e + r_G(e)$ is the least weight spanning tree of $G - e$ containing f. But by lemma 2, this must be the MST of $G - e$. □

Lemma 4. *[17]* *Given a MST T of a graph G, the replacements $r_G(e)$ for all edges e in T can be computed in time $O(m\alpha(m,n))$.*

Note that this bound is never worse than the $O(m \log \beta(m,n))$ time required for constructing the MST of G. The following version of lemma 4 is given in [4].

Lemma 5. *Given a MST T of a planar graph G, the replacements $r_G(e)$ for all edges e in T can be computed in time $O(n)$.*

Lemma 6. *Given a MST T of a graph G, and the values of r_G for each tree edge, in linear time we can compute a set S of $n - k$ tree edges that must be contained in all of the k best spanning trees for G.*

Proof: For each edge e, let $w'(e) = w(r_G(e)) - w(e)$. I.e., w' is the extra weight we have to add to the tree if we remove e from the graph. The values of w' can be computed as above. Then find the edges with the $(k - 1)$ smallest values of w', using a linear time selection algorithm [3]. Let S be all the remaining edges.

Then for each edge e in S there are at least $k - 1$ edges e' with $w(T - e' + r_G(e')) \leq w(T - e + r_G(e))$, and therefore together with T there are at least k trees better than $w(T - e + r_G(e))$, which is in turn (by lemma 3) better than any other spanning tree not containing e. Therefore all the k best spanning trees must contain e. □

Lemma 7. *Given a MST T of a graph G, in time $O(m\alpha(m,n))$ we can find a set of edges S and a graph G' having only k vertices, such that the k best spanning trees of G are exactly the k best spanning trees of G' together with the edges of S. In a planar graph this can be performed in linear time.*

Proof: Let S be the set constructed in lemma 6, and let $G' = G \cdot S$ be the graph formed by contracting each of the edges in S. The contraction can easily be performed in linear time. $\square$

3 Subtracting Useless Edges

Just as tree edges have a single non-tree replacement edge $r_G(e)$, it turns out that non-tree edges have a single tree replacement edge. In particular, define $R_G(e)$ for an edge e connecting vertices x and y to be that edge on the tree path between x and y having the highest weight.

Lemma 8. *If T is the MST of a graph G, then $c_e(T - R_G(e))$ is the MST of $G \cdot e$.*

Proof: We contract tree edges other than $R_G(e)$ one at a time, using lemma 1 to show that they must belong to both MSTs. $\square$

As before, R_G can be computed efficiently for all edges:

Lemma 9. *[17] The replacement edges $R_G(e)$ for each edge e in a graph G, given a MST of G, can be computed in time $O(m\alpha(m,n))$.*

Lemma 10. *The replacement edges $R_G(e)$ for each edge e in a planar graph G can be computed in linear time.*

Proof: Since the minimum spanning tree of a planar graph is the complement of the maximum spanning tree of the dual [10], this follows from lemma 5. $\square$

And as before, this lets us simplify the graph:

Lemma 11. *Given a graph G and its MST T, we can find a set S' of $m - n - k$ non-tree edges such that none of the k best spanning trees contains any edge in S', in time $O(m\alpha(m,n))$.*

Proof: Define $W(e)$ to be $w(e) - w(R_G(e))$; then $W(e)$ is the weight added by including e in a spanning tree. Let f be the non-tree edge having the $(k-1)$st smallest value of W; as before f can be found by a linear time selection algorithm. Then as before, for any e with $W(e) > W(f)$, there are at least k trees better than $T - R_G(e) + e$, and therefore also better than any other tree containing e. Therefore e can never be included in any of the k best spanning trees for G. $\square$

Putting it all together, we have our result:

Theorem 1. *The k least weight spanning trees of a graph can be found in time bounded by $O(m \log \beta(m,n) + k^2)$; for a planar graph they can be found in time $O(n + k^2)$.*

Proof: By results of [13] and [7], the MST of the graph can be found in the given bounds. We can use lemma 7 to reduce the graph to one in which there are k vertices, and therefore $k - 1$ tree edges. Then we can use lemma 11 to reduce the graph to one in which there are $k - 1$ non-tree edges. Therefore the total number of edges in the final reduced graph is $O(k)$. Applying the algorithm of [14] gives the desired total bound. $\square$

4 Euclidean Spanning Trees

Clearly, a Euclidean spanning tree is just a spanning tree in the complete graph with vertices corresponding to the input points in the plane, edges between every pair of vertices, and the weight of an edge equal to the distance between the corresponding vertices. Therefore, from the results above, we have

Theorem 2. *The k best spanning trees for a set of points in the plane can be found in time $O(n^2 + k^2)$.*

To do better than this, we will need to remove many edges quickly from the complete graph. To do this, we will use the standard geometric technique of Voronoi diagrams.

The *order r Voronoi diagram* is a subdivision of the plane into regions. Each region can be denoted $v(S)$ for some set S containing exactly k of the input points (which in this context are called *sites*), and is the locus of points x for which all sites in S are closer to x than any site not in S. A set S corresponds to a Voronoi region $v(S)$ exactly when there is a circle in the plane containing S and no other of the sites; then the center of the circle is one of the points in $v(S)$.

Lemma 12. *If (x,y) is an edge in one of the k best spanning trees, then x and y are both sites in some set S corresponding to a Voronoi region $v(S)$ in the order $(k+1)$ Voronoi diagram of the input points.*

Proof: Let x and y be two arbitrary input points. Let e in the MST T of the points be $R_G((u,v))$, and let the removal of e disconnect T into two components T_1 and T_2. Without loss of generality $x \in T_1$ and $y \in T_2$.

Consider the circle C for which line segment xy is diameter. For each site z other than x and y in this circle, if z is in T_2 then (x,z) connects T_1 and T_2, and is shorter than (x,y); therefore $T - e + (x,z)$ is a better spanning tree than $T - e + (x,y)$. Similarly, if z is in T_1 then $T - e + (z,y)$ is better than $T - e + (x,y)$. But $T - e + (x,y)$ is the best spanning tree containing (x,y); therefore if (x,y) is in any of the k best spanning trees, there can be at most $k - 1$ possible sites z, and C contains at most $k + 1$ sites including x and y. $\square$

Lemma 13. *[1] There are $O(rn)$ pairs of sites (x,y) that belong to a common region of an order r Voronoi diagram. The diagram can be found, and all such pairs can be enumerated, in time $O(k^2n + n\log n)$.*

Theorem 3. *The k best spanning trees for a set of points in the plane can be found in time $O(k^2n + n\log n)$.*

Proof: Lemmas 12 and 13 show that in the time bound we can reduce the problem to a graph problem in a graph with n vertices and $O(kn)$ edges. Then by theorem 1 we can find the k best spanning trees in time $O(kn\log\beta(kn,n) + k^2) = O(kn + n\log n + k^2) = O(k^2n + n\log n)$. □

5   Faster Euclidean Spanning Tree Construction

In theorem 3, we did not make much use of the results of the first two sections, wherein we showed how to reduce the number of edges that need be considered to find the k best spanning trees. Indeed, after the reduction to a graph with $O(kn)$ edges, we could have used the algorithm of [14] instead of theorem 1, and still achieved the same final bound. Therefore it should not come as a great surprise that we can improve this bound, for certain ranges of the parameters k and n.

First, recall that the Euclidean MST of the points can be computed in time $O(n\log n)$. If we could compute $r(e)$ for each edge e of the MST, then by lemma 6 we could show that all but k of the MST edges must remain in all the k best spanning trees. This may be done as follows.

Lemma 14. *For each replacement edge $r(e)$ connecting points x and y, it must be the case that x and y are part of a set S forming a region $v(S)$ in the order-3 Voronoi diagram of the input points.*

Proof: Let e divide the MST T into components T_1 and T_2, with x in T_1 and y in T_2. Construct the circle C with diameter xy. If C contains a site z which is neither x, y, nor an endpoint of e, then as in lemma 12 either $T - e + (x,z)$ or $T - e + (z,y)$ is better than $T - e + (x,y)$, contradicting the assumption that $r(e) = (x,y)$. The only other way that C could contain four sites would be if they were x and y together with the two endpoints of e; but then again we would have a better replacement for e than (x,y). Therefore C contains at most three sites, and x and y belong to the Voronoi region containing the center of C. □

Corollary 1. *We can compute $r(e)$ for each edge e in the MST, in time $O(n\log n)$.*

Proof: The order-3 Voronoi diagram can be constructed in this time by lemma 13. It contains $O(n)$ pairs (x,y) sharing a Voronoi region; therefore by lemma 4 each value of $r(e)$ can be computed in a total time bound of $O(n\alpha(n)) = O(n\log n)$. □

Now as before we can find a set S of all but k edges in the MST, such that the best k spanning trees each contain all the edges in S. Denote the remaining MST edges by $m_1, m_2, \ldots, m_k$. Now we must show how to use this information to reduce the possible non-tree edges we must consider.

Given a point x, and a MST edge m_i, let m_i divide the MST into components T_1 and T_2 with x in T_1; then we define $f_i(x)$ to be the nearest point to x in T_2. We further define

$$F(x) = \min_i w(x, f_i(x)) - w(m_i). \tag{1}$$

Lemma 15. *For each x, $F(x) + w(T)$, where T is the MST, gives the minimum weight of a spanning tree containing all the edges in S and containing an edge not in S adjacent to x.*

Proof: By lemma 8, the spanning tree defined by the lemma must be of the form $T - m_i + (x, y)$ for some i and y. But this is exactly what is minimized by $F(x)$. □

We now show how to compute $F(x)$; this depends on the following well known technique:

Lemma 16. *Given a set A of m points, and a set B of n points, we can compute for each point of B the nearest point in A, in time $O(n \log m)$.*

Proof: We simply construct the Voronoi diagram of A, and use a planar point location procedure [16]. □

Lemma 17. *All values of $F(x)$ can be computed in time $O(nk \log(n/k))$.*

Proof: Removing the edges m_i divides the MST T into $k + 1$ components T_i. We first compute, for each input point, the nearest point in each component. This takes time $O(n \sum \log |T_i|)$; the sum in this bound is maximized when all T_i are equal in size, and the bound then becomes $O(nk \log(n/k))$.

The components T_i, connected by the edges m_i, can be imagined as forming a tree with $(k + 1)$ vertices; there are $2k$ subtrees that can be formed by removing any edge of the tree. For each point, we find the nearest point to it in each of those subtrees; this may easily be done in time $O(k)$ per point by dynamic programming. Therefore this step takes time $O(nk)$.

At this point we have computed $f_i(x)$ for each i and x, as the nearest point to x in the subtree not containing x of the two formed by removing edge m_i from the component tree. From this $F(x)$ may be computed directly from formula 1, in time $O(nk)$. □

Theorem 4. *The k best spanning trees for a set of points in the plane can be computed in time $O(k^2 + kn \log(n/k))$.*

Proof: The algorithm of corollary 1 takes time $O(n \log n)$, which is always dominated by the $O(kn \log(n/k))$ time to compute $F(x)$. By lemma 11, the k best spanning trees together contain only k edges not already in the MST, and (counting the MST edges not in S) $2k$ edges not in S. Therefore only the $4k$ points having the lowest values of $F(x)$ may be endpoints of those edges. By using lemma 17 and a linear time selection algorithm [3] we may find those $4k$ points in time $O(kn \log(n/k))$. These points determine $O(k^2)$ possible non-MST edges. Therefore, the problem becomes one of finding the k best spanning trees in a graph with $O(k^2)$ edges and $O(k)$ vertices; this can be solved in time $O(k^2)$. □

6 Alternate Metrics

All the algorithms described above depend only on some simple properties of the Voronoi diagram, and therefore work just as well for alternate metrics in the plane. However the order r Voronoi diagram for general metrics is not known to take less time than $O(k^2 n \log n)$, and therefore theorem 3 never leads to a better time than the $O(k^2 + kn \log n)$ bound of theorem 4.

However in certain cases we can do better. In particular, for the L_1 (equivalently, L_∞) metric, corresponding to the important case of orthogonal spanning trees, we can achieve a time bound of $O(k^2 + kn \log \log n + n \log n)$; this is always at least as fast as our algorithms for the usual L_2 metric.

The algorithm we use is essentially the same as that of theorem 4. The key difference is in the computation of $f_i(x)$, which was the only step in that theorem requiring time $O(kn \log n)$. Recall that this can be considered to be $O(k)$ computations of the following form: given a partition of the n input points into sets A and B, find the nearest point in B to each point in A. We now show that, after an $O(n \log n)$ preprocessing stage (consisting of sorting the points in various orders) we can solve any such problem in time $O(n \log \log n)$. This will give us the final bound we seek.

Recall that the L_1 distance between points (x, y) and (x', y') is simply $|x - x'| + |y - y'|$. We will solve the following simplified version of the problem:

Lemma 18. Given a subset A containing m of the n input points, and with $O(n \log n)$ preprocessing, we can find in time $O(n \log \log m)$, for each input point (x, y), the point (x', y') in A closest to (x, y) with $x' \leq x$ and $y' \leq y$.

Proof: We process the points in order by their values of x (x'), and within the same value of x in order by y. The preprocessing stage consists of simply sorting the points in this order. At any stage in the processing, corresponding to some particular value of x, we maintain a data structure listing, for each value of y, the nearest point (x', y') in A with $x' \leq x$ and $y' \leq y$. This point must already have been processed by the order of processing. Then to find the nearest neighbor of an input point, we perform a lookup in the data structure; to process a point in A, we update the data structure.

Each point (x', y') in A nearest to any point must correspond to some interval $[y', y]$ in the data structure; this is because if some other point is nearer to a particular value of (x, y) it will be nearer to all points with greater values of y. We may represent this collection of intervals using the *flat tree* integer searching data structure of van Emde Boas [18], indexed by the m possible y coordinates of the points in A. Computing these indices for the n input points can be done by a linear time sweep of the points ordered by their y coordinates.

Then finding the nearest neighbor of an input point (x, y) simply consists of looking up y in the data structure to find which interval contains it. Processing a point (x', y') in A consists of again finding the interval containing y', then splitting that interval at y' to create a new interval for that point, and while each succeeding interval corresponds to a point (x'', y'') farther from (x, y'') than (x', y'), deleting those succeeding intervals and merging them into the new interval for (x', y').

All these data structure operations can be performed in time $O(\log \log m)$. A deletion can be charged to its corresponding insertion, so processing each point requires a constant amortized number of data structure operations. Therefore the whole operation takes time $O(n \log \log m)$. □

We should note that this algorithm is essentially identical to one used by Eppstein et al [9] to compute RNA secondary structure predictions; they also used a more complicated version of the algorithm as part of a method of computing optimal sequence alignments. Our algorithm should also be compared with that of Chew and Fortune [8] which computes the orthogonal Voronoi diagram of a set of points in $O(n \log \log n)$ time; our algorithm differs in simultaneously performing point location within the constructed Voronoi diagram.

Theorem 5. *The k best orthogonal spanning trees for a set of n points in the plane can be found in time $O(k^2 + kn \log \log(n/k) + n \log n)$.*

Proof: The algorithm of lemma 18 may be repeated four times, in four different directions, to solve the problem of, given a set A, finding the nearest point in A to each input point in A. All other steps are the same as in lemma 17 and theorem 4, using the bound of lemma 18 instead of that of lemma 16. □

Acknowledgements

I would like to thank Frances Yao for suggesting the Euclidean version of this problem and directing me to reference [1].

References

[1] A. Aggarwal, H. Imai, N. Katoh, and S. Suri, Finding k Points with Minimum Diameter and Related Problems, J. Algorithms, to appear.

[2] A. Aggarwal and J. Wein, Computational Geometry, MIT LCS Research Seminar Series 3, 1988.

[3] M. Blum, R.W. Floyd, V.R. Pratt, R.L. Rivest, and R.E. Tarjan, Time Bounds for Selection, J. Comput. Syst. Sci. 7, 1972, 448–461.

[4] H. Booth and J. Westbrook, Linear Algorithms for Analysis of Minimum Spanning and Shortest Path Trees in Planar Graphs, Tech. Rep. TR-763, Department of Computer Science, Yale University, Feb. 1990.

[5] R.N. Burns and C.E. Haff, A Ranking Problem in Graphs, 5th Southeast Conf. Combinatorics, Graph Theory and Computing 19, 1974, 461–470.

[6] P.M. Camerini, L. Fratta, and F. Maffioli, The k Shortest Spanning Trees of a Graph, Int. Rep. 73-10, IEEE-LCE Politechnico di Milano, Italy, 1974.

[7] D. Cheriton and R.E. Tarjan, Finding Minimum Spanning Trees, SIAM J. Comput. 5, 1976, 310–313.

[8] L.P. Chew and S. Fortune, Sorting helps for Voronoi diagrams, 13th Symp. Mathematical Progr., Japan, 1988.

[9] D. Eppstein, Z. Galil, R. Giancarlo, and G.F. Italiano, Sparse Dynamic Programming, 1st ACM-SIAM Symp. Discrete Algorithms, San Francisco, 1990, 513–522.

[10] D. Eppstein, G.F. Italiano, R. Tamassia, R.E. Tarjan, J. Westbrook, and M. Yung, Maintenance of a Minimum Spanning Forest in a Dynamic Planar Graph, 1st ACM-SIAM Symp. Discrete Algorithms, to appear.

[11] G.N. Frederickson, Data Structures for On-Line Updating of Minimum Spanning Trees, with Applications, SIAM J. Comput. 14(4), 1985, 781–798.

[12] H.N. Gabow, Two Algorithms for Generating Weighted Spanning Trees in Order, SIAM J. Comput. 6, 1977, 139–150.

[13] H.N. Gabow, Z. Galil, T.H. Spencer, and R.E. Tarjan, Efficient Algorithms for Minimum Spanning Trees on Directed and Undirected Graphs, Combinatorica 6, 1986, 109–122.

[14] N. Katoh, T. Ibaraki, and H. Mine, An Algorithm for Finding k Minimum Spanning Trees, SIAM J. Comput. 10, 1981, 247–255.

[15] E.W. Mayr and C.G. Plaxton, On the spanning trees of weighted graphs, manuscript, 1990.

[16] N. Sarnak and R.E. Tarjan, Planar Point Location using Persistent Search Trees, C. ACM 29(7), 1986, 669–679.

[17] R.E. Tarjan, Applications of Path Compression on Balanced Trees, J. ACM 26, 1979, 690–715.

[18] P. van Emde Boas, Preserving Order in a Forest in Less than Logarithmic Time, 16th IEEE Symp. Found. Comput. Sci., 1975, and Info. Proc. Lett. 6, 1977, 80–82.

The File Distribution Problem for Processor Networks*

Goos Kant   Jan van Leeuwen

Dept. of Computer Science, Utrecht University

P.O. Box 80.089, 3508 TB Utrecht,The Netherlands

Abstract

We investigate the segmentation problem for files in a network, as introduced in [3], for the case of regular interconnection networks. The strict segmentation problem can be reduced to the problem of finding a *strong* coloring of the network with $d + 1$ colors, where d is the degree of the network. We prove that a strong coloring with $d+1$ colors is possible for the d-dimensional hypercube if and only if $d = 2^i - 1$ for some $i > 0$, for the d-dimensional torus of size $l_1 \times \ldots \times l_d$ when $l_i \bmod q = 0, 1 \leq i \leq d$, if $\sqrt[r]{2d+1}|q$ for some $r > 0$, for the Cube-Connected Cycles CCC_d if and only if $d > 2, d \neq 5$, for the directed shuffle-exchange network, for the directed 4-pin shuffle network and for the chordal ring network, for some chord lengths and numbers of nodes. The results show in which instances perfect segmentations are possible for classical regular processor interconnection networks. Some results for other (almost) regular interconnection networks are included.

1 Introduction

In 1988 Bakker and van Leeuwen [3] introduced the segmentation problem for files in a network, which can be described in graph-theoretic terms as follows: "Given a connected network $G = (V, E)$ and a file F, assign to each node $x \in V$ a segment $F_x \subseteq F$ such that for all $x \in V, \bigcup_{\{x,y\} \in E} F_y \cup F_x = F$". The problem derives its significance from the need to distribute data fragments when local storage is at a premium, but communication channels are available to retrieve data from nearby processors. The condition is that in the neighborhood of every node x, a full copy of F can be assembled from the available segments. Interesting versions of the file segmentation problem result when further conditions are imposed on the fragments, e.g. in relation to their size or the maximum redundancy in the various neighbourhoods. By distributing a file the availability of a datafile F is maintained but, if the complete file need not be fully replicated at each node, storage space is saved (see e.g. [6]).

In this paper we consider the strict form of the file distribution problem in which it is required that in every neighbourhood the distributed fragments are free of overlaps, i.e. $\forall (x,y) \in E :$ $F_x \cap F_y = \emptyset$. When there are r different (disjoint) segments of F, this problem is meaningful only for $(r-1)$-regular networks. Consequently we restrict our attention to (almost) regular networks in this paper, which includes many current processor interconnection networks [14].

We now explain the file distribution problem as a coloring problem for graphs in more detail. A *distance-k coloring* of a graph G is any coloring of the vertices such that every two nodes that have distance $\leq k$ are assigned different colors. In this paper we are especially interested in distance-2 colorings, also known as *strong colorings*. Several results for strong colorings of graphs have been summarized in [10], e.g. including the result that for any $k \geq 1$, the DISTANCE-k CHROMATIC NUMBER problem is NP-complete for general graphs ([12]). In our context, we are interested in

---

*This work was supported by the ESPRIT II Basic Research Actions program of the EC under contract No. 3075 (project ALCOM).

strongly coloring regular graphs with exactly r colors, where $r-1$ is the degree of the regular graph. Any such coloring, which we will henceforth refer to as a *perfect coloring*, provides a scheme for the file distribution problem for a processor network whose topology coincides with the regular graph.

Let G be a regular graph of degree $r-1$ and suppose G can be perfectly colored with r colors. Any group of nodes that has the same color in some perfect coloring of G is called a *strong color group*. Whenever colorings are considered, $c(v)$ denotes the color, given to a node v.

Theorem 1.1 ([3]) *If a $(r-1)$-regular graph with $|V| = N$ nodes can be perfectly colored, then $r|N$.*

When G is a cycle, it follows that if $N = 5$ then five colors are required, if $N = 3k$ for some $k > 0$ then three colors are required, otherwise four. When G is a complete graph K_n (which is $(n-1)$-regular), then n colors are required, and a strong coloring with n colors is obtained by assigning each node v_i the color $c(v_i) = i$. It has recently been shown by Bakker [4] that the perfect coloring problem for regular graphs is also NP-complete, even for the case of 3-regular graphs. A property of perfect r-colorable graphs is that they are $(r-1)$-edge colorable, if r is even [10].

In this paper we consider the problem of determining which processor networks admit perfect colorings and, hence, allow for a solution of the file-distribution problem in disjoint (i.e., nonoverlapping) segments. Observe that in the case of directed graphs, a strong coloring of the graph $G = (V, E)$ means that all vertices v_i, for which there is an edge (v_i, w) pointing to a same endpoint w, the colors $c(v_i)$ must have a different color, and the colors of these nodes must be different from the color of the endpoint w. Note that this differs from the undirected case in the sense that now two nodes with distance two may have the same color.

The paper is organized as follows. In section 2 we consider the hypercubes C_d, and we prove that every C_d can be perfectly colored if and only if $d = 2^i - 1$ for some $i > 0$. In section 3 we inspect the d-dimensional torus, and show that a $l_1 \times \ldots \times l_d$ torus can be perfectly colored if $l_i \bmod q = 0, 1 \le i \le d$, with q such that $\sqrt[r]{2d+1}|q$ for some integer $r > 0$. In section 4 we inspect the Cube-Connected Cycles CCC_d, and we prove that we can strongly color these networks, if and only if $d > 2, d \ne 5$. In section 5, some results for the shuffle-exchange and the 4-pin shuffle network are given. In section 6 we inspect the chordal ring network and show for which instances of the chord length and number of nodes a perfect coloring is possible. In section 7 some other regular interconnection networks are considered and some concluding remarks are given. Some familiarity with standard interconnection networks (see e.g. [14]) and coding theory (see e.g. [11]) is assumed throughout. Omitted proofs can be found in the full paper [9].

2 Hypercubes

The first interconnection network we consider is the hypercube network C_n [14]. This network consists of 2^n nodes, $(n \ge 0)$, which we label by the numbers $0, \ldots, 2^n - 1$ (in binary form). There is an edge between node $v = v_{n-1}v_{n-2} \ldots v_1 v_0$ and node $w = w_{n-1}w_{n-2} \ldots w_1 w_0$, if and only if $v_i = \overline{w_i}$, for some $i, (0 \le i < n)$, and $v_j = w_j$, for all $j \ne i$. C_n is a n-regular connected interconnection network.

We will show that there is an intimate connection between the (perfect) strong colorings for C_n and the existence of certain binary codes. Recall that a 1-error correcting code of length n is any set of code words $S \subseteq \{0, 1\}^n$ in which the codewords have Hamming distance ≥ 3 [11].

Lemma 2.1 *Every perfect 1-error correcting code of length n is a strong color group of some perfect coloring of C_n, and vice versa.*

Proof: Let S be a perfect 1-error correcting code, i.e., a 1-error correcting code. $S = \{K_1, \ldots, K_k\}$ with $k = \frac{2^n}{n+1}$. After shifting S over any codeword, we may assume w.l.o.g. that $p_0 = 0 \ldots 0 \in S$. Let $p_i = 0 \ldots 010 \ldots 0$ (with a single "1" in the ith bit position). The cosets $p_i + S$

(with addition componentwise modulo 2) are all disjoint, because for no K_r and K_s one can have $p_i + K_r = p_j + K_s$ for $i \neq j$. (This follows because the $p_i's$ are at Hamming distance ≤ 2 from each other but the codewords have distance ≥ 3.) Together they cover exactly $(n+1).k = 2^n$ nodes, i.e., the entire C_n. Color the nodes of C_n by assigning color 1 to the nodes of S, color 2 to the nodes $p_1 + S$, etc. We claim that the resulting coloring is a strong coloring, hence a perfect coloring of C_n. This follows because for each node q, the nodes $q_0 = q$ and $q_i = $ "q with the ith bit flipped" ($i = 1, \ldots, n$) necessarily belong to different cosets and thus represent a full set of $n + 1$ colors in the neighbourhood of q.

Conversely, let S be a strong color group of some perfect coloring of C_n. We claim that S is a perfect 1-error correcting code. This is clear, as any two elements of S in a distance-2 coloring of C_n must have Hamming distance ≥ 3. By theorem 1.1 we have $|S| = \frac{2^n}{n+1}$ (note that C_n is a n-regular graph) and hence S is a perfect 1-error correcting code. □

The lemma shows that perfect colorings of C_n and, hence, solutions to the file distribution problem for C_n exist if and only if there exists a perfect 1-error correcting code of length n, i.e., a perfect $(n, 3)$ code. This is a well-studied problem in coding theory, which finds an interesting application in this context.

A necessary condition for the existence of a perfect coloring derives from the sphere packing condition for perfect 1-error correcting codes of length $n : n + 1 | 2^n$. Thus $n = 2^i - 1$, for some $i \geq 1$.

Theorem 2.2 *A hypercube C_n can be perfectly colored (i.e., admits a solution to the file distribution problem) if and only if $n = 2^i - 1$ for some $i > 0$.*

Proof: The necessity has just been shown. For the converse, we note that it is well-known that for every $n = 2^i - 1$ there exist perfect (binary) 1-error correcting codes, e.g. the binary Hamming codes (cf. van Lint [11]). □

This completely solves the file distribution problem for hypercubes in well-understood terms.

3 The d-dimensional torus

The second interconnection network we consider is the d-dimensional torus of dimensions $l_1, \ldots, l_d$. The network consists of $\Pi_{i=1}^d l_i$ nodes and there is an edge between node $v = (v_1, \ldots, v_d), 0 \leq v_i < l_i$, and node $w = (w_1, \ldots, w_d)$, if $v_i = (w_i \pm 1) \bmod l_i$ for some $i, 1 \leq i \leq d$, and $v_j = w_j$ for all $j \neq i$. Note that for $l_1 = \ldots = l_d = 2$, this network is equal to the hypercube C_d described in section 2. Note that in all cases this network is regular, and every node has degree $2d - k$, with $k = \#|l_i = 2|$.

Lemma 3.1 *If a d-dimensional torus of dimensions $l_1, \ldots, l_d$ can be perfectly colored, then the d-dimensional torus with dimensions $a_i.l_i$ can be perfectly colored, for every $a_i \in \mathbb{N}$ and i such that $l_i > 2$.*

We now characterize a non-trivial class of tori for which the file distribution problem is solvable. Recall that the Lee weight of a word $b_0 b_1 \ldots b_{n-1}, b_i \in \{0, 1, \ldots, q-1\}$, is defined as the sum of the Lee weights of its digits, where $w_L(b_i) = |b_i|, |b_i| = \pm b_i \bmod q$ and $0 \leq |b_i| \leq \frac{q}{2}$ [5]. A single-Lee-error-correcting code of length n is any set of words $S \subseteq \{0, \ldots, q-1\}^n$ in which the codewords have Lee distance ≥ 3.

Lemma 3.2 *Every perfect single-Lee-error-correcting code of length d is a strong color group in the d-dimensional torus with dimensions $l_1 = l_2 = \ldots = l_d = q$, and vice versa.*

Theorem 3.3 *If $\frac{q}{h} = \sqrt[r]{2d+1}$ (some integers $h, r > 0$) then the d-dimensional torus with dimensions $l_i \bmod q = 0$ ($1 \leq i \leq d$) can be perfectly colored.*

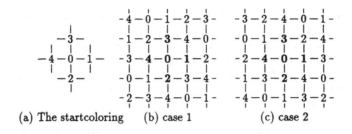

(a) The startcoloring   (b) case 1   (c) case 2

Figure 1: Startcoloring for the 2-dimensional torus

Proof: It is known that if $\frac{q}{h} = \sqrt{2d+1}$ (some integers $h, r > 0$), then there exists a perfect single-Lee-error-correcting code over a q-letter alphabet of length d [2]. Using Lemma 3.1 and Lemma 3.2, the theorem follows.   □

For $r = 1$ in theorem 3.3, a perfect coloring of the d-dimensional torus is given by $c(v) = (\sum_{i=1}^{d} i.v_i) \bmod (2d + 1)$. The construction derives from a result of Golomb and Welch [8] for close-packing the hypertorus with Lee spheres of radius 1.

There are also other d-dimensional tori, for which a perfect coloring is possible. For example, there exist perfect colorings for the $2 \times 3 \times 6$ torus. This leads to the question for which dimensions of a d-dimensional torus, with $\Pi_{i=1}^{d} l_i = \alpha(2d+1-k)$, where $k = \#|l_i = 2|$, a perfect coloring exists. Inspecting this problem for the 2-dimensional torus, we come to the following theorem.

Theorem 3.4 *A 2-dimensional torus of degree r and dimensions l_1 and l_2 can be perfectly colored, if and only if $l_1 = 5k$ and $l_2 = 5l$ or $l_1 = 2$ and $l_2 = 4k$, for some $l, k > 0$.*

Proof: Remark first that $l_1 = l_2 = 2$ gives us a C_4, which needs four colors for a strong coloring. If $l_1 = 2$, then the smallest 2-dimensional torus requiring four colors and $l_2 > 2$ has $l_2 = 4$, and node $v = (v_1, v_2)$ gets color $(2v_1 + v_2) \bmod 4$. If $l_1, l_2 > 2$, then we get a 4-regular network. Starting with some node, we get the startcoloring as described in figure 1. This initial coloring forces the coloring of the remainder of the network, and there are two cases now. The first case leads to the coloring $c(v) = (2v_1 + v_2) \bmod 5$, which is correct by our example. The second case is equal to the first case, by renumbering the colors and swapping the dimensions. Using lemma 3.1, the theorem follows.   □

The Illiac network differs a little from the 2-dimensional torus. It has n nodes arranged in a $\sqrt{n} \times \sqrt{n}$ array (n a square) and node i is adjacent with $(i \pm 1) \bmod n$ and $(i \pm \sqrt{n}) \bmod n$ [14].

Theorem 3.5 *An Illiac network cannot be strongly colored with 5 colors and (hence) is not perfectly colorable.*

Proof: The coloring "starts" in the same way as in the 2-dimensional torus. The vertical "wrapped around"-connections of the Illiac network are equal to those in the 2-dimensional torus, hence the vertical length must be $5k, k > 0$. For a correct sequence in the horizontal dimension, the width must be equal to $5l + 2$ or $5l + 3$ (by swapping the dimensions) for some $l, l \geq 0$. This can never be equal to a number of the form $5k$.   □

In the directed case of the d-dimensional torus, every node $v = (v_1, \ldots, v_d)$ is directly connected to vertex w, if $v_i = (w_i + 1) \bmod l_i$ for some $i, 1 \leq i \leq d$, and $v_j = w_j$ for all $j \neq i$. The in- and outdegree of every node is d.

Theorem 3.6 *If for all $i, 1 \leq i \leq d, l_i \bmod (d+1) = 0$, then the d-dimensional directed torus can be strongly colored with $d+1$ colors.*

Proof: One can easily prove that assigning every node $v = (v_1, \ldots, v_d)$ the color $(\Pi_{i=1}^{d} i.v_i) \bmod (d+1)$ gives a perfect coloring. $\qquad\square$

In the same way one can prove that every 2-dimensional directed torus of dimensions l_1, l_2 can be perfectly colored if and only if $l_1 = 3k$ and $l_2 = 3l$, for some $k, l > 0$. It follows that there is no strong coloring with 3 colors for the directed version of the Illiac network.

4   The Cube-Connected Cycles

The d-dimensional Cube-Connected Cycles (CCC_d) was introduced by Preparata and Vuillemin [13]. The network consists of $d2^d$ nodes and can be described as follows (cf. [7]). Every node v is expressed as a pair of integers (l, p), with $0 \leq l < 2^d$ and $0 \leq p < d$. Node (l, p) is adjacent to nodes $(l, (p \pm 1) \bmod d)$ and $(l + \epsilon^p, p)$, where $\epsilon = 1 - 2\mathrm{bit}_p(l)$ and $\mathrm{bit}_p(l)$ is the p-th bit of the binary representation of l. The CCC_d is 3-regular. Except for some trivial cases, the CCC_d networks all admit a solution to the file distribution problem.

Theorem 4.1 *For every $d > 2, d \neq 5$, there is a perfect coloring of the CCC_d with 4 colors.*

Proof: For $d = 2$, we obtain a C_8, which cannot be strongly colored using 3 colors. For $d = 5$ there exist cycles of length 5 in the network, hence this network cannot be strongly 4-colored. For obtaining a strong 4-coloring in the remaining cases, we employ the following eight strings, where a, b, c and d are the colors we will use.

1. abcd abc	5. dabc dab
2. cdab cda	6. bcda bcd
3. adcb adc	7. dcba dcb
4. cbad cba	8. badc bad

For every string we denote the first four characters together by s_4 and the last three characters by s_3. The idea is to strongly color the cycles of length d in a CCC_d with colors in the sequence $s_4^i o s_3^j$, where i and j are chosen such that $4i + 3j = d$ and o denoting the concatenation of strings for suitable choices of s_4 and s_3. Note that for every $d > 2, d \neq 5$, such "chains" are possible and there is no conflict in any cycle. Now we must assign one of the eight possible strings to each cycle, such that there is no conflict between any two nodes of different cycles. Inspect the string $(abcd)^i o (abc)^j$ and note that the seven other strings are isomorphic to this string, i.e., they can be made by renaming the colors. Inspecting every character of this string, together with its two neighbours, we see that there are six possibilities for connecting it to another cycle. The connections avoiding every conflict are the following.

1.	...cab...	connection with	...bdc...	of string 7
2.	...dab...	connection with	...dcb...	of string 4
3.	...abc...	connection with	...cda...	of string 2
4.	...bca...	connection with	...adb...	of string 8
5.	...bcd...	connection with	...dab...	of string 2
6.	...cda...	connection with	...cba...	of string 3

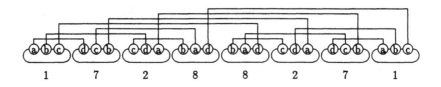

Figure 2: Strong 4 coloring of a CCC_3

The connections between the other strings follow in the same way (by renaming the colors). Suppose that $4i_1 + 3j_1 = d$, and let the string $s_4^{i_1} o s_3^{j_1}$ be assigned to the places $(l,0), \ldots, (l, d-1)$ of a cycle. Then every colored node in this cycle corresponds to one possibility, characterized by its place. The corresponding places of each possibility are given in the first column of the following table of the connections between the strings (with $1 \leq i \leq i_1$ and $1 \leq j \leq j_1$).

poss ibility	places	connection string with ...							
		1	2	3	4	5	6	7	8
1.	$4i_1 + 3j$	7	8	6	5	4	3	1	2
2.	$4i$	4	3	2	1	8	7	6	5
3.	$4i - 3$ or $4i_1 + 3j - 2$	2	1	4	3	6	5	8	7
4.	$4i_1 + 3j - 1$	8	7	5	6	3	4	2	1
5.	$4i - 2$	2	1	4	3	6	5	8	7
6.	$4i - 1$	3	4	1	2	7	8	5	6

This table must be read as follows. Suppose we have assigned string 5 to the nodes $(0,0), (0,1),$ $\ldots, (0, d-1)$. The color for the nodes $(2,0), (2,1), \ldots, (2, d-1)$ are determined by the connection between node $(0,1)$ and $(2,1)$. Node $(0,1)$ is on place $4i-3$ (for $i=1$), which corresponds to possibility 3 in our table. Inspecting this row with string 5, we see in the table that the connection must be made with string 6. Hence string 6 must be assigned to the nodes $(2,0), (2,1), \ldots, (2, d-1)$. Observe that, if for dimension i string s is connected with some other string x, and for dimension j, string s is connected with y and string x is connected with z, then in dimension i, string y is connected with string z. Hence these connections give no conflict. We can color a CCC_d by assigning some string to the nodes $(0,0), (0,1), \ldots, (0, d-1)$. This string determines the strings which must be assigned to all other cycles. There is no conflict between any pair of nodes with distance one or two, and every $CCC_d, d > 2, d \neq 5$ can be perfectly colored with 4 colors. $\square$

An example of a coloring of a CCC is given in figure 2.

5 The shuffle-exchange and the 4-pin shuffle network

The next interconnection network we consider is the shuffle-exchange network SE_n [7]. This network consists of $N = 2^n$ nodes and there are edges from node $v = v_{n-1}v_{n-2} \ldots v_0$ to the nodes corresponding to $v_{n-2}v_{n-3} \ldots v_0 v_{n-1}$ (shuffle) and $v_{n-1}v_{n-2} \ldots \overline{v_0}$ (exchange). This leads to a directed regular network in which every node has in- and outdegree 2. Later in this section we will consider an undirected version of the network and the 4-pin shuffle network.

Definition 5.1 *We call a node v in a colored directed regular network proper, if $|\{c(v)\} \cup \{c(w)| (w,v) \in E\}| = indegree(v) + 1$.*

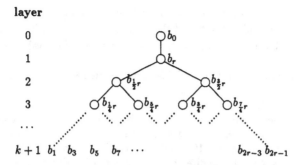

Figure 3: The tree with the connections b_{r+i} with b_i and b_{2i}

Nodes $000\ldots00$ and $111\ldots11$ have selfloops and are somewhat "special". We will try to color the network strongly such that it is proper at each node, except the nodes $000\ldots00$ and $111\ldots11$. For this, we need the following theorem.

Theorem 5.1 *Consider any binary string $b_0 b_1 \ldots b_{2r-1}$, with $r = 2^k$ for some $k \geq 0$, and $b_{r+i} = b_i$ for $0 \leq i < r$ and $b_i = \overline{b_{2i}}$, for $0 < i < r$. Then the bits b_i, with i of the form $i = \frac{r}{2^l}(2j + 1)$ (for some j and l with $0 \leq j < r$ and $0 \leq l \leq \frac{k}{2}$) all have the same value p, $p \in \{0,1\}$, and all other bits have the value $1 - p$.*

Proof: Consider the graph with $2r$ nodes, labelled $b_0, \ldots, b_{2r-1}$, in which b_i is directly connected with $b_{2i}(0 < i < r)$ and b_{r+i} with $b_{2i}(0 \leq i < r)$. Starting with node b_0 as the root, we obtain a tree, as given in figure 3. If the nodes b_i and b_{r+i} are part of layer t, then the node b_{2i} is part of layer $t - 1, t > 1$. When we give node b_0 value $p, p \in \{0,1\}$, then all nodes in even layers must have value p as well, and all nodes in odd layers must have value $1 - p$, according to the condition that $b_{r+i} = b_i(0 \leq i < r)$ and $b_i = \overline{b_{2i}}(0 < i < r)$. The nodes in the odd layers are precisely the nodes b_i, with $i = \frac{r}{2^l}(2j + 1)$, for some $j, 0 \leq j < r$ and $l, 0 \leq l \leq \frac{k}{2}$. $\square$

Theorem 5.2 *Every directed shuffle-exchange network SE_n can be strongly colored with 3 colors, such that every node, except the nodes $000\ldots00$ and $111\ldots11$, is proper.*

Proof: First of all, we relabel the vertices by $w_i, 0 \leq i < 2^n$, such that $w_i \equiv v = v_{n-1} \ldots v_1 v_0$, if $v_{n-1} \ldots v_1 v_0$ is the n-bit binary form of i. Take any string $b_0, \ldots, b_{2r-1}$, with $r = 2^{n-4}$, with the requirements as stated in theorem 5.1 in effect. Assign to the nodes $w_{8i+1}, w_{8i+2}, \ldots, w_{8i+6}$ the values $2, 1, 0, 0, 2, 1$, respectively, to the nodes w_{8i} the value of b_i of the string of theorem 5.1, and to the nodes w_{8i+7} the value $(2 + c(w_{2^n - 8(i+1)})) \bmod 3$, with $0 \leq i < 2^{n-3}$ and p is 1 if n is even and 0 otherwise. We have to prove that this is a correct coloring.

The nodes w_{8i+2} (having color 1) are connected in both directions to the nodes w_{8i+3} (having color 0). There are directed edges from the nodes $w_{\frac{N}{2}+1+4i}$ (having color 2) to the nodes w_{8i+3} and from the nodes w_{1+4i} (having color 2) to the nodes w_{8i+2}, for all $i, 0 \leq i < 2^{n-3}$, hence the nodes w_{8i+2} and w_{8i+3} are proper.

For the nodes w_{8i+1} (having color 2) we note that there are directed edges from w_{8i} and $w_{\frac{N}{2}+4i}$ to those nodes, hence these nodes must be colored 0 or 1, and $c(w_{8i}) \neq c(w_{\frac{N}{2}+4i})$ must hold. For i even is $c(w_{8i}) = b_i$ and $c(w_{\frac{N}{2}+4i}) = b_{n+\frac{i}{2}}$, which values are opposite. For i odd we have $c(w_{\frac{N}{2}+4i}) = 0$, hence $c(w_{8i}) = b_i$ must be 1, which holds in theorem 5.1 for $p = 1$ when n is even and 0 otherwise. Hence the nodes w_{8i+1} are proper. There are directed edges from w_{8i+1} (having color 2) and from

w_{4i} to the nodes w_{8i}, thus $c(w_{8i}) \neq c(w_{\frac{N}{2}+4i})$ must hold and both must be 0 or 1. For i even it follows that $c(w_{4i}) = b_{\frac{i}{2}}$ and $c(w_{8i}) = b_i$ and these two colors are opposite by theorem 5.1. For i odd we have $c(w_{4i}) = 0$, but then is $c(w_{8i}) = b_i = 1$ by our choice of the value of p, hence the nodes w_{8i} are proper.

Checking the nodes $w_{8i+a}, 4 \leq a \leq 7$ can be done in the same way by writing $j = 2^n - 1 - (8i + a)$.

□

In the undirected case we give no direction to the edges, hence we assume that there is an *unshuffle* operation for every node. Now every node has degree 3, except the nodes $000\ldots000$ and $111\ldots111$ (both degree 1), and $1010\ldots10$ and $0101\ldots01$ (both degree 2, only for n even). To solve this problem, we want to color this undirected network strongly with 4 colors. An easy argument shows that this is impossible.

Theorem 5.3 *Every strong coloring of a shuffle-exchange network requires at least 5 colors.*

Proof: Every shuffle-exchange network contains the cycle $(000\ldots001, 000\ldots010, 000\ldots011, 100\ldots001, 100\ldots000)$, which requires 5 colors.

□

The 4-pin shuffle network [7] is another regular network, which can be described as follows. It consists of 2^n nodes and there are edges from node $v = v_{n-1}v_{n-2}\ldots v_0$ to the nodes, corresponding to $v_{n-2}v_{n-3}\ldots v_0 0$ and $v_{n-2}v_{n-3}\ldots v_0 1$. This leads to a regular directed network in which every node has in- and outdegree 2. Since $000\ldots00$ and $111\ldots11$ again have selfloops, we want to color this network strong such that every node, except $000\ldots00$ and $111\ldots11$ are proper.

Theorem 5.4 *Every 4-pin shuffle network can be strongly colored with three colors, such that every node, except the nodes $000\ldots00$ and $111\ldots11$, is proper.*

Proof: First of all, we relabel the vertices by $w_i, 0 \leq i < 2^n$, such that $w_i \equiv v = v_{n-1}\ldots v_1 v_0$, if $v_{n-1}\ldots v_1 v_0$ is the binary form of i. Consider the nodes w_i, with $0 < i < 2^{n-1}$. With k be the greatest number such that $i \geq 2^k$, we color w_i with the color 1 if k is even, and with the color 0, if k is odd. Also we take $c(w_0) = 0$. If n is even, then $c(w_{2^n-i-1}) = c(w_i) + 1$, and $c(w_{2^n-i-1}) = (c(w_i) + 2) \bmod 3$ otherwise $(0 \leq i < 2^{n-1})$. Now we have to prove that this coloring is correct.

Inspect a node w_i, with $0 < i < 2^{n-1}$. There are directed edges from $w_{\lfloor\frac{i}{2}\rfloor}$ and $w_{\lfloor\frac{i}{2}\rfloor+\frac{N}{2}}$ to this node. Let k be the greatest integer, such that $i \geq 2^k$, then it follows that $\lfloor\frac{i}{2}\rfloor \geq 2^{k-1}$, hence $c(w_i) = 1$ and $c(w_{\lfloor\frac{i}{2}\rfloor}) = 0$ or vice versa. Furthermore we note that $i < 2^{n-1}$, thus $\lfloor\frac{i}{2}\rfloor < 2^{n-2}$, hence $2^{n-1} \leq \lfloor\frac{i}{2}\rfloor+2^{n-1} < 2^{n-1}+2^{n-2}$. By our coloring, all these vertices w_j, with $2^{n-1} \leq j < 2^{n-1}+2^{n-2}$ have color 2, hence the vertices $w_i, 0 < i < 2^{n-1}$ are proper.

For a node w_i, with $2^{n-1} \leq i < 2^n$, the proof is similar, because there are directed edges from $w_{2^n-\lfloor\frac{i}{2}\rfloor}$ and $w_{2^n-(\lfloor\frac{i}{2}\rfloor+2^{n-1})}$ to w_i. Thus writing $j = 2^n - i, 0 < j < 2^{n-1}$, the proof is easily completed.

□

In the undirected case, we give no direction to the edges, thus every node has degree 4, except $000\ldots000$ and $111\ldots111$ (both degree 2), and $1010\ldots10$ and $0101\ldots01$ (both degree 3). The following theorem shows that a strong 5-coloring of this network is not possible.

Theorem 5.5 *Every strong coloring of an undirected 4-pin shuffle network requires at least 6 colors.*

56

6 Chordal Ring

Another network we consider is the chordal ring, which can be described as follows (cf. [1]). It consists of a ring of n nodes (n even), labelled $0, 1, \ldots, n-1$ around the ring. There also is some fixed integer w $(0 < w \leq \frac{n}{2})$ called the *chord length*, which is assumed to be odd. Aside from the ring edges, there is an edge between each odd-numbered node i and $(i+w) \bmod n$. Accordingly, there is an edge between each even-numbered node i and $(i-w) \bmod n$. Chordal rings are 3-regular. For characterizing perfectly colorable chordal ring networks, we need the following lemma's.

Lemma 6.1 *Every perfectly colored chordal ring contains a pair of nodes, numbered i and $i+3$ (mod n), colored with the same color.*

Lemma 6.2 *If $n \not\equiv 0$ (mod 4) or $w \equiv 1$ (mod 4) then the chordal ring is not perfectly colorable.*

Proof: The chordal ring is 3-regular, so for a perfect coloring it is necessary that n is divisible by 4 (theorem 2.1).

Assume $w \equiv 1$ (mod 4). From Lemma 6.1 we know that in every perfect coloring of a chordal ring there must be a pair of nodes with distance 3 having the same color. W.l.o.g. we may assume that the nodes $0, 1, 2, 3$ are assigned the colors a, b, c, a respectively (otherwise one can relabel the nodes so this holds). We get the following situation with colors "forced" as shown:

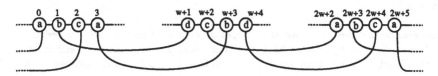

The nodes $k(w+1), \ldots, k(w+1)+3$ (modulo n) must be colored with d, c, b, d respectively if k is odd and with a, b, c, a respectively, if k is even. Color all these nodes for increasing k and note that $n.(w+1) \equiv 0$ (mod n). Inspecting the remaining uncolored nodes (if present), it follows that this coloring implies that *every* node i, with $i \equiv 0$ or 3 (mod 4) must be colored with a or d, and *every* node i, with $i \equiv 1$ or 2 (mod 4) must be colored with b or c. If $w \equiv 1$ (mod 4), then the node $w+1$ must have color b or c. Contradiction with the fact that node $w+1$ has color d. □

From Lemma 6.2 it follows that if a chordal ring is perfectly colorable, then $n = 4l$ for some $l > 0$ and $w = 4p - 1$ for some $p > 0$. We call the nodes $4i, \ldots, 4i+3$ together *block i* $(0 \leq i < k)$. We call the sequence of the four colors a, b, c, a together "color x" and the sequence d, c, b, d together "color y". Coloring block 0 with x implies that the blocks $k.p$ must be colored with x if k is even and with color y otherwise (Lemma 6.2). Note that in a perfect coloring of a chordal ring no two adjacent blocks may have the same color x or y. Let $n + 4t = 4kp$, for some k and t with $0 \leq t < p$.

Lemma 6.3 *If a chordal ring with chord length $4p-1$ $(p > 0)$ and size $4kp - 4t$ $(0 \leq t < p)$ is perfectly colorable, then one of the following cases must occur:*

1. *k and t are even and (if $t > 0$) $\frac{t}{\gcd(t,p)}$ is even, or*

2. *k, $\frac{t}{\gcd(t,p)}$ and $\frac{p}{\gcd(t,p)}$ are odd and $t + p$ is even.*

Proof: First suppose $t = 0$, thus $n = 4kp$. Block 0 is colored with x thus block $kp \equiv$ block 0 (mod n) must be colored x, hence k must be even. If $t > 0$ then we consider the following two cases: (i) k is even, and (ii) k is odd.

Case (i) k is even. Note that block $k.p \equiv$ block t, hence coloring block 0 with x implies that block t $(0 < t < p)$ must be colored with x (k is even), hence all blocks $i.t$ $(i \geq 0)$ must be colored with x. Let $q > 0$ be the smallest integer such that $q.t \bmod p = 0$, thus $q = \frac{p}{\gcd(t,p)}$. Let $r = \frac{q.t}{p}$. If r is odd then block $r.p$ must be colored y, but block $r.p \equiv$ block $q.t$ and, hence, must be colored x. Contradiction, thus r must be even. If t is odd then q must be even (because rp is even). But now $\frac{q}{2}t = \frac{r}{2}p \ (= 0 \bmod p)$, thus q was not the smallest integer as required.

Hence t must be even. We color the blocks as in Lemma 6.2. If p is odd then it follows immediately that, if a block i $(0 \le i < p)$ gets a color x, then i must be even. If a block i $(0 \le i < p)$ gets a color y then i must be odd. Hence there are no two adjacent blocks having the same color. (Note that if p is odd and t is even, then $\frac{t}{\gcd(t,p)}$ is always even.) If p and t are both even, then only (some) even blocks of the chordal ring get a color. So no two adjacent blocks get the same color, and there is no conflict if $r = \frac{t}{\gcd(t,p)}$ is even.

Case (ii) k is odd. Note that block $k.p \equiv$ block t gets color y, block $2t$ gets color x, etc. Block $i.t$ gets color x if i is even and color y otherwise. Let $q > 0$ be the smallest integer such that $q.t \bmod p = 0$, and let $r = \frac{q.t}{p}$. Remark that the blocks $i.p$ are colored x for i even and with y otherwise, thus r and q must be both even or both odd. If q and r are both even then also $\frac{q}{2}t = \frac{r}{2}p$ $(= 0 \bmod p)$ holds, thus q was not the smallest integer as required. Hence q and r must be both odd. If p is odd and t is even (or vice versa) then $r.p$ is odd and $q.t$ is even, which is a contradiction. If p and t are both odd then it follows that if a block i $(0 \le i < p)$ gets color x then i must be even and if a block i $(0 \le i < p)$ gets color y then i must be odd. Hence in the whole ring there are no two adjacent blocks having the same color. If p and t are both even, then it follows that if a block i $(0 \le i < p)$ gets a color, then i must be even, so there are no two adjacent blocks having the same color. □

Now assume that we are in one of the cases of lemma 6.3 and we wish to perfectly color the chordal ring. We start by coloring block 0 with x, block p with y, block $2p$ with x, etc., until we are back in block 0, that is when $4ip \equiv 0 \pmod n$ for some $i > 0$ (which holds when $i = \frac{n}{4.\gcd(n,p)}$).

Lemma 6.4 *Let n and w be as given in Lemma 6.3 and let the blocks ip for $0 \le i < \frac{n}{4.\gcd(n,p)}$ be colored as in lemma 6.2. Let j be the smallest integer > 0 such that block j is colored. Then the distance between every two consecutive colored blocks is j.*

Theorem 6.5 *A chordal ring with chord length $4p - 1$ $(p > 0)$ and $4kp - 4t$ $(0 \le t < p)$ nodes is perfectly colorable if and only if one of the following cases occur:*

1. *k and t are even and (if $t > 0$) $\frac{t}{\gcd(t,p)}$ is even, or*

2. *k, $\frac{t}{\gcd(t,p)}$ and $\frac{p}{\gcd(t,p)}$ are odd and $t + p$ is even.*

Proof: The necessity follows from Lemma 6.3. We show that in each of the cases a perfect coloring is possible. Let the blocks ip, for $0 \le i < \frac{n}{4.\gcd(n,p)}$ be colored as in lemma 6.2. Let $j > 0$ be the smallest integer such that block j is colored. Color every block $ip + 1, \ldots, ip + j - 1$ $(0 \le i < \frac{n}{4.\gcd(n,p)})$ with the colors d, c, b, a respectively if i is even, and with the colors a, b, c, d respectively otherwise. If block j is colored with x then we color the blocks $ip + j - 1$ with y if i is even and with x if i is odd. We claim that this coloring is perfect. Note first that every node is assigned exactly one of the colors a, b, c, d. By case analysis it follows that for every i, the colors of the nodes $i, i+1, i+2$ and $i+w+1$ are all different, hence the described coloring is perfect. □

Corollary 6.6 *If a chordal ring with n nodes and chord length w $(w \le \frac{n}{2})$ is perfectly colorable, then every chordal ring with $a.n$ nodes and chord length w is perfectly colorable.*

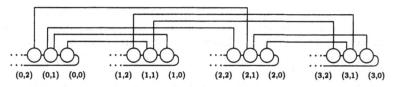

(0,2) (0,1) (0,0)   (1,2) (1,1) (1,0)   (2,2) (2,1) (2,0)   (3,2) (3,1) (3,0)

Figure 4: The graph, which is a subgraph of every binary lens network

7 Other Networks and Final Remarks

In this section we summarize some results and open problems for other processor networks such as the pm2i network, the hexagonal network and the binary lens network. The pm2i interconnection network, introduced by Siegel [14], consists of 2^n nodes ($n \geq 0$), labelled $0, \ldots, 2^n - 1$. Every node v is connected with another node w, if and only if $w = v \pm 2^i$, for some $i, 0 \leq i < n$. This gives a $(2n - 1)$-regular network. The file distribution problem corresponds to the question whether there exists a strong coloring of this network with $2n$ colors. Although this seems easy, this interesting question is still open. From theorem 1.1 it follows that if a pm2i network with 2^n nodes is perfectly colorable, then $n = 2^k$, for some $k > 0$.

Another network we consider is the hexagonal network [14]. It consists of $m \times n$ nodes. Vertex $v = (v_1, v_2)$ is connected with $((v_1 \pm 1) \bmod m, v_2), (v_1, (v_2 \pm 1) \bmod n), ((v_1 + 1) \bmod m, (v_2 + 1) \bmod n)$ and $((v_1 - 1) \bmod m, (v_2 - 1) \bmod n)$. This leads to a 6-regular undirected network.

Lemma 7.1 *Every hexagonal network of size $m \times n$ can be perfectly colored with 7 colors, if and only if $m, n \bmod 7 = 0$.*

The last network we consider in this paper is the binary lens network [13]. It consists of $d2^d$ nodes and every node v is expressed as a pair of integers (l, p), with $0 \leq l < 2^d$ and $0 \leq p < d$. Node (l, p) is adjacent to nodes $(l, (p \pm 1) \bmod d)$, $(l + \epsilon^{p+1}, p + 1)$ and $(l + \epsilon^{p-1}, p - 1)$, where $\epsilon = 1 - 2\mathrm{bit}_p(l)$ and $\mathrm{bit}_p(l)$ is the p-th bit of the binary representation of l. The binary lens network is 4-regular. A simple argument shows that these networks are not perfectly colorable.

Theorem 7.2 *Every strong coloring of a binary lens network with $d2^d$ nodes ($d \geq 3$) requires at least 6 colors.*

Proof: The graph, given in figure 4, is a subgraph of every binary lens network with $d2^d$ nodes ($d \geq 3$) and requires 6 colors. □

There is a nice relationship between parallel algorithms and the file-distribution problem by the following theorem:

Theorem 7.3 *If a $(r - 1)$-regular processor network G is perfectly colorable then every CRCW-program on G can be transformed into an equivalent EREW-program with an extra factor of $O(r)$ in time and no extra space.*

The file distribution problem can be generalised to the more general problem, in which every node must contain a full copy of the file F in its neighbourhood within distance $k, k > 1$. Some of our results can be generalised to this version, e.g. there exists a perfect e-error-correcting code for strings of length n, if and only if there exists a perfect distance-$(2e + 1)$ coloring of the hypercube C_n. Similar results hold for the d-dimensional torus.

We also mention the following version of the file distribution problem: "Given a connected network $G = (V, E)$ and a file F, determine the maximum number of segments K of F such that assigning each node $x \in V$ a segment $F_x \subset F$ out of $\{F_1, F_2, \ldots, F_K\}$ gives an assignment such that for all $x \in V, \bigcup_{\{x,y\} \in E} F_y \cup F_x = F$." The case $K = r$ for $(r-1)$-regular graphs is equal to our perfect coloring problem. The general problem is also NP-complete by observing the equality with the *domatic number* problem, which is known to be NP-complete for $K \geq 3$. Other formulations of this problem are in terms of the "file-allocation"-problems in distributed databases. A survey of these problems is given in [6].

Of course, many other (complicated) versions are possible, but the version studied in this paper seems to be the most practical problem for distributed datastructuring in the present context.

Acknowledgements

The authors wish to thank Hans Bodlaender and Erwin Bakker for some useful suggestions, and J.H. van Lint and K.A. Post for providing some background material on Lee codes.

References

[1] Arden, B.W., and H. Lee, Analysis of chordal ring network, *IEEE Trans. Comput.*, C-30 (1981), pp. 291–295.

[2] Astola, J., *The Theory of Lee–codes*, Manuscript, Lappeenranta University of Technology, Finland.

[3] Bakker, E.M., *The File Distribution Problem*, unpublished manuscript, Utrecht, 1988.

[4] Bakker, E.M., personal communication, 1990.

[5] Berlekamp, E.R., *Algebraic Coding Theory*, McGraw-Hill Book Co., New York, 1968.

[6] Dowdy, L.W., and D.V. Foster, Comparative models of the file assignment problem, *ACM Computing Surveys* 14 (1982), pp. 287–313.

[7] Fishburn, J.P., and R.A. Finkel, Quotient Networks, *IEEE Trans. Comput.*, C-31 (1982), pp. 288–295.

[8] Golomb, S.W., and L.R. Welch, Perfect codes in the Lee metric and the packing of polyominoes, *SIAM J. Appl. Math.*, 18 (1970), pp. 302–317.

[9] Kant, G., and J. van Leeuwen, *The File Distribution Problem for Processor Networks*, Tech. Report RUU-CS-90-16, Dept. of Computer Science, Utrecht University, 1990.

[10] Kant, G., and J. van Leeuwen, *Strong Colorings of Graphs*, Tech. Report RUU-CS-90-15, Dept. of Computer Science, Utrecht University, 1990.

[11] Lint, J.H. van, Coding Theory, *Lecture Notes in Mathematics*, vol. 207, Springer (1971), Berlin.

[12] McCormick, S.T., *Optimal approximation of sparse hessians and its equivalence to a graph coloring problem*, Technical Report, Dept. of Operations Res., Stanford, 1981.

[13] Preparata, F.P., and J. Vuillemin, The Cube-Connected Cycles: A Versatile Network for Parallel Computation, *Comm. ACM*, vol. 24, no. 5 (1981), pp. 300–309.

[14] Siegel, H.J., *Interconnection Networks for Large-Scale Parallel Processing*, Lexington Books, Mass., 1985.

Translating Polygons with Applications to Hidden Surface Removal

Mark de Berg*

Abstract

Let S be a set of polygons in the plane with a total number of n vertices. A translation ordering for S (in direction d) is an ordering of the polygons such that, if the polygons are moved one by one to infinity in direction d according to this ordering, no collisions occur between the polygons. We show that, after $O(n \log n)$ preprocessing using $O(n)$ space, it is possible to determine, for any given d, in $O(\log n)$ time whether such an ordering exists and, if so, to compute an ordering in $O(n)$ time.

Translation orderings correspond to valid orderings for hidden surface removal schemes where objects that are closer to the viewpoint are displayed later than objects that are farther away. Thus our technique can be used to generate displaying orderings for polyhedral terrains. One of the main advantages of our approach is that it can easily be adapted to handle perspective views within the same time and space bounds.

Keywords Computational geometry, separation problems, hidden surface removal, relative convex hulls.

1   Introduction

In its most general form, the *separability problem* can be stated as follows. Given a set of objects in some space, separate them by a sequence of motions. During the motions, the objects should not collide with each other. (A collision between two objects occurs when their interiors have a non-empty intersection.) These problems come in many different flavors, depending on the objects that are considered, the space they are in, and the type of motions that is allowed. Toussaint ([19]) gives an extensive survey of such problems.

We consider the following restricted version of the separability problem. Given a set $S = \{P_1, \ldots, P_m\}$ of non-intersecting polygons in the plane, translate them in some direction d to infinity, one at a time. Thus, every polygon has to be translated in the same direction. The problem now is to determine whether the polygons can be ordered such that, if the polygons are translated according to that ordering, no collisions occur and, if so, to compute such a *translation ordering*.

This problem, which is called the *translation problem*, originated in 1980, when Guibas and Yao ([5]) studied translation orderings for sets of convex polygons. They showed that

---

*Dept. of Computer Science, University of Utrecht, P.O.Box 80.089, 3508 TB Utrecht, the Netherlands. Supported by the Dutch Organisation for Scientific Research (N.W.O.). Partially supported by the ESPRIT II Basic Research Actions Program of the EC under contract No. 3075 (project ALCOM).

a translation ordering always exists for a set of convex polygons and gave an $O(n + m \log m)$ algorithm for computing an ordering. (Here, and in the rest of this paper, m is the number of polygons and $n = \sum_{i=1}^{m} |P_i|$ is the total number of vertices of all polygons.). Since then, their work has been extended in several ways. Ottman and Widmayer ([16]) simplified the method and Nurmi ([9]) adapted the method to arbitrary polygons, achieving a time bound of $O(n \log n)$. Recently, Nussbaum and Sack gave an optimal $O(n + m \log m)$ algorithm for this problem ([10]). Sack and Toussaint ([17]) showed how to compute, in $O(n \log n)$ time, all directions of separability (i.e., directions for which a translation ordering exists) for *two* arbitrary polygons, which was improved to $O(n \log \log n)$ by Toussaint ([20]). Finally, Dehne and Sack ([2]) studied many of these problems when preprocessing is allowed: after $O(m^2(C_S(p)+\log m))$ time and using $O(m^2)$ space, they are able to answer all kinds of questions on translational orderings. (Here each polygon is assumed to have p vertices and $C_S(p)$ is the time needed to determine all directions of separability of two p-vertex polygons, which varies between $O(\log p)$ and $O(p \log \log p)$ depending on the type of the polygons.) Although their method is very efficient when the number of polygons is small, it becomes very costly when there are many polygons. When all polygons have constant size, for example, their preprocessing takes time $O(n^2 \log n)$ and space $O(n^2)$ and computing an ordering for a given direction still takes $O(n^2)$ time.

In this paper it is shown that a set of (arbitrary) polygons can be preprocessed in time $O(n \log n)$ and space $O(n)$, such that it is possible to determine, for any given direction d, in time $O(\log n)$ whether there exists a translation ordering and to compute one (if it exists) in time $O(n)$. We also show that all directions of separability can be computed in $O(n \log n)$ time. Hence, when the number of polygons is large, this improves the results of Dehne and Sack considerably.

One of the main applications of the translation problem is in computer graphics. To render a realistic picture of a scene, hidden surface removal has to be performed. One way to do this is to display the objects in the scene in a 'back to front' (with respect to the viewpoint) order. This way the objects in the front are painted over the objects in the back, thereby achieving the desired overlaying effect. A moment's thought will make it clear that a valid displaying order for this so-called *painter's algorithm* corresponds to a translation ordering for the objects in the direction perpendicular to the viewing plane. However, computing translation orderings in three dimensions efficiently is difficult. Fortunately, for an important class of three dimensional scenes, the so-called *terrains* (polyhedral scenes in which the projections of the faces of the objects on the xy–plane do not intersect), solutions to the two dimensional translation problem can be used.

The translation ordering for the set of polygonal faces of a scene corresponds to a parallel view of the scene. This is of course unwanted. One of the advantages of our method is that it can easily be adapted to yield an valid displaying order for perspective views within the same bounds. Thus we can preprocess a terrain consisting of convex polygonal faces with a total number of n vertices in time $O(n \log n)$ using $O(n)$ space, such that for any viewpoint a displaying order for the faces can be found in time $O(n)$. Notice that this gives a better space bound than the $O(n \log n)$ needed in the binary partition scheme of Paterson and Yao ([12]).

The sequel of this paper is organised as follows.

We start in section 2 by presenting some preliminary results on relative convex hulls.

In section 3, our solution to the translation problem is presented. First, we consider sets of convex polygons to illustrate the main idea of our method, namely triangulating

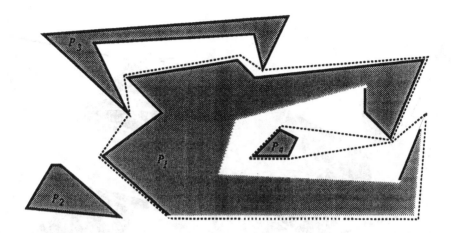

Figure 1: The dashed line is the boundary of the convex hull of P_1 relative to $\{P_2, P_3, P_4\}$. Note that the relative convex hull is not a simple polygon, since there is a vertex that is used twice.

the area *in between* the polygons and translating the set of polygons augmented with the triangles of the triangulation. Then we show how to handle arbitrary polygons, using the concept of relative convex hulls.

In section 4, we discuss the application to hidden súrface removal. It is shown how perspective views can be handled and how to treat viewpoints above the terrain.

We conclude with a brief summary of our results and by mentioning some open problems in section 5.

2   Preliminaries

In this section we present some results on *relative convex hulls*, a generalisation of convex hulls introduced by Toussaint ([20]). Relative convex hulls are defined as follows. Define a polygonal circuit to be a closed polygonal path without (proper) self-crossings.

Definition 1 *Let P be a polygon and V a set of polygons. The* convex hull of P relative to V, *denoted $CH(P|V)$, is the polygon whose boundary is the shortest polygonal circuit that includes P but excludes V, i.e., $int(P) \subseteq int(CH(P|V))$ and $int(P') \subseteq ext(CH(P|V))$ for each $P' \in V$.*

(Thus our polygons are a slight generalization of a simple polygons, where we allow some edges and vertices to be used more than once.) Intuitively, if we release an elastic band that is wrapped around P, then it tries to take the shape of the convex hull of P but it can be stopped by the other polygons and it takes the shape of the relative convex hull of P. An example is given in Figure 1.

Let S be a set of polygons. For a polygon $P \in S$ we define $p^* = CH(P|S - \{P\})$ and we define $S^* = \{P^*|P \in S\}$. In the remainder of this section it is shown how S^* can be computed efficiently. Toussaint ([20]) has shown how to do this for a set of two polygons. Using ideas similar to his, we show how this can be done for larger sets of polygons.

63

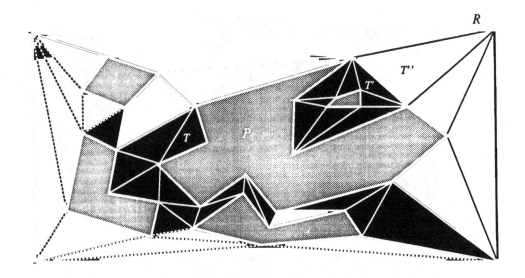

Figure 2: The triangles around P form the sleeve of P. Triangle T is a dead end and does not belong to $sleeve(P)$. Observe that triangles T' and T'' occur two times in $sleeve(P)$.

The idea is to compute first an area around each polygon P, called the *sleeve* of P, that contains the boundary of its relative convex hull. Let R be a large rectangle that contains S properly. Triangulate $R - S$, the area inside R between the triangles. $sleeve(P)$ is constructed by concatenating the triangles of the triangulation as they are encountered during a clockwise walk along the boundary of P starting at the leftmost vertex v_0 of P. However, we do not concatenate the last triangle thus encountered to the first one. Then we compute a shortest circuit that starts at v_0, goes 'around' P and returns to v_0. This shortest circuit is the shortest path from (the copy of) v_0 in the first triangle of the sleeve to (the copy of) v_0 in the last triangle of the sleeve. This is done using algorithms of Chazelle ([1]) or Lee and Preparata ([8]). They have shown that if the dual tree of the triangulation of a simple polygon is a chain, then a shortest path between two points in such a polygon can be computed in time linear in the number of vertices of the polygon. To ensure that the dual tree of $sleeve(P)$ is indeed a chain, we have to remove certain triangles that are 'dead ends' from the sleeve. The fact that $sleeve(P)$ is not a simple polygon (some triangle may occur more than once in a sleeve) is no real problem, as Toussaint ([19]) observed. If we embed the sleeve onto a Riemann surface of several layers ([6]), so that if a triangle occurs for the second or third time in a sleeve it lies 'above' its previous occurence, so to speak, the algorithm that computes the shortest path won't know any better than that it is working with a simple polygon. See Figure 2 for an illustration.

Since a triangle only occurs in $sleeve(P)$ when it shares a vertex with P, the total complexity of all sleeves is $O(n)$, and we get:

Lemma 1 S^* *can be computed in* $O(n \log n)$ *time and* $O(n)$ *space.*

3   Translating polygons

A translation ordering for a set S of polygons (in direction d) is defined as an ordering such that if the polygons are moved one at a time (in direction d) to infinity according to this ordering, no collisions occur. The computation of translation orderings involves computing some sort of dominance relation between the polygons. A polygon P dominates another polygon P' if P' collides with P when it is moved before P is moved. Thus a translation ordering exists iff the dominance relation between the polygons is free of cycles. It has been shown by Guibas and Yao ([5]) that it is not necessary to compute all (possibly $\Theta(m^2)$) dominances explicitly, but that it suffices to compute the *immediate* dominances. (P immediately dominates P' if, when P' is moved, some portion of P' intersects some portion of P before it intersects some other polygon.)

This immediate dominance relation changes radically, however, when the direction of translation d changes. Hence, if we want to do preprocessing to speed up the computation of a translational ordering for any given d, we have to take a different approach. We will show that a triangulation of the area in between the polygons gives us all the information we need to compute a translation ordering for any given direction.

3.1   Translating convex polygons

Let $S = \{P_1, \ldots, P_m\}$ be a set of convex polygons. The convex hull of S is denoted by $CH(S)$. Furthermore let $T = \{T_1, \ldots, T_k\}$ be a triangulation of $CH(S) - S$, the area in between the polygons of S. The idea is to translate the set $S \cup T$. Observe that this new set still contains only convex polygons and, hence, it can still be translated. Surprisingly, translating $S \cup T$ is an easier task than translating S, as follows from the lemma given next. First we define d-neighbours, a concept that is crucial in our method.

Definition 2 *Let Q and Q' be two polygons. Q is a d-neighbour of Q' iff*
(i) Q and Q' share an edge e
(ii) there is a ray in direction d that intersects $int(Q')$ just before it intersects e and $int(Q)$ just after it intersects e.

Notice that if two polygons Q and Q' share an edge e, then either Q is a d-neighbour of Q', or Q' is a d-neighbour of Q, or e is parallel to d. The following lemma shows the importance of this concept.

Lemma 2 *A polygon $Q \in S \cup T$ (possibly a triangle) can be translated to infinity in direction d without collisions if and only if all its d-neighbours already have been translated without collisions.*

Lemma 2 immediately leads to the following simple scheme.

The preprocessing just consists of computing a triangulation T of $CH(S) - S$ and the dual graph $G(S \cup T)$ of $S \cup T$. (The nodes in this graph correspond to the polygons in $S \cup T$ and there is an arc between two nodes iff the corresponding polygons share an edge.)

Now, given a query direction d, we proceed as follows. First we turn $G(S \cup T)$ into a directed graph G_d. Let a be an arc in $G(S \cup T)$ connecting nodes corresponding to polygons Q and Q'. If Q is a d-neighbour of Q' then the arc in G_d corresponding to a, denoted a_d, is directed from Q to Q'. If Q' is a d-neighbour of Q then a_d is directed from Q' to Q. Otherwise (the edge shared by Q and Q' is parallel to d) a has no corresponding arc in G_d. Thus (a node corresponding to) some polygon has incoming arcs from all its d-neighbours and outgoing arcs to all polygons for which it is a d-neighbour.

From Lemma 2 and the definition of G_d it easily follows that a topological ordering of the nodes in G_d corresponds to a translation ordering in direction d for the polygons in $S \cup T$. (Note that the fact that $S \cup T$ can be translated guarantees that G_d is acyclic.) Clearly, if the triangles of T are omitted from this ordering, we get the desired translation ordering for S. We thus have:

Lemma 3 *A set S of convex polygons can be preprocessed in $O(n \log n)$ time and $O(n)$ space such that, given a direction d, a translation ordering for S in direction d can be computed in time $O(n)$.*

Proof: The convex hull of S as well as the triangulation (and its dual graph) can be computed in time $O(n \log n)$ and $O(n)$ space ([13]). (Note that the total number of edges in $S \cup T$ (and therefore the number of nodes and arcs in $G(S \cup T)$ as well) is $O(n)$).

Since we can decide in constant time for an arc a in $G(S \cup T)$ what the direction of its corresponding arc a_d in G_d will be, the construction of G_d takes only linear time. Topologically sorting a directed (acyclic) graph can also easily be done in linear time (see, e.g., Knuth ([7])). □

3.2 Translating arbitrary polygons

To apply the same idea to a set of arbitrary polygons, some extra preprocessing has to be done. The problem is that if there are non-convex polygons, the triangles of the triangulation might prevent a translation ordering, i.e., it is possible that a translation ordering for S exists, but not for $S \cup T$. This problem can be overcome by using the concept of relative convex hulls (see section 2).

In the remainder we shall need the following lemma, proved by Toussaint in [19].

Lemma 4 ([19]) *There exists a translation ordering for a set of polygons if and only if there exists a translation ordering for every pair of polygons in the set.*

Now we are ready to show that the method of the previous section can be used if we first replace every polygon by its convex hull relative to the other polygons. Let $S = \{P_1, \ldots, P_m\}$ be a set of polygons. For a polygon $P \in S$ we define $P^* = CH(P|S - \{P\})$ and we define $S^* = \{P^*|P \in S\}$. Any ordering on S naturally corresponds to a unique ordering on S^* (and vice versa) and this correspondence is also preserved when restricted to translation orderings, as follows from [20]:

Lemma 5 *An ordering on S is a translation ordering (in direction d) for S if and only if the corresponding ordering on S^* is a translation ordering (in direction d) for S^*.*

Once we have replaced the polygons in S by their relative convex hulls, it is safe to triangulate the region between the polygons and add the triangles to the set to be translated. Let T be a triangulation of $CH(S^*) - S^*$, then we have:

Lemma 6 *There exists a translation ordering in direction d for S^* if and only if there exists a translation ordering in direction d for $S^* \cup T$.*

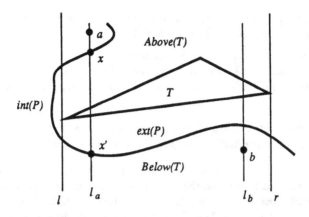

Figure 3: Illustration of the proof of Lemma 6

Proof: The "if"–part is trivial. The proof of the "only if"–part makes use of Lemma 4. Since there exists (by assumption) a translation ordering for every pair of polygons in S^*, and there exists a translation ordering for every pair of triangles (they are convex), it remains to prove that there exists a translation ordering for every pair $P^* \in S^*$, $T \in \mathcal{T}$.

Suppose for a contradiction that some pair P^*, T cannot be ordered and assume w.l.o.g. that d is vertically upward. Let l and r be the two vertical lines tangent to T and denote the area between l and r and above T by $Above(T)$ and the area below T by $Below(T)$ (see Figure 3). If there is no ordering for P^* and T then $Above(T) \cap P^* \neq \emptyset$ and $Below(T) \cap P^* \neq \emptyset$. Let a be a point in the first intersection and b a point in the second intersection and let l_a and l_b denote the vertical lines through a and b. At least one of l_a and l_b, say l_a, must intersect P^* above T as well as below T. Consider x, the first intersection of l_a with P^* above T, and x', the first intersection point below T. Now $\overline{xx'}$ cuts off some portion of $ext(P^*)$ (that has a non-empty intersection with T). Clearly, the part of the boundary of P^* that borders this portion is longer than $\overline{xx'}$. But this means that there must be (a part of) a polygon $P' \in S - \{P\}$ lying in this portion of $ext(P^*)$, contradicting the fact that P^* and $(P')^*$ can be ordered. $\qquad \Box$

We thus arrive at the following scheme for translating a set S of arbitrary polygons. As a preprocessing step, S^* and a triangulation $\mathcal{T}$ of $CH(S^*) - S^*$ (together with its dual graph $G(S^* \cup \mathcal{T})$) are computed in $O(n \log n)$ time and $O(n)$ space. Then, given a direction d, we construct G_d as in the previous section. (Note that there might be more than one arc between two nodes in G_d, because two polygons can share more than one edge.) If S^* can be translated in direction d, then G_d does not contain a cycle and we can compute a topological ordering of the nodes in G_d. This ordering corresponds to a translation ordering for S^* (omitting the triangles of the triangulation) which, by Lemma 5, corresponds to a translation ordering for S. (Note that Lemma 2 is true for non-convex polygons too.) If G_d contains a cycle, then $S^* \cup \mathcal{T}$ cannot be translated. By Lemma's 5 and 6 we can then conclude that no translation ordering for S exists either.

Lemma 7 *A set S of arbitrary polygons can be preprocessed in $O(n \log n)$ time and $O(n)$ space such that, given a direction d, a translation ordering for S in direction d can be computed in time $O(n)$ (if it exists).*

We now show how all directions of separability (i.e., all directions for which a translation ordering exists) can be computed in $O(n \log n)$ time.

Toussaint has shown in [20] that there exists a translation ordering for two polygons in direction d if and only if their relative convex hulls are monotonic in direction $d + \frac{1}{2}\pi$. Lemma 4 implies that the fact stated above for two polygons also holds for larger sets of polygons:

Lemma 8 *There exists a translation ordering in direction d for a set of polygons if and only if the relative convex hulls of the polygons are monotonic in direction $d + \frac{1}{2}\pi$.*

A polygon is monotonic in direction d if and only if it has no reflex vertex v such that the two edges incident to v lie on the same side of the line through v with slope d. For a reflex vertex v of some $P^* \in S^*$, let $I_v \subset [0 : 2\pi]$ be the interval such that the two edges incident to v lie on the same side of a line through v with slope d if and only if $d \in I_v$ (in fact, I_v can consist of two disjoint intervals, one starting at 0, the other ending at 2π). Given the two edges incident to v, I_v is easily computed in constant time. Thus, in $O(n)$ time, we can compute $I(S^*) = \{I_v | v$ is a reflex vertex of a $P^* \in S^* \}$. By Lemma 8, a translation ordering for S in direction d exists iff $d \notin \bigcup I(S^*)$. In other words, the set D of directions for which a translation ordering exists is the set $[0 : 2\pi] - \bigcup I(S^*)$. Using a simple line sweep algorithm, D can be computed in time $O(n \log n)$.

Theorem 1 *All directions for which a translation ordering exists for a given set S of polygons with a total number of n vertices can be determined in $O(n \log n)$.*

Observe that D consists of $O(n)$ disjoint intervals. Hence, D can be stored in a search tree which can be built in $O(n \log n)$ time and uses $O(n)$ space. With this tree it can be decided, for a given direction d, in time $O(\log n)$ if $d \in D$ and thus if there exists a translation ordering in direction d.

We now state our main theorem, which summarizes the results of this section.

Theorem 2 *A set S of polygons, with a total number of n vertices, can be preprocessed in $O(n \log n)$ time and $O(n)$ space, such that, given any direction d, it can be decided in time $O(\log n)$ if there exists a translation ordering for S in direction d and, if so, an ordering can be computed in time $O(n)$.*

4   Application to hidden surface removal

One of the most important applications of translation orderings is in computer graphics. When an object of a scene is displayed onto a screen, it is painted over the objects that already have been displayed. Therefore, the objects must be displayed in a 'back to front' order. This order corresponds to a translation order perpendicular to the projection plane. Translation orderings for polygons in the plane can be used to obtain displaying orders for so-called (polyhedral) *terrains*. A terrain is a set of polygonal faces in 3–space that do not intersect when projected onto the xy–plane[1]. Observe that this is a very general definition of a terrain: we do not require the scene to be 'connected' (as, e.g., is necessary for the hidden surface removal algorithm of Reif and Sen ([15])).

We will now show how our translation algorithm can be used to generate displaying orders for the most general type of views for terrains consisting of convex faces: a

---

[1]In this paper, all projections onto the xy–plane are orthogonal.

perspective view from an arbitrary point above the terrain. Let $F = \{f_1, \ldots, f_m\}$ be a set of convex polygons in 3–space, the faces of the terrain, and let $\overline{F} = \{\overline{f_1}, \ldots, \overline{f_m}\}$ be the (non–intersecting) set of projections of these faces onto the xy–plane. Let h be the viewing plane and let X be the viewpoint. Thus we want to project the faces of F onto h as seen by an observer at position X. Again, we permit ourselves a preprocessing of $O(n \log n)$ to compute a triangulation T of $CH(\overline{F}) - \overline{F}$ and its dual graph $G(\overline{F} \cup T)$. After this, given a viewpoint X and a viewing plane h, a correct displaying order can be calculated in linear time, as is shown in the remainder of this section.

Let us assume that $\overline{X}$, the projection of X onto the xy–plane, does not lie in (the interior of) the convex hull of $\overline{F}$. (This can easily be accomplished by splitting $\overline{F} \cup T$ into two sets with a line through $\overline{X}$.)

To find a valid displaying ordering for the faces of F that corresponds to a perspective view all that we have to change in the algorithms of the previous section is the concept of neighbourhood.

Definition 3 *Let Q and Q' be two polygons and $\overline{X}$ be a point in the plane. Q is a $\overline{X}$-neighbour of Q' iff*
(i) Q and Q' share an edge e
(ii) There is a ray starting at $\overline{X}$ and intersecting e that intersects $int(Q')$ just before intersecting e and $int(Q)$ just after intersecting e.

The analogon of Lemma 4 is as follows.

Lemma 9 *A face $f \in F$ can be safely displayed (onto h, perspective w.r.t. X) if all the faces corresponding to $\overline{X}$-neighbours of $\overline{f}$ in $\overline{F} \cup T$ already have been displayed safely.*

Of course, there are no real faces corresponding to the triangles that were added when $CH(\overline{F}) - \overline{F}$ was triangulated, and displaying a face corresponding to such a triangle is just a dummy statement. Also (the parts of) the faces of F that lie on the same side of h as X should not be displayed. Note that, since all faces are convex, we always find an ordering. We conclude:

Theorem 3 *A terrain F consisting of convex polygonal faces with a total number of n vertices can be preprocessed in $O(n \log n)$ time and $O(n)$ space such that, given a viewpoint X and a projection plane h, a valid displaying order for the faces of F to obtain a perspective view can be determined in $O(n)$ time.*

Remark: If the terrain contains non-convex faces, we can always cut up these faces into convex parts without changing the complexity of the scene. The restriction to convex faces is necessary because if there are non-convex faces it is possible that there is a valid displaying order for the faces of the terrain, but no translation order for the corresponding 2–dimensional problem.

5 Concluding remarks

In this paper, we have presented an efficient solution to the translation problem. It was shown that, after $O(n \log n)$ preprocessing using $O(n)$ space, a translation ordering for a set of polygons in the plane can be determined in linear time (if it exists). One of the main advantages of our method is that it can easily be adapted to yield a valid displaying order for *perspective* views of a terrain (consisting of convex polygonal faces). It should be stressed that the preprocessing of the terrain as well as the algorithm that yields the displaying order are conceptually very simple and a good candidate for efficient implementations.

Using the the methods of [4], edges (or polygons of constant size) can be inserted and deleted in linear time in our structure. An open problem is whether a method exists with better update times. (Note that it takes linear time to compute a translation ordering anyway, so a linear update time is in fact not that bad.)

As yet, little work has been done on the 3–dimensional problem. Nurmi ([9]) has considered 3–dimensional translation orderings, but his results are theoretically not very strong, since the time complexity of his algorithm depends of the number of intersections in the viewing plane. Nussbaum and Sack ([10]) have considered the problem of determining all directions of separability for two polyhedra. Egyed ([3]) has tried to find efficient ways to cut up a scene such that the 2–dimensional translation algorithms can be used. However, efficient methods to solve the general problem are still lacking.

Acknowledgement

I would like to thank Peter Egyed for proposing the problem and for useful discussions on the problem. Also, I thank Mark Overmars for giving valuable comments.

References

[1] Chazelle, B., A Theorem on Polygon Cutting with Applications, *Proc. 23rd Annual IEEE Symp. on Foundations of Computer Science*, 1982, pp. 339-349.

[2] Dehne, F., and J.-R. Sack, Separability of Sets of Polygons, *Proc. 12th International Workshop on Graph-Theoretic Concepts in Computer Science*, 1986, pp. 237-251.

[3] Egyed, P., Hidden Surface Removal in Polyhedral–Cross–Sections, *The Visual Computer* 3 (1988), pp. 329-343.

[4] El Gindy, H.A., and G.T. Toussaint, Efficient Algorithms for Inserting and Deleting Edges from Triangulations, *Proc. International Conf. on Foundations of Data Organization*, 1985.

[5] Guibas, L.J., and F.F. Yao, On Translating a Set of Rectangles, in: F. P. Preparata (Ed.), *Advances in Computing Research Vol. I: Computational Geometry*, JAI Press Inc., 1983, pp. 61-77.

[6] Kahn, J., M. Klawe and D. Kleitman, Traditional Galleries Require Fewer Watchmen, *SIAM J. Alg. Disc. Meth.* 14 (1983), pp. 194-206.

[7] Knuth, D.E., *Fundamental Algorithms: The Art of Computer Programming I*, Addison-Wesley, Reading, Mass., 1968.

[8] Lee, D.T., and F.P. Preparata, Euclidean Shortest Paths in the Presence of Rectilinear Barriers, *Networks* 14 (1984), pp. 393-410.

[9] Nurmi, O., On Translating a Set of Objects in Two– and Three–dimensional Space, *Computer Vision, Graphics and Image Processing* 36 (1986), pp. 42-52.

[10] Nussbaum, D., and J.-R. Sack, Translation Separability of Polyhedra, *manuscript*, presented at the 1st Canadian Conf. on Computational Geometry.

[11] Nussbaum, D., and J.-R. Sack, Disassembling Two-dimensional Composite Parts Via Translations, *Proc. Int. Conf. on Optimal Algorithms*, 1989.

[12] Paterson, M.S., and F.F. Yao, Binary Partitions with Applications to Hidden-Surface Removal and Solid Modelling, *Proc. 5th Annual ACM Symp. on Computational Geometry*, 1989, pp. 23-32.

[13] Preparata, F.P., and M.I. Shamos, *Computational geometry, an introduction*, Springer-Verlag, New York, 1985.

[14] Preparata, F.P., and K.J. Supowit, Testing a Simple Polygon for Monotonicity, *Inform. Proc. Letters* 12 (1981), pp.161-164.

[15] Reif, J.H., and S. Sen, An Efficient Output-Sensitive Hidden-Surface Removal Algorithm and its Parallelization, *Proc. 4th Annual ACM Symp. on Computational Geometry*, 1988, pp. 193-200.

[16] Ottman, T., and P. Widmayer, On Translating a Set of Line Segments, *Computer Vision, Graphics and Image Processing* 24 (1983), pp. 382-389.

[17] Sack, J.-R., and G.T. Toussaint, Translating Polygons in the Plane, *Proc. 2nd Annual Symp. on Theoretical Aspects of Computer Science*, 1985, pp. 310-321.

[18] Tarjan, R.E., and C.J. van Wyk, An $O(n \log \log n)$ Time Algorithm for Triangulating Simple Polygons, *SIAM J. Comput.* 17 (1988), pp. 143-178.

[19] Toussaint, G.T., Movable Separability of Sets, in: G.T. Toussaint (Ed.), *Computational Geometry*, North Holland, 1985, pp. 335-376.

[20] Toussaint, G.T., On Separating Two Simple Polygons by a Single Translation, *Techn. Rep. SOCS-88.8*, McGill University, 1988.

[21] Yao, F.F., On the Priority Approach to Hidden-Surface Algorithms, *Proc. 21st Annual IEEE Symp. on Foundations of Computer Science*, 1980, pp. 301-307.

Output-Sensitive Generation of the Perspective View of Isothetic Parallelepipeds

(extended abstract)

Franco P. Preparata[*]
University of Illinois
Urbana, IL 61801, USA

Jeffrey Scott Vitter[†]
Brown University
Providence, RI 02912, USA

Mariette Yvinec[‡]
Ecole Normale Supérieure
75320 Paris, France

Abstract

We present a practical and new hidden-line elimination technique for displaying the perspective view of a scene of three-dimensional isothetic parallelepipeds (3D-rectangles). We assume that the 3D-rectangles are totally ordered based upon the dominance relation of occlusion. The perspective view is generated incrementally, starting with the closest 3D-rectangle and proceeding away from the viewpoint. Our algorithm is scene-sensitive and uses $O((n + d)\log n \log \log n)$ time, where n is the number of 3D-rectangles and d is the number of edges of the display. This improves over the heretofore best known technique. The primary data structure is an efficient alternative to dynamic fractional cascading for use with augmented segment and range trees when the universe is fixed beforehand. It supports queries in $O((\log n + k)\log \log n)$ time, where k is the size of the output, and insertions and deletions in $O(\log n \log \log n)$ time, all in the worst case.

1 Introduction

Substantial attention has been devoted to the generation of the planar display of a three-dimensional scene, a problem that is of central importance in areas such as computer graphics, animation, and flight simulation [9,16]. In this paper we consider output-sensitive approaches, which are more attractive than intersection-sensitive approaches,

[*]Support was provided in part by NSF research grant CCR–8906419. Part of this research was done while visiting Ecole Normale Supérieure in Paris, France.

[†]Support was provided in part by NSF Presidential Young Investigator Award CCR–8947808 with matching funds from an IBM research contract and by NSF research grant DCR–8403613. Part of this research was done while visiting Ecole Normale Supérieure in Paris, France.

[‡]Support was provided by CNRS.

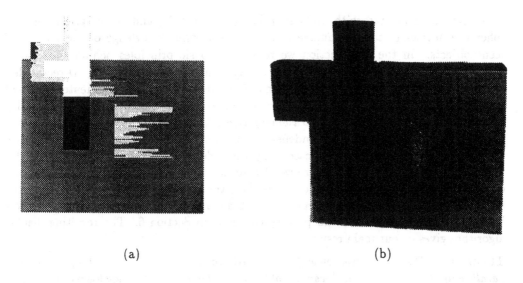

Figure 1: (a) The axial view of a set of isothetic 3D-rectangles (b) The perspective view

since the size of the display is never larger than the size of the underlying intersection graph and is frequently much smaller.

Güting and Ottmann [11] considered the case of the axial view of a collection of n isothetic rectangles in space (that is, the view from infinity along the z-axis of rectangles whose sides are parallel to the x- and y-axes), as shown in Figure 1(a). Their algorithm makes use of augmented segment trees and requires $O((n + d) \log^2 n)$ execution time, where d is the complexity of the resulting display. It uses a "priority" sweep, that is, the rectangles are processed starting with the rectangle closest to the point of view and proceeding away from the viewpoint. Such a total ordering of the rectangles exists for this case and also, in fact, for the case when the objects are isothetic parallelepipeds (hereafter called *3D-rectangles*) considered in the axial view.

The solution to the problem of the axial view of 3D-rectangles was recently improved upon by Preparata, Vitter, and Yvinec [19] with an $O(n \log^2 n + d \log n)$-time algorithm. The algorithm also uses a priority sweep, but with a very practical and efficient implementation of semidynamic augmented segment trees, based upon "contracted binary trees." Bern [3] and Näher, Mehlhorn, and Uhrig [15] obtained the same time bound using a radically different technique; their algorithms perform a planar sweep of the display and use the depth (z-coordinate) to determine the visible portions of the 3D-rectangles being scanned. Fractional cascading can be used to improve their running times to $O(n \log n \log \log n + d \log n)$, and a different preprocessing technique yields $O((n + d) \log n)$. Atallah and Goodrich proposed an $O(n^{3/2} + d)$-time algorithm [1], and Atallah, Goodrich, and Overmars provide a continuous spectrum between $O(n^{3/2} + d)$ time and $O((n + d) \log n)$ time [2].

The next natural question to consider is how to generate the perspective view, as shown in Figure 1(b). We make the preliminary assumption that the object-occlusion relation is acyclic [10]; this restriction does not exclude the very important case where the objects have disjoint projections on a plane, a subcase that occurs, for example, when all objects rest on a ground plane.

In this paper we start with the basic priority sweep of [11] and introduce appropriate coherence devices to take advantage of the available prior knowledge of the individual scene objects. In the next section we review the basic principles behind the display algorithms of [11] and [19] for generating the axial display of a scene of n 3D-rectangles. Our data structures are efficient semidynamic implementations of augmented segment trees and range trees, in which the secondary data structure used to store the catalog at each node consists of a *contracted binary tree* (CBT) and a *contracted stratified tree* (CST). The data structures are of independent interest and can be thought of as practical alternatives to dynamic fractional cascading: the code is simpler to implement and faster to execute, and the time bounds are worst-case, not amortized. In Section 3 we give an overview of the data structures and discuss how they can be used to generate the axial view. In Sections 4 and 5 we introduce the CBT and CST data structures. We show how to generate the parallel and perspective views in Section 6. The resulting display algorithm gives us our main result:

Theorem 1 *The perspective view from an arbitrary viewpoint of a set of n isothetic parallelepipeds (3D-rectangles) can be obtained in $O((n + d) \log n \log \log n)$ time using $O((n + d) \log n \log \log n)$ storage space, where d is the size of the display.*

2 The Display Algorithm

The 3D-rectangles are processed successively starting close to the viewpoint and proceeding away from the viewpoint. For each 3D-rectangle r, the first two steps generate the contribution of r to the display, while the subsequent steps update the silhouette of the 3D-rectangles already processed.

Step 1. For each side s of the current 3D-rectangle r, we use the data structure storing the silhouette edges to intersect s with the current silhouette and obtain an ordered list of intersections.

Step 2. Using the list of intersections from Step 1, we generate the contribution of the current 3D-rectangle r to the display and insert into our data structures the edges and vertices contributed by r to the silhouette.

Step 3. We update the silhouette edges cut by r, which we determined in Step 1. This involves deletions from and insertions in our data structures.

Step 4. We use the data structure storing the silhouette vertices to find the edges of the silhouette that are internal to r. We delete these edges from the silhouette and update our data structures.

In the next section we give an overview of the data structures we use to store the vertices and edges of the silhouette, and in the remaining sections we discuss their implementation and analysis. The data structures we use to store the vertices and edges require $O((n + d) \log n \log \log n)$ storage space, where d is the size of the final display. By the analysis we present in Theorems 3 and 4, Step 1 requires $O((\log n + k) \log \log n)$ time and Steps 2–4 each take $O((k + 1) \log n \log \log n)$ time, where k is the number of items updated at each step. Theorem 1 follows from these bounds, since the sum of k taken over all 3D-rectangles is $O(d)$.

3   Data Structures for the Axial View

In this section we give an overview of data structures we use to store the edges and vertices of the silhouette for the special case of generating the axial view, in which the image of each 3D-rectangle is a (two-dimensional) rectangle. The extensions to general parallel and perspective views will be discussed in Section 6.

In the case of axial views, we use two *augmented segment tree* data structures T_x and T_y to store the edges of the silhouette; T_x maintains the set of horizontal edges of the silhouette and is queried in Step 1 with the vertical sides of the processed rectangles, while T_y maintains the vertical edges of the silhouette and is queried in Step 1 with horizontal sides of the rectangles. We store the vertices of the silhouette in an *augmented range tree* data structure $\mathcal{R}_x$. In Step 4, given a query rectangle $[x_1, y_1) \times [x_2, y_2)$, we can find the silhouette vertices it contains by a range search using $\mathcal{R}_x$. The reader is referred to [18] for further background on segment trees and range trees.

In order to insert and delete edges and perform intersection queries in an efficient way, each secondary structure attached to a segment tree node embeds at the same time two related substructures, the *contracted binary tree* (CBT) and the *contracted stratified tree* (CST). The CBT handles insertions and deletions very efficiently. Each edge has to be inserted into or deleted from the catalogs of $O(\log n)$ segment tree nodes. The main idea is to provide bridges that link together entries of different catalogs, thus customizing to our purpose the principles behind fractional cascading, which was introduced by Chazelle and Guibas [6], building on the works of [5,7,8,12], and adapted by Mehlhorn and Näher [14] to fully dynamic operations. The bridging between different catalogs is obtained by adding dummy elements (called *witnesses*) to the catalogs.

While witnesses are crucial to the efficiency of insertions and deletions, they unfortunately pollute the answers to intersection queries, and filtering them out can drastically slow down the handling of these queries. The CST substructure attached to each segment tree node is designed to remedy this shortcoming. The CST is inspired by the priority queue of van Emde Boas et al [4].

The CBT and CST data structures are efficient, easy-to-implement, and practical realizations of augmented segment trees and augmented range trees in a dynamic setting. We explore their structures and properties in detail in Sections 4 and 5.

4   The Contracted Binary Tree (CBT)

A *contracted binary tree* (CBT) [19] is a tree structure that implements a dynamic dictionary L of distinct elements belonging to a finite totally ordered universe U known *a priori*. In our case, we use $U = \{0, 1, \ldots, 2n - 1\}$. The CBT is best described in terms of its underlying hypothetical "shadow structure" $\mathcal{SS}$, which is a balanced binary tree with the leaves $0, 1, \ldots 2n - 1$. The CBT $\mathcal{B}_L$ is a binary tree built on the "active" nodes of $\mathcal{SS}$. (See Figure 2.)

Lemma 1 *If L_1 and L_2 are two subsets of the universe such that $L_1 \subseteq L_2$, then for each node in $\mathcal{B}_{L_1}$ there is a node with the same label in $\mathcal{B}_{L_2}$.*

Two other notions are useful for characterizing our algorithms. Let y be an element in U that is not in L; y corresponds to a non-active leaf of $\mathcal{SS}$. The *neighbor* of y in $\mathcal{B}_L$ is

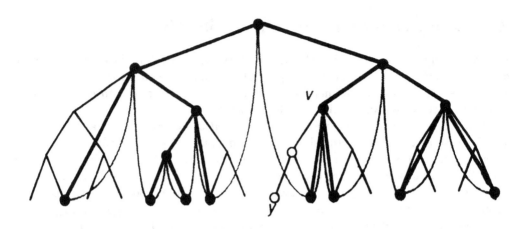

Figure 2: A contracted binary tree (CBT) and its underlying shadow structure. The active nodes are represented by black circles. The *parent*, *left*, and *right* links are shown as heavy lines, and the *next* and *prev* pointers are shown as dashed lines. The "companion nodes" v and u are used to insert a new leaf y in constant time.

the leaf in $\mathcal{B}_L$ whose label has the longest common prefix with the binary representation of y. Let z be an element in U. The first *companion node* v is the lowest active ancestor of z in $\mathcal{SS}$. If z is in v's left subtree, then the second companion node u is defined to be the lowest active right ancestor of z in $\mathcal{SS}$; otherwise, u is defined to be the lowest active left ancestor of z in $\mathcal{SS}$.

Lemma 2 *[19] The CBT $\mathcal{B}_L$ uses $O(|L|)$ space and supports the operations of insert(y), delete(y), and search(y) in $O(\log n)$ time, where $2n$ is the size of the universe. In addition delete(y) can be performed in constant time if we are given a pointer to the leaf storing y in the CBT, and insert(y) can be done in constant time if we are given a pointer to y's neighbor in $\mathcal{B}_L$ or if we are given the neighbor's companion nodes.*

4.1 CBTs in Segment Trees

Each catalog $A(V)$ is stored in a CBT $\mathcal{B}_{A(V)}$. When a horizontal edge e is added to the silhouette, an entry for this edge is added to the catalog $A(V)$ of each primary node V allocated to e. To insert these entries efficiently, we add to the data structure some links between entries in different catalogs. The links are analogous to the bridges between different catalogs used in the dynamic implementation of fractional cascading [14]. Our system of linked CBTs is however notably simpler than general dynamic fractional cascading, because it takes full advantage of the fact than the universe is finite and fixed in advance. In particular, no rebalancing of the structure is required when insertions or deletions are performed. The result is a simple implementation and fast performance.

Another original aspect of our data structure is that the idea of fractional cascading appears implicitly in the insertion or deletion of an edge. Intuitively we link together catalogs whose primary nodes are likely to appear together in the allocation set. We achieve this as follows: Starting from the segment tree, we construct a new structure,

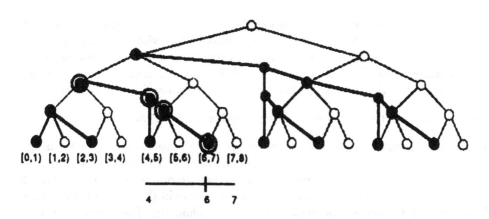

$[0,1)$ $[1,2)$ $[2,3)$ $[3,4)$ $[4,5)$ $[5,6)$ $[6,7)$ $[7,8)$

4   6   7

Figure 3: The left allocation forest G_l. For edge $e = (4, 7; y)$, the left allocation path includes the nodes allocated for e (which span x-ranges $[4, 6)$ and $[6, 7)$) plus a dummy node plus the extra node that spans the x-range $[0, 4)$.

called the *allocation forest* G (see also [17]), which is a forest of binary trees whose nodes are the segment tree nodes (less the root) and additional "dummy" nodes. Graph G consists of two entirely symmetrical components, which we call the *left allocation forest* G_l (pictured in Figure 3) and the *right allocation forest* G_r (defined symmetrically). The edges of the allocation forest G define the linking structure between entries of catalogs.

We extend the notion of primary node to include the dummy nodes, and we store the catalog of each dummy node in a CBT as well. We also modify our definition of "recording an edge" so that each edge $e = [x_1, x_2; y)$ is recorded not only at the nodes allocated to e in $\mathcal{T}_x$, but also at the ancestors of these nodes in G, as shown in Figure 3. This makes Lemma 1 applicable. We say that e is a *witness* in catalog $A(V)$ if V is not an allocation node of e.

Theorem 2 *The total time required for the insertion or deletion of an edge in the CBTs attached to the primary nodes on its left and right allocation paths is $O(\log n)$.*

Proof Sketch: The basic idea is that edge e is inserted into the catalogs of each of the $O(\log n)$ primary nodes on its left and right allocation paths in constant time (by Lemma 2) by using its companion nodes. At the root V of the left allocation path, the companion nodes in $\mathcal{B}_{A(V)}$ can be found in $O(\log n)$ time. Given the companion nodes of e in $\mathcal{B}_{A(V)}$, the companion nodes of e in $\mathcal{B}_{A(V')}$, where V' is a child of V in G_l, can be found by walking upward in $\mathcal{SS}$. Each walk starts where the previous one finishes. Since the height of $\mathcal{SS}$ is $\log n + 1$, the total time to find the companion nodes for all the insertions is $O(\log n)$. Insertion into the right allocation path is identical. $\qquad\square$

A separate bridge structure is needed to speed up queries. Let us recall that when we process a query $e = [x; y_1, y_2)$ we traverse a path in $\mathcal{T}_x$ from the root to the leaf associated with x; at each segment tree node V we search the catalog $A(V)$ and return all edges whose y-coordinates are in the range $[y_1, y_2)$. In order to accelerate the searching, we need extra witnesses that link together corresponding entries in the CBTs along this search path.

The solution is as follows: Whenever we record an edge e at a primary node V using the process outlined above in Theorem 2, we also record e as a witness at all ancestors of V in T_x. Theorem 2 remain valid even with these extra witnesses, since the crucial subset property remains valid. The following lemma shows that the total number of witnesses introduced to realize the two bridge structures described above is not too large:

Lemma 3 *Each inserted edge causes $O(\log n)$ witnesses to be introduced.*

Given a query vertical rectangle side $e = [x; y_1, y_2)$, we process the query by using the companion nodes of y_1 at each node V on the search path. We use a method similar to that used for Theorem 2 to amortize the cost of finding the companion nodes at each step.

Unfortunately, this algorithm does not perform queries in $O(\log n + k)$ time, where k is the number of intersected horizontal edges of the silhouette. The reason is that there might be many witnesses returned as the result of a query. Since witnesses are really "dummy" entries, they have to be filtered out during the query process. The resulting time bound is $O(\log n + k + w)$, where w, the numbered of witnesses encountered, might be significantly greater than $\log n + k$. The solution is to use a complementary data structure at each segment tree node that allows us to "jump over" witnesses and find the truly recorded edge e having the next-larger y-coordinate. The contracted stratified tree (CST) data structure discussed in Section 5 allows us to do the jumping in $O(\log \log n)$ time (see Theorem 6). This gives us the following performance:

Theorem 3 *Given a query vertical side, the horizontal edges of the silhouette that it intersects can be found in $O((\log n + k) \log \log n)$ time, where k is the number of intersected edges. The time required to insert or delete an edge is $O(\log n \log \log n)$.*

The time for insertion and deletion increases from the $O(\log n)$ bound quoted in Theorem 2 to $O(\log n \log \log n)$ because of the updating required for the CST.

4.2 CBTs in Range Trees

We augment the primary nodes of the range tree $\mathcal{R}_x$ used to store the silhouette vertices in a dual way to that for the segment trees $\mathcal{T}_x$ and $\mathcal{T}_y$ in Section 4.1. We use bridges as before also. An analysis similar to that in Section 4.1 gives us the following performance bounds:

Theorem 4 *Given a query rectangle, the vertices in the silhouette that it contains can be found in $O((\log n + k) \log \log n)$ time, where k is the number of interior vertices. The time required to insert or delete a vertex is $O(\log n \log \log n)$.*

5 The Contracted Stratified Tree (CST)

The bridge data structures described in Section 4 enable us to find the elements within a specified range of the accessed expanded catalogs. By expanded catalog, we mean the catalog consisting of witnesses as well as truly recorded entries. Let us refer to the truly recorded entries as *marked*. In order to filter out the witnesses during the query, we need to be able to quickly "jump over" the unmarked entries in a list of entries and find the next-larger marked entry. More formally, we define the *expanded catalog A* to be a subset of a totally ordered universe U, and we define the *catalog* (or *marked set*) to be a subset C of A. Both A and C are dynamic sets. Modifications to A are called *insert* and *delete*, and modifications to C are called *mark* and *unmark*. Given $x \in A$, a search (called *find*(x)) returns the smallest element y in C that is $\geq x$.

The prototypical solution is offered by the stratified tree structure of van Emde Boas et al [4]. Similar ideas are used in the priority queue data structure of Johnson [13], which implements a different set of primitives. The data structure is semidynamic, in that it is designed for a universe known *a priori*, as is our set U. However, the stratified tree of [4] is fully initialized and uses $O(n \log \log n)$ storage space irrespective of the number $|A|$ of items it actually stores. This does not suit our needs, since each node in the segment tree requires such an auxiliary structure, many of which store relatively few items. A structure inspired by [4], but with memory requirement $O(|A| \log \log |A|)$ rather than $O(|U| \log \log |U|)$, has been recently proposed by Mehlhorn and Näher [14]. It is designed to solve a generalized version of guided insertion, where the totally ordered set U need not be finite, and as a result there is high overhead.

Our proposed data structure, called the *contracted stratified tree* (CST) combines in a nontrivial and efficient way the features of the two data structures mentioned above. It performs the *insert*, *delete*, and *find* operations in $O(\log \log n)$ worst-case time using $O(|A| \log \log n)$ space. The motivation behind the CST is germane to that of the CBT. We start with the same underlying shadow structure SS, described in Section 4, which is a (fixed) balanced binary tree whose leaves correspond to the elements of U. We implement only those nodes of SS that are necessary in the execution of the desired operations. The set of nodes implemented in the CST for A will certainly contain all the nodes of the CBT for A, along with an additional set of nodes used to efficiently support the operations *find*, *mark*, and *unmark*.

As before, we assume that $|U|$ is a power of 2; let $|U| = 2^K$. The shadow structure SS is the balanced binary tree with 2^K leaves. The *level* of a node u of SS (denoted *level*(u)) is the number of edges in the root-to-leaf paths of the subtree rooted at u. Leaves have level 0, for example. The levels of SS are organized hierarchically, as shown in Figure 4 for the case $K = 9$. Let $\mathcal{L}$ be a balanced (but otherwise arbitrary) segment tree on $\{0, 1, \ldots, K\}$. An edge joining a node of SS at level i to a node of SS at level j, where $i < j$, can be viewed as an interval $[i, j)$; this interval (and the corresponding edge) is uniquely partitioned by $\mathcal{L}$ according to the well-known segment tree mechanism. Any subtree of SS with its root at level j and its leaves at level i is called a *canonical subtree* of SS with *span* $[i, j)$. The *order* of a canonical subtree T is the distance (in terms of number of edges in $\mathcal{L}$) from the root to the node of $\mathcal{L}$ whose span is the span of T. In particular, the root has order 0, its children have order 1, and so on.

Definition 1 Let $\mathcal{B}_A$ be the CBT for catalog A, augmented with the root if it is not already present. Each edge (u, w) of $\mathcal{B}_A$ is partitioned by $\mathcal{L}$ and is replaced by a chain where a node (called *buffer* node) is inserted at each cut-point. The resulting nodes, which we call *active nodes*, comprise the node set of the CST $\mathcal{S}_A$.

The nodes of the CST $\mathcal{S}_A$ are exactly the "active" nodes in the stratified tree of [4]. We now show that the implantation of the buffer nodes upon the CBT provides the support for the pointers needed to implement the operations *find*, *mark*, and *unmark*. A canonical

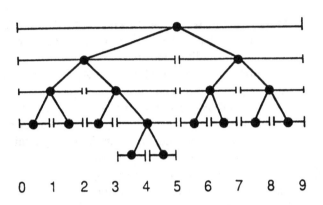

Figure 4: A segment tree $\mathcal{L}$ with base $\{0, 1, \ldots, K\}$. Each node of $\mathcal{L}$ is shown with its associated range.

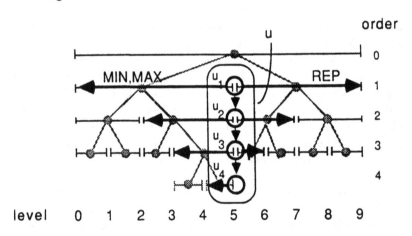

Figure 5: The subnodes of a given node, with their respective pointers

subtree of $\mathcal{SS}$ is called a *canonical subtree* of $\mathcal{S}_A$ if its root and at least one of its leaves are nodes of $\mathcal{S}_A$. For each canonical subtree T_p of $\mathcal{S}_A$ of order p having node u of $\mathcal{S}_A$ as a leaf, u has a pointer $rep_p[u]$ to the root of T_p. Conversely, for each canonical subtree T_p of $\mathcal{S}_A$ of order p having u as its root, u has pointers $min_p[u]$ and $max_p[u]$ to its marked leaves with smallest and largest values, respectively. (If none of its leaves are marked, we have $min_p[u] = max_p[u] = $ **nil**.) It is convenient to view node u of $\mathcal{S}_A$ as a list of *subnodes* $u_h, u_{h+1}, \ldots, u_{\max\{k,l\}}$, such that $u_{i+1} = succ(u_i)$, for $h \leq i \leq \max\{k, l\} - 1$. Each subnode is equipped with the appropriate fields *rep*, *min*, and *max*. All subnodes of u are given the same label as u; from the implementation point of view, all subnodes have a pointer to a common instantiation of the label of u. This interpretation is illustrated in Figure 5, where a node u of $\mathcal{S}_A$ at level 5 is assumed to have its pointers fully utilized and is replaced by the chain of subnodes u_1, u_2, u_3, and u_4. In Figure 6 we show an example of a CST $\mathcal{S}_A$, complete with the specification of its *min* and *max* pointers. Each pointer is expressed with as many bits as in the label of its destination.

We now analyze the storage used by the pointer structure introduced above. Since

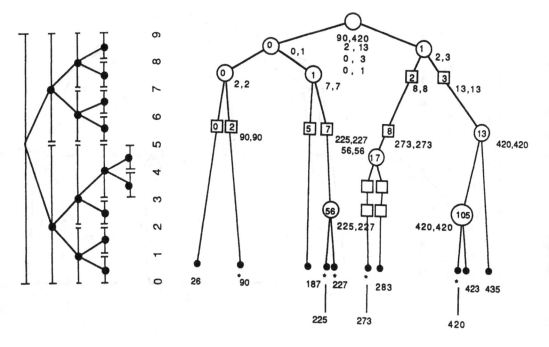

Figure 6: The CST S_A for $A \subseteq \{0, 1, \ldots, 511\}$. The marked elements C are designated by "*". Each node is labeled with the decimal interpretation of its bit string. Branching nodes are depicted as circles and buffer nodes as squares. Only the *min* and *max* pointers that are not nil are shown; each pointer is represented by the label value it points to. Shown on the left is the underlying segment tree $\mathcal{L}$.

S_A has $|A|$ leaves, the number of branching nodes in the CST is either $|A| - 1$ or $|A|$ (it is $|A|$ when the root has degree 1) and the total number of nodes is $O(|A| \log \log |U|)$. Each node of S_A can have as many as $O(\log \log |U|)$ *min*, *max*, and *rep* pointers, however collectively the following holds:

Theorem 5 *The storage used by CST S_A is $O(|A| \log \log |U|)$.*

We now consider the performance of the other operations on S_A. By analogy with the preceding analysis, the deletion of a leaf x of A and the updating of the pertinent nodes and pointers can be done in time $O(\log \log |U|)$.

Let us consider next the management of the marked set C and the execution of the operations *find*, *mark*, and *unmark*. If $y = find(x)$ is defined (that is, if $y \neq$ **nil**) there are two paths in S_A issuing, respectively, from leaves x and y (with x to the left of y) and converging at their lowest common ancestor $lca(x, y)$. The objective of *find* is to "home in" on $lca(x, y)$, specifically, on the unique maximal (that is, having minimum order) canonical substructure T' of S_A having $lca(x, y)$ as its root. This search may be viewed as the traversal of a sequence of nested canonical substructures of S_A, starting with the entire CST S_A and ending with T' (see Figure 7 for an example). Each of these substructures corresponds to a node of $\mathcal{L}$, and the history of a *find* operation is correctly

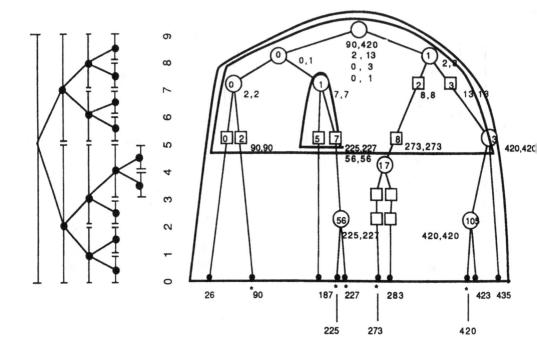

Figure 7: The sequence of nested canonical substructures traversed by *find*(187). Note that *lca*(187, 225) is the node labeled 1 at level 7.

interpreted as the traversal of a path in $\mathcal{L}$ starting at the root. The advancing mechanism of *find* traverses $O(\log \log |U|)$ canonical substructures, taking constant time per step, so that the overall running time of the procedure is $O(\log \log |U|)$ in the worst case. The recursive procedures *mark* and *unmark* similarly take $O(\log \log |U|)$ running time, in the worst case. Details appear in the full paper. Combining this with Theorem 5, we get the following:

Theorem 6 *The CST S_A for catalog A and universe U uses space $O(|A| \log \log |U|)$ and can perform the operations insert, delete, find, mark, and unmark in time $O(\log \log |U|)$, in the worst case.*

6   The Parallel and Perspective Views

First we show how to implement our display algorithm for the case of the parallel view, in which the point of view is at infinity, but no longer along one of the axes. In this case the two-dimensional image of each 3D-rectangle is a hexagon, whose opposite sides are parallel. All the hexagons share the same three orientations ℓ_1, ℓ_2, and ℓ_3. We denote by $\overline{\ell_i}$ the direction orthogonal to ℓ_i, for each i, as shown in Figure 8. We let S_i denote the set of ℓ_i-oriented edges of the current silhouette.

We replace the segment tree data structures $\mathcal{T}_x$ and $\mathcal{T}_y$ used for the axial view by six analogous segment trees $\mathcal{T}_{1,2}$, $\mathcal{T}_{1,3}$, $\mathcal{T}_{2,1}$, $\mathcal{T}_{2,3}$ $\mathcal{T}_{3,1}$, and $\mathcal{T}_{3,2}$. For each $1 \leq i, j \leq 3$,

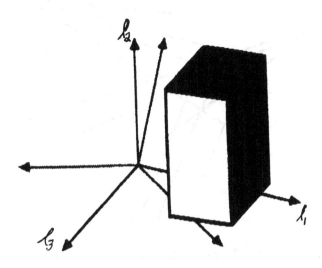

Figure 8: In the parallel view from infinity, each 3D-rectangle appears as a hexagon with pairs of parallel sides oriented along three directions ℓ_1, ℓ_2, and ℓ_3. Its image can be naturally partitioned into three parallelograms for purposes of finding interior edges of the silhouette.

$i \neq j$, segment tree $T_{i,j}$ is used for finding intersections between a query segment in the ℓ_i direction and the stored edges in the ℓ_j direction; $T_{i,j}$ stores the edges of S_j projected onto the $\overline{\ell_i}$ axis, using the $\overline{\ell_j}$ dimension for discriminating in the secondary data structure. The edges of S_j that are stored in $T_{i,j}$ might not start or end at $\overline{\ell_i}$ ordinates. In such cases, the edge is "trimmed" at each end to the nearest ordinate and only the remainder of the edges is stored in $T_{i,j}$; the pieces that are trimmed off cannot generate any intersections with query segments in the ℓ_i direction.

In a similar way we replace the range tree data structure $\mathcal{R}_x$ used for the axial view by three range trees $\mathcal{R}_{1,2}$, $\mathcal{R}_{1,3}$, and $\mathcal{R}_{2,3}$. For each $1 \leq i < j \leq 3$, range tree $\mathcal{R}_{i,j}$ is used for finding the vertices of the silhouette that are interior to a query parallelogram oriented in the ℓ_i and ℓ_j directions; $\mathcal{R}_{i,j}$ stores the vertices of the silhouette projected onto the $\overline{\ell_i}$ axis, using the $\overline{\ell_j}$ dimension for discriminating between vertices in the secondary data structure.

In the parallel view from infinity, the sides of each hexagon belong to lines forming three distinct pencils, in the ℓ_1, ℓ_2, and ℓ_3 directions. The pole of each of these pencils is a point lying on the line at infinity. Assuming that not all three pencil poles are at infinity transforms the parallel view into a perspective view, as shown in Figure 9. The technique described in the previous section requires only small changes to adapt to the more general case of the perspective view. Specifically, segment trees and range trees are now built on sets of polar angles rather than on sets of cartesian coordinates. Details are in the full paper.

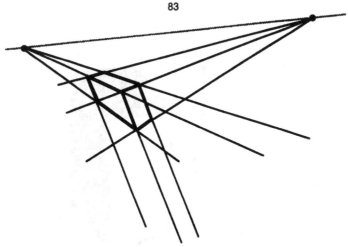

Figure 9: The perspective view of a 3D-rectangle

7 Conclusions

We have presented a new algorithm for hidden-line elimination in the perspective view of a scene of three-dimensional isothetic parallelepipeds (3D-rectangles). It uses $O((n + d) \log n \log \log n)$ time, where n is the number of 3D-rectangles and d is the number of edges of the display. It is scene-sensitive, in that its running time depends on the complexity of the resulting display and not on the potentially much larger set of underlying intersections. The main component of our algorithm is a fast and simple implementation of augmented segment and range trees in a semidynamic setting. The principles of fractional cascading appear, but not the customary overhead.

We made the assumption in this paper that the 3D-rectangles in the scene can be ordered in some way consistent with the dominance relation based on occlusion; in other words, we assumed that the dominance relation is acyclic. An interesting open problem is whether there is an $o(n^1 2)$ algorithm to determine if the dominance relation on a set of 3D-rectangles is acyclic. If the dominance relation is not acyclic, then we can split some 3D-rectangles until it becomes acyclic. Another open problem is whether finding a minimal set of 3D-rectangles to split is NP-hard.

Another open problem is to consider more general polyhedral scenes. Reif and Sen consider hidden-line elimination in polyhedral terrains. [20]. Polyhedral terrains are more general in some respects and more restrictive in others than the scenes we consider in this paper. It would be interesting to combine our techniques with those of [20] to enlarge the class of scenes we can handle.

References

[1] M. J. Atallah & M. T. Goodrich, "Output-sensitive Hidden Surface Elimination for Rectangles," Johns Hopkins University, Technical Report CS–88–13, 1988.

[2] M. J. Atallah, M. T. Goodrich & M. Overmars, "New Output-Sensitive Methods for Rectilinear Hidden-Surface Removal," Department of Computer Science, The Johns Hopkins University, Technical Report 89-09, 1989.

[3] M. Bern, "Hidden Surface Removal for Rectangles," *Journal of Comp. Sys. Sci.* (to appear 1990).

[4] P. van Emde Boas, R. Kaas & E. Zijlstra, "Design and Implementation of an Efficient Priority Queue," *Mathematical Systems Theory* 10 (1977), 1977.

[5] B. Chazelle, "Filtering Search: A New Approach to Query-Answering," *24th IEEE Symposium on Foundations of Computer Science*, Tucson, Arizona (November 1983).

[6] B. M. Chazelle & L. J. Guibas, "Fractional Cascading I: A Data Structuring Technique," *Algorithmica* 1 (1986), 133–162.

[7] R. Cole, "Searching and Storing Similar Lists," *Journal of Algorithms* 7 (1986), 202–220.

[8] H. Edelsbrunner, L.J. Guibas & J. Stolfi, "Optimal Point Location in a Monotone Subdivision," *SIAM Journal on Computing* 15 (1986), 317–340.

[9] J. D. Foley, A. van Dam, S. K. Feiner & J. F. Hughes, *Computer Graphics: Principles and Practice*, Addison-Wesley, Reading, MA, 1990.

[10] L. J. Guibas & F. F. Yao, "Translating a Set of Rectangles," in *Advances in Computing Research, Volume 1: Computational Geometry*, F. P. Preparata, ed., JAI Press Inc., 1983, 61–78.

[11] R. H. Güting & T. H. Ottmann, "New Algorithms for Special Cases of the Hidden Line Elimination Problem," *Computer Vision, Graphics, and Image Processing* 40 (1987).

[12] H. Imai & T. Asano, "Dynamic Orthogonal Segment Intersection Search," *Journal of Algorithms* 8 (1987), 1–18.

[13] D. B. Johnson, "A Priority Queue in Which Initialization and Queue Operations Take $O(\log \log D)$ Time," *Mathematical Systems Theory* 15 (1982), 295–309.

[14] K. Mehlhorn & S. Näher, "Dynamic Fractional Cascading," preprint 1987.

[15] S. Näher, K. Mehlhorn & C. Uhrig, Universität des Saarlandes, Technical Report A 02/90, 1990.

[16] W. M. Newman & R. E. Sproull, *Principles of Interactive Computer Graphics*, McGraw-Hill, New York, 1979.

[17] F. P. Preparata & D. T. Lee, "Parallel Batch Planar Point Location on the CCC," *Information Processing Letters* (to appear).

[18] F. P. Preparata & M. I. Shamos, *Computational Geometry*, Springer-Verlag, New York, 1985.

[19] F. P. Preparata, J. S. Vitter & M. Yvinec, "Computation of the Axial View of a Set of Isothetic Parallelepipeds," *ACM Transactions on Graphics* (to appear 1990).

[20] J. H. Reif & S. Sen, "An Efficient Output-Sensitive Hidden-Surface Removal Algorithm and its Parallelization," *Proceedings of the Fourth Annual ACM Symposium on Computational Geometry* (June 1988).

GRAPHICS IN FLATLAND REVISITED

Michel Pocchiola[*]

(pocchiol@dmi.ens.fr)

Abstract

Let S be a set of n non-intersecting convex objects in the plane. A view of S from a given viewpoint p is the circular sequence of objets intersected by a semi-infinite line rotating continuously around the viewpoint p. We show how to preprocess the set S in time $O(n^2 \cdot \alpha(n))$ where $\alpha(n)$ is a pseudo-inverse of Ackermann's function, so that the view from a given point in the plane can be computed in time $O(n \cdot 2^{\alpha(n)})$. Then we show how, with the same preprocessing time, the view can be maintained in $O(\log n)$ time per change while advancing the viewpoint along a given connected curve lying in the complement of the convex hull of S. Both algorithms use $O(n^2)$ storage. In the process we show how to reduce by a $\log n$ factor the preprocessing time of the ray-shooting problem and we investigate a preprocessing version of the edge visibility problem. This work also serves to demonstrate that the geometric duality between the plane and the set of oriented straight lines of the plane is a "good" framework in dealing with objects more general then the classical points and segments which we usually meet in the context of computational geometry.

1 Introduction

In this paper we investigate the following problems. Given a set S of n disjoint convex objects like line segments, polygons, circles, etc (an exact definition of the objects we consider can be founded in [EOW83]): (1) compute the first object intersected by a given semi-infinite line, then enumerate in order the succeeding intersected objects, (2) compute a view of S from a given point, (3) maintain the view while walking along a given connected curve, and (4) compute the region of the plane illuminated by a given luminous neon. We assume that S is fixed once and for all, so we are allowed to preprocess S.

The first problem is known in the literature as the "ray-shooting problem" or "segment intersection problem" [HDM82], [Cha86]. Chazelle [Cha86] studies this problem in the case where S is a set of non-intersecting segments. His algorithm uses an $O(n^2)$ data structure computable in $O(n^2 \log n)$ time and runs in an optimal $O(\log n)$ time for the first intersection, then an optimal $O(1)$ time for each succeeding intersection. The main tools used by [Cha86] are the point-line duality and a data structure, the hive graph which is related to the technique of filtering search. Sarnak and Tarjan [ST86] have shown that the Persistent Search Trees provide a simpler alternative to the hive graph. By the amortized properties of red-black trees, we show here that the preprocessing time

---

[*]Laboratoire d'Informatique de l'Ecole Normale Supérieure, URA 1327, CNRS, 45 rue d'Ulm, 75230 Paris Cédex 05, France.

86

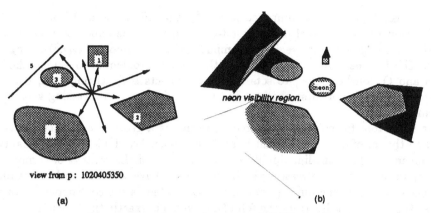

view from p: 1020405350

(a)   (b)

Figure 1: a view and a neon visibility region

can be reduced by a $\log n$ factor; furthermore, we introduce the duality between point and oriented straight line to deal with more general objects. At last we mention that tradeoffs between space and query time can be found in [Aga89].

The second problem is known as "the point visibility problem". In the case where S is the set of edges of a simple polygon the problem is solved in linear time (see [EA81],[Mor88],[ORo87]). In the general case the problem is solved in $O(n \log n)$ time, which is proved to be optimal by reduction from the problem of sorting (see [EOW83], [SO86]). What can we do if we are allowed to preprocess the set S ? Edelsbrunner *et al* [EOW83] consider a restricted version of the problem where *the viewpoint lies on a given connected curve*. In this case they compute a view in $O(n)$ time using an $O(n^2 \log n)$ data structure. Our algorithm uses an $O(n^2)$ data structure computable in $O(n^2 \alpha(n))$ time and computes a view in $O(n 2^{\alpha(n)})$ time for *every viewpoint in the plane*, which is quasi-optimal in the worst-case. It should be noted that the number of distinct views is in $O(n^4)$.

The third problem, which we call "the visibility maintenance problem", is also studied in [EOW83]. They maintain the view in $O(n^2 + k \log n)$ total time using $O(\frac{n^2}{\log n})$ space, where k is the number of intersections between the curve and the common tangents to pair of objects of S. Our algorithm uses an $O(n^2)$ data structure, computable in $O(n^2 \alpha(n))$ time; it runs in: (1) $O(\log n)$ time each time the viewpoint crosses a common tangent to two objects of S; the total running time is then $O(k \log n)$, (2) $O(\log n)$ time per change in the view, while the viewpoint moves along a given connected curve lying in the complement of the convex hull of S; the total running time is then $O(l \log n)$ where $l \leq k$ is the number of change in the view. It should be noted that the data structure used in our algorithm is independent of the curve on which the viewpoint lies.

The fourth problem, which we call the "neon visibility problem", has been studied in the following two cases: (1) S is a simple polygon and the neon is an edge of the polygon (in this case the region illuminated is a simple polygon called the "edge visibility polygon"), (2) S is a set of non-intersecting line segments and the neon is one of these line segments (in this case the region illumined is called the "edge visibility region" (see [ORo87])). The edge visibility polygon is computed in $O(n \log n)$ time in [CG88] [LL6a] [ElG85] [ORo87]. Using the $O(n \log \log n)$ Tarjan-Van Wyk [TW88] triangulation algorithm the "edge visibility polygon" is computed in $O(n \log \log n)$ time (see [GHL*86]).

The edge visibility region is computed in $O(n^4)$ time by Suri and O'Rourke in [SO86], which is shown to be optimal in the worst-case. In the same paper Suri and O'Rourke show that the edge visibility region is computable as the union of $O(n^2)$ triangular regions in $O(n^2)$ time. Our contribution to this fourth problem is to extend the results of Suri and O'Rourke [SO86] to sets of convex objects such as polygons, circles, etc, (the neon is then one of these objects) and to investigate a preprocessing version of the problem.

Our main tool for solving these four problems is a geometric duality between the plane and the set of oriented straight lines of the plane. We think that this duality is an improvement to the point-line duality commonly used in the context of computational geometry for at least three reasons; firstly we do not have to suppose that the lines are not vertical; secondly the dual space can be interpreted as the continuous set of all the views as the viewpoint describes the full plane, which is exactly the "answers's space" to visibility questions; thirdly we deal easily with a larger class of objects then the classical points and segments. This duality was suggested to us by the work of Plantinga and Dyer [PD87]. In this paper Plantinga and Dyer investigate the problem of computing the continuous set of all the parallel views of a set of n plane convex polygons in the 3D space.

2 Duality

2.1 The space of oriented straight lines

Consider $\mathcal{A}$, the topological space of oriented straight lines of an Euclidean plane P, and let p be a point of P and O a subset of P. We use the following definitions:

$$\sigma(p) = \{d \in \mathcal{A} \mid p \in d\} \ (\textit{i.e the lines trough } p)$$
$$\sigma(O) = \bigcup_{p \in O} \sigma(p) \ (\textit{i.e the lines intersecting } O)$$

Let now (I, x, y) be a coordinate system for the plane P. We identify the spaces $\mathcal{A}$ and $\mathbb{R}/2\pi Z \times \mathbb{R}$ using the bijection which maps the pair (θ, u) on the straight line $y \cos \theta - x \sin \theta - u = 0$ oriented by the angle θ; this bijection is well known in the context of geometric probability [San67],[Sol78]. It is straightforward that if O is connected then $\sigma(O) = \sigma(\textit{convex hull of } O)$. The following theorem describes $\sigma(O)$

Theorem 2.1 *Let O be a compact convex subset of P. There exists two functions λ and μ such that*
$$\sigma(O) = \{(\theta, u) \mid \mu(\theta) \leq u \leq \lambda(\theta)\}.$$
Moreover $\mu(\theta) = -\lambda(\theta + \pi)$.

Proof. The two functions λ and μ are defined by the following condition: the straight lines $y \cos \theta - x \sin \theta - \lambda(\theta) = 0$ and $y \cos \theta - x \sin \theta - \mu(\theta) = 0$ oriented by the angle θ are the two tangents of direction θ to the convex O (see figure 2).$\Box$

For example if O is a line segment whose endpoints are $m_i = (a_i, b_i)$ for $i = 1, 2$, then $\lambda = max(F_1, F_2)$ and $\mu = min(F_1, F_2)$ where $F_i(\theta) = b_i \cos \theta - a_i \sin \theta$, while if O is the ellipse $x = x_0 + a\cos\phi$, $y = y_0 + b\sin\phi$, $\phi \in [0, 2\pi]$ then $\lambda(\theta) = y_0\cos\theta - x_0\sin\theta + a\cos^2\theta + b\sin^2\theta$.

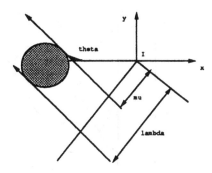

Figure 2: The λ and μ functions

Let now O_1 and O_2 be two convex compact subsets (convex objects in short) of P. The corresponding curves $\lambda_1, \mu_1, \lambda_2, \mu_2$ have the following property which will be useful in the sequel

Lemma 2.1 *The curves associated to two convex objects intersect two by two in at most two points.*

Proof. The points of intersection are exactly the common tangents of the two objects. But as the two objects are convex they have at most 4 common tangents. We recall now that the lines are oriented to complete the proof of our lemma.□

2.2 The dual of a set of n convex objects

Consider a set $S = \{O_1, O_2, \ldots, O_n\}$ of n disjoint convex compact subsets of the plane P and let $AR(S)$ be the arrangement of the $2n$ connected curves λ_i, μ_i of Theorem 2.1. We recall that the arrangement of a set of curves is the cell decomposition of the plane whose faces are the maximal connected components of the complement of these curves, the vertices are the intersection points of the curves and the edges are the connected components of these curves minus the vertices. As two curves meet in at most two points the arrangement can be constructed in $O(n^2 \cdot 2^{\alpha(n)})$ time and $O(n^2)$ space by an incremental method; this result is based on the Zone theorem of [EGP*88] and on the remarkable upper bound for the maximal length of an $(n, 4)$ Davenport-Schinzel Sequence (see [Aga89]). Here we can do slightly better.

Theorem 2.2 *The arrangement $AR(S)$ can be constructed in $O(n^2\alpha(n))$ time and $O(n^2)$ space.*

Proof (Outline). Introduce the curves in increasing u-coordinate for a given θ-coordinate. □

For every oriented straight line a let $\beta(a)$ be the sequence of elements of S which are intersected by the oriented line a. If the line a is tangent to the convex object O_i then we write $O_i{}^t$ in place of O_i in the sequence $\beta(a)$. Now we say that two oriented straight lines a and b are *equivalent*, denoted $a \sim b$, if we can move continuously from a to b under the condition that $\beta(a)$ remains constant. Then we have the following obvious result.

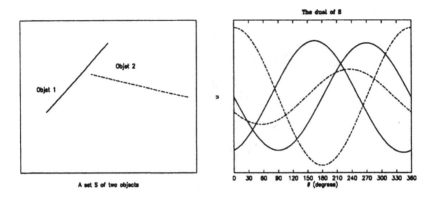

Figure 3: The dual of a set of two segments

Theorem 2.3 *The faces, edges and vertices of the arrangement $AR(S)$ are exactly the elements of the quotient-space $\mathcal{A}/\sim$.*

Definition 2.1 *Let S be a set of n convex compact subsets of the plane P. The arrangement $AR(S)$ where each face, edge and vertex is labelled by the common value $\beta(.)$ is called the dual of S; we denote it S^*.*

The figure 3 shows the dual of a set of two segments.

3 Applications

In the present section we use various representation of the dual S^* of S to solve the four problems defined in the introduction.

3.1 The ray-shooting problem

The problem here is to compute the first object, among a set of convex objects, intersected by a given semi-infinite line. The following result appears in the case of a set of segments in [Cha86].

Theorem 3.1 *Given a set of n convex objects there exists an algorithm that preprocesses the set in time $O(n^2 \log n)$ using $O(n^2)$ storage, such that the first intersection of a given semi-infinite line and the set can be computed in $O(\log n)$ time and such that each succeeding intersection can be computed in $O(1)$ time.*

Proof (Outline). Let $(p, \vec{u})$ be a semi-infinite line and $a \in \mathcal{A}$ the straight line oriented by $\vec{u}$ passing through p. The intersections of the set of objects and the semi-infinite line is a part of the sequence $\beta(a)$. So we must: (1) determine the sequence $\beta(a)$, (2) determine in $\beta(a)$ the first object intersected by the semi-infinite line, and (3) enumerate the succeeding intersections.

The first task is accomplished by an optimal planar point location in the arrangement $AR(S)$, the second task by a binary search in the sequence $\beta(a)$ represented as a search tree; the third task is then straightforward. To achieve the time and space preprocessing announced we compute the arrangement $AR(S)$ in quasi-optimal time $O(n^2\alpha(n))$ and $O(n^2)$ space and store the $O(n^2)$ sequences $\beta(.)$ using the Persistent Search Trees of Sarnak and Tarjan [ST86][DSST86], during a traversal of the arrangement $AR(S)$ from faces to adjacent faces.□

We observe that the above structure allows to compute the k intersections of a given segment with the set of objects in $O(k + \log n)$ time; this justifies the term "segment intersection problem".

As we have noticed in the introduction, Chazelle [Cha86] obtains the same bounds for a set of disjoint segments using the structure of hive graph. Let us now explain how we can reduce the preprocessing time by a $\log n$ factor.

Theorem 3.2 *Given a set of n disjoints segments there exists an algorithm that pre-processes the set in time $O(n^2)$ using $O(n^2)$ storage, such that the first intersection of a given semi-infinite line and the set can be computed in $O(\log n)$ time and such that each succeeding intersection can be computed in $O(1)$ time.*

Proof (Outline). In the case where the objects are segments, the arrangement $AR(S)$ is equivalent to an arrangement of lines; hence it can be constructed in $O(n^2)$ time and a location can be done in $O(\log n)$ after a preprocessing in $O(n^2)$ if we use the method of Kirkpatrick [Kir83]. It remains to show how we can store in a persistent structure the $O(n^2)$ sequences $\beta(.)$ in $O(n^2)$ time. For that we observe that the cost of an insertion or deletion in a persistent red-black tree can be decomposed in three parts: (1) $O(\log n)$ for searching the insertion or deletion node, (2) $O(1)$ amortized for color changes and (3) $O(1)$ amortized for nodes copying and pointer changes. Then we see that the $\log n$ factor comes from the time to search the node where we do an insertion or a deletion. It follows that if we were able to search the insertion (or deletion) node in $O(1)$ time then, using bidirectional pointers, we could completed the operation in $O(1)$ amortized time and the sequences $\beta(.)$ are stored in $O(n^2)$ time. Actually we can do that in $O(1)$ time by encoding in each edge of the arrangement the following information

Let f_1 and f_2 be two adjacent faces of the arrangement $AR(S)$. The label $\beta(f_2)$ is obtained from the label $\beta(f_1)$ by inserting (or deleting) object O_i between objects O_l and O_k.

Such an encoding can be computed by a vertical topological sweep of the arrangement $AR(S)$ in time $O(n^2)$ using only $O(n^2)$ extra storage. We postpone the proof to the next section. This completes the proof of our theorem.□

3.2 The point visibility problem

The problem here is to compute quickly a view from a given point after preprocessing the set of objects. Before providing our results we observe firstly that without preprocessing, a view can be computed en $O(n \log n)$ time by sorting the objects around the viewpoint and storing in a dictionary the set of objects intersected by a semi infinite line rotating continuously around the viewpoint (furthermore this bound is optimal), and secondly that the number of distinct views is in $O(n^4)$. Indeed a view can change each time the viewpoint crosses a tangent common to two objects of the scene. As the number of common tangents is in $O(n^2)$, their arrangement has a complexity in $O(n^4)$. This bound is attained for n points on a circle by example.

Theorem 3.3 *Given a set of n convex objects there exists an algorithm that preprocesses the set in time $O(n^2\alpha(n))$ using $O(n^2)$ storage, such that the view from a given point in the plane can be computed in $O(n2^{\alpha(n)})$.*

Proof (outline). Let p the viewpoint. The problem is equivalent to computing the value of the ray shooting problem for a particular direction, then to maintain the value of the ray shooting problem while the direction $\vec{u}$ of the semi-infinite line $(p, \vec{u})$ rotates continuously from 0 and 2π. Let $\delta(\theta)$ be the oriented straight line trough p of direction θ and let $d(\theta)$ the semi-infinite line whose origin is p and direction is θ. We define

- $f_1, \cdots, f_{2n+1}$ the $2n + 1$ faces of $AR(S)$ intersected by the line $\theta = 0$. These faces are computable in $O(n \cdot \alpha(n))$ time by insertion of the line $\theta = 0$ in the arrangement $AR(S)$.

- β_i the sequence of the face f_i; we represent β_i by a double linked list. These lists can be computed in $O(n^2)$ time.

- $pt[\cdot, \cdot]$ an array of pointers indexed by the $2n + 1$ faces f_i and the n objects of the scene; the pointer $pt[f_i, O_j]$ stores the address of the object O_j in the list β_i.

Such a structure can be built in time and space $O(n^2)$. The first point of our strategy is obtained by a double binary search; one for computing the face f_i at which $\delta(0)$ belongs and the second for computing in the list β_i the first object intersected. The second point of our strategy consists of sweeping the arrangement $AR(S)$ by the point $\delta(\theta)$ as θ increases and maintaining the value of the first object intersected by the semi infinite line $d(\theta)$. The sweep can be done in $O(n \cdot 2^{\alpha(n)})$ time [EGP*88]; while the keeping up phase can be done in $O(1)$ time by encoding in each edge of the arrangement the following information

> Let $f1$ and $f2$ be two adjacent faces of the arrangement $AR(S)$. The label $\beta(f2)$ is obtained from the label $\beta(f1)$ by inserting (or deleting) object O_i between adjacent objects O_k and O_l.

The array $pt[., .]$ allows us to maintain the value of $\beta(\delta(\theta))$ in constant time as the sweep crosses an edge of the arrangement $AR(S)$. It remains to explain how to encode the edges. The encoding can be computed by a vertical topological sweep of the arrangement $AR(S)$ in time $O(n^2)$; during the sweep we maintain the structure (lists β_i, faces f_i, array $pt[., .]$). We observe that when the vertical topological line crosses a vertex of the arrangement there is a unique list, say β_i, whose value is changed; but the modification to be done on β_i (insertion or deletion of two objects) is completely determinated by the two sequences adjacent to β_i. The array $pt[., .]$ allows us to make the modifications in constant time. The proof of our theorem is now completed.□

We should acknowledge the fact that the array technique used in the above proof has been used by McKenna [McK87] in developing a worst-case optimal hidden-surface removal.

3.3 The visibility maintenance problem

The problem here is to maintain the view while walking along a given connected curve. Our main result is that the preprocessing time is independent of the curve on which the viewpoint lies.

Theorem 3.4 *Given a set S of n convex objects there exists an algorithm that prepro-*
cesses the set in time $O(n^2\alpha(n))$ using $O(n^2)$ storage, such that the view can be main-
tained in $O(\log n)$ time per change while advancing the viewpoint along a given simple
connected curve lying in the complement of the convex hull of S.
If the given connected curve do not lies in the complement of the convex hull of S then
with the same preprocessing time and space, the view can be maintained in $O(\log n)$ time
each time the viewpoint crosses a common tangent to two object of S.

Proof. First we examine the case where the curve lies in the complement of the convex
hull of the scene. We call ASP the planar subgraph of $AR(S)$ obtained by deleting every
edge whose two incident faces have the same first element in their labels β. The ASP can
be easily computed in $O(n^2)$ time with the algorithm described in the previous section.
Let p be a point lying in the complement of the convex hull of S. We observe that

> the intersection of the ASP and the curve $\sigma(p)$ is exactly the view of S from
> the viewpoint p.

Let $p(t)$ be the viewpoint at time t. The problem is equivalent to maintaining the
intersection between the curve $\sigma(p(t))$ and the ASP as t varies, because the viewpoint is
in the complement of the convex hull of S. The view changes each time the curve $\sigma(p(t))$
passes through a vertex of the ASP. In addition, the next vertex of the ASP "visited"
by the curve $\sigma(p(t))$ after time, say t_1, is one of the endpoints of the edges of the ASP
intersected by the curve $\sigma(p(t_1))$. So we have to maintain the set of these endpoints
ordered by the times where these points will lie on the curve $\sigma(p(t))$. This can be done
in time $O(\log n)$ per change if we are able to solve in t the equation *"the vertex v lies on*
the curve $\sigma((p(t))$" in $O(\log n)$ time. This condition induces a restriction on the family
of connected curves we consider. A similar restriction is supposed in [EOW83].

Let us now examine the case where the curve do not lies in the complement of the
convex hull of S. We have seen in the previous subsection that a view of S from a
given point p is obtained by maintaining the value of the ray-shooting problem while the
direction $\vec{u}$ of the semi-infinite line $(p, \vec{u})$ rotates continuously from 0 to 2π. If we think
in the dual space then the view from p is exactly the intersection between the curve $\sigma(p)$
and the arrangement $AR(S)$ augmented for each face traversed by $\sigma(p)$ by the value
of the first object intersected. Let us call this augmented intersection the *intersection*
between the dual S^* and the curve $\sigma(p)$. Clearly the visibility maintenance problem is
equivalent to the maintenance of the intersection between S^* and $\sigma(p)$ as p varies. But
this intersection varies each time the curve $\sigma(p(t))$ sweeps a vertex of $AR(S)$. As in
the first case we can maintain the set of swept vertices in $O(log n)$ time each time the
viewpoint crosses a commun tangent to two object O and O' of the scene. Now we have
to explain how we can maintain the value of the first object intersected. But this is clear
because the new value belongs to a set of at most four objects. Namely the old value, O
,O' and the first object O'' intersected after O' provided that the tangent hits O before
O' (see figure 4). Hence the new value can be computed in constant time if during the
preprocessing time we have computed for each tangent the value of O''; this can be easily
done by similar technique we have developed in the point visibility problem. The proof
is now completed.□

This notion of ASP is the 2D version of the *aspect representation or asp* of Plantinga
and Dyer [PD87].

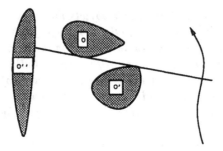

the view point crosses a commun tangent

Figure 4: illustrate the proof of theorem 3.4

3.4 The neon visibility problem

Consider B the set of semi-infinite lines of the plane P, let N be a subset of B and $b = (o, \vec{u}) \in B$. Define $\rho(b)$ to be the maximal line segment $[o, m]$ of b which does not intersect the objects of S, and $\rho(N) = \bigcup_{b \in N} \rho(b)$. The set $\rho(N)$ is the region of the plane illuminated by the "neon" N.

Let S be a set of n disjoint convex objects and O an object of S. Define $D(O)$ to be the set of semi-infinite lines "starting" from a point of O, and $D^*(O)$ the set of semi-infinite lines of $D(O)$ tangent to an object of S (which includes O). Then the illuminated regions by the neon $D(O)$ and the neon $D^*(O)$ are the same

Lemma 3.1 $\rho(D(O)) = \rho(D^*(O))$

Proof. Let p be a point of $\rho(D(O))$ and let $b \in D(O)$ such that $p \in \rho(b)$. Let a be the oriented supporting line of b and let a_1 be one of the points of intersection between $\sigma(p)$ and the frontier of the face of $AR(S)$ at which a belongs. This oriented line a_1 intersects the object O in o_1 and o_2, goes trough p and is tangent to an object of the scene (possibly O itself). But it is now clear that the semi-infinite line starting from o_2 trough p is a point of $D^*(O)$; hence it follows that $p \in \rho(D^*(O))$ (see figure 5) which completes the proof of our lemma. $\square$

We call $ASP(O)$ the planar subgraph of $AR(S)$ obtained by deleting every face, edge or vertex whose label β does not contain the element O or O^t and by merging every pair of incident faces having the same element succeeding the element O in their labels β. The $ASP(O)$ is a representation of the partition of the set of semi-infinite lines starting from a point of O under the relation "*intersect the same first object of S*". The planar graph ASP introduced in the previous subsection is obtained for O a circle including the whole scene.

Lemma 3.2 $\rho(D(O)) = \bigcup_{edge \in ASP(O)} \rho(edge)$

Proof. The proof is similar to the proof of the previous lemma. $\square$

We call $\rho(edge)$ a *pseudo-triangular region* (see figure 5) because in the case where the objects of S are line segments, $\rho(edge)$ is the triangular region of Suri and O'Rourke [SO86].

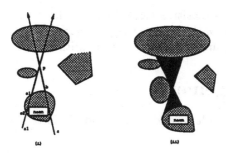

Figure 5: (i) illustrate the lemma 3.1 and (ii) pseudo-triangular region

Theorem 3.5 *Given a set S of n disjoint convex objects there exists an algorithm that computes in $O(n^2\alpha(n))$ time the region illuminated by one object O of S as the union of $O(n^2)$ pseudo-triangular regions.*

Proof. The previous lemma 3.2 shows that it is sufficient to compute the $O(n^2)$ edges of $ASP(O)$. But the graph $ASP(O)$ is clearly computable in $O(n^2\alpha(n))$ time by an algorithm similar to the one we have describe in the previous subsection.□

We observe that the time to compute l'ASP (O) is in $\Omega(n^2)$; this observation motivates the following theorem

Theorem 3.6 *Given a set S of n disjoint convex objects there exists an algorithm that preprocesses the set in time $O(n^2\alpha(n))$ using $O(n^2)$ storage, such that the region illuminated by a new object O lying in the complement of the convex hull of S can be computed in $O(n \cdot 2^{\alpha(n)} + k)$ time as the union of $k = O(n^2)$ pseudo-triangular regions.*

Proof. We observe that the illuminated region R by a neon O lying in the complement of the convex hull of the scene S is a subset of the illumined region by a circle including the whole scene. More precisely R is the union of the $\rho(\sigma(O) \cap edge)$ where *edge* describes the edges of the graph ASP. The preprocessing consists then in computing the graph ASP. We know that this can be done in $O(n^2\alpha(n))$ time and $O(n^2)$ space. The illuminated region R is obtained by introducing the curves λ and μ associated to the object O in the graph ASP (this can be done in $O(n2^{\alpha(n)})$ time), then enumerating the edges of the graph ASP lying between these two curves. The computing time is then $O(k + n2^{\alpha(n)})$ where k is the number of pseudo- triangular regions.□

4 Conclusion

We have shown how the duality between the plane and the set of oriented lines of the plane enables us to consider in a likeable manner visibility questions upon convex objects. In particular we have interpreted the dual of the scene as the continuous set of all views as the viewpoint describes the full plane. This interpretation provides a uniform treatment of the basic visibility questions in graphics: ray-shooting, view, coherence and illumination.

The design of output sensitive algorithms for the point visibility and the visibility maintenance problem without increasing the quadratic preprocessing space required should be an interesting improvement to our algorithms.

5   Acknowledgements

I wish to thank Richard Cole, Claude Puech and Mariette Yvinec for useful discussions.

References

[Aga89]   Pankaj K. Agarwal. *Intersection and Decomposition Algorithms for Arrangements of Curves in the Plane.* Technical Report Robotics Report No. 207, New York University Courant Institute of Mathematical Sciences, August 1989.

[CG88]   Bernard Chazelle and Leonidas J. Guibas. *Visibility and Intersection Problems in Plane Geometry.* Technical Report CS-TR-167-88, Princeton University Departement of Computer Science, June 1988.

[Cha86]   Bernard Chazelle. Filtering search: a new approach to query-answering. *SIAM J. Comput.*, 15(3):703–724, august 1986.

[DSST86]   Driscoll, Sarnak, Sleator, and Tarjan. Making data structures persitents. In *Proceedings of the 18th Annual ACM Symposium on Theory of Computing*, pages 669–679, May 1986.

[EA81]   H. ElGindy and D. Avis. A linear algorithm for computing the visibility polygon from a point. *J.Algorithms*, 2:186–197, 1981.

[EGP*88]   Edelsbrunner, Guibas, Pach, Pollack, Seidel, and Sharir. Arrangements of curves in the plane - topology, combinatorics and algorithms. In *ICALP*, pages 214–229, Springer-Verlag, 1988.

[ElG85]   H. ElGindy. *Hierarchical decomposition of polygon with applications.* PhD thesis, NcGill Univ., 1985.

[EOW83]   Herbert Edelsbrunner, Mark H. Overmars, and Derick Wood. Graphics in flatland: a case study. *Advances in Computing Research*, 1:35–59, 1983.

[GHL*86]   L.J. Guibas, J. Hersberger, D. Leven, M. Sharir, and R.E. Tarjan. Linear time algorithms for visibility and shortest path problems inside simple polygons. In *Proceedings of the Second Annual ACM Symposium on Computational Geometry*, pages 1–13, 1986.

[HDM82]   H.Edelsbrunner, D.G.Kirkpatrick, and H.A. Maurer. Polygonal intersection searching. *Information processing letters*, 14:83–91, 1982.

[Kir83]   David Kirkpatrick. Optimal search in planar subdivision. *SIAM J. Comput.*, 12(1):28–35, February 1983.

[LL6a]   D.T. Lee and A.K. Lin. Computing visibility polygon from an edge. *Comput. Vision, Graphics, and Image Proc.*, 34:1–19, 1986a.

[McK87]   M. McKenna. Worst-case optimal hidden-surface removal. *ACM Trans. on graphics*, 6:19–28, January 1987.

[Mor88]   J.-M. Moreau. A simple linear algorithm for polygon visibility. In ACM Siggraph France, editor, *proceedings of first annual Conference on Computer Graphics in Paris, PIXIM 88*, pages 151–161, 1988.

[ORo87]   Joseph O'Rourke. *Art Gallery Theorems and Algorithms*. Oxford University Press, 1987.

[PD87]   Harry Plantinga and Charles Dyer. *The Aspect Representation*. Technical Report 683, University of Wisconsin-Madison, January 1987.

[San67]   L.A. Santaló. *Integral Geometry and Geometric Probability*. Addison-Wesley, 1967.

[SO86]   Subhash Suri and Joseph O'Rourke. Worst-case optimal algorithms for constructing visibility polygons with holes. In *proceedings of the second ACM Symposium on Computational Geometry*, pages 14–23, 1986.

[Sol78]   Herbert Solomon. *Geometric Probability*. Society for Industrial and Applied Mathematics, 1978.

[ST86]   Neil Sarnak and Robert E. Tarjan. Planar point location using persistent search trees. *Communications of the ACM*, 29, July 1986.

[TW88]   Robert E. Tarjan and Christopher J. Van Wyk. An $O(n \log \log n)$- time algorithm for triangulating a simple polygon. *Siam J. Comput.*, 1:143–178, february 1988.

The Visibility Diagram: a Data Structure for Visibility Problems and Motion Planning

Gert Vegter

Dept. of Computing Science, University of Groningen

P.O.Box 800, 9700 AV Groningen, The Netherlands

1 Introduction

The visibility diagram $VIS(S)$ of a set S of line segments in the plane—which don't intersect except possibly at their boundary points—is a subdivision (cell complex) of a certain two-dimensional space whose points correspond to lines in the plane. This geometric object may be considered as a generalization of the visibility graph of the set S of obstacle edges. We recall that the boundary points of segments in S form the set of vertices of the visibility graph $G_S = (V_S, E_S)$ of S, while for two boundary points p and q the pair $\{p, q\}$ is an edge of G_S if and only if the points p and q can see each other, i.e. the line segment (p, q) is disjoint from all segments in S. $VIS(S)$ implicitly represents the visibility graph: its 0-cells correspond to edges of G_S. The additional topological information, viz. the 1- and 2-dimensional cells, makes this data structure suitable for solving some geometric problems concerning visibility and motion planning. We consider the computation of the visibility polygon of a query point and the problem of planning a feasible motion of a rod amidst the set of obstacles S. In the latter problem both the length of the rod and its initial and final position are parameters of the query.

The computation of the visibility polygon of a query point has been considered in [AAGHI], where an algorithm is presented that solves the problem in $O(n)$ time with $O(n^2)$ space and $O(n^2)$ preprocessing time, where $n = |S|$. The time bound achieved in that paper is actually $\theta(n)$, even when the size k of the visibility polygon satisfies $k = o(n)$. Conceptually speaking the algorithm for answering a visibility polygon query consists of a rotational sweep of a line about the query point q. During the sweep the subsegments visible from q along the sweep-line are collected. The sweep-line passes a number of critical positions in which it contains at least one boundary point—of a segment in S—that is *visible from q*. These critical positions will be computed on the fly in polar order with respect to q. In fact this more or less obvious approach is similar to that of [AAGHI]. However in the latter paper the critical positions correspond to situations in which the sweep-line contains a boundary point, *irrespective of whether it is visible from q or not*. The algorithm to be described has output sensitive time complexity in the sense that it has an $O(k \log(n/k)))$ bound for the query time, which coincides with the result of [AAGHI] in the worst case $k = \theta(n)$.

Motion planning for a rod—line segment—with three degrees of freedom amidst polygonal obstacles has been considered in several papers, e.g. in [LeSh]. In that paper a subdivision of the space of free positions is constructed, yielding a 3-dimensional cell complex. The incidence structure of this cell complex is called the *connectivity graph*. The algorithm uses $O(n^2)$ space and $O(n^2 \log n)$ time for finding a motion of the rod.

In [ShSi] an algorithm is presented that requires $O(h \log n)$ time for planning a motion, where h is the number of pairs of obstacle-edges whose distance is at most the length of the rod. Note that $h = \theta(n^2)$ in the worst case.

We construct a 2-dimensional cell complex which is a subdivision of the set of lines supporting the rod in a free position. The incidence structure is obtained by modifying the visibility diagram of S. We show how $VIS(S)$ is used to plan in $O(|E_S|)$ time a motion of the rod consisting of a sequence of $O(|E_S|)$ elementary motions, or to decide that no such motion exists. Our algorithm yields a motion in which the rod may *touch* a line segment of S. The method can however easily be adapted to yield a *free* motion of the rod.

The computation of $VIS(S)$ requires $O(n^2)$ time. The space needed in preprocessing is $O(n^2)$, while the space needed to store $VIS(S)$ is proportional to the size of the visibility graph. Since in [KeO] a quadratic lower bound has been derived for planning the motion of a rod, the algorithm presented in this paper is optimal.

2 The Visibility Diagram

Introduction of $VIS(S)$

In this section we define the visibility diagram $VIS(S)$. For the sake of convenience we assume throughout the paper that the segments in S are disjoint, and that no three boundary points are collinear. These restrictions can easily be removed without increasing the—asymptotic—complexity of the methods.

First we introduce some terminology. The set $Bd(S)$ consists of all boundary points of segments of S. The subset $Obst \subset \mathbb{R}^2$ is the union of the line segments in S.

$\mathcal{L}$ is the set of directed lines in the plane. The set $\mathcal{L}$ is a 2-dimensional topological space. We shall consider its topology in more detail in section 2.2. For a line segment s the set $\mathcal{L}_s$ consists of all directed lines intersecting s. Similarly for a point p the set of directed lines containing p is denoted by $\mathcal{L}_p$. We introduce a circle s_∞—the 'blue sky'—surrounding S with infinite radius and finite center. The augmented set $S \cup \{s_\infty\}$ is denoted by S^+. By definition $\mathcal{L}_{s_\infty} = \mathcal{L}$.

With each $s \in S^+$ we associate the *visibility function* $f_s : \mathcal{L}_s \to S^+$ in the following way. If $s \in S$, $l \in \mathcal{L}_s$, then $f_s(l)$ is the segment of S visible from s along l, or s_∞ if no such segment exists. If $l \in \mathcal{L}_{s_\infty}$ is disjoint from all segments in S then by definition $f_{s_\infty}(l) = s_\infty$, otherwise $f_{s_\infty}(l)$ is the segment of S that is uniquely determined by the property that it is not visible along l from any other segment of S.

The idea of our method is to precompute for each $s \in S^+$ the maximal connected regions of $\mathcal{L}_s$ on which f_s is locally constant (see [ChG] for a similar approach). These 2-dimensional regions turn out to be convex in a sense to be made more precize later. Therefore for a query point q the intersection of $\mathcal{L}_q$ and the boundary of a 2-cell can be determined efficiently. Each of these intersections corresponds to a situation in which the sweep-line contains a point that is *visible from* q. To describe this subdivision of $\mathcal{L}_s$ formally we introduce the notion of a *semi-free* segment, which is a—possibly unbounded—

line segment *not piercing* the set *Obst*. It is allowed to *touch* the set *Obst*, however (cf. [LeSh]). For $l \in \mathcal{L}_s$ the subsegment $SF(s,l)$ of l is the closure of the maximal semi-free segment containing the open interval joining $l \cap s$ and $l \cap f_s(l)$, cf. Figure 1.

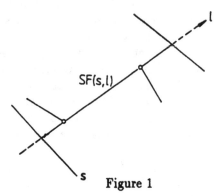

Figure 1

Next we characterize the cells of the subdivision $\mathcal{V}is(s)$ of $\mathcal{L}_s$.

Definition 2.1 *Let* $s \in S$. *A face (edge,vertex) of the subdivision* $\mathcal{V}is(s)$ *of* $\mathcal{L}_s$ *is a maximal connected subset* c *of* $\mathcal{L}_s$ *with the property that for all* $l \in c$ *the segment* $SF(s,l)$ *contains no (exactly one, at least two) points of* $Bd(S)$.

Two cells c_0 and c_1 of a subdivision are called *incident* —upon each other—if their dimensions differ by one, while c_0 is contained in the boundary of c_1 or c_1 is contained in the boundary of c_0.

Consider a line $l \in e$, where e is an edge of $\mathcal{V}is(s)$. Translating l slightly into its left or right half-plane yields a line belonging to a face of $\mathcal{V}is(s)$ incident upon e, which is denoted by $e \cdot left$ or $e \cdot right$, respectively. See Figure 2.

It is obvious from definition 2.1 that f_s is constant on each cell of the subdivision $\mathcal{V}is(s)$. The value of f_s on c is denoted by $c \cdot vis$. For a cell c of $\mathcal{V}is(s)$, $s \in S^+$, the segment s is denoted by $c \cdot seg$.

The algorithm that computes the visibility polygon maintains the segment *from which* the query point is visible along the sweep-line. To update this information efficiently when passing a critical position, we associate with each edge e an edge $e \cdot ref$ in the following way. Let e be an edge of $\mathcal{V}is(s)$ then for any line $l \in e$ the segment $SF(s,l)$ contains exactly one point p of $Bd(S)$.

If p is not an end-point of s, let s' be the segment containing p. Then l, considered as an element of $\mathcal{L}_{s'}$, is also on an edge of $\mathcal{V}is(s')$, which is denoted by $e \cdot ref$, see Figure 2.

If p is an end-point of s, let s' be the segment from which p is visible along l. Again l, considered as an element of $\mathcal{L}_{s'}$, is also on an edge of $\mathcal{V}is(s')$, which is denoted by $e \cdot ref$. Note that $e' = e \cdot ref$ iff $e = e' \cdot ref$.

Definition 2.2 *The* Visibility Diagram VIS(S) *of* S *is the graph* $(\mathcal{V}_S, \mathcal{E}_S)$ *such that*
$\mathcal{V}_S$ *is the set of cells of* $\{\mathcal{V}is(s) \mid s \in S^+\}$
$\{c_0, c_1\} \in \mathcal{E}_S$ *iff one of the following conditions holds :*
$c_0 \cdot seg = c_1 \cdot seg$ *and* c_0 *and* c_1 *are incident;*
c_0 *and* c_1 *are edges (of* $\mathcal{V}is(c_0 \cdot seg)$ *and* $\mathcal{V}is(c_1 \cdot seg)$, *respectively) with* $c_0 \cdot ref = c_1$.

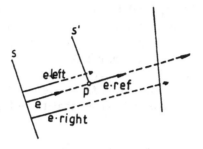

Figure 2

The Subdivision $\mathcal{V}is(s)$

We now introduce coordinates on the set $\mathcal{L}_s$ using the well known duality transform. Under such a transform almost all of $\mathcal{L}_s$ is mapped onto a subset of the Euclidean plane, and the subdivision $\mathcal{V}is(s)$ is mapped onto a straight-line subdivision of the plane having convex faces.

To make these ideas more precise let l_s^+ and l_s^- be the directed lines supporting s. Fixing an orientation on $\mathbb{R}^2$ we can distinguish the connected components of $\mathcal{L}_s \setminus \{l_s^+, l_s^-\}$: $\mathcal{L}_s^\pm$ is the set of directed lines with the property that for all $l \in \mathcal{L}_s^\pm$ the pair $(l, l_s^\pm)$ is *positive* with respect to the orientation on $\mathbb{R}^2$.

For $l_0, l_1 \in \mathcal{L}_s^+$ $(\mathcal{L}_s^-)$ the—closed—s-interval $[l_0, l_1]$ is defined as the subset of $\mathcal{L}_s^+$ $(\mathcal{L}_s^-)$ obtained by rotating a line from l_0 to l_1 about the point of intersection of l_0 and l_1. Whether the rotation is clockwise or counter-clockwise is determined by the fact that all lines of the interval $[l_0, l_1]$ must intersect s. A subset c of $\mathcal{L}_s \setminus \{l_s^+, l_s^-\}$ is called s-*convex* if for all $l_0, l_1 \in c$ the s-interval $[l_0, l_1]$ is a subset of c. An s-*polygon* is a simply connected region in $\mathcal{L}_s \setminus \{l_s^+, l_s^-\}$ whose boundary is the union of finitely many s-intervals, which don't intersect except possibly at their boundary points.

To embed $\mathcal{L}_s^\pm$ in the plane introduce Cartesian coordinates x_1, x_2 matching the orientation on $\mathbb{R}^2$ such that the x_2-axis is along l_s^+. Then any $l \in \mathcal{L}_s^+(\mathcal{L}_s^-)$ is directed according to *increasing (decreasing)* x_1-coordinate. The *duality transform* $\mathcal{D}_s^\pm$ maps a non-vertical line l with equation $x_2 = h_1 x_1 + h_2$ onto the point $\mathcal{D}_s^\pm(l) = (h_1, \pm h_2)$, and the point $p = (p_1, p_2)$ onto the line $\mathcal{D}_s^\pm(p)$ with equation $x_2 = \mp p_1 x_1 + p_2$. There are several duality transforms, cf. [Me]. Ours has the convenient feature that for an edge $e \subset \mathcal{L}_s^\pm$ the dual image of the face $e \cdot left$ lies to the left of the dual image of e. The map $\mathcal{D}_s : \mathcal{L}_s \setminus \{l_s^+, l_s^-\} \to \mathbb{R}^2$ is defined by $\mathcal{D}_s | \mathcal{L}_s^\pm = \mathcal{D}_s^\pm$. Note that the dual image $\mathcal{D}_s^\pm(\mathcal{L}_s^\pm)$ is the horizontal strip bounded by the lines $x_2 = \pm a$ and $x_2 = \pm b$. Here $(0, a)$ and $(0, b)$ are the boundary points of s. Obviously $\mathcal{D}_s$ maps an s-convex set onto a convex subset of the plane.

Lemma 2.3 *Let $s \in S$. Any edge of $\mathcal{V}is(s)$ is an open s-interval. Any face of $\mathcal{V}is(s)$ is an s-polygon, which is s-convex.*

Proof: Omitted from this version. The proof is contained in [V]. $\square$

A subset c of $\mathcal{L}_{s,\infty} = \mathcal{L}$ is called *weakly convex* if there is a segment s, not necessarily in S, such that $c \subset \mathcal{L}_s \setminus \{l_s^+, l_s^-\}$, while c is convex as a subset of $\mathcal{L}_s$. Similarly c is called a *weakly convex polygon* if moreover c is an s-convex polygon in $\mathcal{L}_s$.

This definition makes sense in view of the following fact: if c is a subset of both $\mathcal{L}_{s_0} \setminus \{l_{s_0}^+, l_{s_0}^-\}$ and $\mathcal{L}_{s_1} \setminus \{l_{s_1}^+, l_{s_1}^-\}$, then c is s_0-convex iff c is s_1-convex.

The subdivision $\mathcal{V}is(s_\infty)$ has two *exceptional* faces c_+ and c_-, determined by the property that $l \in c_+$ (c_-) if and only if the convex hull of the set S is a subset of the right (left) half-plane of l. It is easy to see that these faces are not simply connected, and hence not convex.

Lemma 2.4 *Any cell c of $\mathcal{V}is(s_\infty)$ different from c_+ and c_- is weakly convex.*

Proof: Again the proof is omitted from this version. See [V]. $\qquad\qquad\qquad\qquad\square$

An obvious consequence of lemma 2.3 and lemma 2.4 is

Corollary 2.5 *Let c be a face of $\mathrm{VIS}(S)$ different from $c_\pm$. Then for any $q \in \mathbb{R}^2$ the intersection of $\mathcal{L}_q$ and the boundary of c can be determined in $O(\log n_c)$ time (after $O(n_c)$ preprocessing time), where n_c is the number of edges incident upon c.*

The construction of $VIS(S)$ is a trivial modification of Welzl's algorithm for computing the visibility graph of S, cf. [We]. We refer to [V] for the proof of the following result.

Theorem 2.6 *The visibility diagram $\mathrm{VIS}(S)$ can be computed in $O(n^2)$ time. It can be stored in $O(|E_S|)$ space.*

3 Answering a Visibility Polygon Query

As outlined in the Introduction the visibility polygon of a query point q is computed during a rotational sweep of a directed line l about q, starting in vertically downward position. We assume that q is in general position, i.e. the maximal semi-free subsegment of l containing q contains at most one point of $Bd(S)$. In [V] this restriction is removed without increasing the time complexity.

To sketch the algorithm we introduce variables s and c satisfying
Q: s is the segment of S^+ *from which q is visible along the sweep-line l in a non-critical position and c is the face of $\mathcal{V}is(s)$ containing l.*

We first describe how s and c are initialized and subsequently how the invariant Q is restored after the sweep-line has passed a critical position.

Initialization

Consider the *vertical adjacency map* $M(S)$ of S obtained by extending from each point of $Bd(S)$ a vertical line in both directions, until a point of an other segment in S is hit, cf. [PS]. $M(S)$ generates a subdivision of the plane into $O(n)$ regions. With each region r corresponds a unique cell $c(r)$ of $VIS(S)$ such that all points $p \in r$ are visible from the segment $c(r) \cdot seg$ along the vertical line $l(p)$ through p which is directed downward, while moreover $l(p) \in c(r)$. With each region r of $M(S)$ we associate the cell $c(r)$ of $VIS(S)$, which therefore can be determined in $O(1)$ time. Using an optimal point-location structure for $M(S)$—which can be built in $O(n \log n)$ preprocessing time and

$O(n)$ space—the region of $M(S)$ containing the query point q can be determined in $O(\log n)$ time, cf. [Me]. Therefore the initial value of s and c can be found in $O(\log n)$ time.

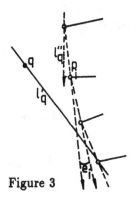

Figure 3

Remark 3.1 If the vertical line l_q through q is disjoint from the convex hull of S the initial value of c is one of the non-convex faces $c_\pm$ of $Vis(s_\infty)$. To determine the intersection l'_q of $\mathcal{L}_q$ and the boundary ∂c of c let l''_q be the vertical line bounding the region of $M(S)$ containing q. Walking along ∂c towards l''_q we pass the edges $e_1, \ldots, e_h$, with $e_i \subset \mathcal{L}_{p_i}$ for certain points $p_i \in Bd(S)$, $1 \le i \le h$, see Figure 3. Obviously l''_q is tangent to $\text{conv}(S)$ and p_i is visible from q, for $1 \le i \le h$. Consequently $h \le k$, where k is the number of edges of the visibility polygon of q, so the intersection of $\mathcal{L}_q$ and ∂c can be determined in $O(k)$ time. Therefore the initialisation time does not dominate the overall time complexity.

Transition at a Critical Direction

Consider the transition at the position l_{crit}. In view of our genericity assumption l_{crit} lies on an edge e of $Vis(s)$. Let p be the unique point of $Bd(S)$ on the semi-free segment $SF(s, l_{crit})$, so in particular $e \subset \mathcal{L}_p$. We distinguish two cases.

Case 1: $c = e \cdot right$, see Figure 4.

In this case p is visible from q along l_{crit}. Any line l_1, obtained by rotating l_{crit} slightly about q in counter-clockwise direction can also be obtained by translating a suitable line $l_0 \in e$ into its left half-plane. Therefore the new value of c is $e \cdot left$.

Case 2: $c = e \cdot left$.

In this case q is visible from p along l_{crit}. There are two cases, according to whether p is an end-point of the segment $c \cdot seg$ or not. In both cases one easily checks that any line, obtained by rotating l_{crit} slightly about q in counter-clockwise direction can also be obtained by translating a suitable line $l_0 \in e \cdot ref$ into its right half-plane. Therefore the new value of c is $e \cdot ref \cdot right$.

The visibility polygon is built during the sweep process using the values $c \cdot seg$ and $c \cdot vis$. The—trivial—details are omitted.

Let $\mathcal{C}$ be the set of faces of $VIS(S)$ over which the variable c ranges in the algorithm. Note that $|\mathcal{C}| = k$. For $c \in \mathcal{C}$ the number of edges of the face c is denoted by n_c. Since each transition takes constant time, we derive from Corollary 2.5 and Remark 3.1

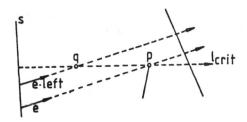

Figure 4

that the query time is $O(\sum_{c \in C} \log n_c)$. A rough estimate for the query time therefore is $O(k \log n))$. However one can prove:

Lemma 3.2 $\sum_{c \in C} n_c = O(n)$.

The proof of this lemma is rather intricate and therefore omitted from this version. We refer to [V] for details.

In view of the previous lemma $\sum_{c \in C} n_c \leq an$, for some positive constant a. Since the geometric means of the numbers $n_c, c \in C$, is bounded above by the arithmetic means, it follows that $\prod_{c \in C} n_c \leq (an/k)^k$. Therefore $\sum_{c \in C} \log n_c \leq k \log(an/k)$. Hence we have the following result:

Theorem 3.3 *Given the visibility diagram VIS(S) of a set S of n line segments that don't intersect except possibly at their boundary points, the visibility polygon of a query point q can be computed in $O(k \log(n/k))$ time, where k is the number of edges of the visibility polygon of q.*

4   Planning the Motion of a Rod

We describe an efficient method to determine a motion for a rod of length d avoiding the set of obstacles *Obst*. To give a precize setting for this problem we introduce the set $SFP_d(S)$ of *d-semi-free positions*—with respect to S—which is the subset of $\mathbb{R}^2 \times S^1$ defined by: $(p, v) \in SFP_d(S)$ iff the segment $\{p + \lambda v | 0 \leq \lambda \leq d\}$ is semi-free. A *feasible d-motion* is a continuous path $\alpha : [0, 1] \to SFP_d(S)$. A *motion planning query* comes as a triple $(d, (p_0, v_0), (p_1, v_1))$ with $d \in \mathbb{R}$ and $(p_i, v_i) \in SFP_d(S)$, $i = 0, 1$, and asks to determine a feasible d-motion from initial position (p_0, v_0) to final position (p_1, v_1), or to report that no such motion exists.

To bound the complexity of a motion we introduce the concept of *elementary d-motion*, which is a feasible d-motion of one of the following types.

Type 1. A jump.

This motion corresponds to a translation of the rod along the line supporting it, i.e. a motion of the form $\alpha(t) = (p + tv, v)$, $t_0 \leq t \leq t_1$, with $(p, v) \in \mathbb{R}^2 \times S^1$. In many cases the points $p + t_i v$, $i = 0, 1$, lie on segments of S, in which case the rod 'jumps' from one segment to the other.

Type 2. A sliding rotation.
This motion is of the form $\alpha(t) = (p(t), v(t))$, where the line $p(t) + Rv(t)$ supporting the rod rotates about a fixed point p, while the tail of the tail $p(t)$ of the rod lies on a fixed line segment of S.

A feasible d-motion $\alpha : [0,1] \to SFP_d(S)$ consists of k elementary d-motions if there is a partition $0 = t_0 < t_1 < \ldots < t_n = 1$ of the parameter interval such that $\alpha|[t_i, t_{i+1}]$ is an elementary d-motion for $0 \leq i < n$. It is reasonable to consider k as the complexity of the motion. The main result of this section is

Theorem 4.1 *Consider a set S of n line segments which don't intersect, except possibly at their end-points. There is an algorithm for answering a motion planning query for a rod of query-length d amidst the set of obstacles S in $O(|E_S|)$ time. The algorithm determines whether such a motion exists and, if so, returns a motion consisting of $O(|E_S|)$ elementary motions. It requires $O(n^2)$ preprocessing time.*

The preprocessing step consists of the construction of the visibility diagram of S. In the next section we indicate how $VIS(S)$ is used to answer a motion planning query. Although d is a parameter of a query we suppress it henceforth in our notation, writing $SFP(S)$ in stead of $SFP_d(S)$, etc.

A common approach to motion planning problems is the construction of a 'simple' low dimensional set Σ of semi-free positions. One then shows that:
there is a map $\pi : SFP(S) \to \Sigma$ such that there is a simple feasible motion from $X \in SFP(S)$ to $\pi(X)$;
there is a feasible motion from initial position X_0 to final position X_1 iff there is a path in Σ from $\pi(X_0)$ to $\pi(X_1)$.
Computing a path in Σ usually boils down to a standard graph traversal. We follow this approach by first mapping a semi-free position to an element of $L(S) \overset{\text{def}}{=} \cup_{s \in S^+} \mathcal{L}_s$. Subsequently we construct a graph Σ of size $O(|E_S|)$ corresponding to a 'simple' one-dimensional subset of $L(S)$. We then show that any path in Σ corresponds to a finite sequence of elementary motions.

A simple case

To explain our method we first consider the situation in which any two distinct segments in S are at least distance d apart. So in particular any $l \in \mathcal{L}_s$, $s \in S$, corresponds to a semi-free position in which the tail of the rod is on s while the direction of the rod is along l. Let $\Pi : \mathbb{R}^2 \times S^1 \to L(S)$ map a position (p, v) onto the line $p + Rv \in \mathcal{L}_s$, where $s \in S^+$ is the segment *from which p is visible* along $p + Rv$.

In this case Σ is the subgraph of $VIS(S) = (\mathcal{V}_S, \mathcal{E}_S)$ obtained by removing from $\mathcal{V}_S$ all nodes corresponding to 2-cells and from $\mathcal{E}_S$ the corresponding incidences. It is easy to check that Σ is connected. We show that this is sufficient for planning a feasible motion from any initial position $X_0 \in SFP(S)$ to any final position $X_1 \in SFP(S)$ which consists of $O(|E_S|)$ elementary motions.
Step 1. Determine cells c_0 and c_1 of $VIS(S)$ containing $\Pi(X_0)$ and $\Pi(X_1)$, respectively;
Step 2. If c_i is a *face* or a *vertex* of $VIS(S)$ determine an interval I_i connecting $\Pi(X_i)$ to some edge e_i incident upon c_i, otherwise set $e_i = c_i$, $i = 0, 1$;
Step 3. Determine a path $\gamma_0 \ldots \gamma_m$ in Σ of length $m + 1 = O(|E_S|)$ with $\gamma_0 = e_0$ and

$\gamma_m = e_1$.

Step 4. Transform the path found in step 3 into a sequence of $O(|E_S|)$ elementary feasible motions. Append the sliding rotation corresponding to the intervals I_0 and I_1 at the head and the tail of this sequence. Finally append—at the head—a jump *from* position X_0 to the segment $c_0 \cdot \text{seg}$. Append a similar jump *to* position X_1 at the tail.

Steps 1, 2 and 3 take $O(|E_S|)$ time. With regard to step 4 we may assume that any three consecutive elements of the path $\gamma_0 \ldots \gamma_m$ are different. Therefore any triple of consecutive elements of this path is of the form:

1. v, e, v', where v an v' are vertices of $VIS(S)$ incident upon edge e;
2. $v, e, e \cdot ref$—or $e \cdot ref, e, v$—where v is a vertex incident upon edge e;
3. e, v, e', where e and e' are edges of $VIS(S)$ incident upon vertex v.

In case 1 walking along edge e from v to v' corresponds to a sliding rotation about the point $p \in Bd(S)$ for which $e \subset \mathcal{L}_p$, while the rod is in contact with $e \cdot \text{seg}$, see Figure 5. In case 2 let v' be the vertex succeeding $e \cdot ref$ on the path $\gamma_0 \ldots \gamma_m$, and let l be any element of e. This part of the path corresponds to the following sequence of elementary motions:

a sliding rotation corresponding to a walk along e from v to l;

a jump from $l \in e$ to $l \in e \cdot ref$;

a sliding rotation corresponding to a walk along $e \cdot ref$ from l to v'.

Case 3 is treated similarly. Therefore step 4 can be executed in $O(|E_S|)$ time.

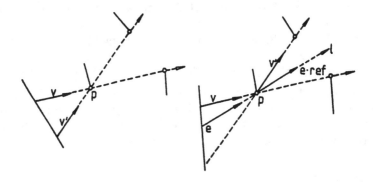

Figure 5

The general case

In the general situation a position X is feasible iff for $l = \Pi(X) \in \mathcal{L}_s$ the length of $SF(s,l)$ is at least d. So for each face f of $VIS(S)$ we consider the set $FREE(f)$ consisting af all $l \in f$ for which the length of $SF(s,l)$ is at least d, where $s = f \cdot \text{seg}$. The boundary of $FREE(f)$ consists of an *interior* and an *exterior* part, defined by $\partial_{int}(f) = \partial(FREE(f)) \cap cl(f)$ and $\partial_{ext}(f) = \partial(f) \setminus \partial_{int}(f)$. Here cl denotes the closure operator. Our aim is to construct a one-dimensional subset Σ_f of $FREE(f)$, which contains $\partial_{ext}(f)$ and has the properties:

for any $l \in FREE(f)$ there is an interval in $FREE(f)$ connecting l to some point of Σ_f;

if two points on $\partial_{ext}(f)$ are in the closure of a connected component of $FREE(f)$ they can be connected by a path in Σ_f.

To this end we consider $FREE(f)$ in more detail. Introduce Cartesian coordinates x_1, x_2 on the plane such that no $l \in \text{cl}(f)$ is parallel to the x_2-axis. Let $\mathcal{D}$ be the duality transform mapping a line $x_2 = h_1 x_1 + h_2$ onto the point (h_1, h_2). Then $\mathcal{D}(f)$ is a *bounded convex polygon*.

Lemma 4.2 *(i) The dual image $\mathcal{D}(\partial_{int}(f))$ is a smooth curve (not necessarily connected);*
(ii) $\partial_{int}(f)$ intersects each edge incident upon f in at most three points, which can be determined in constant time.

Proof: (i) Exploit the freedom of choice of Cartesian coordinates by taking the coordinate axes along the bisectors of the angles formed by the lines l_0 and l_1 supporting the segments $f \cdot \text{seg}$ and $f \cdot \text{vis}$, respectively, while moreover any $l \in f$ intersects the positive x_2-axis from left to right, see Figure 6.

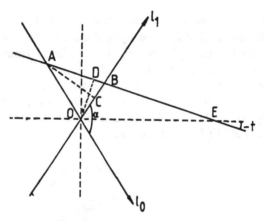

Figure 6

Consider the set L_0 of lines intersecting the positive x_2-axis, such that l_0 and l_1 cut a segment $\overline{AB}$ of length d from l. Obviously $\partial_{int}(f) = f \cap L_0$. Let α be the angle from l_0 to l_1, and let $t \in (-\alpha/2, \alpha/2)$ be the angle of a line $l \in L_0$ and the positive x_1-axis. Let C be the perpendicular projection of A on l_1, while D is the perpendicular projection of the origin O on l. Referring to Figure 6 we derive $\angle OAB = \pi - \angle AOE - \angle OEA = \pi - (\pi - \alpha/2) - (-t) = \alpha/2 + t$, and $\angle OBA = \angle BOE + \angle OEB = \alpha/2 - t$. Hence $\overline{AC} = \overline{AB} \sin \angle OBA = d \sin(\alpha/2 - t)$, and $\overline{OD} = \overline{OA} \sin \angle OAB = (\overline{AC}/\sin \alpha) \sin(\alpha/2 + t)$. Therefore $\overline{OD} = d \sin(\alpha/2 - t) \sin(\alpha/2 + t)/(\sin \alpha) = c(\cos \alpha - \cos 2t)$, where $c = -d/(2 \sin \alpha)$. The equation of the line $l(t)$ through A and B is therefore $x_2 = x_1 \tan t + c(\cos \alpha - \cos 2t)/(\cos t)$, with $-\alpha/2 < t < \alpha/2$. In these coordinates a parametrization of the curve $\mathcal{D}(L_0)$, which contains $\mathcal{D}(\partial_{int}(f))$, is:
$(\xi(t), \eta(t)) = (\tan t, c(\cos \alpha - \cos 2t)/(\cos t))$, with $-\alpha/2 < t < \alpha/2$.
(ii) We shall prove that in the dual plane each line intersects the curve $\mathcal{D}(L_0)$ in at most three points. Since the dual image of an edge is a line segment this obviously implies the lemma. We easily derive, after some calculations:
$\xi'(t) \eta''(t) - \xi''(t) \eta'(t) = c(3 \cos 2t + \cos \alpha)/(2 \cos^3 t)$.
Therefore the curvature of $\mathcal{D}(L_0)$ has fixed sign if $\cos \alpha \geq 0$. When $\cos \alpha < 0$ there

are two inflection points—points of zero curvature—at $t = \pm t_0$, where $0 < t_0 < \alpha/2$ satisfies $3\cos 2t_0 + \cos\alpha = 0$, see Figure 7. Since $\mathcal{D}(L_0)$ is symmetric in the sense that $(\xi(-t), \eta(-t)) = (-\xi(t), \eta(t))$ each line intersects $\mathcal{D}(L_0)$ in at most three points, which can be determined in constant time using the expressions above for $\xi(t)$ and $\eta(t)$. $\quad\square$

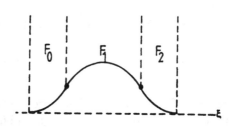

Figure 7

We call $l \in \partial_{int}(f)$ *convex* (*concave*) if in the dual plane the line tangent to the curve $\mathcal{D}(\partial_{int}(f))$ in the point $\mathcal{D}(l)$ is locally contained in the exterior (interior) of $\mathcal{D}(FREE(f))$. Since convexity (concavity) is invariant under linear transformations this definition does not depend on the particular choice of $\mathcal{D}$.

An *f-connecting edge* is an open s-interval (l_0, l_1) with
(i) $l_0, l_1 \in \partial_{int}(f) \cap \partial(f)$, and l_0, l_1 are *convex* elements of $\partial_{int}(f)$;
(ii) l_0 and l_1 are in the boundary of a connected component of $\partial(f) \setminus \partial_{ext}(f)$;
(iii) $(l_0, l_1) \subset FREE(f)$, see Figure 8.

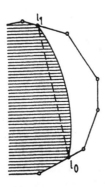

Figure 8

The set of f-connecting edges is denoted by $CE(f)$.

Lemma 4.3 $CE(f)$ *can be determined in* $O(n_f)$ *time. (*n_f *is the number of edges incident upon face* f.)

Proof: This is an obvious consequence of lemma 4.2(ii). It also follows from the proof of that lemma that condition (iii) in the definition of f-connecting edge can be checked in constant time. $\quad\square$

Let Σ_f be the set $\partial_{ext}(f) \cup (\cup_{e \in CE(f)} e)$. For a connected component c of $FREE(f)$ we define $\Sigma_c = \Sigma_f \cap cl(c)$. The crucial properties of Σ_c are collected in the following result.

Lemma 4.4 *Let c be a connected component of $FREE(f)$.*
(i) c is (homeomorphic to) a disc;
(ii) Σ_c is connected;
(iii) There is a map $\pi : c \to \Sigma_c$ such that the interval connecting $l \in c$ to $\pi(l)$ is contained in c; moreover $\pi(l)$ can be determined in $O(n_f)$ time.

Proof: We use the same coordintes and notation as in the proof of lemma 4.2.
(i) $\mathcal{D}(c)$ intersects each vertical line $\xi = \xi_0$ in a segment that varies continuously with ξ_0. Therefore $\mathcal{D}(c)$ is a disc.
(ii) Assume $\cos \alpha < 0$, the proof for the other case being trivial. Note that $\mathcal{D}(f)$ is a convex polygon contained in the region $F \stackrel{\text{def}}{=} \{(\xi, \eta) | - \tan \alpha/2 \leq \xi \leq \tan \alpha/2, \eta \geq 0 \}$, since any $l \in f$ has slope inbetween $-\alpha/2$ and $\alpha/2$, while any such l intersects the *positive* x_2-axis. Let F_0, F_1 and F_2 be the *unbounded* closed subsets of F bounded by the curve $\mathcal{D}(L_0)$ and the vertical lines through the inflection points, cf. Figure 7. Obviously F_0 and F_1 are convex. Moreover $\mathcal{D}(FREE(f)) \subset F_0 \cup F_1 \cup F_2$.

Suppose $\mathcal{D}(c) \cap F_0 \neq \emptyset$. Then $\mathcal{D}(c) \cap F_0 = \mathcal{D}(f) \cap F_0$. We shall first prove that $\mathcal{D}(\Sigma_c) \cap F_0$ is connected. If $\mathcal{D}(f)$ intersects $F_0 \cap \mathcal{D}(L_0)$ in at most two points then obviously $\mathcal{D}(\Sigma_c) \cap F_0$ is connected. If $\mathcal{D}(f)$ intersects $F_0 \cap \mathcal{D}(L_0)$ in at least three points, then it does not intersect $F_2 \cap \mathcal{D}(L_0)$. This follows from the convexity of $\mathcal{D}(f)$ and from the fact—requiring some trivial calculations—that the right-most inflection point $(\xi(t_0), \eta(t_0))$ lies *below* the tangent to $\mathcal{D}(L_0)$ in its left-most point $(\xi(-\alpha/2), \eta(-\alpha/2))$. Consider $l_0, l_1 \in \partial f$ such that $\mathcal{D}(l_0)$ and $\mathcal{D}(l_1)$ are successive—with respect to ξ-coordinate—points of intersection of $\mathcal{D}(f)$ and $\mathcal{D}(L_0)$, lying in F_0. Then l_0, l_1 are boundary points of a connected component of either $\partial_{ext}(f)$ or $\partial f \setminus \partial_{ext}(f)$. In the first case these points are connected in Σ_f via $\partial_{ext}(f)$, in the second case by the f-connecting edge (l_0, l_1), since $(\mathcal{D}(l_0), \mathcal{D}(l_1)) \subset F_0$, i.e. $(l_0, l_1) \subset FREE(f)$. Therefore $\mathcal{D}(\Sigma_c) \cap F_0$ is connected. One similarly proves that $\mathcal{D}(\Sigma_c) \cap F_1$ and $\mathcal{D}(\Sigma_c) \cap F_2$ are connected.
(iii) For $l \in c$ let $I \subset f$ be the interval whose dual image $\mathcal{D}(I)$ connects $\mathcal{D}(l)$ to the point of $\mathcal{D}(\Sigma_c)$ lying vertically above $\mathcal{D}(l)$. Then obviously $\mathcal{D}(I) \cap \mathcal{D}(L_0) = \emptyset$, i.e. $I \subset c$. $\square$

Each edge e of $VIS(S)$ is subdivided by points of intersection of e and $\partial_{int}(e \cdot left)$ and $\partial_{int}(e \cdot right)$—as far as the latter sets are defined. These points will be called *primary obstruction vertices*. To incorporate 'jumps' we subdivide e even further by introducing a vertex at $l \in e$ iff l correponds to a primary obstruction vertex of $e \cdot ref$. These vertices will be called *secondary obstruction vertices*, see Figure 9. In this way we associate with each sub-edge e' of e a unique sub-edge of $e \cdot ref$, which we denote by $e' \cdot ref$.

A *feasible edge* is either:
(i) a sub-edge e' of an edge e of $VIS(S)$ incident upon at least one of the sets $FREE(e \cdot left)$ or $FREE(e \cdot right)$, or:
(ii) an f-connecting edge, for some face f of $VIS(S)$.

A *feasible vertex* is a vertex—of $VIS(S)$, or a primary or secondary obstruction vertex—incident upon a feasible edge. The graph $\Sigma = (V_\Sigma, E_\Sigma)$ is defined by:
V_Σ is the set of feasible edges and vertices;
$\{c_0, c_1\} \in E_\Sigma$, for $c_0, c_1 \in V_\Sigma$, iff one of the following conditions holds:
- c_0 and c_1 are incident;
- c_0 and c_1 are feasible edges with $c_0 \cdot ref = c_1$.

Lemma 4.5 *(i) $|V_\Sigma| + |E_\Sigma| = \Theta(|E_S|)$;*

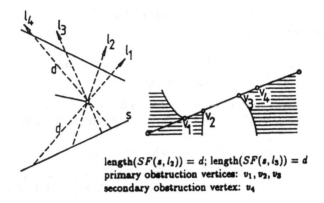

length$(SF(s, l_2)) = d$; length$(SF(s, l_3)) = d$
primary obstruction vertices: v_1, v_2, v_3
secondary obstruction vertex: v_4

Figure 9

(ii) Σ *can be constructed from* VIS(S) *in* $O(|E_S|)$ *time.*

Proof: This is an obvious consequence of Lemmas 4.2 and 4.3. ☐

We finally show that answering a motion planning query boils down to finding a path in Σ.

Proof of Theorem 4.1: First construct Σ from $VIS(S)$ in $O(|E_S|)$ time. Let X_0 and X_1 be feasible initial and final positions. In $O(|E_S|)$ time we determine cells f_0 and f_1 of $VIS(S)$ containing $\Pi(X_0)$ and $\Pi(X_1)$, respectively. In $O(n_{f_0})$ time we determine a simple motion corresponding to an interval in $FREE(f_0)$ connecting $\Pi(X_0)$ to a feasible edge e_0 contained in an edge of $VIS(S)$ incident upon f_0, cf. Lemma 4.4. We similarly associate an edge e_1 with $\Pi(X_1)$. A standard deformation argument using Lemma 4.4 shows that there is a feasible motion from X_0 to X_1 iff there is a path in Σ from e_0 to e_1. We refer to [Y] for details. In $O(|\Sigma|)$ time such a path can be determined, or it is detected that no such path exists. With this path we associate a feasible motion consisting of $O(|E_S|)$ elementary motions as in Section 4. ☐

References

[AAGHI] T.Asano, T.Asano, L.Guibas, J.Hershberger and H.Imai: "Visibility of disjoint polygons." *Algorithmica*, 1 (1986), pp. 49-63.

[ChG] B.Chazelle and L.Guibas: "Visibility and intersection problems in plane geometry." *Proceedings of the ACM Symposium on Computational Geometry*, Baltimore, 1985, pp. 135-146.

[KeO] Y. Ke and J. O'Rourke: "Lower bounds on moving a ladder in two and three dimensions", *Discrete and Computational Geometry*, 3 (1987), pp. 197-218.

[LeSh] D.Leven and M.Sharir: "An efficient and simple motion-planning algorithm for a ladder moving in 2-dimensional space amidst polygonal barries", *J. Algorithms*, 8 (1987), pp. 192-215.

[Me]   K. Mehlhorn: "Data Structures and Algorithms", Vol.3, Springer Verlag, 1984.

[PS]   F.Preparata and I.Shamos: "Computational Geometry, an Introduction", Springer Verlag, 1985.

[ShSi] S. Sifrony and M. Sharir: "A New Efficient Motion-Planning Algorithm for a Rod in Two-Dimensional Polygonal Space", *Algorithmica*, 2 (1987), pp. 367-402.

[V]   G. Vegter: "The Visibility Diagram: a Data Structure for Visibility and Motion Planning Problems, Part 1: Application to the Visibility Polygon Query Problem", Techn. Report CS8913, Univ. of Groningen (1989).

[Y]   C.K.Yap: "Algorithmic Motion planning", in: Advances in Robotics, ed. J.T. Schwarz and C.K. Yap (1987), Lawrence Erlbaum Ass., Publ., Hillsdale NJ, London.

[We]   E.Welzl: "Constructing the visibility graph for n line segments in the plane", *Inf. Proc. Letters*, 20 (1985), pp.167-171.

Fast Updating of Well-Balanced Trees

Arne Andersson
Lund University
Sweden

Tony W. Lai
University of Waterloo
Canada

Abstract

We focus on the problem of maintaining binary search trees with an optimal and near-optimal number of incomplete levels. For a binary search tree with one incomplete level and a height of exactly $\lceil \log(n+1) \rceil$, we improve the amortized insertion cost to $O(\log^3 n)$. A tree with 2 incomplete levels and a near-optimal height of $\lceil \log(n+1) + \epsilon \rceil$ may be maintained with $O(\log^2 n)$ amortized restructuring work per update. The amount of restructuring work is decreased to $O(\log n)$ by increasing the number of incomplete levels to 4, while the height is still kept as low as $\lceil \log(n+1) + \epsilon \rceil$. This yields an improved amortized bound on the dictionary problem.

Trees of optimal and near-optimal height may be represented as a pointer-free structure in an array of size $O(n)$. In this way we obtain an array implementation of a dictionary with $O(\log n)$ search cost and $O(\log^2 n)$ update cost, allowing interpolation search to improve the expected search time.

1 Introduction

The binary search tree is a fundamental and well studied data structure, commonly used in computer applications to implement the abstract data type *dictionary*. In a comparison-based model of computation, the lower bound on the three basic operations *insert*, *delete* and *search* is $\lceil \log(n+1) \rceil$ comparisons per operation. This bound may be achieved by storing the set in a binary search tree of optimal height.

Definition 1 *A binary tree has optimal height if and only if the height of the tree is $\lceil \log(n+1) \rceil$.*

A special case of a tree of optimal height is an *optimally balanced* tree, as defined below.

Definition 2 *A binary tree is optimally balanced if and only if the difference in length between the longest and shortest paths is at most one.*

In the literature, there are a number of algorithms presented for maintenance of optimally balanced trees, all having an amortized cost of $\Theta(n)$ per update [7, 8, 9, 15]. Although this cost is high, it is regarded as affordable when updates are rare and when a low search cost is essential.

In this paper we improve the tradeoff between balance and maintenance cost for binary search trees. Starting with optimal balance, that is, at most one incomplete level and a height of $\lceil \log(n+1) \rceil$, we improve the amortized insertion time from linear to $O(\log^3 n)$. Ending with at most 4 incomplete levels and a near-optimal height of $\lceil \log(n+1) + \epsilon \rceil$, we achieve an optimal amortized update cost of $O(\log n)$ per insertion and deletion.

The paper is organized as follows. In Section 2, we show that a tree, defined by a simple balance criterion similar to the general balanced tree [3, 5], may be maintained with at most 2 incomplete levels and a height of $\lceil \log(n+1) + \epsilon \rceil$ at an amortized cost of $O(\log^2 n)$ per update. This result is analogous to the bound for 2-complete trees and weight-balanced trees, obtained by Lai and Wood [11, 12].

In Section 3, we turn to the problem of maintaining a tree with at most one incomplete level and an optimal height of exactly $\lceil \log(n+1) \rceil$. Based on the result in Section 2, we give an insertion algorithm with an amortized cost of $O(\log^3 n)$, provided that no deletions are made. Just as there is an efficient algorithm for insertions only into an optimally balanced tree, there is an efficient algorithm for deletions only.

In Section 4, we show that by allowing at most 4 incomplete levels, we can maintain a tree with near-optimal height $\lceil \log(n+1) + \epsilon \rceil$ at a cost of $O(\log n)$ per update. This result is achieved by introducing *dense trees* in which we replace the leaves of a tree of size $\Theta(n/\log n)$ by perfectly balanced subtrees of size $\Theta(\log n)$.

The algorithms presented in Sections 2 and 3 use partial rebuilding for maintenance. This fact, together with the low height, implies that a tree of optimal or near-optimal height may be represented as a pointer-free structure in an array of size $O(n)$. In this way we obtain an array implementation of a dictionary with $O(\log^2 n)$ update cost, allowing interpolation search to improve the expected search time.

The number of leaves in the subtree rooted at node v is called the *size* of v, denoted $|v|$. this implies that the size of a tree containing n elements is $n + 1$. We assume that rebuilding the subtree rooted at v to perfect balance requires time proportional to the size of v. Linear algorithms for balancing a binary tree can be found in [7, 8, 15].

2 Maintaining Near-Optimal Balance

If we allow the height of a binary search tree to be an additive constant larger than the optimal height, we have a tree of *near-optimal* height. In this section we show that a near-optimal height of $\lceil \log(n+1) + \epsilon \rceil$ may be maintained at an amortized cost of $O(\log^2 n)$ per update. We also show how to bound the length of the shortest path to $\lfloor \log(n+1) - \epsilon \rfloor$. In this way, we are able to maintain a tree with at most two incomplete levels.

We use the simplest possible balance criterion, allowing the tree to take any shape as long as the height is at most $\lceil \log(n+1) + \epsilon \rceil$ for a given constant ϵ. As we will show, this simple balance criterion is sufficient to maintain the tree at an amortized cost of $O(\log^2 n)$ per update.

Theorem 1 *A binary search tree of height $\lceil \log(n+1) + \epsilon \rceil$ may be maintained with an amortized cost of $O\left(\frac{\log^2 n}{\epsilon}\right)$ per update, for any constant $\epsilon > 0$.*

Proof: We maintain the tree by partial rebuilding. The entire tree is rebuilt when the number of updates since the last global rebuilding equals half the size of the tree. The cost of each of these global rebuildings is amortized over $\Theta(n)$ updates, which gives an amortized cost of $O(1)$ per update.

We still have to show the amortized cost of the updates between these global rebuildings. To prove the cost of updating we define a potential function $\Phi(v)$ for each node v. The function is chosen in such a way that the decrease in potential during a rebuilding corresponds to the cost of the rebuilding. The increased potential during an update corresponds to the amortized cost of the update. By showing that the potential is always positive or zero, we prove that the amortized cost covers the cost of restructuring.

Let $\delta(v)$ denote the difference in sizes between v's two subtrees and n_0 denote the size of the entire tree at the latest global rebuilding. The cost of rebuilding the subtree rooted at v

to perfect balance is $R|v|$ for some constant R, where $|v|$ denotes the number of leaves in the subtree v. The potential $\Phi(v)$ is chosen as

$$\Phi(v) = \begin{cases} 0 & \text{if } v \text{ is perfectly balanced} \\ \frac{4R\log n_0}{\epsilon}\delta(v) & \text{otherwise.} \end{cases} \tag{1}$$

A single update changes the potential by $O\left(\frac{\log n_0}{\epsilon}\right)$ at a logarithmic number of nodes, which implies that the amortized cost per update is $O\left(\frac{\log^2 n}{\epsilon}\right)$.

Left to show is that there is a maintenance algorithm with a restructuring cost which is covered by the potential function Φ. When a single insertion or deletion causes the height of the tree to exceed $\lceil \log(n+1) + \epsilon \rceil$, we make a partial rebuilding at the lowest node v which satisfies

$$\text{height}(v) > \left\lceil \left(1 + \frac{\epsilon}{\log(n+1)}\right)\log|v|\right\rceil, \tag{2}$$

Clearly, at least one such node v exists, since Eq. (2) is satisfied by the root. Let v_1 denote v's highest child. Since v is the lowest node satisfying Eq. (2), we know that

$$\text{height}(v_1) \le \left\lceil \left(1 + \frac{\epsilon}{\log(n+1)}\right)\log|v_1|\right\rceil \tag{3}$$

and

$$\text{height}(v) = \text{height}(v_1) + 1. \tag{4}$$

Combining Eqs. (2), (3), and (4) gives

$$\left\lceil \left(1 + \frac{\epsilon}{\log(n+1)}\right)\log|v|\right\rceil < \text{height}(v)$$
$$= \text{height}(v_1) + 1$$
$$\le \left\lceil \left(1 + \frac{\epsilon}{\log(n+1)}\right)\log|v_1|\right\rceil + 1 \tag{5}$$

which implies that

$$|v_1| > 2^{-1/\left(1+\frac{\epsilon}{\log(n+1)}\right)}|v|. \tag{6}$$

The cost of a rebuilding at the node v is $R|v|$. This cost has to be covered by the decrease in v's potential. After the rebuilding the subtree v is perfectly balanced and $\Phi(v) = 0$. From Eq. (6) we get the potential immediately before the rebuilding.

$$\Phi(v) = \frac{4R\log n_0}{\epsilon}\delta(v)$$
$$= \frac{4R\log n_0}{\epsilon}(|v_1| - (|v| - |v_1|))$$
$$= \frac{4R\log n_0}{\epsilon}(2|v_1| - |v|)$$
$$> \frac{4R\log n_0}{\epsilon}\left(2^{1-1/\left(1+\frac{\epsilon}{\log(n+1)}\right)} - 1\right)|v|$$
$$> \frac{2R\log n_0}{\epsilon}\frac{\epsilon|v|}{\log(n+1)}$$
$$> R|v|. \tag{7}$$

The fact that

$$2\left(2^{1-1/\left(1+\frac{\epsilon}{\log n}\right)} - 1\right) > \frac{\epsilon}{\log n}, \quad \epsilon < \log n \tag{8}$$

is straightforward to show by the substitution

$$t = \frac{1}{1 + \frac{\epsilon}{\log n}}. \tag{9}$$

From Eq. (7) follows that the decrease in potential covers the restructuring cost, which completes the proof. □

As a matter of fact, the algorithm described above may be viewed as the maintenance algorithm for a general balanced tree or GB(c)-tree [3, 5], where we let the value of the tuning parameter c vary such that $c = 1 + \frac{\epsilon}{\log(n+1)}$.

In the same way as we may bound the longest path to be near-optimal, we may bound the shortest path to be close to maximum.

Theorem 2 *A binary search tree with a shortest path of at least $\lfloor \log(n+1) - \epsilon \rfloor$ may be maintained with an amortized cost of $O\left(\frac{\log^2 n}{\epsilon}\right)$ per update, for any constant $0 < \epsilon \leq 1$.*

Proof: The proof is similar to the proof of Theorem 1. Let $\text{short}(v)$ denote the length of v's shortest path. Each time a path becomes too short, we make a partial rebuilding at the lowest node v which satisfies

$$\text{short}(v) < \left\lfloor \left(1 - \frac{\epsilon}{\log(n+1)}\right) \log |v| \right\rfloor. \tag{10}$$

Let v_1 denote v's child on the shortest path. Since v is the lowest node satisfying Eq. (10) we know that

$$\text{short}(v_1) \geq \left\lfloor \left(1 - \frac{\epsilon}{\log(n+1)}\right) \log |v_1| \right\rfloor \tag{11}$$

and

$$\text{short}(v) = \text{short}(v_1) + 1. \tag{12}$$

Combining Eqs. (10), (11), and (12) gives

$$\left\lfloor \left(1 - \frac{\epsilon}{\log(n+1)}\right) \log |v| \right\rfloor > \text{short}(v)$$
$$= \text{short}(v_1) + 1$$
$$\geq \left\lfloor \left(1 - \frac{\epsilon}{\log(n+1)}\right) \log |v_1| \right\rfloor + 1 \tag{13}$$

which implies that

$$|v_1| < 2^{-1/\left(1 - \frac{\epsilon}{\log(n+1)}\right)} |v|. \tag{14}$$

As in the proof of Theorem 1, we may choose a potential function $\Phi(v)$ to be $\Theta\left(\frac{\log n}{\epsilon} \delta(v)\right)$, where $\delta(v)$ denotes the difference in sizes between v's two subtrees. From Eq. (14) follows that when v is to be rebuilt, the value of $\delta(v)$ will be large enough to make the potential cover the cost of the rebuilding. □

By combining the results of Theorem 1 and Theorem 2, we may restrict both the longest and the shortest path in the tree.

Theorem 3 *A binary search tree with a maximum height of $\lceil \log(n+1) + \epsilon \rceil$ and a shortest path of length $\lfloor \log(n+1) - \epsilon \rfloor$ may be maintained at a cost of $O\left(\frac{\log^2 n}{\epsilon}\right)$ per update.*

Proof: By combining the two algorithms described in the proofs of Theorem 1 and Theorem 2 we obtain the bounds stated above. □

3 Maintaining Optimal Balance

A consequence of Theorem 3 is that when the value of ϵ is small, the height of the tree will be $\lceil \log(n+1) \rceil$ and the shortest path will be of length $\lfloor \log(n+1) \rfloor$, except when n is close to a power of 2. Thus, the number of incomplete levels in the tree will be one in most cases and two at most.

This fact can be used to obtain an efficient algorithm for insertion into an optimally balanced tree. The cost of this algorithm is $O(\log^3 n)$, which is a significant improvement compared to previous linear algorithms [7, 8, 9, 15]. The idea is to vary the value of ϵ in Theorem 3 in such a way that the height of the tree is optimal at any time.

Theorem 4 *An optimally balanced binary search tree may be maintained with a restructuring work of $O(\log^3 n)$ per insertion, provided that no deletions are made.*

Proof: It is sufficient to prove the theorem for insertions when $2^{k-1} \leq n + 1 \leq 2^k$ for any positive integer k. The case when $n \geq 0$ follows immediately. We maintain the tree as in the proof of Theorem 1 with the difference that we use a varying value of the tuning parameter ϵ. Each time $n + 1 = 2^k - 2^{k-i+1}$, $i = 2, 3, 4 \ldots k$, we rebuild the tree and change the value of ϵ to $\log \frac{2^i}{2^i - 1}$. In other words, the value of ϵ is set to $\log \frac{4}{3}$ when the lowest level is empty, $\log \frac{8}{7}$ when the level is half-filled, $\log \frac{16}{15}$ when it is $\frac{3}{4}$-filled, and so forth. Suppose that $n + 1 < 2^k - 2^{k-i}$. Then, the height is given by

$$\text{height}(T) \leq \lceil \log(n+1) + \epsilon \rceil$$

$$\leq \left\lceil \log\left(2^k - 2^{k-i}\right) + \log \frac{2^i}{2^i - 1} \right\rceil$$

$$= \left\lceil \log \frac{\left(2^k - 2^{k-i}\right) 2^i}{2^i - 1} \right\rceil$$

$$= k. \tag{15}$$

The fact that $k = \lceil \log(n+1) \rceil$ proves that the height is optimal. Let ϵ_i be the value of ϵ after $n + 1 = 2^k - 2^{k-i+1}$. The total cost for filling the lowest level is given by

$$O\left(\sum_{i=2}^{k} (\text{number of insertions for this value of } i) \cdot \frac{\log^2 n}{\epsilon_i} \right)$$

$$= O\left(\sum_{i=2}^{k} \frac{n}{2^i} \cdot \frac{\log^2 n}{\epsilon_i} \right)$$

$$= O\left(n \log^2 n \cdot \sum_{i=2}^{k} \frac{1}{2^i \log \frac{2^i}{2^i - 1}} \right)$$

$$= O\left(n \log^2 n \cdot \sum_{i=2}^{k} 1 \right)$$

$$= O\left(n \log^3 n \right). \tag{16}$$

This cost is amortized over $\Theta(n)$ updates, which gives a cost of $O(\log^3 n)$ per update. Thus, the proof is completed. $\square$

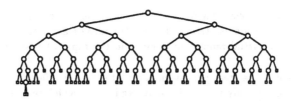

Figure 1: *A tree which requires rebuilding*

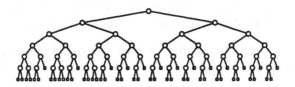

Figure 2: *The tree of Figure 1 after a partial rebuilding*

Example: Figure 1 shows a tree which has become unbalanced by an insertion. The number of leaves in the tree is 53, the value of i is 3, thus $\epsilon = \log \frac{8}{7}$. A rebuilding is made at the lowest node v which satisfies

$$\text{height}(v) > \left\lceil \left(1 + \frac{\log\left(\frac{8}{7}\right)}{\log 53}\right) \log |v| \right\rceil. \tag{17}$$

Eq. (17) is satisfied by the left subtree of the root. A rebuilding is made at this node. The result of this rebuilding is illustrated in Figure 2. $\square$

In the same way that we may handle insertions by bounding the longest path, we may handle deletions by bounding the shortest path.

Theorem 5 *An optimally balanced binary search tree may be maintained with a restructuring work of $O(\log^3 n)$ per deletion, provided that no insertions are made.*

Proof: The proof is analogous to the proof of Theorem 4. $\square$

4 Dense Trees

Although the bounds for maintenance of trees with optimal and near-optimal height obtained in the preceding sections are low, there is still a gap between these bounds and the optimal of $O(\log n)$ per update. In this chapter we show how to overcome this gap by giving logarithmic algorithms for maintenance of trees of near-optimal height. In this way we obtain a near-optimal amortized solution to the dictionary problem.

The maintenance algorithms presented in this section differ from the algorithms in the previous sections in the way that we use a mixture of partial rebuilding and local restructuring. The basic idea is to maintain a tree of near-optimal height in which the leaves are replaced by subtrees of size $\Theta(\log n)$. We refer to the tree as a *dense tree* consisting of a *topmost tree* and *leaf-subtrees*. Both the topmost tree and the leaf-subtrees are maintained by partial rebuilding. However, when a partial rebuilding is made in the topmost tree, the leaf-subtrees are not affected.

The advantage of dense trees is that most updates do not affect the topmost tree. Changes in the topmost tree occur only when a leaf-subtree is split, which results in an insertion into the topmost tree, or when two leaf-subtrees are joined with a subsequent deletion from the topmost tree. Provided that $\Theta(\log n)$ updates are required between two such operations, the amortized cost of updates in the topmost tree may be reduced by a factor of $\Theta(\log n)$. The price we pay for this improvement in time is that the number of incomplete levels is increased to 4. However, by making global rebuildings often and choosing the sizes of the leaf-subtrees carefully, the height may still be kept very low. The result is given in Theorem 6 below.

Theorem 6 *A binary search tree of height $\lceil \log(n+1) + \epsilon \rceil$ and at most 4 incomplete levels may be maintained at an amortized update cost of $O\left(\frac{\log n}{\epsilon^2}\right)$, where $0 < \epsilon \le 1/2$.*

Proof: We use a dense tree, where the topmost tree T has a near-optimal height of $\lceil \log(|T| + 1) + \epsilon/4 \rceil$ and the leaf-subtrees are perfectly balanced trees of size between $2^{\lceil \log \log n_0 \rceil}(1 - \epsilon/4)$ and $2^{\lceil \log \log n_0 \rceil + 1} - 1$, where n_0 denotes the size of the tree the last time the entire tree was rebuilt.

To keep the value of n_0 up-to-date, we rebuild the entire tree completely after $\epsilon^2 n_0/32$ updates. At each global rebuilding we choose the size of each leaf-subtree to be $(2^{\lceil \log \log n_0 \rceil + 1} - 1)(1 - \epsilon/4)$.

Updates are performed by adding or removing nodes in leaf-subtrees. When a leaf-subtree gets too small or too large, it is joined with another leaf-subtree or split. Only then an update is made in the topmost tree.

Theorem 1 implies that the amortized cost per update in the topmost tree is $O\left(\frac{\log^2 n}{\epsilon}\right)$. However, the way the sizes of the leaf-subtrees are restricted guarantees that each update in the topmost tree is preceded by $\Omega(\epsilon \log n)$ updates in the entire tree. Therefore, the cost of maintenance of the topmost tree is

$$\frac{O\left(\frac{\log^2 n}{\epsilon}\right)}{\Omega(\epsilon \log n)} = O\left(\frac{\log n}{\epsilon^2}\right). \tag{18}$$

The amortized cost of an update is the sum of the amortized cost of rebuilding the entire tree, the amortized cost of updating the topmost tree, and the amortized cost of updating the leaf-subtrees. This cost is

$$O(1) + O\left(\frac{\log n}{\epsilon^2}\right) + O(\log n) = O\left(\frac{\log n}{\epsilon^2}\right), \tag{19}$$

which proves the maintenance cost.

Next, we compute the maximum height of a dense tree. Due to the way the sizes of the leaf-subtrees are chosen, the size of the topmost tree immediately after a global rebuilding will be at most

$$\frac{n_0}{(2^{\lceil \log \log n_0 \rceil + 1} - 1)(1 - \epsilon/4)}. \tag{20}$$

Every update in the topmost tree is preceded by at least $(2^{\lceil \log \log n_0 \rceil + 1} - 1)\epsilon/4$ updates in the entire tree. Thus, when the tree is rebuilt the next time at most

$$\frac{\epsilon^2 n_0/32}{(2^{\lceil \log \log n_0 \rceil + 1} - 1)\epsilon/4} = \frac{\epsilon n_0/8}{(2^{\lceil \log \log n_0 \rceil + 1} - 1)} \tag{21}$$

updates have been performed in the topmost tree. Thus, the size of the topmost tree is at most

$$\frac{n_0}{(2^{\lceil \log\log n_0 \rceil+1} - 1)(1 - \epsilon/4)} + \frac{\epsilon n_0/8}{(2^{\lceil \log\log n_0 \rceil+1} - 1)} \leq \frac{(1 + \epsilon/8)n_0}{(2^{\lceil \log\log n_0 \rceil+1} - 1)(1 - \epsilon/4)}. \tag{22}$$

Since we make a global rebuilding after $\epsilon^2 n_0/2$ updates, we have

$$n \geq (1 - \epsilon^2/32)n_0$$
$$n_0 \leq \frac{n}{1 - \epsilon^2/32}. \tag{23}$$

Thus, from Eqs. (22) and (23) we get that the height of the topmost tree is at most

$$\left\lceil \log\left(\frac{(1 + \epsilon/8)n}{(2^{\lceil \log\log n_0 \rceil+1} - 1)(1 - \epsilon/4)(1 - \epsilon^2/32)} + 1\right) + \frac{\epsilon}{4} \right\rceil$$

$$= \left\lceil \log(n + 2^{\lceil \log\log n_0 \rceil+1}) + \log\left(1 + \frac{\epsilon}{8}\right) - \log\left(1 - \frac{\epsilon}{4}\right) - \log\left(1 - \frac{\epsilon^2}{32}\right) - \lceil \log\log n_0 \rceil - 1 + \frac{\epsilon}{4} \right\rceil$$

$$\leq \lceil \log(n + 1) + \epsilon \rceil - \lceil \log\log n_0 \rceil - 1, \tag{24}$$

provided that $\epsilon \leq 1/2$ and n is sufficiently large. The maximum size of a leaf-subtree is $2^{\lceil \log\log n_0 \rceil+1} - 1$; thus, its height is at most $\lceil \log\log n_0 \rceil + 1$ since it is perfectly balanced. The maximum height of the entire tree is given by

$$\text{height}(T) \leq \lceil \log(n + 1) + \epsilon \rceil - \lceil \log\log n_0 \rceil - 1 + \lceil \log\log n_0 \rceil + 1$$
$$= \lceil \log(n + 1) + \epsilon \rceil. \tag{25}$$

In the same way that we restrict the height of the topmost tree, we may restrict its shortest path to have length at least

$$\lfloor \log(n + 1) - \epsilon \rfloor - \lceil \log\log n_0 \rceil - 1. \tag{26}$$

The shortest path of a leaf-subtree has a length of at least

$$\left\lfloor 2^{\lceil \log\log n_0 \rceil}(1 - \epsilon/4) \right\rfloor \geq \lceil \log\log n_0 \rceil - 1. \tag{27}$$

By adding the two lengths together, we see that the length of the shortest path of the entire tree is at least

$$\lfloor \log(n + 1) - \epsilon \rfloor - \lceil \log\log n_0 \rceil - 1 + \lceil \log\log n_0 \rceil - 1 = \lfloor \log(n + 1) - \epsilon \rfloor - 2. \tag{28}$$

Thus, the maximum number of incomplete levels is given by

$$\lceil \log(n + 1) + \epsilon \rceil - (\lfloor \log(n + 1) - \epsilon \rfloor - 2) \leq 4. \tag{29}$$

The proof follows from Eqs. (19), (25), and (29). □

Thus, we have shown that an almost optimally balanced binary search tree may be efficiently maintained. An example of a dense tree is given in Figure 3.

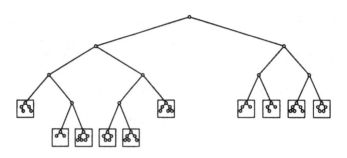

Figure 3: *A dense tree. The leaf-subtrees are marked by rectangles.*

5 Comments

We have presented improvements in the tradeoff between balance and maintenance cost for binary search trees.

Starting with one incomplete level, we have improved the amortized cost of insertions from $\Theta(n)$ to $O(\log^3 n)$. An analogous algorithm is also shown for deletions. These significant improvements make an optimally balanced tree much more attractive for the maintenance of a dynamic set than before.

Two incomplete levels may be maintained at an amortized cost of $O(\log^2 n)$ per update. This has also been shown by Lai and Wood [12] by introducing *level-layered trees* and a variant of weight-balanced trees. Compared to their results, the balance criteria used in this paper are simpler in the same sense that general balanced trees [3, 5] are simpler than other classes of trees, such as weight-balanced trees and AVL-trees.

Finally, four incomplete levels and a height of $\lceil \log(n+1) + \epsilon \rceil$ can be maintained with an optimal cost of $O(\log n)$ per update. This result allows us to improve the amortized upper bound on the dictionary problem.

Corollary 1 *For any value of ϵ, $\epsilon > 0$, there is a solution to the dictionary problem with the following costs:*

search: $\lceil \log(n+1) + \epsilon \rceil$ comparisons in the worst case;

update: $\Theta\left(\frac{\log n}{\epsilon^2}\right)$ amortized.

Thus, dense trees are superior (with respect to the search cost) to trees defined by weaker balance criteria [1, 6, 13, 14]. The only competitive tree presented so far is the ϵ-tree by Andersson [4, 5]. Compared to the ϵ-tree, however, the dense tree has simpler maintenance algorithms. A dense tree is also more balanced in the sense that we can guarantee at most four incomplete levels.

An interesting consequence of Theorem 1 is that we can store a binary search tree without any pointers in an array of size $O(N)$, where N is the maximum number of elements in the set, guaranteeing a search cost of at most $\lceil \log(n+1) + \epsilon \rceil$ and an amortized update cost of $O\left(\frac{\log^2 n}{\epsilon}\right)$. This is due to the fact that a partial rebuilding can be made in an implicitly stored tree, since no pointer movements are required. Furthermore, the tree in Theorem 1 may be implemented without any balance information in the nodes in the same way as a general balanced tree.

Corollary 2 *For any value of $\epsilon > 0$, a set of n elements, $n \leq N$, may be maintained in an array of size $O(N)$ with the following costs:*

search: $\lceil \log(n+1) + \epsilon \rceil$ *comparisons in the worst case;*

update: $O\left(\frac{\log^2 n}{\epsilon}\right)$ *amortized.*

The ability to store the tree implicitly allows random access to a node at any level of the tree which makes *interpolation search* [10, 16] among the elements possible. If the elements are smoothly distributed, this will give an expected search cost of $O(\log \log n)$.

We are also able to improve the behaviour of a well-known sorting algorithm, namely *treesort* [2]. Using a dense tree we may sort n elements in $O(n \log n)$ time, using $n \log n + O(n)$ comparisons.

References

[1] G. M. Adelson-Velskii and E. M. Landis. An algorithm for the organization of information. *Dokladi Akademia Nauk SSSR*, 146(2), 1962.

[2] A. V. Aho, J. E. Hopcroft, and J. D. Ullman. *Data Structures and Algorithms*. Addison-Wesley, Reading Mass, 1983.

[3] A. Andersson. Improving partial rebuilding by using simple balance criteria. In *Proc. Workshop on Algorithms and Data Structures, WADS '89, Ottawa*, 1989.

[4] A. Andersson. Optimal bounds on the dictionary problem. In *Proc. Symposium on Optimal Algorithms, Varna*, 1989.

[5] A. Andersson. *Efficient Search Trees*. Ph. D. Thesis, Lund University, Sweden, 1990.

[6] R. Bayer. Symmetric binary B-trees: Data structure and maintenance algorithms. *Acta Informatica*, 1(4), 1972.

[7] H. Chang and S. S. Iynegar. Efficient algorithms to globally balance a binary search tree. *Communications of the ACM*, 27(7), 1984.

[8] A. C. Day. Balancing a binary tree. *Computer Journal*, 19(4), 1976.

[9] T. E. Gerasch. An insertion algorithm for a minimal internal path length binary search tree. *Communications of the ACM*, 31:579–585, 1988.

[10] G. H. Gonnet. *Interpolation and Interpolation-Hash Searching*. Ph.D. Thesis, University of Waterloo, Canada, 1977.

[11] T. W. Lai and D. Wood. Updating approximately complete trees. Technical Report CS-89-57, University of Waterloo, 1989.

[12] T. W. Lai and D. Wood. Updating almost complete trees or one level makes all the difference. In *Proceedings of the 7th Symposium on Theoretical Aspects of Computer Science*, 1990.

[13] J. Nievergelt and E. M. Reingold. Binary trees of bounded balance. *SIAM Journal on Computing*, 2(1), 1973.

[14] H. J. Olivie. A new class of balanced search trees: Half-balanced binary search trees. *R.A.I.R.O. Informatique Theoretique*, 16:51–71, 1982.

[15] Q. F. Stout and B. L. Warren. Tree rebalancing in optimal time and space. *Communications of the ACM*, 29:902–908, 1986.

[16] A. C.-C. Yao and F. F. Yao. The complexity of searching an ordered random table. In *Proc. Foundations of Computer Science, Houston*, 1976.

How to Update a Balanced Binary Tree with a Constant Number of Rotations

Thomas Ottmann

Institut für Informatik
Universität Freiburg
Rheinstraße 10-12
D-7800 Freiburg
WEST GERMANY

Derick Wood

Data Structuring Group
Department of Computer Science
University of Waterloo
Waterloo, Ontario N2L 3G1
CANADA

Abstract

We provide a unifying framework for balanced binary trees in which we show how to ensure that insertions and deletions require a constant number of rotations or promotions. At the same time, the updating algorithms are also logarithmic in the worst case. We say that the updating algorithms have *constant linkage cost*.

The result provides insight into the constant linkage cost updating algorithms for red-black, red-h-black, and half-balanced trees. Moreover, it enables us to design new constant linkage cost updating algorithms for these as well as for other classes of trees. Specifically, we give constant linkage cost updating algorithms for α-balanced trees.

1 Introduction

Binary search trees have been extended to maintain not only the ordering of records by their primary keys, but also, simultaneously, orderings on secondary keys. The best known example of such a search structure are the priority search trees of McCreight [8]. They are search trees with respect to their primary keys and priority trees with respect to their secondary keys; thus, enabling them to answer queries of the form: return all records whose primary key lies in some range and whose secondary key is greater than some bound. Other example structures are the persistent search trees of Sarnak and Tarjan [11] and the dynamic contour trees of Frederickson and Rodger [5].

Extended search trees introduce new maintenance problems; both orderings have to be maintained when updates are performed. If the underlying tree is not balanced, the problems are minor; however, we normally use balanced trees to ensure $O(\log n)$ behavior in the worst case. In this case, we use promotions (or rotations) to restructure the given tree after an update. As is well known, promotions preserve the binary search ordering, but not, in general, the secondary key ordering. For example, in a priority search tree each promotion can cause one "trickle down" operation to re-establish the secondary key ordering. Since the height of the tree is $O(\log n)$, a trickle down operation requires $O(\log n)$ time in the worst case. If an update produces p promotions, then, in the worst case, it takes $O(p \log n)$ time. Moreover, if $p = O(\log n)$, then an update takes $O(\log^2 n)$ time in the worst case, which is unacceptable. For this reason, McCreight did not base priority search trees on AVL trees, since deletion requires $O(\log n)$ promotions in the worst case. Instead he chose the half-balanced trees of Olivié [10], since they need at most three promotions in the worst case for both insertion and deletion. We call such a class of trees and the associated updating algorithms *constant linkage cost* (*CLC*, for short).

The half-balanced trees were the first class demonstrated to be constant linkage cost. As was shown later by Tarjan [12], the half-balanced trees are the same as the symmetric binary B-trees of

Bayer [3], but the updating algorithms in [3] are not CLC. Indeed, they are also equivalent to the red-black trees of Guibas and Sedgewick [6], but the updating algorithms given in [6] are also not CLC. The updating algorithms for red-black trees given in Sarnak and Tarjan [11] are, however, CLC.

But why are red-black trees CLC? Are AVL trees or weight-balanced trees CLC? In examining the CLC demonstrations in [10, 11, 12] no underlying principles are exposed; the algorithms appear just as the rabbit appears from the proverbial magician's hat. In this paper, we provide the principle that underlies the CLC red-black updating algorithms. For this purpose we introduce a general framework for binary trees related to the stratified trees of [13] and to the dichromatic framework of [6]. We define classes of binary trees to be made up of strata; the trees appearing in the strata are chosen from a given finite set; the boundary nodes of each stratum are colored black; and the interior nodes of each stratum are colored red. The coloring is used solely to identify the strata, nothing more.

Within the general framework, we show that CLC updating operations cause some boundary nodes to cross their boundaries—a push-up or a pull-down effect—and, under appropriate conditions, boundary crossings do not cause any link changes. The usefulness of this general result on CLC updating operations is demonstrated by showing that the red-black updating algorithms of [11] fall within this framework; CLC updating algorithms for the red-h-black trees of Icking et al. [7] (independently discovered by Andersson [2]) are obtained; and, finally, CLC updating algorithms for the classes of α-balanced trees of Olivié [9] are derived, thereby solving (at least partially) a problem left open by him.

In Section 2 we briefly recall the definitions in [13] and adapt them to our current needs. Because of space limitations we present only a generic deletion algorithm in Section 3 and show in Section 4 that the deletion algorithm for red-black trees and red-h-black trees are special cases of the generic procedure. Once the right spectacles are worn, it is easy to obtain algorithms for new classes of balanced trees which require a constant number of link changes for each insertion or deletion and which are tunable to performance. In the remainder of the section, we prove that none of our classes of trees is AVL or k-height balanced, for any $k > 0$. Thus, it is an open problem whether or not there is a subset of AVL trees that is CLC.

2 Stratification

Though it is not necessary for our theory, we assume for simplicity that all trees are binary. We are interested, then, in binary search trees. The *height* and *weight* of a tree are defined in the usual way. The nullary nodes of a tree are usually called *external nodes* or *leaves*, while the other nodes are said to be *internal nodes*.

Now, *stratification* means that we view a tree as consisting of a number of *strata* or *layers*. The tree can be decomposed into a small "irregular" part of bounded size at the top, the *apex*, and a number of strata which are glued together at their borders. Each stratum consists of small trees of a prespecified set of *stratum trees*.

Let Z be a finite set of trees, $l_Z = min(\{weight(T) : T \in Z\})$, and $h_Z = max(\{weight(T) : T \in Z\})$; we call Z a *stratum set*. Observe that we do not require that all trees in Z are of the same height. But usually we require Z to be *nontrivial*; that is, $1 < l_Z < h_Z$. An *apex* set is simply a nonempty finite set of trees. To define a class of stratified trees inductively, we use a tree constructor. Let T_0 be a tree with weight t, let $T_1, \ldots, T_t$ be trees, and let T_0's external nodes be enumerated from left to right from 1 to t. Then, we denote by $T_0[T_1, \ldots, T_t]$ the tree obtained by replacing, for all i, $1 \leq i \leq t$, the i-th external node of T_0 with T_i.

So far we do not require any specific properties of Z and A except that both sets must be finite. We define the class of (A, Z)-*stratified* trees to be the smallest class of trees such that

(i) each tree in A is said to be (A, Z)-stratified, and

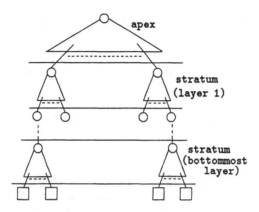

Figure 1: Decomposition of a stratified tree

(ii) if T_0 is (A, Z)-stratified and has weight t, then $T_0[T_1, \ldots, T_t]$ is (A, Z)-stratified, for all $T_1, \ldots, T_t$ in Z.

The class of (A, Z)-stratified trees is denoted by $S(A, Z)$. Because we did not restrict the possible apexes and trees in stratum sets so far, there may be different decompositions for a given tree $T \in S(A, Z)$. We assume, however, that for each tree $T \in S(A, Z)$ its decomposition into an apex from A and strata with trees chosen from Z is explicitly at hand. We can, therefore, define the *stratum height*, denoted by $sh(T)$, as follows: $sh(T) = 0$ if $T \in A$ and $sh(T) = 1 + sh(T_0)$ if $T = T_0[T_1, \ldots, T_t]$.

We can consider a tree $T \in S(A, Z)$ as a multiway tree if we "collapse" the trees in the apex and in the stratum set, respectively, into single nodes. All internal collapsed nodes except possibly the root have degree d where $l_Z \leq d \leq h_Z$, $l_Z = min(\{weight(T) : T \in Z\})$, and $h_Z = max(\{weight(T) : T \in Z\})$. By definition, all leaves of this corresponding multiway tree have the same distance to the root, namely $sh(T)$. If Z is nontrivial, we have $l_Z \geq 2$ and, therefore, $sh(T) = O(\log N)$, for $T \in S(A, Z)$, where N is the total number of collapsed nodes in T. Clearly, N differs from the total number n of nodes in T by at most a constant factor. Similarly, expanding collapsed nodes back to trees in Z and A, respectively, increases the height by at most a constant factor. This gives the following theorem.

Theorem 2.1 *Let Z be a nontrivial stratum set and let A be an apex set. Then, for each tree $T \in S(A, Z)$, $sh(T) = O(\log n)$ and $height(T) = O(\log n)$, where $n = weight(T)$.*

If $sh(T) > 0$, all leaves of the apex of T and all leaves of stratum trees except those at the bottommost level are identified with roots of stratum trees chosen from Z. Throughout this paper we visualize decompositions of stratified trees as shown in Figure 1. That is, we draw lines just above the leaves of the apex and above the leaves of the trees in the stratum set.

Example 2.1 Let A and Z be identical sets of trees containing exactly the four trees shown in Figure 2. Then $S(A, Z)$ coincides with the class of symmetric binary B-trees [3] which equals the class of half-balanced trees [10] and the class of red-black trees [6]. Figure 3 shows a decomposition of a tree $T \in S(A, Z)$ with stratum height 2.

We say that a tree *complete* if its leaves appear on at most two different levels. The stratum and apex trees of Example 2.1 as well as the tree shown in Figure 3 are complete trees.

Example 2.2 Let A_h be the set of all complete trees of height at most $h + 1$ and let Z_h be the set

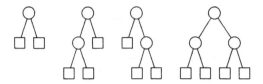

Figure 2: Stratum and apex sets defining the class of symmetric binary B-trees

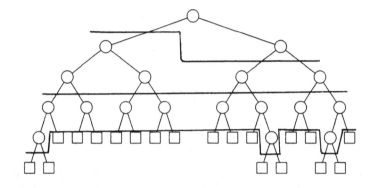

Figure 3: Decomposition of a tree of stratum height 2

which contains the complete tree of weight 2^h and height h and, furthermore, all complete trees of weight j, for $2^h + 1 \leq j \leq 2^{h+1}$, and height $h + 1$. For $h = 1$ we obtain once more the sets A and Z of Example 2.1. The class $S(A_h, Z_h)$ is identical with the class of red-h-black trees introduced in [7].

Van Leeuwen and Overmars [13] imposed additional constraints on apex and stratum sets to ensure that there exist updating procedures for (A, Z)-stratified trees that can be performed in $O(\log n)$ steps. One of their conditions is that all trees in the stratum set Z must be of the same height. Hence, the stratum trees of Examples 2.1 and 2.2 do not fit into their framework. We will see that it is not necessary to require stratum trees to be of equal height in order to obtain updating procedures which take $O(\log n)$ time. If we want to achieve updating procedures that require only constant link changes for each insertion or deletion, the condition of van Leeuwen and Overmars is even prohibitive. This will become clear in Section 3.

3 Updating stratified trees

We want to design generic CLC-updating algorithms for (A, Z)-stratified trees. This requires that we look at the structure of the apex and stratum trees in some detail. We cannot simply argue from the weights of trees and ignore their structure as van Leeuwen and Overmars [13] do. However, the general flavor of the algorithms are similar and resemble the well known maintenance algorithms for B-trees [4]. We assume that the decomposition of each tree into an apex chosen from A and strata (or layers) of trees chosen from Z is explicitly at hand. For example, bordering nodes are colored black and non-bordering nodes are colored red. We describe a generic deletion algorithm and, simultaneously, derive conditions that ensures it can be carried out with constant linkage cost.

Let $l_Z = min(\{weight(T) : T \in Z\})$, $h_Z = max(\{weight(T) : T \in Z\})$, and $1 < l_Z < h_Z$. We call a stratum set *gap-free* if for each t with $l_Z \leq t \leq h_Z$, there is a tree $T \in Z$ with $weight(T) = t$.

In order to delete an item from an (A, Z)-stratified tree, we first locate the item by a search

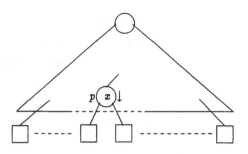

Figure 4: Deletion of an item

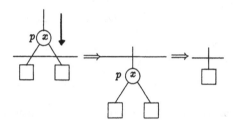

Figure 5: Deletion at the leaves

starting at the root. We may assume that the item is stored in an internal node just above the bottommost stratum border; that is, in a node which has only leaves as its children. For, by taking successors or predecessors in symmetric order, we can always reduce a deletion to this case. An internal node just above a stratum border has to be pulled down below it. In other words, the stratum tree containing this node loses one of its internal nodes and one of its leaves; see Figure 4. We assume that the stratum set Z is nontrivial and gap-free and that the apex set A contains a tree T of weight t, for each t with $1 \le t \le max(\{h_Z, 2l_Z - 1\})$. We say that Z is *weight-deletion closed* or *wd-closed*, for short, and A is *wd-closed with respect to* Z if they satisfy the above conditions.

A call of *pull_down(p)* may lead to a local structural change and termination or may lead to a recursive call for a node one layer higher up in the tree. We derive conditions for Z which ensure that no structural changes precede the recursive call. When *pull_down(p)* is called for the first time, p is a node immediately above a stratum border. We will ensure that the same condition holds for every further call of *pull_down*. A call of *pull_down(p)* triggers the following actions.

We locally change the stratum border just below p and make p a node immediately below the border. When this transformation has been carried out for the first time at the bottommost level, we replace p by a leaf and delete the item at p; see Figure 5. Note that the local change of the stratum border does not involve any structural change in the tree. We must, however, ensure on higher levels that the tree with root p is a stratum tree in order to maintain the whole tree in $S(A, Z)$.

Let p be a node of stratum tree T and t be its weight. By pulling p down below the stratum border we have changed T into a tree T_p^- of weight $t - 1$. If $t > l_Z$, we can replace T_p^- by a Z-tree and halt. This requires a constant number of link changes. If $t = l_Z$, then we consider all siblings of T_p^- in the same stratum. If at least one of them is of weight greater than l_Z or if there are more than $l_Z - 1$ siblings, then we can redistribute the elements such that all the, at least l_Z, subtrees become trees in Z. It requires the restructuring of at most $h_Z + 1$ subtrees and takes only a constant

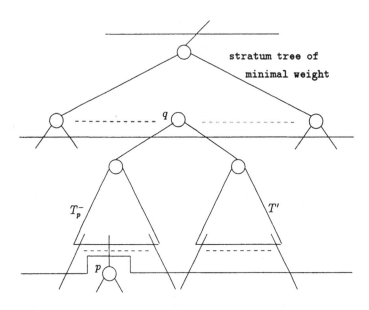

Figure 6: Merging of two minimally filled stratum trees

number of steps. If there are only $l_Z - 1$ neighboring siblings and they all have minimal weight l_Z, then we must call *pull_down* recursively, on the parent of of the root of T_p^-. However, we want to ensure that no structural changes precede the recursive call. This goal leads naturally to the following definition.

A stratum set Z is *structurally deletion closed* or *sd-closed*, for short, if it is wd-closed and satisfies the following conditions (i) and (ii).

(i) Each tree $T \in Z$ with $weight(T) = l_Z$ has the property that each leaf of T has a leaf node as its sibling. (In other words, each parent of a leaf has only leaves as its children.)

(ii) For each tree $T' \in Z$ with $weight(T') = l_Z$ and, for each tree T^- obtained from a tree T in Z, of weight l_Z, by replacing an arbitrary internal node with two leaf children by a leaf, the trees (u, T', T^-) and (u, T^-, T') also belong to Z, where u is a new root node.

Let us now assume that Z is sd-closed. Then, the situation to be dealt with can be depicted as shown by Figure 6. Our assumptions ensure that there is a node q just above the next stratum border which has T_p^- and a minimally filled stratum tree T' as its subtrees. Therefore, we can call *pull_down(q)*. The trees (q, T_p^-, T') and (q, T', T_p^-) are stratum trees by the assumption that Z is sd-closed. The procedure is repeated in this way until it either finishes or reaches the apex. The trees in the first layer and the apex can always be rebuilt in a constant number of steps to form a tree in $S(A, Z)$. As long as T_p^- has at least one sibling T' on the first layer with $weight(T') > l_Z$, rebuilding does not decrease the stratum height. If T' has only l_Z leaves, T' is the only sibling of T_p^- on the first layer, and if both are subtrees of a node q, then the tree (q, T_p^-, T') or (q, T', T_p^-) can be rebuilt to form a new apex of weight $2l_Z - 1$. In this last case the stratum height decreases by one.

The general approach to deletion is this. Initially, the procedure *pull_down(p)* is called for the node p which has the item to be deleted. Then, *pull_down* is called recursively for a sequence of nodes in minimally-filled stratum trees that have the minimal number of siblings, all minimally-filled, on the same stratum. As soon as a node in a stratum tree is reached which does not have this

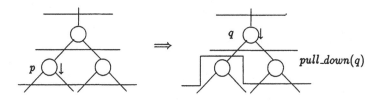

Figure 7: Recursive call of *pull_down* in the deletion algorithm for symmetric binary B-trees

property, finite restructuring terminates the deletion. It is clear that the total amount of work to be done is $O(\log n)$ and that only a constant number of linkage changes are required if the stratum set is sd-closed. Hence, we have proved the following theorem.

Theorem 3.1 *Let Z be a wd-closed stratum set and A be an apex set which is wd-closed with respect to Z. Then, deletion of an item from an arbitrary (A, Z)-stratified tree can be carried out in time $O(\log n)$. Furthermore, if Z is sd-closed, deletion is CLC.*

4 Applications

4.1 Symmetric binary B-trees (SBB-trees)

Consider the apex and stratum sets of Example 2.1. It is easy to check that Z is sd-closed. The apex set A is wd-closed with respect to Z. Therefore, deletions can be carried out in $O(\log n)$ time and they are CLC. It is interesting to specialize the generic updating algorithms in this case and to compare them with others in the literature. Deletion requires a recursive call of *pull_down* if and only if the situation shown by Figure 7 or its symmetric variant applies. In all other cases, we can locally rebuild trees in two adjacent strata or in the topmost stratum and the apex such that an SBB-tree is obtained. Figure 8 shows two of the many possible cases. Note that there is still some freedom in the choice of restructuring.

4.2 Red-h-black trees and α-balanced binary trees

Consider the apex and stratum sets A_h and Z_h of Example 2.2. For each $h \geq 1$, A_h is wi-closed and wd-closed with respect to Z_h. The stratum sets Z_h are both si-closed and sd-closed for all $h \geq 1$. Therefore, it is clear that all classes $S(A_h, Z_h)$ have $O(\log n)$ time CLC-updating algorithms. This property is utilized in [7] to build external priority search trees with trees from $S(A_h, Z_h)$ as underlying search trees.

For a node p in a tree $T \in S(A_h, Z_h)$, the difference between the length l_p of a longest path from node p to a leaf and the length s_p of a shortest path from node p to a leaf can become arbitrarily large. But it should be obvious that the asymptotic value of the quotient s_p/l_p is bounded from below by $h/(h+1)$. For, Z_h has only trees of heights h and $h+1$ and the worst we can do is to append repetitively stratum trees of height h in one subtree of p and stratum trees of height $h+1$ in the other.

In his thesis [9], Olivié takes the quotient of the two path lengths as a balance criterion and requires that quotient be between certain limits. More precisely, let α be a real value such that $0 \leq \alpha \leq 1$. A binary tree T is an *α-balanced binary tree* or an *αBB tree*, for short, if for each node p of T, the length s_p of a shortest path from p to a leaf and the length l_p of a longest path from p to a leaf satisfy the following conditions.

(i) $0 \leq \alpha \leq s_p/l_p \leq 1$, if $l_p \geq \frac{1}{1-\alpha}$, and

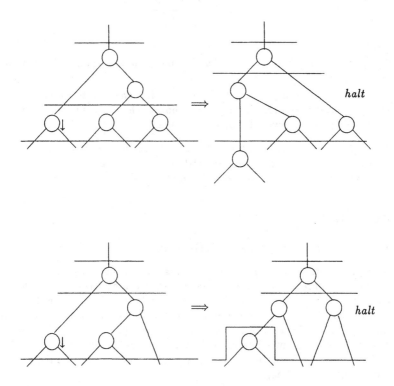

Figure 8: Restructuring of adjacent strata during the deletion of an item from a symmetric binary B-tree

(ii) $l_p - 1 \leq s_p$, if $l_p < \frac{1}{1-\alpha}$.

For $\alpha = 0$ we get all binary trees, for $\alpha = 1$ we get the complete binary trees (then only condition (ii) applies), and for $\alpha = 1/2$ we get the half-balanced binary trees [10]. For $0 \leq \alpha \leq 1/2$, condition (i) is the only condition to be considered, as $1/(1 - \alpha) \leq 2$. We say a node p α-balanced if the subtree with root p is α-balanced.

If the apex set A_h in Example 2.2 contained only complete binary trees, where all leaves have the same distance to the root, we could immediately infer that $S(A_h, Z_h)$ is a class of α-balanced trees with $\alpha = h/(h + 1)$. Because this is not the case, we obtain a slightly weaker result.

Theorem 4.1 *For each $h \geq 1$, the stratified class of red-h-black trees $S(A_h, Z_h)$ is a class of α-balanced trees with $\alpha = h/(h + 2)$.*

For some specific values of α, Olivié has shown that α-balanced trees have $O(\log n)$ time CLC-updating algorithms. The case $\alpha = 1/2$ is contained in [10] and the case $\alpha = 1/3$ is presented in [9] and it is left as an open problem whether there exist other classes of α-balanced trees with CLC-updating procedures. Our results show that there are an infinite number of these classes with α in the range $1/3 \leq \alpha \leq 1$. In particular, one can choose α to be arbitrarily close to 1 and still obtain CLC-updating algorithms. However, a note of caution is on order. We do not claim that for arbitrarily chosen values of α, $1/3 \leq \alpha \leq 1$, there exist $O(\log n)$ time CLC-updating algorithms for the class of α-balanced trees. Rather, we have shown that there is a subclass of any class of α-balanced trees that has the desired property. The subclass is, in general, a proper subclass that it is closed under insertions and deletions. For example, consider the class $S(A_1, Z_1)$. From Theorem 4.1 we infer that this class is $1/3$-balanced. But we know already, see Example 2.1, that $S(A_1, Z_1)$ coincides with the class of $1/2$-balanced trees.

4.3 Height balanced trees

The attentive reader may have noticed already that none of our example classes of (A, Z)-stratified trees is height balanced. This is not accidental as the next theorem shows. For an integer $k \geq 1$, we call a tree *k-height balanced* if, for each node p in the tree, the heights of p's left and right subtrees differ by at most k. The 1-height balanced trees are also called *height balanced* or AVL-trees [1].

Theorem 4.2 *Let Z be a set of stratum trees which is si-closed or sd-closed and let A be a nonempty apex set. Then, for all $k \geq 1$, $S(A, Z)$ is not a class of k-height balanced trees.*

Acknowledgements

The work of the first author was supported under Grant No. Ot 64/5-2 from the Deusche Forschungsgemeinschaft and that of the second author under a Natural Sciences and Engineering Research Council of Canada Grant No. A-5692.

References

[1] G.M. Adel'son-Vel'skii and Y.M. Landis. An algorithm for the organization of information. *Doklady Akademi Nauk*, 146:263–266, 1962.

[2] A. Andersson. Binary search trees of almost optimal height. Technical Report LU-CS-TR: 88:41, Department of Computer Science, Lund University, Lund, Sweden, 1988.

[3] R. Bayer. Symmetric binary B-trees: Data structure and maintenance algorithms. *Acta Informatica*, 1:290–306, 1972.

[4] R. Bayer and E.M. McCreight. Organization and maintenance of large ordered indexes. *Acta Informatica*, 1:173–189, 1972.

[5] G. Frederickson and S. Rodger. A new approach to the dynamic maintenance of maximal points in the plane. In *Proceedings of the 25th Annual Allerton Conference on Communication, Control, and Computing*, pages 879–888, Urbana-Champaign, Illinois, 1987.

[6] L.J. Guibas and R. Sedgewick. A dichromatic framework for balanced trees. In *Proceedings of the 19th Annual Symposium on Foundations of Computer Science*, pages 8–21, 1978.

[7] Chr. Icking, R. Klein, and Th. Ottmann. Priority search trees in secondary memory. In *Graphtheoretic Concepts in Computer Science (WG '87), Staffelstein, Lecture Notes in Computer Science 314*, pages 84–93, 1987.

[8] E.M. McCreight. Priority search trees. *SIAM Journal on Computing*, 14:257–276, 1985.

[9] H. J. Olivié. *A Study of Balanced Binary Trees and Balanced One-Two Trees*. PhD thesis, Departement Wiskunde, Universiteit Antwerpen, Antwerp, Belgium, 1980.

[10] H. J. Olivié. A new class of balanced search trees: Half-balanced search trees. *RAIRO Informatique théorique*, 16:51–71, 1982.

[11] N. Sarnak and R. E. Tarjan. Planar point location using persistent search trees. *Communications of the ACM*, 29:669–679, 1986.

[12] R.E. Tarjan. Updating a balanced search tree in O(1) rotations. *Information Processing Letters*, 16:253–257, 1983.

[13] J. van Leeuwen and M. Overmars. Stratified balanced search trees. *Acta Informatica*, 18:345–359, 1983.

Ranking Trees Generated by Rotations

Samuel W. Bent
Department of Mathematics and Computer Science
Dartmouth College
Hanover, NH 03755

1. Introduction

In this paper, we develop ranking and unranking algorithms for certain systematic lists of all binary trees of a given size. The lists we consider have the property that consecutive trees differ by one rotation; such lists were proved to exist by Lucas [2]. The ranking problem is to determine the position of a given tree in the list; the unranking problem is to produce the tree that occupies a given position.

The problem of listing and ranking n-node binary trees has been extensively studied. Most approaches involve representing trees by codewords of some sort, such as balanced parentheses [4] ballot sequences [7], or stack-sortable permutations [1], and listing the codewords in lexicographic order. Typically, the listing algorithms produce each new codeword in $O(1)$ amortized time, and the ranking and unranking algorithms perform $O(n)$ arithmetic operations given a precomputed table of size $O(n^2)$. (The arithmetic operands and table entries can be quite large; the number of n-node trees is well known to be the n-th Catalan number $C_n = \binom{2n}{n}/(n+1)$, whose binary representation requires roughly $2n$ bits, and the intermediate results also require $O(n)$ bits. If bit operations are important, the bounds on space and time should be multiplied by $O(n)$.)

One difficulty with these methods is that consecutive codewords often correspond to vastly different trees. Proskurowksi and Ruskey list balanced parentheses in a "Gray code"-like order, in which consecutive strings differ in only two positions [5, 6]. But even two nearly identical parenthesis strings may correspond to trees with different structure.

Lucas showed how to list n-node trees so that consecutive trees differ by a single rotation of an edge [2]. Following Sleator, Tarjan, and Thurston [8], she defined the rotation graph $RG(n)$ to be the graph on C_n vertices, one for each tree, in which two vertices are connected by an edge if the corresponding trees can be transformed into each other by rotating one tree edge. By describing the structure of $RG(n)$ in terms of "d-dimensional stacks of height h", she was able to find Hamiltonian paths in $RG(n)$. She did not address the ranking and unranking problems for these lists.

The present paper contributes a more direct proof of the existence of Hamiltonian paths in $RG(n)$, using the more familiar and intuitive language of lattice paths. The ideas in the proof are actually the same as in Lucas' proof, but the surface presentation is quite different. Using the insight from this proof, we develop ranking and unranking algorithms. The straightforward implementation of these algorithms perform $O(n^2)$ arithmetic operations using a precomputed table of size $O(n^2)$. (As usual, multiply by $O(n)$ if bit operations are important.) Next we show how to reduce the time to $O(n^2)$ operations on $O(\log n)$-bit numbers and only $O(n)$ operations on $O(n)$-bit numbers, given a precomputed table of size $O(n^2)$. Finally we show how to compute the table on the fly, using only $O(n)$ memory cells.

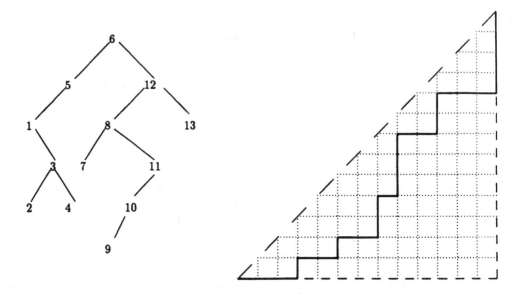

$$\mathbf{x} = (3, 2, 2, 0, 1, 0, 0, 2, 0, 3, 0, 0, 0, 0)$$

$$\mathbf{d} = (0, 2, 3, 4, 3, 3, 2, 1, 2, 1, 3, 2, 1, 0)$$

Figure 1.

A tree, its lattice path, and its vectors

2. Encoding trees by lattice paths

We begin by exhibiting a correspondence between binary trees and lattice paths from $(0,0)$ to (n,n) that never go above the main diagonal. The representation of trees by ballot sequences or balanced parentheses gives such a correspondence, but ours is different. A symmetric version of our correspondence was studied by Zerling [9].

Let T be an n-node binary tree. The *right path* of T is the set of nodes reachable from the root by following right child pointers. Number the nodes in symmetric order. For any node v, let $lm(v)$ be the leftmost descendant, let $la(v)$ be the closest left ancestor of v, and let $rpa(v)$ be the closest ancestor to v on the right path. Formally:

$$lm(v) = \min\{w \mid w \text{ is a descendant of } v\}$$
$$la(v) = \max\{w \mid w \text{ is an ancestor of } v \text{ and } w < v\}$$
$$rpa(v) = \max\{w \mid w \text{ is an ancestor of } v \text{ and } w \text{ is on the right path}\}$$

If v is on the right path, then $rpa(v) = v$.

We dissect the tree into vertex-disjoint paths, namely the right path itself plus all maximal paths that follow only left child pointers and do not touch the right path. We record the size of each path at its bottommost node. Formally, let

$$x_v = \begin{cases} 0, & \text{if } v \text{ has non-empty left subtree} \\ \min\big(d(v, la(v)),\ d(v, rpa(v))\big), & \text{otherwise} \end{cases}$$

and let x_0 be the number of nodes on the right path.

Form a lattice path starting at $(0,0)$ by placing x_v horizontal steps at height v, for $0 \le v \le n$. The vector $(x_0, \ldots, x_n)$ completely describes the path, and the tree can be uniquely recovered from the vector. An alternative description of the path will be convenient; instead of describing how many horizontal steps there are at each height, we describe how close the path gets to the main diagonal. Let $d_0 = 0$, and for $0 \le v < n$ let $d_{v+1} = d_v + x_v - 1$ be the distance from the main diagonal to the path at height $v + 1$. Figure 1 illustrates these definitions.

It is easy to produce the x and d vectors from the tree in time $O(n)$. The algorithm ENCODE of Figure 2 does the job by successively rotating all the vertices on each maximal left-pointing path onto the right path.

procedure ENCODE(T, n)
  { determine the vectors $(x_0, \ldots, x_n)$ and $(d_0, \ldots, d_n)$ corresponding to tree T }
    $v \leftarrow root(T)$
    $nodecount \leftarrow 0$
    for $k \leftarrow 1$ **to** n **do**
        { count how many rotations it takes to get k onto the right path }
        $x_k \leftarrow 0$
        while $v.left \ne null$ **do**
            $w \leftarrow v.left$
            rotate (v, w) right
            $x_k \leftarrow x_k + 1$
            $v \leftarrow w$
        $v \leftarrow v.right$
        $nodecount \leftarrow nodecount + x_k$
    $x_0 \leftarrow n - nodecount$
    $d_0 \leftarrow d_n \leftarrow 0$
    for $k \leftarrow 0$ **to** $n - 1$ **do**
        $d_{k+1} \leftarrow d_k + x_k - 1$
end ENCODE

Figure 2.
Encoding a tree by a lattice path

Using the fact that the maximal paths are disjoint, and that only the path above the leftmost vertex in a subtree can leave that subtree, the following properties of our encoding are easily verified:

$$0 \le d_v \le n - v, \qquad \text{for } 0 \le v \le n \qquad (1)$$

$$d_v \le d_{v+1} + 1, \qquad \text{for } 0 \le v < n \qquad (2)$$

$$\sum_{j \le k \le v} x_k < v - j, \qquad \text{for } 1 \le v \le n \text{ and } lm(v) < j \le v \qquad (3)$$

$$\sum_{lm(v) \le k \le v} x_k \ge v - lm(v), \qquad \text{for } 1 \le v \le n \qquad (4)$$

$$\sum_{j \le k \le v} x_k < v - j, \qquad \text{for } v \text{ on the right path and } 0 < j \le v \qquad (5)$$

$$\sum_{0 \le k \le v} x_k \ge v, \qquad \text{for } v \text{ on the right path} \qquad (6)$$

If we rotate an edge in T, the effect on the corresponding lattice path is surprisingly simple. Figure 3 illustrates a right rotation of an edge $e = (v, w)$. Let $a = lm(v)$. If w is not on the right path, the rotation simply decreases x_a and increases x_{v+1} by 1, leaving all other x_k's unchanged, thus "pulling" the part of the lattice path between height a and $v + 1$ to the left by one unit. By (3), a diagonal dropped from $(v + 1 + d_{v+1}, v + 1)$ lies strictly above the lattice path until height a. At this point the path crosses strictly through the diagonal, by (4) and the fact that v is a left child of w.

If w is on the right path, the rotation decreases x_a and increases x_0 by 1, thus "pushing" the lattice path one unit to the right between height 0 and a. Because v is a left child of w, a diagonal dropped from $(a + d_a, a)$ leaves the lattice path from the left end of a horizontal step; by (5) and (6) it next touches the path at height 0.

Left rotations have the opposite effect. Rotating an edge not on the right path corresponds to pushing a piece of the lattice path, and rotating an edge on the right path corresponds to pulling. For any lattice path, we can determine which paths arise by performing one rotation as follows:

From every point (x_2, y_2) on the path, drop a diagonal and let (x_1, y_1) be the next point of contact with the path. If the path strictly crosses the diagonal, we can "pull at y_2", moving the part of the path between height y_1 and y_2 one unit to the left. If the path is horizontal at (x_2, y_2), we can "push at y_2", moving the partial path one unit to the right. (If the diagonal lies under the path, neither case applies.)

For $0 \le h \le w$, let $\text{LP}(w, h)$ be the graph whose vertices correspond to lattice paths from $(0, 0)$ to (w, h) that never go above the main diagonal. Two paths are connected by an edge if one can be changed into the other by pushing or pulling.

Let $p(w, h)$ be the number of paths in $\text{LP}(w, h)$. From the well-known reflection principle [3], we have $p(w, h) = \binom{w+h}{h} - \binom{w+h}{h-1}$. Of course $p(n, n) = \binom{2n}{n}/(n+1) = C_n$, the n-th Catalan number.

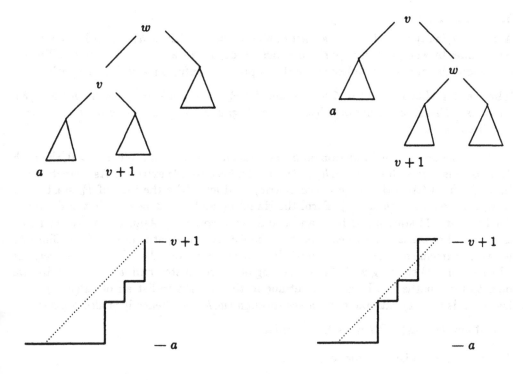

Figure 3.

The effect of rotations on lattice paths

Call a path *h-extreme* if every horizontal segment at height h or below touches the main diagonal. Formally, a path described by $(x_0, \ldots, x_n)$ and $(d_0, \ldots, d_n)$ is h-extreme if for $0 \le k \le h$ either $x_k = 0$ or $d_k = 0$. If $d_{h+1} > 0$, then the diagonal from $(h + 1 + d_{h+1}, h + 1)$ will touch an h-extreme path at a horizontal segment that strictly crosses the diagonal; in other words, we can always pull an h-extreme path at $h + 1$ if $d_{h+1} > 0$. Of course, we can push whenever $x_{h+1} > 0$ (this is true for any path).

3. Listing trees

A *Hamiltonian list* of $LP(w, h)$ is a list of all the paths in $LP(w, h)$ in which consecutive paths differ by a single pull or push. In other words, it's a Hamiltonian path in $LP(w, h)$. (We avoid the phrase 'Hamiltonian path' to prevent confusion with 'lattice path'.)

Theorem 1 (Lucas). *Let $0 \leq h \leq w$ and let P_0 be any h-extreme path in $LP(w, h)$. There is a Hamiltonian list of $LP(w, h)$ that begins with P_0 and ends with some other h-extreme path.*

Proof. The proof is by induction on h. The case $h = 0$ is trivial. When $h > 0$, the path P_0 must pass through either (h, h) or $(w, h - 1)$, because it's extreme. Assume it passes through (h, h) (the other case is symmetric), and let P_0' be the part of P_0 starting at $(0, 0)$ and ending at $(h, h - 1)$. Form the Hamiltonian list as follows: First, inductively list $LP(h, h - 1)$ starting with P_0' and ending with some P_1'. Append one vertical step and $w - h$ horizontal ones to each path in the list, creating paths in $LP(w, h)$. Since P_1' is $(h-1)$-extreme and $x_h = w - h$, push P_1 at height h to form P_2. Next, inductively list $LP(h + 1, h - 1)$ starting with P_2', appending one vertical step and $w - h - 1$ horizontal ones to form paths in $LP(w, h)$. Continue in this fashion to list all of $LP(w, h)$. The last path is $(h - 1)$-extreme and passes through $(w, h - 1)$, hence it is also h-extreme.

Corollary (Lucas). *$RG(n)$ is Hamiltonian.*

Proof. $RG(n)$ is isomorphic to $LP(n, n)$.

Ranking

Let P_0 be an n-extreme path in $LP(n, n)$, corresponding to tree T_0. The proof of Theorem 1 constructs a Hamiltonian list $P_0, P_1, \ldots, P_{C_n}$ of paths in $LP(n, n)$. Given a tree T with corresponding path P, the algorithm RANK of Figure 4 finds the unique integer $r = rank(T)$ with $P_r = P(T)$.

The first phase of the algorithm sets dir_h to $+1$ if the Hamiltonian list is working left-to-right at height h, from (h, h) to $(h + d_h, h)$, when P appears. It sets dir_h to -1 if it is working right-to-left from $(h + d_h, h)$ to (h, h). To do this, it determines how many times the list has reversed direction at height h, by solving the rank problem for P with respect to the shortened lattice with bottom-left corner at (h, h).

The second phase computes the rank. If $dir_h = +1$, we increase the rank by the number of paths that agree with P at height $h + 2$ and above, yet pass through height h to the left of P; these are exactly the paths listed while working from (h, h) to $(h + d_h, h)$. Since the latter parts of these paths are fixed, we can count them by counting the number of paths from $(0, 0)$ to (h, h), $(h + 1, h)$, $\ldots$, $(h + d_h - 1, h)$. This number is

$$\sum_{0 \leq k < d_h} p(h + k, h) = \sum_{0 \leq k < d_h} \left[\binom{2h + k}{h} - \binom{2h + k}{h - 1} \right]$$
$$= \binom{2h + d_h}{h + 1} - \binom{2h + d_h}{h}.$$

Similarly, the number of paths agreeing with P at height $h + 2$ and above, yet passing through height h to the right of P, is

$$
\sum_{d_h < k \le d_{h+1}+1} p(h+k, h) = \sum_{d_h < k \le d_{h+1}+1} \left[\binom{2h+k}{h} - \binom{2h+k}{h-1} \right]
$$

$$
= \left[\binom{2h+d_{h+1}+2}{h+1} - \binom{2h+d_{h+1}+2}{h} \right] -
$$

$$
\left[\binom{2h+d_h+1}{h+1} - \binom{2h+d_h+1}{h} \right].
$$

The upper bound on the range of summation follows from (2).

> **procedure** RANK(T, T_0, n)
> { find the rank of T in the Hamiltonian list of n-node trees that starts with T_0 }
>   call ENCODE(T, n) to determine $(x_0, \ldots, x_n)$ and $(d_0, \ldots, d_n)$
>   call ENCODE(T_0, n) to determine $(y_0, \ldots, y_n)$ and $(e_0, \ldots, e_n)$
>   **for** $k \leftarrow 0$ **to** n **do**
>   **if** $e_k = 0$ **then** $dir_k \leftarrow +1$ **else** $dir_k \leftarrow -1$
>
>   { phase 1: determine the sweep direction at each height }
>   **for** $h \leftarrow n - 1$ **downto** 0 **do**
>   **for** $k \leftarrow n - 1$ **downto** $h + 1$ **do**
>   **if** $dir_k = +1$ **then**
>   **if** IsOdd($A[k - h, d_k]$) **then** $dir_h = -dir_h$
>   **else**
>   **if** IsOdd($A[k - h, d_{k+1} + 2]$) **then** $dir_h = -dir_h$
>   **if** IsOdd($A[k - h, d_k + 1]$) **then** $dir_h = -dir_h$
>
>   { phase 2: compute the rank }
> $rank \leftarrow 0$
>   **for** $h \leftarrow n - 1$ **downto** 0 **do**
>   **if** $dir_h = 1$ **then**
> $rank \leftarrow rank + A[h + 1, d_h]$
>   **else**
> $rank \leftarrow rank + A[h + 1, d_{h+1} + 2] - A[h + 1, d_h + 1]$
> **end** RANK

Figure 4.
Finding the rank of a tree

Algorithm RANK finds these numbers in a table defined by

$$
A(a, b) = \binom{2a + b - 1}{a} - \binom{2a + b - 1}{a - 1}.
$$

The straightforward implementation of our algorithm assumes that A has been precomputed. The algorithm performs $O(n^2)$ arithmetic operations and uses a precomputed table of size $O(n^2)$.

Unranking is done similarly. We determine the path from top to bottom; for each height h we find whether the Hamiltonian list is sweeping to the right or to the left at height h when the tree of the desired rank is listed, again by solving a rank problem in a shortened lattice. Then we compute how far the height h sweep has progressed by skipping as many paths as possible without skipping the desired rank. The details are given in Figure 5.

```
procedure UNRANK(rank, T₀, n)
    { find the tree of given rank in the list of n-node trees that starts with T₀ }
        call ENCODE(T₀, n) to determine (y₀,...,yₙ) and (e₀,...,eₙ)
        for k ← 0 to n do
            if eₖ = 0 then dirₖ ← +1 else dirₖ ← −1

    dₙ ← 0
    r ← 0
    for h ← n − 1 downto 0 do
        { determine dₕ }
        for k ← n − 1 downto h + 1 do
            { find the sweep direction at height h }
            if dirₖ = +1 then
                if IsOdd(A[k − h, dₖ]) then dirₕ = −dirₕ
            else
                if IsOdd(A[k − h, dₖ₊₁ + 2]) then dirₕ = −dirₕ
                if IsOdd(A[k − h, dₖ + 1]) then dirₕ = −dirₕ

        { sweep, but don't skip tree with desired rank }
        if dirₕ = +1 then
            dₕ ← 0
            while r + A[h + 1, dₕ] < rank do dₕ ← dₕ + 1
            r ← r + A[h + 1, dₕ]
        else
            dₕ ← dₕ₊₁ + 1
            while r + A[h+1, dₕ₊₁+2] − A[h+1, dₕ+1] < rank do dₕ ← dₕ−1
            r ← r + A[h + 1, dₕ₊₁ + 2] − A[h + 1, dₕ + 1]
end UNRANK
```

Figure 5.
Finding the tree with given rank

4. Saving time

It's easy to verify that $A(a, b)$ satisfies a simple recurrence:

$$
\begin{aligned}
A(a, 0) &= 0, & &\text{for } a \geq 0, \\
A(0, b) &= 1, & &\text{for } k \geq 1, \\
A(a, b) &= A(a, b-1) + A(a-1, b+1), & &\text{for } a, b \geq 1.
\end{aligned} \tag{7}
$$

The first phase doesn't need the actual values in the table; it suffices to know whether they are even or odd. The next lemma shows that the parity of the (a, b)-th entry can be computed with $O(1)$ operations on $O(\log n)$-bit numbers.

Lemma 3. *Let* $B(a, b) = A(a, b) \bmod 2$. *The* $B(a, b) = 1$ *if and only if* $b > 0$ *and* $(2a + b)\&a = a$. *(Here, '&' denotes bitwise logical AND.)*

Proof. The B table satisfies recurrence (7) with arithmetic done modulo 2. Let 2^r be the largest power of two not exceeding a, so $a = 2^r + x$ with $0 \leq x < 2^r$; if a is not a power of two, let 2^s be the largest power of two not exceeding x, so $a = 2^r + 2^s + y$ with $r > s$ and $0 \leq y < 2^s$.

From Figure 6, we see that B has the following properties:

i) Row $2^r + x$ is periodic with period 2^{r+1}.
ii) The period in row 2^r consists of 2^r zeros followed by 2^r ones.
iii) Row $2^r + 2^s + y$ contains the first 2^r elements of row $2^s + y$, offset from the right of the period by distance 2^{s+1}.
iv) Outside of this region, the period of row $2^r + 2^s + y$ contains only zeros.

These facts may be verified by induction.

Write a in binary as

$$
10^{d_1} 10^{d_2} 1 \cdots 10^{d_k}
$$

where '0^d' means "contatenate d zeros", and align b under a. Fact (i) means that we can ignore any bits of b lying to the left of a. Fact (iv) means that $B(a, b) = 0$ unless the part of b lying under 10^{d_1} must represent a number x_1 in the range $2^{d_1} - 1 \leq x_1 < d^{d_1 + 1} - 1$. Fact (iii) means that we can strip off the leading $d_1 + 1$ bits from both a and b, and repeat the argument with the remaining bits. Thus, for $1 \leq i < k$, the part of b lying under 10^{d_i} must represent a number x_i in the range $2^{d_i} - 1 \leq x_i < d^{d_i + 1} - 1$. Finally, fact (ii) means that the part of b lying under 10^{d_k} must represent a number x_k in the range $2^{d_k} \leq x_k < d^{d_k + 1}$.

On the other hand, $(2a + b)\&a = a$ exactly when $2a + b$ has the form

$$
y_0 1 y_1 1 y_2 1 \cdots 1 y_k
$$

where y_i is an arbitrary d_i-bit number (counting leading zeros) for $i > 0$, and y_0 is any positive number. Subtracting $2a$ from this and examining the result from right to left, this is equivalent to saying that b can be written as

$$
x_0 x_1 x_2 \cdots x_k
$$

```
0     0 1 1 1 1 1 1 1 1 1 1 1 1 1 1 1
1     0 (1)
2     0 0 (1 1)
3     (0 1) 0 0
4     0 0 0 0 (1 1 1 1)
5     0 0 (0 1 0 1) 0 0
6     (0 0 1 1) 0 0 0 0
7     (0 1 0 0) 0 0 0 0
8     0 0 0 0 0 0 0 0 (1 1 1 1 1 1 1 1)
9     0 0 0 0 0 0 (0 1 0 1 0 1 0 1) 0 0
10    0 0 0 0 (0 0 1 1 0 0 1 1) 0 0 0 0
11    0 0 0 0 (0 1 0 0 0 1 0 0) 0 0 0 0
12    (0 0 0 0 1 1 1 1) 0 0 0 0 0 0 0 0
13    (0 0 0 1 0 1 0 0) 0 0 0 0 0 0 0 0
14    (0 0 1 1 0 0 0 0) 0 0 0 0 0 0 0 0
15    (0 1 0 0 0 0 0 0) 0 0 0 0 0 0 0 0
```

Figure 6.

The table $B(a, b)$

where x_k is a $(d_k + 1)$-bit number in the range $2^{d_k} \leq x_k < d^{d_k+1}$, x_i is a $(d_k + 1)$-bit number in the range $2^{d_i} - 1 \leq x_i < d^{d_i+1} - 1$ for $1 \leq i < k$, and $x_0 \geq 0$ is arbitrary. This proves that $B(a, b) = 1$ exactly when $(2a + b)\&a = a$. ∎

This suggests an implementation of the ranking algorithm in which phase 1 performs $O(n^2)$ operations on $O(\log n)$-bit numbers. If the A table is precomputed, phase 2 can get by with $O(n)$ operations on $O(n)$-bit numbers, for a grand total of $O(n^2 \log n)$ bit operations. In practice, we usually represent large numbers by sequences of $O(\log n)$-bit "words", and perform operations on an entire word in constant time. Under this interpretation, the ranking algorithm does $O(n^2)$ word operations.

5. Saving space

Using the ideas of the previous section, the first phase doesn't need to consult the table at all. The h-th iteration of the second phase needs only row $h + 1$ of the table, so we could store one row at a time and use recurrence (7) to create the next row when needed, discarding the old row. This gives a substantial savings, reducing the space from $O(n^2)$ entries to $O(n)$, and from $O(n^3)$ bits to $O(n^2)$.

6. References

1. Gary D. Knott, "A numbering system for binary trees." *CACM* **20** (1977), 113–115.

2. Joan M. Lucas, "The rotation graph of binary trees is Hamiltonian." *J. Algorithms* **8** (1987), 503–535.

3. Gopal Mohanty, *Lattice Path Counting and Applications.* Academic Press, 1979.

4. Andrzej Proskurowski, "On the generation of binary trees." *JACM* **27** (1980), 1–2.

5. Andrzej Proskurowski and Frank Ruskey, "Binary tree gray codes." *J. Algorithms* **6** (1985), 225–238.

6. Andrzej Proskurowski and Frank Ruskey, "Generating binary trees by transpositions." *Proceedings of the First Scandanavian Workshop on Algorithm Theory*, published as *Lecture Notes in Computer Science* **318**, 199–207. Springer-Verlag, 1988.

7. Doron Rotem and Y. L. Varol, "Generation of binary trees from ballot sequences." *JACM* **25** (1978), 396–404.

8. Daniel D. Sleator, Robert E. Tarjan, and William P. Thurston, "Rotation distance, triangulations, and hyperbolic geometry." *J. American Mathematical Society* **1** (1988), 647–681.

9. David Zerling, "Generating binary trees using rotations." *JACM* **32** (1985), 694–701.

Expected Behaviour Analysis of AVL Trees

Ricardo Baeza-Yates

Gaston H. Gonnet

Depto. de Cs. de la Computación
Universidad de Chile
Santiago, Casilla 2777
Chile

Department of Computer Science
University of Waterloo
Waterloo, Ontario N2L3G1
Canada

Nivio Ziviani

Depto. de Cíencia da Computação
Universidade Federal de Minas Gerais
Belo Horizonte, Minas Gerais
Brazil *

Abstract

In this paper we improve previous bounds on expected measures of AVL trees by using fringe analysis. A new way of handling larger tree collections that are not closed is presented. An inherent difficulty posed by the transformations necessary to keep the AVL tree balanced makes its analysis difficult when using fringe analysis methods. We derive a technique to cope with this difficulty obtaining the exact solution for fringe parameters even when unknown probabilities are involved. We show that the probability of a rotation in an insertion is between 0.37 and 0.73, that the fraction of balanced nodes is between 0.56 and 0.78, and that the expected number of comparisons in a search seems to be at most 12% more than in the complete balanced tree.

1 Introduction

Balanced tree structures are efficient ways of storing information. They provide an excellent solution for the dictionary data structure problem. For N elements the operations find, insert, and delete can be done in $O(\log N)$ units of time. The most popular, for main memory, are AVL trees (also called Height Balanced trees).

AVL trees were introduced by Adel'son-Vel'skii and Landis in 1962 [AVL62]. A binary search tree is AVL if the height of the subtrees at each node differ by at most one. A balance field in each node can indicate this with two bits: +1, higher right subtree; 0, equal heights; - 1, higher left subtree.

---

*The work of the first author was also supported by the the Institute for Computer Research of the University of Waterloo, the second author by a Natural Sciences and Engineering Research Council of Canada Grant No. A-3353, and the third by a Brazilian Coordenação do Aperfeiçoamento de Pessoal de Nível Superior Contract No. 4799/77 and by the University of Waterloo.

The first valuable attempt to analyze a balanced search tree in the average case was performed by Yao [Yao78]. In his work Yao presented a method which he used to obtain a partial analysis of 2-3 trees and B-trees. The method used by Yao was later used by Brown [Bro79] to obtain a partial analysis of AVL trees. In his analysis Brown considered the collection of AVL subtrees with three or less leaves and called it the fringe of the AVL tree. By analyzing the fringe of large AVL trees Brown was able to derive bounds on the expected number of balanced nodes in the whole tree.

An improvement on Brown's results for AVL trees was obtained by Mehlhorn [Meh79], through the study of 1-2 brother trees [OS76]. The main technical contribution of Mehlhorn's paper is a method for analyzing the behaviour of 1-2 brother tree schemes where the rebalancing operations require knowledge about the brother of a node. Using the close relationship between 1-2 brother trees and AVL trees [OW80], Mehlhorn was able to improve the bounds on the expected number of balanced nodes in AVL trees. Also, Mehlhorn [Meh82] presented a fringe analysis of AVL trees under random insertions and deletions.

Consider an AVL tree T with N keys and consequently $N + 1$ external nodes. These N keys divide all possible key values into $N + 1$ intervals. An insertion into T is said to be a *random insertion* if it has an equal probability of being in any of the $N + 1$ intervals defined above. A *random AVL tree* with N keys is an AVL tree constructed by making N successive random insertions into an initially empty tree. In this paper we assume that all trees are random trees.

We now define certain complexity measures:

- Let $\overline{b}(N)$ be the expected number of balanced nodes in an AVL tree after the random insertion of N keys into an initially empty tree;

- Let $r(N)$ be the expected number of rotations required during the insertion of the $(N+1)^{st}$ key into a random AVL tree with N keys;

- Let $\overline{C}(N)$ be the expected number of comparisons in a successful search in an AVL tree with N keys.

- Let $\overline{f}(N)$ be the expected number of nodes in the fringe of an AVL tree after the random insertion of N keys into an initially empty tree;

- Let $\overline{h}(N)$ be the expected height of a random AVL tree with N elements;

- Let $\overline{c}(N)$ be the expected number of balance changes in an AVL tree during the N random insertions into an initially empty tree.

Table 1 shows the summary of our main results related to AVL trees.

2 Fringe Analysis

In this section we survey briefly fringe analysis, and we formalize fringe properties to deal with AVL-tree collections. We do so by following an example related to AVL trees.

Let us define a tree collection as a finite collection $C = \{T_1, \cdots, T_m\}$ of trees. The collection of AVL trees with three leaves or fewer forms a tree collection, as shown in Figure 1.

The fringe of a tree consists of one or more subtrees that are isomorphic to members of a tree collection C. Typically, the fringe will contain all subtrees that meet this definition; for example

Tree Collection		$\overline{f}(N)$	$r(N)$	$\frac{\underline{b}(N)}{N}$	$\frac{\overline{c}(N)}{N}$
Size	Kind				
2 [Bro79]	closed	$0.57N$		$[0.47, 0.86]$	
3 (Ours)	closed ambiguous	$0.66N$	$[0.29, 0.86]$	$[0.51, 0.86N]$	$[1.43, 2.34]$
3	weakly-closed ambiguous	$[0.66N, 0.69N]$	$[0.29, 0.77]$ [MT86]	$[0.51, 0.81]$ [Meh79]	$[1.47, 2.25]$ [MT86]
4 (Ours)	weakly-closed ambiguous	$[0.747N, 0.753N]$	$[0.37, 0.73]$	$[0.56, 0.78]$	$[1.59, 2.15]$

Table 1: Summary of AVL tree results

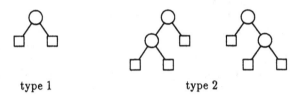

type 1   type 2

Figure 1: Tree collection of AVL trees with three leaves or fewer

the fringe of an AVL tree that corresponds to the tree collection of Figure 1 is obtained by deleting all nodes at a distance greater than 2 from the leaves. Figure 2 shows an instance of an AVL tree with eleven keys in which the fringe that corresponds to the tree collection of Figure 1 is delimited by a dashed line.

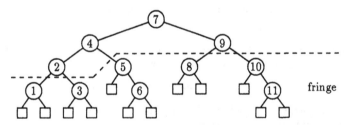

Figure 2: An AVL tree and its fringe

The composition of the fringe can be described in several ways. One possible way is to consider the probability that a randomly chosen leaf of the tree belongs to each of the members of the corresponding tree collection [EZG+82]. In other words, the probability p is

$$p_i(N) = \frac{Expected \ number \ of \ leaves \ of \ type \ i \ in \ a \ N-key \ tree}{N+1} \quad (1)$$

We now introduce some concepts about the fringes of search trees.

Definition: A tree collection $C = \{T_1, \ldots, T_m\}$ is **weakly-closed** if for all $j \in [1, \ldots, m]$ an insertion into T_j always leads to one or more T_i, $i \in [1, \ldots, m]$. ∎

Definition: A tree collection C is **closed** when (i) C is weakly-closed and (ii) the effect of an insertion on the composition of the fringe is determined only by the subtree of the fringe where the insertion is performed. ∎

The tree collection of Figure 1 is an example of a closed tree collection (proved by Brown [Bro79]). On the other hand the collection of AVL trees with more than 2 and fewer than 6 leaves (see Figure 7) is not closed. This is because an insertion into a type 2 tree of Figure 10, when the type 2 tree is part of the fringe of an AVL tree, may cause a rotation higher in the tree, and the composition of the fringe depends on this rotation at the higher level. Figure 3 shows an instance of an AVL tree where an insertion into a type 2 tree does not lead to a type 3 tree as expected.

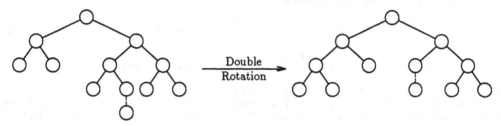

Figure 3: Example of an insertion that unexpectedly changes the fringe of an AVL tree
(dashed edge shows the point of insertion)

Definition: A tree collection C is **ambiguous** when a tree in C appears as a subtree of another tree in C. Figure 4 shows an AVL tree collection that is ambiguous, since a tree of type 1 is a subtree of trees of type 3. ∎

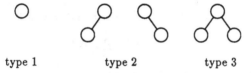

type 1   type 2   type 3

Figure 4: Tree collection of AVL trees with more than 1 and less than 5 leaves
(leaves not shown)

Definition: A tree collection C is **open** if it is not weakly-closed. ∎

The transitions between trees of a tree collection can be used to model the insertion process. In an insertion of a key into the type 1 tree shown in Figure 1 two leaves of type 1 are lost and three leaves of type 2 are obtained. In an insertion of a key into the type 2 tree three leaves of the type 2 are lost and four leaves of the type 1 tree are obtained.

Definition: A fringe defined by a tree collection C and a set of transitions between members of C is **ambiguous**, if is not possible, to determine which subtrees belongs to the fringe. ∎

To improve the results obtained by Brown we need larger tree collections. A new tree collection for AVL trees composed of trees with four leaves or fewer is shown in Figure 4. In Brown's paper, the fringe is defined statically; from the tree itself we can decide how to break it up into the different subtree types without knowing the sequence of insertions that built it. For this class, the fringe is defined dinamically; that is, every insertion, in addition to building a new tree, defines the fringe of the new tree as a function of the fringe of the old tree and the transformations caused by the insertion.

THEOREM 2.1 *The AVL tree collection shown in Figure 4 is closed.*

Proof: Similar to the proof by Brown [Bro79]. ∎

The results obtained with this fringe are better than Brown's results, and some are as good as Mehlhorn's results. They are shown in Table 1.

Consider the tree collection of AVL trees shown in Figure 1. It is possible to see by studying the transitions of each type of subtree that

$$\vec{P}(N) = \left(I + \frac{H}{N+1} \right) \vec{P}(N-1) \tag{2}$$

where $\vec{P}(N) = [p_1(N), p_2(N)]$, $H = \begin{bmatrix} -3 & 4 \\ 3 & -4 \end{bmatrix}$ is the transition matrix, and I is the identity matrix. In general, the same recurrence type holds for a tree collection with m types.

Formal details of the development of this theory can be found in Ziviani [Ziv82], or in Eisenbarth, Ziviani, Gonnet, Mehlhorn and Wood [EZG+82]. The following theorem [EZG+82, BYG89] is basic for deriving our new results.

THEOREM 2.2 *Let H be the $m \times m$ transition matrix of a connected fringe analysis problem, as in Eq. 2. Let $\lambda_1, \ldots, \lambda_m$ be the eigenvalues of H, then*

$$\lambda_1 = 0 > \mathrm{Re}\lambda_2 \geq \mathrm{Re}\lambda_3 \geq \cdots \geq \mathrm{Re}\lambda_m .$$

Let t be the multiplicity of λ_2. Then for every vector $\vec{P}(0)$

$$\vec{P}(N) = \vec{P}(\infty) + O(\log^t N \ N^{\mathrm{Re}\ \lambda_2})$$

where $\vec{P}(N)$ is defined by Eq. 1, and $\vec{P}(\infty)$ is the unique solution to the system of equations

$$H\vec{P}(\infty) = 0 \quad \text{and} \quad \sum_i p_i(\infty) = 1 .$$

Let $A_i(N)$ be the expected number of trees of type i in a random search tree with N keys. Let L_i be the number of leaves of the type i tree. Note that Eq. 1 can be written as

$$p_i(N) = \frac{A_i(N)L_i}{N+1} . \tag{3}$$

3 Basic Relations

In this section we show how to relate the fringe of an AVL tree to the defined measures. To simplify notation $p_i(N)$ is written as p_i throughout the remainder of this paper.

Lemma 3.1 *The expected number of rotations in a random AVL tree with N keys during the $(N+1)^{st}$ insertion is bounded above by*

$$(i) \quad r(N) = 1 - Pr\{no\ rotation\}$$

and

$$(ii) \quad r(N) \leq r(N) \text{ in the fringe } + \text{ rotations outside the fringe } .$$

Proof: For case (i) it is known that the maximum number of rotations per insertion in an AVL tree is 1. For case (ii) $r(N)$ must be less than or equal to the number of rotations per insertion in the fringe plus all possible rotations per insertion that may occur outside the fringe. ∎

Lemma 3.2 *The expected number of nodes in the fringe of an AVL tree with N keys that corresponds to a tree collection $C = T_1, \ldots, T_m$ is*

$$\overline{f}(N) = \sum_{i=1}^{m} \frac{p_i(L_i - 1)}{L_i}(N+1)$$

Proof: Each subtree type has $L_i - 1$ nodes. Then, $\overline{f}(N) = \sum_{i=1}^{m}(L_i - 1)A_i(N)$. Using Eq. 3 we have the desired result. ∎

Lemma 3.3 *The expected number of balanced nodes in a random AVL tree with N keys is bounded above by*

$$(i) \quad \overline{b}(N) = N - \overline{u}(N) \text{ for } N \geq 1$$

and

$$(ii) \quad \overline{b}(N) \leq \overline{b}(N) \text{ in the fringe} + [N - \overline{f}(N)] \text{ for } N \geq 1,$$

where $\overline{u}$ is the expected number of unbalanced nodes.

Proof: For case (i) $\overline{b}(N) + \overline{u}(N) = N$. For case (ii) $\overline{b}(N)$ must be less than or equal to the number of balanced nodes in the fringe plus all nodes outside the fringe. ∎

THEOREM 3.1 *The expected number of balance changes during N random insertions in an initial empty AVL tree is bounded by*

$$N(2 + r(N)) - \overline{b}(N) - 1.44\log_2 N + O(1) \leq \overline{c}(N) \leq N(2 + r(N)) - \overline{b}(N) - \log_2 N + O(1).$$

Proof: Mehlhorn and Tsakalidis show that [MT86]

$$\overline{c}(N) = \overline{u}(N) + R(N) + N - \overline{h}(N),$$

where $R(N) = \sum_{i=1}^{N} r(i)$. Using this definition, Lemma 3.5, and the fact that [Knu73]

$$\log_2(N) \leq \overline{h}(N) \leq 1.44\log_2(N).$$

∎

4 Weakly-closed AVL Tree Collections

If the effect of an insertion on the composition of the fringe is determined not only by the subtree of the fringe where the insertion is performed, but by some other transformation that may happen outside the fringe, then the tree collection is weakly-closed (Definition 2). We will show that the tree collection of AVL trees with four or less leaves shown in Figure 4 is not closed if the fringe is not ambiguous.

Lemma 4.1 *If the trees shown in Figure 4 form the fringe of a random AVL tree with N keys and $N \to \infty$, then an insertion into a leaf of a type 3 subtree (i) decreases by one the number of type 3 subtrees and increases by one the number of type 1 and type 2 subtrees; or (ii) decreases by one the number of type 1 subtrees and increases by one the number of type 2 subtrees.*

Proof: By Mehlhorn [Meh79]. ∎

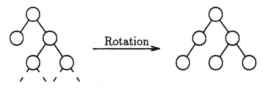

Figure 5: AVL transformation with changes (symmetric transformations occur)
(dashed edges shows the point of insertion)

Lemma 4.1 tells us that any AVL tree collection that contains the type 3 shown in Figure 4 is not closed, if the fringe is not ambiguous. (i.e. it is weakly-closed.) In fact it is not difficult to show that every AVL tree type that contains more than one internal node and has its root node balanced suffers from the same type of misbehaviour that occurs with type 3 (i.e. consider the AVL tree with six nodes). Consequently an AVL tree collection that contains a tree type with the root node balanced and has more than two types is weakly-closed, if the fringe is not ambiguous. (cf. Def. 2).

We know from Lemma 4.1 that an insertion into the type 3 tree shown in Figure 4, when it belongs to a fringe of an AVL tree with N keys, produces a transition that is not well defined: the transition depends on the unknown probability t_N which also depend on N. First of all let us give a more precise meaning to t_N. Let I be the expected number of leaves in an AVL tree with N keys such that an insertion in one of the I leaves causes the transformations show in Figure 5. Thus

$$t_N = \frac{I}{N+1} \ .$$

Although the probability t_N is unknown it cannot assume arbitrary values between 0 and 1.

Lemma 4.2 *The probability t_N is bounded by $0 \le t_N \le \min(2p_1(N), p_3(N))$.*

Proof: We have that the number of type 1 subtrees $(A_1(N))$ must be greater or equal than the number of type 3 subtrees that have a brother of type 1 $(t_N(N+1)/4)$. This is because not all the type 1 subtrees are paired with type 3 subtrees and in the worst case all the type 3 subtrees have a brother of type 1. This give $t_N \le 2p_1(N)$. Also, the t_N leaves are a subset of p_3 leaves. ∎

The probability t_N models an absolute event. On the other hand, Mehlhorn [Meh79] used a conditional probability of the brother of a type 1 subtree being a type 3 subtree. The main problem is that he used $A_3(N)/A_1(N)$ to bound this probability. However, in general, $E[x/y] \neq E[x]/E[y]$ where x and y are random variables, and $E[x]$ is the expected value of x. Our approach avoids this problem.

Table 2 shows the exact value of t_N for $N \leq 20$.

From the results of Lemma 4.1 we can examine the insertion process and obtain

$$
\vec{P}(N) = \left(I + \frac{1}{N+1} \begin{bmatrix} -3 & 0 & 2 \\ 3 & -4 & 3 \\ 0 & 3 & -5 \end{bmatrix} \right) \vec{P}(N-1) + \frac{t_{N-1}}{N+1} \begin{bmatrix} -4 \\ 0 \\ 4 \end{bmatrix} \qquad (4)
$$

$$
= (I + \frac{H}{N+1})\vec{P}(N-1) + \frac{t_{N-1}}{N+1}\vec{U}
$$

where t_N depend on N. Figure 6 shows how the values of the third column of H and the value of $\vec{U}$ were obtained.

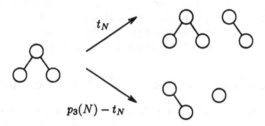

Figure 6: Transitions for type 3

The following theorem shows that the bounds for p_1 and p_3 are independent of the convergence of t_N.

THEOREM 4.1 *The solution of Eq. 5 is bounded by*

$$
\frac{8}{91} \leq p_1 \leq \frac{8}{35}, \quad p_2 = \frac{3}{7}, \quad \frac{12}{35} \leq p_3 \leq \frac{44}{91}
$$

for $N \geq 12$ with the condition $p_1 + p_3 = \frac{4}{7}$.

Corollary 4.1 $t_N \leq 16/91$ *applying Lemma 4.2.*

Proof: We have that the number of subtrees with two leaves is 2/7 and with three leaves is 1/7 for $N \geq 6$ [Bro79]. This implies

$$
\frac{p_1}{L_1} + 2\frac{p_3}{L_3} = \frac{2}{7} \quad \text{and} \quad p_2 = \frac{3}{7} \qquad (5)
$$

for $N \geq 6$. Introducing this in Eq. 5, we obtain for $N \geq 6$

$$
p_1(N) = p_1(N-1)(1 - \frac{5}{N+1}) + \frac{8}{7(N+1)} - \frac{4}{N+1}t_{N-1}
$$

and

$$p_3(N) = p_3(N-1)(1 - \frac{5}{N+1}) + \frac{12}{7(N+1)} + \frac{4}{N+1}t_{N-1}$$

Using Lemma 4.2 $p_1(N)$ is bounded by

$$p_1(N-1)(1 - \frac{13}{N+1}) + \frac{8}{7(N+1)} \leq p_1(N) \leq p_1(N-1)(1 - \frac{5}{N+1}) + \frac{8}{7(N+1)} . \quad (6)$$

The solution of a recurrence of the form

$$x_N = x_{N-1}(1 - \frac{a}{N+1}) + \frac{b}{N+1}$$

is $x_N = \frac{b}{a}$ for $N \geq (a-1)$ if a is an integer.

Solving Eq. 6 gives $8/91 \leq p_1 \leq 8/35$ for $N \geq 12$. In the same way, ussing Lemma 4.2 and relations 5 we obtain $12/35 \leq p_3 \leq 44/91$ for $N \geq 12$. ∎

The results for this tree collection are shown in Table 1, and are similar to Mehlhorn [Meh79] and Mehlhorn and Tsakilidis [MT86] results (although using a different formulation). The lower bounds are the minimum values and the upper bounds the maximum values to hold for any value of p_1 and p_3 in the given ranges, with the constraint that $p_1 + p_3 = 4/7$.

5 Larger Weakly-closed AVL Tree Collections

In Section 4 we showed that any AVL tree collection that contains a tree type with its root node balanced and has more than two types is weakly-closed. This happens because every AVL tree type that contains more than one internal node and has its root node balanced suffers from the same type of misbehaviour that occurs with type 3 of Figure 4, as described in Lemma 4.1.

It is easy to prove a lemma similar to Lemma 4.1 for the tree collection shown in Figure 7. The only difference in the proof of such lemma is that now the trees shown in Figure 8 adds another unknown probability that we call s_N, and is divided in two cases: $s1_N$ and $s2_N$. In this cases the number of type 3 trees decrease by one and the number of type 1 and type 2 trees increases by one. The recurrence relation corresponding to the tree collection shown in Figure 7 involves the two unknown probabilities t_N and s_N, as follows

$$\vec{P}(N) = \left(I + \frac{1}{N+1} \begin{bmatrix} -3 & 0 & 0 & 6/5 \\ 3 & -4 & 4 & 12/5 \\ 0 & 4 & -5 & 12/5 \\ 0 & 0 & 5 & -6 \end{bmatrix} \right) \vec{P}(N-1) + \frac{1}{N+1} \begin{bmatrix} -2 & 2 \\ 3 & 3 \\ 4 & 0 \\ -5 & -5 \end{bmatrix} \begin{bmatrix} t_{N-1} \\ s_{N-1} \end{bmatrix}$$

$$(7)$$

where t_N and s_N depend on N. Figure 9 shows the transitions in which t_N and s_N appears.

For trees of size $N \leq 20$ we are able to obtain s_N exactly. Table 2 shows the values for $s1_N$ (cases (a) and (b) of Fig. 8), $s2_N$ (case (c) of Fig. 8) and $s_N = s1_N + s2_N$.

Lemma 5.1 We have that $t_N = 2p_1(N)$.

Proof: A type 1 subtree is always paired with a type 3 subtree, otherwise will be a type 3 or a type 4 subtree. Then, the number of type 3 subtrees with a type 1 subtree as a brother is $A_1(N)$. ∎

152

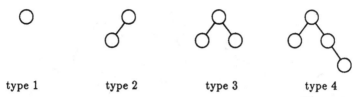

| type 1 | type 2 | type 3 | type 4 |

Figure 7: Tree collection of AVL trees with five or less leaves
(leaves not shown)

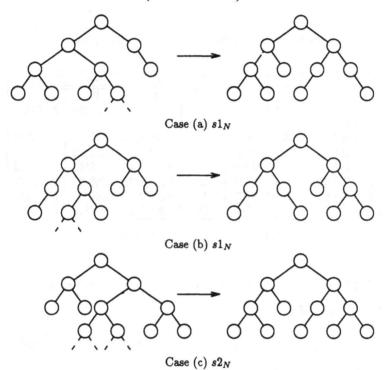

Case (a) $s1_N$

Case (b) $s1_N$

Case (c) $s2_N$

Figure 8: Transformations in the fringe with changes (symmetric transformations occur)
(dashed edges shows the points of insertion)

Now the relation between this collection to Brown's collection [Bro79] give us the following
equations:

$$\frac{p_2}{3} + \frac{p_4}{5} = \frac{1}{7} \quad \text{and} \quad p_1 + p_3 + \frac{2}{5}p_4 = \frac{4}{7}. \tag{8}$$

Using this equations and Lemma 5.1, the new matrix recurrence relation is reduced for $N \geq 6$
to

$$\vec{P}(N) = \left(I + \frac{1}{N+1} \begin{bmatrix} -7 & -2 \\ 9 & -8 \end{bmatrix} \right) \vec{P}(N-1) + \frac{1}{N+1} \left(\begin{bmatrix} 6/7 \\ 12/7 \end{bmatrix} + s_{N-1} \begin{bmatrix} 2 \\ 3 \end{bmatrix} \right) \tag{9}$$

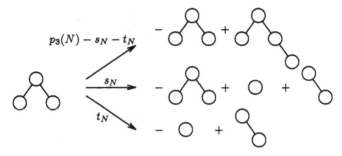

$$p_3(N) - s_N - t_N$$

Figure 9: Transitions for type 2 that involves t_N and s_N

N	t_N	N	t_N	N	$s1_N$	$s2_N$	s_N
5	0.4	13	0.09376	10	0.045455	0.	0.045455
6	0.	14	0.08948	11	0.	0.038961	0.038961
7	0.	15	0.08914	12-14	0.	0.	0.
8	0.09524	16	0.09074	15	0.003207	0.	0.003207
9	0.11429	17	0.09299	16	0.005758	0.001983	0.007741
10	0.10390	18	0.09490	17	0.007114	0.003889	0.011003
11	0.11039	19	0.09613	18	0.007560	0.005117	0.012677
12	0.10739	20	0.09672	19	0.007428	0.005705	0.013133
				20	0.007752	0.005830	0.013582

Table 2: Exact values for t_N and s_N

with $\vec{P}(N) = [p_1(N), p_2(N)]$, and $p_3(N)$ and $p_4(N)$ are obtained from Eqs. 8.

Lemma 5.2 s_N is bounded by

$$0 \leq s_N \leq \frac{p_3(N) - 2p_1(N)}{3} = \frac{4}{21} - p_1(N) - \frac{2}{15}p_4(N) = \frac{2}{21} + \frac{2}{9}p_2(N) - p_1(N).$$

Corollary 5.1 $0 \leq s_N \leq \frac{4}{21}$.

Proof: In the worst case all the type 3 subtrees without a type 1 subtree as a brother belongs to trees associated with s_N. Using Lemma 5.1 this number is $A_3(N) - A_1(N)$. If all are associated with $s1_N$ we have that $s_N \leq 2(A_3(N) - A_1(N))/2$ and if are all associated with $s2_N$ we have that $s_N \leq 4(A_3(N) - A_1(N))/3$. Using relations 8 we obtain the last equalities. ■

Although we have reduced again the problem to one unknown probability, s_N, how we solve recurrence 9 in general?. The next theorem, one of the most important of our results, gives the exact solution to this type of recurrences.

THEOREM 5.1 *Let t_N^i be m unknown probabilities. The solution to a recurrence of the form*

$$\vec{P}(N) = (I + \frac{1}{N+1}H)\vec{P}(N-1) + \frac{1}{N+1}\vec{F} + \frac{1}{N+1}\sum_{i=1}^{m} t_{N-1}^i \vec{G}_i,$$

where $\vec{F}$ and $\vec{G}_i$ $(i = 1..m)$ are constant vectors, is

$$\vec{P}(N) = -H^{-1}\vec{F} + \frac{(-1)^{N+1}}{(N+1)!}R^{\underline{N+1}}\,\vec{C}$$

$$+ \sum_{i=1}^{m}\sum_{k\geq N_0}^{N-1}(-1)^{N-k}t_k^i \sum_{j=0}^{n-k-1}\frac{R^{\underline{j}}\,H^{\underline{N-k-j-1}}}{(N-j+1)(N-k-j-1)!j!}\vec{G}_i$$

where $R = -H - I$, and $H^{\underline{n}} = (H - (n-1)I)\cdots(H-I)H$ denotes descendent factorials over matrices, and with $\vec{C}$ obtained from the initial condition

$$\vec{P}(N_0) = [1, 0, 0, ..., 0]^T \, ,$$

where N_0 is the number of elements in the smallest subtree type of the fringe collection.

Corollary 5.2 If $t_N^i \leq t^i$ for all $N \geq N_1$, then we have

$$\vec{P}(N) \leq -H^{-1}(\vec{F} + \sum_{i=1}^{m}t^i\vec{G}_i) \, + \, O(1/N)$$

for some $N_1 > N_0$. A similar relation holds if $t_N^i \geq t^i$.

Proof: Introducing the generating function

$$\vec{P}(z) = \sum_{n\geq 0}\vec{P}(N)z^N \, ,$$

in the matrix recurrence, we obtain the following first order non-linear differential equation

$$\frac{d\vec{P}(z)}{dz} = \left(\frac{2z-1}{z(1-z)}I + \frac{1}{1-z}H\right)\vec{P}(z) \, + \, \frac{1}{z(1-z)^2}\vec{F} \, + \, \frac{1}{1-z}\sum_{i=1}^{m}t^i(z)\vec{G}_i,$$

where $t^i(z)$ is the generating function associated to t_N^i. The solution of the previous equation is (this can be checked by simple substitution)

$$\vec{P}(z) = \frac{1}{z}e^{R\ln(1-z)}\vec{C} \, - \, \frac{1}{z(1-z)}H^{-1}\vec{F} \, - \, \sum_{i=1}^{m}\frac{1}{z}e^{R\ln(1-z)}\int z\,t^i(z)e^{H\ln(1-z)}dz\,\vec{G}_i$$

where $\vec{C}$ is obtained from the initial condition.
For $N \geq N_0$, we have

$$\vec{P}(N) = [z^N]\vec{P}(z) = -H^{-1}\vec{F} \, + \, \sum_{k\geq 0}R^k[z^{N+1}]\frac{\ln^k(1-z)}{k!}\vec{C} \, - \, \sum_{i=1}^{m}[z^N]\vec{U}_i(z)$$

where $[z^n]P(z)$ denotes the coefficient in z^n of $P(z)$, and

$$\vec{U}_i(z) = \frac{1}{z}e^{R\ln(1-z)}\int z\,t^i(z)e^{H\ln(1-z)}dz\,\vec{G}_i$$

$$= \frac{1}{z}e^{R\ln(1-z)}\sum k \geq N_0 \sum_{j\geq 0}\int z\,t_k^i z^{k+j+1}H^j\frac{\ln^j(1-z)}{j!}dz\,\vec{G}_i \, .$$

But

$$[z^n]\frac{\ln^k(1-z)}{k!} = \frac{(-1)^n}{n!}\mathcal{S}_n^{(k)} \ ,$$

for $n \geq k$ where $\mathcal{S}_n^{(k)}$ denotes Stirling numbers of the first kind [Knu69]. Using the previous formula and that

$$\sum_{k=0}^{n+1} \mathcal{S}_{n+1}^{(k)} H^k = H^{\underline{n+1}}$$

we obtain the desired result. This simplifies if t_i^k is bounded by a constant, giving the corollary. ∎

Applying this solution to the recurrence for the collection of Section 4 we obtain, for $N > 5$,

$$p_1(N) = \frac{8}{35} - 4\sum_{j>4}^{N-1} \frac{(j+1)^{\underline{4}}}{(N+1)^{\underline{5}}} t_j$$

Using Lemma 5.2 and Corollary 5.2 (chosing $N_1 = 20$) we can bound the set of probabilities p_i. Boot-strapping this result in Lemma 5.2 we obtain the following improved bounds for t_N and s_N.

Lemma 5.3 *The probabilities t_N and s_N are bounded by*

$$\frac{24}{259} \approx 0.09266 \leq t_N \leq \frac{1168}{9583} \approx 0.1218 \quad \text{and} \quad 0 \leq s_N \leq \frac{4}{37} \approx 0.1081 \ .$$

The previous lemma gives the following bounds: $12/259 \leq p_1 \leq 584/9583$, $69/259 \leq p_2 \leq 3099/9583$, $108/259 \leq p_3 \leq 4220/9583$, and $240/1369 \leq p_4 \leq 10/37$. The next lemma gives the relation between the current tree collection and a larger tree collection that is used for the final results.

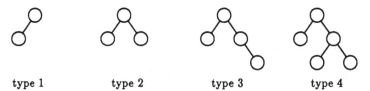

type 1   type 2   type 3   type 4

Figure 10: Tree collection of AVL trees with more than 2 and less than 7 leaves
(leaves not shown)

Lemma 5.4 *The probabilities, q_i, of the tree collection showed in Figure 10 are*

$$q_1 = p_2, \quad q_2 = p_3 - 2p_1, \quad q_3 = p_4, \quad \text{and} \quad q_4 = 3p_1.$$

Proof: Using Lemma 5.1. ∎

The following results are for the collection of Figure 10 using Lemma 5.3 and the bounds on the p_i's.

THEOREM 5.2 *The expected number of rotations in a random AVL tree with N keys during the $(N+1)^{st}$ insertion is bounded by*

$$2\frac{q_1}{L_1} + 2\frac{q_3}{L_3} + 4\frac{q_4}{L_4} + s_N \leq r(N) \leq 2\frac{q_1}{L_1} + 2\frac{q_3}{L_3} + 4\frac{q_4}{L_4} + q_2$$

Corollary 5.3 $\frac{14}{37} \approx 0.3784 \leq r(N) \leq \frac{994}{1369} \approx 0.7261$ *for* $N \geq 20$.

Proof: The above expression can be obtained by observing Figure 10. and using Lemma 3.1. ∎

CONJECTURE 5.1 *The expected number of comparisons in a successful search is upper bounded by*

$$\overline{C}_N \leq \frac{60}{37}H_N + O(1) \approx 1.117\log_2 N \ ,$$

where $H_N = \sum_{i=1}^{N} 1/i$ is the N-th harmonic number.

Proof: For each input of N elements (permutation) we model the tree process as building a random binary search tree (BST) and then applying a transformation to the corresponding AVL tree. This transformation is based in each rotation performed in the AVL tree. Each rotation in the fringe decreases the internal path length in at least one (in fact, only the rotation in the case (a) of Figure 8 the path lenght is decreased by two). We assume that the efect of all rotations above the fringe does not increase the internal path length (this is not true in general for a single rotation). This seems to be true, but we wre not able to prove it. If a fringe rotation was done in the i-th insertion, that rotation decreases the internal path length of i elements. Then

$$C_N^{AVL} \leq C_N^{BST} - \sum_{i=1}^{N-1} \frac{Rot_i}{i}$$

where Rot_i is 0 or 1 depending if for that input there it was a fringe rotation in the i-th insertion. Taking the expected value over all inputs, we have

$$\overline{C}_N^{AVL} \leq \overline{C}_N^{BST} - \sum_{i=1}^{N-1} \frac{r(i)}{i} \ .$$

It is known that $\overline{C}_N$ for a random binary search tree is $2H_N + O(1)$. 1. In the i-th insertion, a fringe rotation happens with probability 14/37 (previous theorem). Thus, $\overline{C}_N \leq 2H_N - \sum_{i=N_1}^{N-1} \frac{14}{37}\frac{1}{i} + O(1)$ which gives the desired result. This conjecture also holds for unsuccessful searches (C'_N), by using the relation $C_N = (1 + 1/N)C'_N - 1$. ∎

Lemma 5.5 *The expected number of nodes in the fringe of an AVL tree with N keys corresponding to the tree collection of Figure 10 is $\frac{7159}{9583} \approx 0.7471(N+1) \leq \overline{f}(N) \leq \frac{195}{259} \approx 0.7529(N+1)$ for $N \geq 20$.*

Proof: The above expression can be obtained by observing Figure 10 and by using Lemma 3.1. ∎

From the previous lemma, and taking a balanced tree and the tallest tree outside the fringe, and adding the average height of the fringe, we obtain the following bounds

$$\log_2(N) + 0.3413 \leq \overline{h}(N) \leq 1.44\log_2 N - 0.4464 \ .$$

Lemma 5.6 *The expected number of unbalanced nodes outside the fringe defined by the current tree collection, of a random AVL tree with N keys, is at least $\frac{s_N}{4}(N+1)$.*

Proof: The above expression is obtained as follows: a tree associated with s_N always have one unbalanced node (the root) outside the fringe. In the worst case all these trees belongs to case (c) of Fig. 8. ∎

Lemma 5.7 *The expected number of balanced nodes outside the fringe defined by the current tree collection, of a random AVL trees with N keys, is at least*

$$\left(\frac{q_1}{L_1} + \frac{q_2}{L_2} - \frac{q_3}{L_3} - \frac{q_4}{L_4}\right)\frac{(N+1)}{3}.$$

Proof: Type 1 or 2 subtrees may be paired with type 3 or type 4 subtrees. In that case we have $A_1(N)+A_2(N)$ subtrees of height 3. In the worst case for balanced nodes, the $A_3(N)+A_4(N)-A_1(N)$ $A_2(N)$ remaining subtrees of height 2 are paired with the subtrees of height 3, i.e. $A_1(N)+A_2(N)$ pairs. Any 3 of the others $(A_3(N) + A_4(N) - A_1(N) - A_2(N))$ must generate at least one balanced node (see Figure 11). ∎

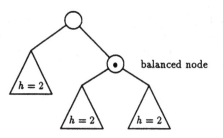

Figure 11: Balanced nodes outside the fringe

THEOREM 5.3 *The expected number of balanced nodes in a random AVL tree with N keys is bounded by*

$$\left(\frac{4q_1}{3L_1} + \frac{10q_2}{3L_2} + \frac{5q_3}{3L_3} + \frac{11q_4}{3L_4}\right)(N+1) \le \bar{b}(N) \le N - \left(\frac{q_1}{L_1} + 2\frac{q_3}{L_3} + \frac{q_4}{L_4} + \frac{s_N}{4}\right)(N+1)$$

Corollary 5.4 $\frac{146}{259} \approx 0.5637 \le \frac{\bar{b}(N)}{N} \le \frac{202}{259} \approx 0.7799$ *for $N \ge 20$.*

Proof: The left hand side of the above expression is obtained by observing Figure 10, Lemma 5.7 and by using Eq. 3. The right hand side is obtained by using Lemmas 3.3, 5.5, and 5.6. ∎

THEOREM 5.4 *The expected number of balance changes during N random insertions in an initial empty AVL tree is bounded by*

$$\frac{414}{259}N + o(N) \approx 1.5985N \le \bar{c}(N) \le \frac{61718}{28479} + o(N) \approx 2.1468N$$

for $N \ge 20$.

Proof: Similar to Theorem 3.1. ∎

Experimental results show that $r(N) \approx 0.47$ [ZT82], $\bar{b}(N) \approx 0.68N$ [Knu73], and $\bar{c}(N) \approx 1.78N$ [MT86].

Table 3 shows simulation results for larger trees for t_N and s_N, obtained with a 95% confidence interval. For example, from Table 3 the value of $s2_N$ seems to converge to $12/2500$ when N is large, but we are not able to prove it. Moreover $s2_N$ may oscillate smoothly, in such a way that simulations cannot detect. (e.g. consider $s2_N = 12/2500 + \cos(\ln N)/10000$.)

Tree Size	Number of Trees	t_N	s_N (percent)		
			Total	$s1_N$	$s2_N$
20	50000	0.0968 ± 0.0011	1.367 ± 0.036	0.789 ± 0.023	0.00578 ± 0.00029
50	10000	0.0964 ± 0.0016	1.062 ± 0.047	0.565 ± 0.028	0.496 ± 0.038
100	5000	0.0961 ± 0.0016	1.094 ± 0.048	0.605 ± 0.030	0.490 ± 0.039
500	5000	0.09524 ± 0.00070	1.110 ± 0.022	0.614 ± 0.013	0.496 ± 0.018
1000	2000	0.09559 ± 0.00078	1.090 ± 0.023	0.593 ± 0.015	0.497 ± 0.019
2500	2000	0.09611 ± 0.00048	1.086 ± 0.015	0.6114 ± 0.0096	0.474 ± 0.012
5000	1000	0.09603 ± 0.00048	1.097 ± 0.015	0.6029 ± 0.0095	0.494 ± 0.012
10000	1000	0.09565 ± 0.00035	1.091 ± 0.010	0.6050 ± 0.0066	0.4859 ± 0.0085
15000	500	0.09582 ± 0.00043	1.098 ± 0.013	0.6078 ± 0.0077	0.490 ± 0.010
20000	500	0.09550 ± 0.00037	1.101 ± 0.011	0.6019 ± 0.0064	0.4986 ± 0.0087

Table 3: Simulation results for t_N and s_N

Lemma 5.8 *If* $\lim_{N\to\infty} t_N = t$ *and* $\lim_{N\to\infty} s_N = s$ *exist, then* $t = \frac{24}{259} + \frac{10}{37}s$.

Proof: Replacing s_{N-1} by s in Eq. 9, solving it, and using Lemma 5.1 and Theorem 2.2. ∎

We conjecture that t_N and s_N converges. In fact, the simulation results agrees with Lemma 5.8. For example, if we assume that s_N for large N is approximately 0.011, then using Lemma 5.8 the value for t_N is 0.0956. The simulation value is 0.0955.

6 Conclusions

We have formalized the concept of weakly-closed and ambiguous tree collections, in relation to AVL-trees. We present a new closed AVL tree collection that allows to obtain almost all the results obtained by Mehlhorn using a weakly-closed collection [Meh79]. In Section 4 we model weakly-closed tree collections, giving the exact solution to the new fringe analysis recurrence in Section 5. Our last tree collection improves all previous known results for the expected case of AVL-trees. By using this exact solution it is possible to analyze larger tree collections. In fact, a tree collection with subtrees from height 2 to 4 includes 10 types and at least 8 unknown probabilities.

Like AVL trees, weight-balanced trees are balanced by single and double rotations [Knu73, Sec. 6.2.3]. For this reason only small tree collections of weight-balanced trees are closed. For large tree collections we find the same type of difficulties showed for AVL trees. Consequently, the technique presented for the analysis of AVL trees is also suitable for the analysis of weight-balanced trees.

References

[AVL62] G.M. Adel'son-Vel'skii and E.M. Landis. An algorithm for the organization of information. *Dokladi Akademia Nauk SSSR*, 146(2):263–266, 1962. English translation in Soviet Math. Doklay 3 ,1962, 1259-1263.

[Bro79] M.R. Brown. A partial analysis of random height-balanced trees. *SIAM J on Computing*, 8(1):33–41, Feb 1979.

[BYG89] R.A. Baeza-Yates and G.H. Gonnet. Solving matrix recurrences with applications. Technical Report CS-89-16, Department of Computer Science, University of Waterloo, May 1989.

[EZG+82] B. Eisenbarth, N. Ziviani, Gaston H. Gonnet, Kurt Mehlhorn, and Derick Wood. The theory of fringe analysis and its application to 2-3 trees and B-trees. *Information and Control*, 55(1):125–174, Oct 1982.

[Knu69] D.E. Knuth. *The Art of Computer Programming: Fundamental Algorithms*, volume 1. Addison-Wesley, Reading, Mass., 1969.

[Knu73] D.E. Knuth. *The Art of Computer Programming: Sorting and Searching*, volume 3. Addison-Wesley, Reading, Mass., 1973.

[Meh79] K. Mehlhorn. A partial analysis of height-balanced trees. Technical Report Report A 79/13, Universitat de Saarlandes, Saarbrucken, West Germany, 1979.

[Meh82] Kurt Mehlhorn. A partial analysis of height-balanced trees under random insertions and deletions. *SIAM J on Computing*, 11(4):748–760, Nov 1982.

[MT86] Kurt Mehlhorn and A. Tsakalidis. An amortized analysis of insertions into AVL-trees. *SIAM J on Computing*, 15(1):22–33, Feb 1986.

[OS76] Th. Ottmann and H.W. Six. Eine neue klasse von ausgeglichenen binarbaumen. *Augewandte Informartik*, 9:395–400, 1976.

[OW80] Thomas Ottmann and Derick Wood. 1-2 brother trees or AVL trees revisited. *Computer Journal*, 23(3):248–255, Aug 1980.

[Yao78] A.C-C. Yao. On random 2-3 trees. *Acta Informatica*, 9(2):159–170, 1978.

[Ziv82] N. Ziviani. *The Fringe Analysis of Search Trees*. PhD thesis, Department of Computer Science, University of Waterloo, 1982.

[ZT82] N. Ziviani and F.W. Tompa. A look at symmetric binary B-trees. *Infor*, 20(2):65–81, May 1982.

Analysis of the Expected Search Cost in Skip Lists*

Thomas Papadakis   J. Ian Munro
Department of Computer Science
University of Waterloo
Waterloo, Ont N2L 3G1
Canada

Patricio V. Poblete
Departamento de Ciencias de la Computación
Universidad de Chile
Casilla 2777
Santiago, Chile

Abstract

Skip lists, introduced by W. Pugh, provide an alternative to search trees. The exact value of the expected search cost is derived (in terms of previously studied functions), as well as an asymptotic expression for it. The latter suggests that Pugh's upper bound is fairly tight for the interesting cases.

1   Introduction

The skip list, recently introduced by W. Pugh [5], is an interesting and practical alternative to search trees. The approach is to store each of its n $(n \geq 1)$ elements in one or more of a set of sorted linear linked lists. All elements are stored in sorted order on a linked list denoted as level 1, and each element on the list at level i $(i = 1, 2, \ldots)$ is included with (independent) probability p in the list at level $i + 1$. A header contains the references to the first element in each list (see Figure 1). The height of the data structure, that is, the number of linked lists, is also stored.

A search for an element begins at the header of the highest numbered list. This list is scanned, until it is observed that its next element is greater than or equal to the one sought (or the reference is null). At that point, the search continues one level below until it terminates at level 1 (see the search for the 6th element in Figure 1). We have adopted

---

*This research was supported in part by the Natural Science and Engineering Research Council of Canada under grant No. A-8237, the Information Technology Research Centre of Ontario, and FONDECYT (Chile) under grant 90-1263.

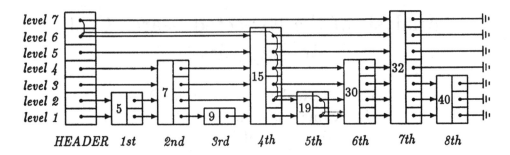

Figure 1: A skip list of 8 elements, and the search path when searching for the 6th element

the convention that an equality test is done only at level 1 as the last comparison. This is the usual choice in a standard binary search and avoids two tests (or a three way branch) at each step.

The search cost is defined as the number of pointer inspections excluding the last one for the equality test. For the search in Figure 1, this is 9.[1] Of principal interest is $C_p(m,n)$ $(m = 1, 2, \ldots, n + 1)$, which, assuming that the 0th element is $-\infty$ and the $(n+1)$st element is $+\infty$, is defined as the average search cost when performing a successful search for the mth element, or an unsuccessful search for an element between the $(m-1)$st and the mth in a list of n elements. This average is taken over all possible skip lists, for fixed m and n.

It is worthy of note at this early stage that the expected search cost will depend on the relation between the element sought and the other elements in the skip list. The expected cost of a search for the element that turns out to be the mth smallest, will be an increasing function of m. In particular, searching for the smallest element, or one less than the smallest, will have cost equal to the height of the structure, while searching for other elements will have the additional cost of following links horizontally. It is clear that the height of a skip list is a lower bound on the (expected) search cost for any element. The spirit of the structure is that p be a probability of moderate size, say $\frac{1}{2}$ or $\frac{1}{3}$.

Insertions and deletions (as defined by Pugh) are very straightforward. A new element is inserted where a search for it terminated at level 1. As it is put in list i $(i = 1, 2, \ldots)$, it is inserted, with probability p $(0 < p < 1)$, where its search terminated at level $i + 1$. This continues until, with probability $q = 1 - p$, the choice is not to insert. The counter for the height of the data structure is increased, if necessary. Clearly, this suggests the notion of nodes with i pointer fields and a data field. Deletions are completely analogous to insertions. An element to be deleted is removed from the lists in which it is found. The height of the data structure is updated by scanning the header's pointers by decreasing level until a non-null pointer is found.

At this point we can make an observation on the adopted "scan the list until the next list element is greater than or equal to the one sought" criterion to drop down a level during the search. Had it been simply "greater than the one sought", then the search path for the mth element in this case would have been identical to the search path for

---

[1]Note that our definition of the search cost is 1 greater than Pugh's definition. Our references to his results are translated into our terms.

the $(m+1)$st element in our case. This implies that the expected search cost in this case would have been higher. Moreover, our approach makes the deletion a simple extension of the search.

An immediate property of skip lists is that the deletion of an element produces the structure that would have existed had it never been inserted. This infers, through Pugh's upper bound on the expected search cost, that skip lists maintain the expectation of logarithmic search and update cost, even after a long sequence of updates. This is in sharp contrast with the results such as those of Culberson and Munro [2, 3], indicating that the usual update algorithms in binary search trees lead to degeneration in behaviour, although the notions of Aragon and Seidel [1] show how to maintain the "history independent" property in (nearly) standard binary search trees.

A skip list, as defined above, can have an unbounded number of levels. It is "fairly obvious" though, that it should be expected to have about $\log_{\frac{1}{p}} n$ levels. Difficulties related to the difference between this expected height and the actual height caused Pugh to propose [5, initial version] cutting the structure off at level $\lceil \log_{\frac{1}{p}} n \rceil$. The *upper bound* on the cost for a search of an element greater than the last that he gave for this data structure, was later [5] extended for an unboundly high skip list. Finally, he showed [6] that in general the expected cost for a search of the mth element (or for an unsuccessful search for an element between the $(m-1)$st and the mth) is *bounded above* by

$$\log_{\frac{1}{p}} n + \frac{q}{p} \log_{\frac{1}{p}} (m-1) + \frac{1}{q} + 1. \tag{1}$$

In this paper, we give an analysis of the expected search cost *itself* for an unboundly high skip list. This analysis yields the expected search cost

$$\log_{\frac{1}{p}} n + \frac{q}{p} \log_{\frac{1}{p}} (m-1) + \frac{1}{p}\left(\frac{\gamma}{\ln \frac{1}{p}} - \frac{1}{2}\right) + 1 + w\left(\frac{1}{p}, n\right) + \frac{q}{p} w\left(\frac{1}{p}, m-1\right) + O\left(\frac{1}{m}\right),$$

where the term $w\left(\frac{1}{p}, n\right) + \frac{q}{p} w\left(\frac{1}{p}, m-1\right)$ not only can be computed for any p, m and n, but its absolute value is bounded by a very small function of p. In particular, it is essentially a periodic function in m and n of absolute value at most $.31 \times 10^{-5}$ for $p = \frac{1}{2}$.

Throughout the paper, n, m, p, q and $C_p(m,n)$ are used only as described above, l is used for either n or $m-1$, and x is used for $\frac{1}{p}$.

2 Expected Search Cost: Exact Value

In order to compute the expected search cost $C_p(m,n)$, we divide the steps on the search path into vertical steps (that is, one less than the height of the skip list) and horizontal steps, and compute the expected value of each separately.

To this end, define the random variables H and L as

$$H = \text{height of a skip list of } n \text{ elements}$$

and

$$L = \text{number of full horizontal steps on the path followed from}$$
$$\text{the header to the } (m-1)\text{st element.}$$

We use the term *full* horizontal steps to emphasize actually moving to the next node in the list, rather than simply inspecting it and dropping down a level. H will be a positive integer, while L will be nonnegative. For example, for the skip list of Figure 1 we have $n = 8$, $m = 6$, $H = 7$ and $L = 2$, while for any skip list $m = 1$ implies $L = 0$ and $m = 2$ implies $L = 1$.

Clearly, the expected search cost is given by

$$C_p(m, n) = E(H) + E(L), \qquad (2)$$

and expressions for $E(H)$ and $E(L)$ are given in the next two lemmata.

Lemma 1 *The expected height of a skip list of n elements is*

$$A_n \overset{\text{def}}{=} E(H) = \sum_{k=1}^{\infty} k \left((1 - p^k)^n - (1 - p^{k-1})^n \right). \qquad (3)$$

Proof: Straightforward from

$$E(H) = \sum_{k=1}^{\infty} k \times P[\text{height of a skip list is } k]. \qquad \blacksquare$$

Lemma 2 *The expected number of full horizontal steps on the search path followed from the header to the $(m-1)$st element $(m = 1, 2, \ldots, n+1)$ is*

$$B_{m-1} \overset{\text{def}}{=} E(L) = \begin{cases} 1 + \dfrac{q}{p} \displaystyle\sum_{i=1}^{m-2} \sum_{k=1}^{\infty} p^k (1 - p^k)^i & \text{if } m = 2, 3, \ldots, n+1 \\ 0 & \text{if } m = 1 \end{cases} \qquad (4)$$

(where, by convention, $B_1 = 1$).

Proof: Following the search path backwards, consider the $m - 2$ indicator random variables:

$$I_i = \begin{cases} 1 & \text{if } i\text{th element is on path from header to } (m-1)\text{st element} \\ 0 & \text{otherwise} \end{cases} \qquad i = 1, \ldots, m-2.$$

(For example, for the search path of Figure 1, we have $I_1 = I_2 = I_3 = 0$ and $I_4 = 1$). Clearly,

$$L = \begin{cases} 1 + \displaystyle\sum_{i=1}^{m-2} I_i & \text{if } m = 2, 3, \ldots, n+1 \\ 0 & \text{if } m = 1 \end{cases}. \qquad (5)$$

If H_i is a random variable denoting the height of the ith element $(i = 1, 2, \ldots, n)$, then

$$P[I_i = 1] = P[H_i \geq H_{i+1}, \ldots, H_{m-1}], \qquad i = 1, 2, \ldots, m-2,$$

or, since H_i are independent,

$$P[I_i = 1] = q \sum_{k=1}^{\infty} p^{k-1} (1 - p^k)^{m-i-1}, \qquad i = 1, 2, \ldots, m-2.$$

The last, together with (5) and the fact that $E(I_i) = P[I_i = 1]$, completes the proof. ∎

Having computed $C_p(m, n)$, we now derive an expression for it, in terms of the previously studied function [4, section 5.2.2, ex. 50]

$$U_{x,l} \stackrel{\text{def}}{=} \sum_{j=0}^{l-2} \binom{l}{j+2} (-1)^j \frac{1}{x^{j+1} - 1}, \qquad \text{for } x \in \mathbb{R}, \quad x > 1, \quad \text{and} \quad l = 0, 1, 2, \ldots, \qquad (6)$$

where, by convention, $U_{x,0} = U_{x,1} = 0$. The key result (although not the most technically difficult to prove) towards this end, is next lemma, which relates the "expected number of vertical and horizontal steps".

Lemma 3
$$A_l - \frac{p}{q}(B_l - 1) = \frac{1}{q}, \qquad l = 1, 2, \ldots.$$

Proof: It is easy to verify that B_l can be rewritten as

$$B_l = 1 + \frac{q}{p} \sum_{k=1}^{\infty} \left\{ (1 - p^k) - (1 - p^k)^l \right\}, \qquad l = 1, 2 \ldots,$$

which implies that it suffices to prove

$$\sum_{k=1}^{\infty} k \left\{ (1 - p^k)^l - (1 - p^{k-1})^l \right\} - \sum_{k=1}^{\infty} \left\{ (1 - p^k) - (1 - p^k)^l \right\} = \frac{1}{1 - p}, \qquad l = 1, 2 \ldots.$$

But this is equivalent to

$$\lim_{a \to \infty} \left(\sum_{k=1}^{a} \left\{ (k+1)(1 - p^k)^l - k(1 - p^{k-1})^l \right\} - \sum_{k=1}^{a} 1 + \sum_{k=1}^{a} p^k \right) = \frac{1}{1 - p}, \qquad l = 1, 2, \ldots,$$

and, by applying the telescopic property and doing some algebra, the last is equivalent to

$$\lim_{a \to \infty} \left((a+1)(1 - p^a)^l - a + \frac{1 - p^{a+1}}{1 - p} - 1 \right) = \frac{1}{1 - p}, \qquad l = 1, 2, \ldots.$$

which can be shown to be true, since $p < 1$. ∎

It now suffices to express either $E(H)$ or $E(L)$ in terms of $U_{\frac{1}{p}, n}$ or $U_{\frac{1}{p}, m}$. We choose the latter.

Lemma 4
$$B_l = \frac{q}{p} \left(U_{\frac{1}{p}, l+1} - U_{\frac{1}{p}, l} \right), \qquad l = 0, 1, 2, \ldots.$$

Proof: Consider the inner sum of the double sum of B_l, that is

$$D_i \stackrel{\text{def}}{=} \sum_{k=1}^{\infty} p^k (1 - p^k)^i, \qquad i = 1, 2, \ldots.$$

Expanding the $(1 - p^k)^i$ term and interchanging the two summations, we get

$$D_i = \sum_{j=0}^{i} \binom{i}{j} (-1)^j \frac{p^{j+1}}{1 - p^{j+1}}, \qquad i = 1, 2, \ldots.$$

Well-known properties of the binomial coefficients allow us now to rewrite this as

$$
D_i = \sum_{j=0}^{i-2}\left\{\binom{i}{j+2}(-1)^j \frac{1}{\left(\frac{1}{p}\right)^{j+1}-1}\right\} - 2\sum_{j=0}^{i-1}\left\{\binom{i+1}{j+2}(-1)^j \frac{1}{\left(\frac{1}{p}\right)^{j+1}-1}\right\}
$$
$$
+ \sum_{j=0}^{i}\left\{\binom{i+2}{j+2}(-1)^j \frac{1}{\left(\frac{1}{p}\right)^{j+1}-1}\right\}, \qquad i = 1, 2, \ldots,
$$

which implies

$$
D_i = U_{\frac{1}{p},i} - 2U_{\frac{1}{p},i+1} + U_{\frac{1}{p},i+2}, \qquad i = 1, 2, \ldots.
$$

Therefore, B_l can be rewritten as

$$
B_l = \begin{cases} 1 + \dfrac{q}{p}\left(\displaystyle\sum_{i=1}^{l-1}\left(U_{\frac{1}{p},i} - U_{\frac{1}{p},i+1}\right) + \sum_{i=1}^{l-1}\left(U_{\frac{1}{p},i+2} - U_{\frac{1}{p},i+1}\right)\right) & \text{if } l = 1, 2, \ldots \\ 0. & \text{if } l = 0 \end{cases}
$$

Application of the telescopic property, together with the fact that $U_{\frac{1}{p},1} = 0$ and $U_{\frac{1}{p},2} = \frac{p}{q}$, completes the proof. ∎

We can now state the first of our two main results.

Theorem 1 *The expected search cost of a successful search for the mth element, or an unsuccessful search for one between the $(m-1)$st and the mth, in a skip list of n elements, is*

$$
C_p(m, n) = U_{\frac{1}{p},n+1} - U_{\frac{1}{p},n} + \frac{q}{p}\left(U_{\frac{1}{p},m} - U_{\frac{1}{p},m-1}\right) + 1.
$$

Proof: Lemmata 3 and 4 imply that

$$
A_n = E(H) = U_{\frac{1}{p},n+1} - U_{\frac{1}{p},n} + 1, \qquad n = 1, 2, \ldots.
$$

Since we also have

$$
B_{m-1} = E(L) = \frac{q}{p}\left(U_{\frac{1}{p},m} - U_{\frac{1}{p},m-1}\right), \qquad m = 1, 2, \ldots, n+1
$$

from Lemma 4, the theorem follows immediately from (2). ∎

3 Expected Search Cost: Asymptotic Value

Theorem 1 gives the expected search cost in terms of a function that "someone" has studied. In this section, we provide an approximation to the expected search cost in terms of functions that "everyone" has studied.

In Theorem 2 we give an expression within $O(\frac{1}{m})$ of the exact expected search cost, in terms of well-known constants, logarithms and a small term involving an infinite sum of the real part of rotations of the complex Gamma function. This small term is unpleasant, but definitely *can* be computed. Moreover, this small term is a function oscillating around

0, and it takes small values for reasonable values of p. Its range is tightly (for interesting values of p) bounded by an easily computable function of p (i.e. it is independent of m and n), as we shall see in (17). This leads to the observation that the expected search cost can be expressed as falling within a small constant of a very simple form.

Our first requirement is an expression for the asymptotic value of $U_{x,l}$. Knuth has shown [4, section 5.2.2, ex. 50] that

$$U_{x,l} = l\log_x l + l\left(\frac{\gamma-1}{\ln x} - \frac{1}{2} + f_{-1}(x,l)\right) + \frac{x}{x-1} - \frac{1}{2\ln x} - \frac{1}{2}f_1(x,l) + O\left(\frac{1}{l}\right), \quad x \in \mathbb{R}, \ x > 1,$$

$$(7)$$

where

$$f_s(x,l) \stackrel{\text{def}}{=} \frac{2}{\ln x} \sum_{k=1}^{\infty} \Re\left(\Gamma\left(s - i\frac{2\pi k}{\ln x}\right) e^{i\,2\pi k\log_x l}\right), \quad \text{for } s = \pm 1. \tag{8}$$

Both $f_1(x,l)$ and $f_{-1}(x,l)$ are absolutely convergent series. This is a consequence of the fact that $|e^{i2\pi ky}| = 1$ for all $y \in \mathbb{R}$, and of the properties

$$|\Gamma(1+iy)|^2 = \frac{y\pi}{\sinh(\pi y)} \quad \text{for all } y \in \mathbb{R} \tag{9}$$

and

$$|\Gamma(-1+iy)|^2 = \frac{\pi}{y(1+y^2)\sinh(\pi y)} \quad \text{for all } y \in \mathbb{R} \tag{10}$$

of the Gamma function.

In order to proceed, we require a bound on the range of the difference of two consecutive (with respect to l) terms of $f_1(x,l)$, and a bound of the range of the function

$$w(x,l) \stackrel{\text{def}}{=} (l+1)\,f_{-1}(x,l+1) - l\,f_{-1}(x,l). \tag{11}$$

These bounds are presented in the next two lemmata, and the well-known relation

$$\log_x(l+1) - \log_x l = \frac{1}{(l+1)\ln x} + O\left(\frac{1}{l^2}\right) \tag{12}$$

will be used in their proofs.

Lemma 5 $\qquad |f_1(x,l+1) - f_1(x,l)| = O\left(\frac{1}{l}\right).$

Proof: Since the series $f_1(x,l)$ converges absolutely for all l, its definition implies

$$|f_1(x,l+1) - f_1(x,l)| \le \frac{2}{\ln x} \sum_{k=1}^{\infty} \left|\Re\left(\Gamma\left(1 - i\frac{2\pi k}{\ln x}\right)\left(e^{i\,2\pi k\log_x(l+1)} - e^{i\,2\pi k\log_x l}\right)\right)\right|.$$

Since now $\Re(z) \le |z|$ for all $z \in \mathbb{C}$, and $|z_1 - z_2| \le |\arg(z_1) - \arg(z_2)|$ for all z_1, z_2 on the unit circle, the last inequality implies

$$|f_1(x,l+1) - f_1(x,l)| \le \frac{2}{\ln x} \sum_{k=1}^{\infty} \left\{\left|\Gamma\left(1 - i\frac{2\pi k}{\ln x}\right)\right| |2\pi k\log_x(l+1) - 2\pi k\log_x l|\right\},$$

and so, using (12),

$$|f_1(x,l+1) - f_1(x,l)| \leq \frac{2}{\ln x} \sum_{k=1}^{\infty} \left\{ \left| \Gamma\left(1 - i\frac{2\pi k}{\ln x}\right) \right| \left(\frac{2\pi k}{(l+1)\ln x} + O\left(\frac{1}{l^2}\right) \right) \right\}.$$

Application now of (9) yields

$$|f_1(x,l+1) - f_1(x,l)| \leq \frac{1}{l+1} \left(\frac{2}{\ln x}\right)^{5/2} \pi^2 \sum_{k=1}^{\infty} \left(\frac{k^3}{\sinh\left(\frac{2\pi^2 k}{\ln x}\right)}\right)^{1/2} + O\left(\frac{1}{l^2}\right),$$

which completes the proof. ■

Lemma 6 $\quad |w(x,l)| \leq W(x) + O\left(\dfrac{1}{\sqrt{l}}\right),$

where

$$W(x) \stackrel{\text{def}}{=} \sqrt{\frac{2}{\ln x}} \sum_{k=1}^{\infty} \left(\frac{1}{k \sinh\left(\frac{2\pi^2 k}{\ln x}\right)}\right)^{1/2}. \tag{13}$$

Proof: Reasoning as in the proof of Lemma 5, it is easy to see that

$$|w(x,l)| \leq \frac{2}{\ln x} \sum_{k=1}^{\infty} \left\{ \left| \Gamma\left(-1 - i\frac{2\pi k}{\ln x}\right) \right| T \right\}, \tag{14}$$

where

$$T \stackrel{\text{def}}{=} \left| (l+1)e^{i2\pi k \log_x(l+1)} - l e^{i2\pi k \log_x l} \right|.$$

The well-known complex numbers identities $e^{iy} = \cos y + i \sin y$ and $|y+iz| = \sqrt{y^2 + z^2}$ and the trigonometric identities $\cos^2 y + \sin^2 y = 1$ and $\cos(y-z) = \cos y \cos z + \sin y \sin z$, for all $y, z \in \mathbb{R}$, allow us to rewrite the last as

$$T = \sqrt{(l+1)^2 + l^2 - 2l(l+1) \cos\{2\pi k(\log_x(l+1) - \log_x l)\}},$$

which, since $\cos y \geq 1 - \frac{y^2}{2}$ for all $y \in \mathbb{R}$, implies

$$T \leq \sqrt{1 + l(l+1)4\pi^2 k^2 \{\log_x(l+1) - \log_x l\}^2},$$

or, according to (12),

$$T \leq \sqrt{1 + \frac{4k^2\pi^2}{\ln^2 x}} + O\left(\frac{1}{\sqrt{l}}\right). \tag{15}$$

The lemma now follows from (14), (10) and (15). ■

We are now ready to state the second of our main results.

Theorem 2 *The expected search cost of a successful search for the mth element, or an unsuccessful search for one between the $(m-1)$st and the mth, in a skip list of n elements, is*

$$C_p(m,n) = \log_{\frac{1}{p}} n + \frac{q}{p} \log_{\frac{1}{p}}(m-1) + \frac{1}{p}\left(\frac{\gamma}{\ln\frac{1}{p}} - \frac{1}{2}\right) + 1$$

$$+ w\left(\frac{1}{p}, n\right) + \frac{q}{p} w\left(\frac{1}{p}, m-1\right) + O\left(\frac{1}{m}\right), \tag{16}$$

and moreover

$$\left| w\left(\frac{1}{p}, n\right) + \frac{q}{p}\, w\left(\frac{1}{p}, m-1\right) \right| \leq \frac{1}{p}\, W\left(\frac{1}{p}\right) + O\left(\frac{1}{\sqrt{m}}\right). \tag{17}$$

Proof: According to Theorem 1, it suffices to find the asymptotic value of $U_{x,l+1} - U_{x,l}$. Since (12) implies

$$(l+1)\log_x(l+1) - l\log_x l \;=\; \log_x l + \frac{1}{\ln x} + O\left(\frac{1}{l}\right),$$

the asymptotic expression for $U_{x,l}$ in (7) yields

$$\begin{aligned}
U_{x,l+1} - U_{x,l} \;=\;& \log_x l + \frac{\gamma}{\ln x} - \frac{1}{2} \\
&+ (l+1)\,f_{-1}(x, l+1) - l\,f_{-1}(x,l) - \frac{1}{2}\{f_1(x, l+1) - f_1(x, l)\} \\
&+ O\left(\frac{1}{l}\right).
\end{aligned}$$

According to Lemma 5, this can be rewritten as

$$\begin{aligned}
U_{x,l+1} - U_{x,l} \;=\;& \log_x l + \frac{\gamma}{\ln x} - \frac{1}{2} \\
&+ (l+1)\,f_{-1}(x, l+1) - l\,f_{-1}(x, l) + O\left(\frac{1}{l}\right),
\end{aligned}$$

which proves (16). (17) follows from Lemma 6. ∎

The $w\left(\frac{1}{p}, n\right) + \frac{q}{p}\, w\left(\frac{1}{p}, m-1\right)$ term in the above theorem, although not very nice, can be computed to an arbitrary accuracy for any value of p, m and n (from the definitions of w and f in (11) and (8)). From Table 1 we observe that for reasonable values of p (i.e. between .25 and .75) its range is rather small, and the bound given is very tight. The computation of $W\left(\frac{1}{p}\right)$ from (13) is *very fast*, since this series converges very rapidly. On the other hand, the computation of the given $\liminf_{l\to\infty}\left\{w\left(\frac{1}{p}, l\right)\right\}$ and $\limsup_{l\to\infty}\left\{w\left(\frac{1}{p}, l\right)\right\}$ was very time consuming. It was based on the fact that if a maximum/minimum of $w(x, l)$ occurs at $l = l_0$, then the next maximum/minimum will occur around $x l_0$.

A quick glance at the top half of the table shows the terms $\liminf_{l\to\infty}\left\{w\left(\frac{1}{p}, l\right)\right\}$ and $\limsup_{l\to\infty}\left\{w\left(\frac{1}{p}, l\right)\right\}$ to be equal in absolute value up to the precision computed. This, together with the fact that $w\left(\frac{1}{p}, l\right)$ is in the limit multiplicatively periodic in l, gives rise to the hypothesis that $w\left(\frac{1}{p}, l\right)$ is sinosoidal in $\log l$. Indeed, this may be a rather good approximation, but hopes for such a simple form are dashed by the values in the latter portion of the table.

4   Concluding Remarks

The main results of this paper are the computation of the exact value for the expected search cost in terms of previously studied functions (Theorem 1), and an asymptotic

p	$W\left(\frac{1}{p}\right)$	$\liminf\limits_{l\to\infty}\left\{w\left(\frac{1}{p},l\right)\right\}$	$\limsup\limits_{l\to\infty}\left\{w\left(\frac{1}{p},l\right)\right\}$
$1-10^{-6}=.999999$	$.5178767957\times10^{-4,286,309}$	$-.5178767957\times10^{-4,286,309}$	$.5178767957\times10^{-4,286,309}$
$1-10^{-5}=.999990$	$.2959332278\times10^{-428,626}$	$-.2959332278\times10^{-428,626}$	$.2959332278\times10^{-428,626}$
$1-10^{-4}=.999900$	$.1981128348\times10^{-42,858}$	$-.1981128348\times10^{-42,858}$	$.1981128348\times10^{-42,858}$
$1-10^{-3}=.999000$	$.4262909886\times10^{-4,282}$	$-.4262909886\times10^{-4,282}$	$.4262909886\times10^{-4,282}$
$1-10^{-2}=.990000$	$.6534524334\times10^{-425}$	$-.6534524334\times10^{-425}$	$.6534524334\times10^{-425}$
$9/10=.900000$	$.1280348461\times10^{-39}$	$-.1280348461\times10^{-39}$	$.1280348461\times10^{-39}$
$8/9=.888889$	$.2365238603\times10^{-35}$	$-.2365238603\times10^{-35}$	$.2365238603\times10^{-35}$
$7/8=.875000$	$.4350709920\times10^{-31}$	$-.4350709920\times10^{-31}$	$.4350709920\times10^{-31}$
$6/7=.857143$	$.7962533306\times10^{-27}$	$-.7962533306\times10^{-27}$	$.7962533306\times10^{-27}$
$5/6=.833333$	$.1448664720\times10^{-22}$	$-.1448664720\times10^{-22}$	$.1448664720\times10^{-22}$
$4/5=.800000$	$.2617946479\times10^{-18}$	$-.2617946479\times10^{-18}$	$.2617946479\times10^{-18}$
$3/4=.750000$	$.4699899518\times10^{-14}$	$-.4699899518\times10^{-14}$	$.4699899518\times10^{-14}$
$7/10=.700000$	$.3217123526\times10^{-11}$	$-.3217123526\times10^{-11}$	$.3217123526\times10^{-11}$
$2/3=.666667$	$.8427531658\times10^{-10}$	$-.8427531658\times10^{-10}$	$.8427531658\times10^{-10}$
$1-1/e=.632121$	$.1334403619\times10^{-8}$	$-.1334403619\times10^{-8}$	$.1334403618\times10^{-8}$
$3/5=.600000$	$.1137467089\times10^{-7}$	$-.1137467088\times10^{-7}$	$.1137467084\times10^{-7}$
$1/2=.500000$	$.1573158429\times10^{-5}$	$-.1573158209\times10^{-5}$	$.1573157193\times10^{-5}$
$2/5=.400000$	$.00004386535361$	$-.00004386429002$	$.00004386511484$
$1/e=.367879$	$.0001034501560$	$-.0001034426044$	$.0001034501401$
$1/3=.333333$	$.0002393726280$	$-.0002393501022$	$.0002393677843$
$3/10=.300000$	$.0005019537889$	$-.0005018376596$	$.0005018745792$
$1/4=.250000$	$.001375397941$	$-.001375151586$	$.001374070362$
$1/5=.200000$	$.003428651367$	$-.003428631258$	$.003418133199$
$1/6=.166667$	$.006072944165$	$-.006069926234$	$.006041295507$
$1/7=.142858$	$.009029623921$	$-.009012113876$	$.008967892454$
$1/8=.125000$	$.01211860233$	$-.01206734939$	$.01202313730$
$1/9=.111111$	$.01523225636$	$-.01512320708$	$.01510368073$
$1/10=.100000$	$.01830841225$	$-.01811564976$	$.01814992208$
$10^{-2}=.010000$	$.1193488286$	$-.1019027505$	$.1188307423$
$10^{-3}=.001000$	$.2211622651$	$-.1705439068$	$.2107083553$
$10^{-4}=.000100$	$.3037564564$	$-.2193982501$	$.2714772985$
$10^{-5}=.000010$	$.3708513182$	$-.2550544167$	$.3115810658$
$10^{-6}=.000001$	$.4268099774$	$-.2820941271$	$.3395665486$

Table 1: The "bound" $W\left(\frac{1}{p}\right)$ of $\left|w\left(\frac{1}{p},l\right)\right|$ (from Lemma 6), and the actual $\liminf\limits_{l\to\infty}\left\{w\left(\frac{1}{p},l\right)\right\}$ and $\limsup\limits_{l\to\infty}\left\{w\left(\frac{1}{p},l\right)\right\}$

expression for the expected search cost in terms of well-known functions, along with the upper bound on the "wiggling" term in this asymptotic expression (Theorem 2). This upper bound can be computed very quickly and is an excellent estimate of the range of this "wiggling" term for all values of p appropriate for skip lists. It is also worth reiterating that despite the presence of this "wiggling" term, the expected search cost remains a strictly increasing function of m and n.

As a corollary, we can compute the "expected cost" for an insertion or a deletion. The expected cost of an insertion is equal to the cost of a search, plus $\frac{2}{q}$ to update the pointers, since $\frac{1}{q}$ is the average height of an element. The expected cost for a deletion is equal to the cost of a search, plus $\frac{1}{q}$ to update the pointers, plus (see (3)) $1+A_n-A_{n-1}$, to discover any update in the height of the structure.

As another corollary, we can compute the expected search cost counting *only the comparisons between elements*, that is, ignoring null pointer tests. Under this model, the search cost will be reduced by the average number of null pointers encountered on the search path. This reduction will be $A_n - A_{n-m}$.

It is appropriate to ask how close our asymptotic expression, or indeed Pugh's upper bound, is to the exact value of the expected cost, in the range of n for which calculating the exact value is feasible. For simplicity, we consider the case of $m = n + 1$. Adopting the notation $C_p^*(m, n)$ for our approximation to $C_p(m, n)$ and $C_p^+(m, n)$ for Pugh's upper bound, our results from Theorems 1 and 2 become

$$C_p(n+1,n) = \frac{1}{p}\left(U_{\frac{1}{p},n+1} - U_{\frac{1}{p},n}\right) + 1 \tag{18}$$

$$= C_p^*(n+1,n) + O\left(\frac{1}{n}\right),$$

where

$$C_p^*(n+1,n) \stackrel{\text{def}}{=} \frac{1}{p}\log_{\frac{1}{p}} n + \frac{1}{p}\left(\frac{\gamma}{\ln\frac{1}{p}} - \frac{1}{2}\right) + 1 + \frac{1}{p}w\left(\frac{1}{p},n\right), \tag{19}$$

and Pugh's upper bound from (1) becomes

$$C_p(n+1,n) \le C_p^+(n+1,n) \stackrel{\text{def}}{=} \frac{1}{p}\log_{\frac{1}{p}} n + \frac{1}{q} + 1. \tag{20}$$

Values of $C_p(n+1,n)$, $C_p^*(n+1,n)$ and $C_p^+(n+1,n)$, computed for $n = 50, 500, 1000$ and 5000 and for several p, are given in Tables 2 and 3. The computation of $C_p(n+1,n)$ from (18) was *very* time consuming, despite the fact that the faster formula

$$U_{x,n+1} - U_{x,n} = \sum_{j=0}^{\frac{n}{2}-2}\left\{\binom{n}{j+1}(-1)^j\left(\frac{1}{x^{j+1}-1} + \frac{1}{x^{n-j-1}-1}\right)\right\}$$

$$+ \binom{n}{\frac{n}{2}}(-1)^{\frac{n}{2}-1}\frac{1}{x^{\frac{n}{2}}-1} - \frac{1}{x^n-1}, \qquad \text{for even } n$$

was used. Indeed, the computation of the values shown in these tables, which were computed to an accuracy of 10^{-10}, took approximately 70 hours on a MIPS M2000 (a 20 mip machine), with the computation of $C_p(5001,5000)$ alone taking over 65 hours. The computation of $C_p^*(n+1,n)$ from (19) did not create problems, although the computation of $w\left(\frac{1}{p},n\right)$ took longer than the computation of its bound $W\left(\frac{1}{p}\right)$. Finally (and obviously), the computation of $C_p^+(n+1,n)$ from (20) was *extremely* fast.

Comparing our determination of the exact value of the expected search cost and Pugh's upper bound indicates that Pugh's upper bound is a reasonably good approximation for the interesting values of p. Comparing our exact and asymptotic search cost indicates that our asymptotic cost (is not only much faster than the exact cost to compute, but also) approaches the exact cost very rapidly.

Finally, we turn to the issue of choosing p. It seems natural to choose $p = \frac{1}{2}$. However, as Pugh has noted [5], his formula (20) for a search for an element greater than the last one in a large structure is minimized when $p = \frac{1}{e}$. Our tables indicate that for moderate values of n, making p a bit smaller, say $\frac{1}{3}$, is a slightly better choice.

p	$C_p^*(51,50)$	$C_p(51,50)$	$C_p^+(51,50)$	$C_p^*(501,500)$	$C_p(501,500)$	$C_p^+(501,500)$
$1-10^{-6}=.999999$	4,489,241.414950	4,499,208.087933	4,912,025.961441	6,791,827.659238	6,792,827.326405	7,214,612.205729
$1-10^{-5}=.999990$	448,926.611666	449,923.283449	491,205.256571	679,186.272268	679,286.239434	721,464.917172
$1-10^{-4}=.999900$	44,895.131460	44,994.803123	49,123.186229	67,922.133778	67,932.130945	72,150.188547
$1-10^{-3}=.999000$	4,491.984661	4,501.956317	4,914.980648	6,795.722007	6,796.722174	7,218.717994
$1-10^{-2}=.990000$	451.682311	452.684003	494.174760	683.101794	683.202265	725.594243
$9/10=.900000$	47.787066	47.892172	52.255419	72.069671	72.080214	76.538025
$8/9=.888889$	43.316285	43.411482	47.365533	65.309336	65.318884	69.358583
$7/8=.875000$	38.850686	38.935988	42.481890	58.557856	58.566412	62.189060
$6/7=.857143$	34.392807	34.468238	37.607569	51.819582	51.827148	55.034344
$5/6=.833333$	29.947172	30.012770	32.748067	45.102276	45.108855	47.903170
$4/5=.800000$	25.522706	25.578537	27.914273	38.421270	38.426869	40.812837
$3/4=.750000$	21.139814	21.186007	23.131233	31.811711	31.816345	33.803130
$7/10=.700000$	18.266226	18.306145	20.001954	27.488649	27.492653	29.224377
$2/3=.666667$	16.857737	16.894609	18.472354	25.376048	25.379746	26.990664
$1-1/e=.632121$	15.692456	15.726831	17.210901	23.634103	23.637550	25.152548
$3/5=.600000$	14.813670	14.846188	16.263726	22.326296	22.329557	23.776352
$1/2=.500000$	12.953202	12.981956	14.287712	19.597063	19.599947	20.931569
$2/5=.400000$	11.998489	12.025736	13.340198	18.280666	18.283390	19.622552
$1/e=.367879$	11.843807	11.870873	13.215958	18.103231	18.105960	19.475033
$1/3=.333333$	11.758132	11.785077	13.182630	18.045994	18.048705	19.470340
$3/10=.300000$	11.761368	11.788659	13.259445	18.137274	18.140042	19.634409
$1/4=.250000$	11.957875	11.987661	13.621046	18.596813	18.599693	20.264902
$1/5=.200000$	12.455814	12.488350	14.403383	19.598352	19.601433	21.556766
$1/6=.166667$	13.016668	13.048090	15.300050	20.778362	20.782136	23.010633
$1/7=.142858$	13.590076	13.619582	16.239342	21.926362	21.929905	24.522404
$1/8=.125000$	14.176568	14.205798	17.193140	23.041936	23.044990	26.051615
$1/9=.111111$	14.781571	14.812913	18.148946	24.194017	24.197057	27.580510
$1/10=.100000$	15.406099	15.441777	19.100811	25.410007	25.413665	29.100811
$10^{-2}=.010000$	40.942305	41.003221	86.958601	105.240219	105.270740	136.958601
$10^{-3}=.001000$	48.858488	49.844421	568.324336	395.128712	395.121431	901.657669
$10^{-4}=.000100$	40.592053	50.882696	4,249.425111	488.146396	488.779542	6,749.425111
$10^{-5}=.000010$	−34.797804	50.988252	33,981.400097	491.574486	499.759568	53,981.400097
$10^{-6}=.000001$	−667.547550	50.998825	283,163.667390	429.041064	500.875771	449,830.334057

Table 2: A comparison of our estimate $(C_p^*(n+1,n))$, the exact value $(C_p(n+1,n))$, and Pugh's bound $(C_p^+(n+1,n))$, of the expected search cost in searching for a value greater than the nth element in skip lists of size $n=50$ and of size $n=500$

References

[1] Aragon, C. R., Seidel R. G. "Randomized Search Trees". *Proceedings of the 30th Annual IEEE Symposium on Foundations of Computer Science*, Research Triangle Park, NC, Oct. 1989, pp. 540–545.

[2] Culberson, J., Munro, J. I. "Explaining the Behaviour of Binary Search Trees Under Prolonged Updates: A Model and Simulations". *The Computer Journal*, vol. 32, no. 1, Feb. 1989, pp. 68–75.

[3] Culberson, J., Munro, J. I. "Analysis of the Standard Deletion Algorithms in Exact Fit Domain for Binary Search Trees". *Algorithmica*, in press.

[4] Knuth, D. E. *The Art of Computer Programming*, Volume 3: Sorting and Searching. Addison-Wesley, Reading, MA, 1973.

[5] Pugh, W. "Skip Lists: A Probabilistic Alternative to Balanced Trees". To appear in *Communications of the ACM*. (Initial version appeared in *Lecture Notes in Computer Science, no. 382, Proceedings of the Workshop of Algorithms and Data Structures*. Ottawa, Canada, August 1989, pp. 437–449.)

p	$C_p^*(1001,1000)$	$C_p(1001,1000)$	$C_p^+(1001,1000)$	$C_p^*(5001,5000)$	$C_p(5001,5000)$	$C_p^+(5001,5000)$
$1-10^{-6}=.999999$	7,484,975.186372	7,485,475.103288	7,907,759.732863	9,094,413.903525	9,094,513.900242	9,517,198.450016
$1-10^{-5}=.999990$	748,501.336900	748,551.328817	790,779.981805	909,445.932869	909,455.932586	951,724.577774
$1-10^{-4}=.999900$	74,853.952186	74,858.951603	79,082.006955	90,949.136097	90,950.136113	95,177.190866
$1-10^{-3}=.999000$	7,489.216050	7,489.716217	7,912.212037	9,099.459353	9,099.559400	9,522.455340
$1-10^{-2}=.990000$	752.766000	752.816244	795.258449	914.521277	914.531327	957.013726
$9/10=.900000$	79.379464	79.384736	83.847818	96.352277	96.353332	100.820631
$8/9=.888889$	71.929904	71.934678	75.979151	87.302386	87.303341	91.351633
$7/8=.875000$	64.490305	64.494584	68.121509	78.265026	78.265881	81.896230
$6/7=.857143$	57.065564	57.069348	60.280326	69.246358	69.247115	72.461119
$5/6=.833333$	49.664417	49.667707	52.465311	60.257380	60.258038	63.058274
$4/5=.800000$	42.304124	42.306925	44.695692	51.319834	51.320394	53.711401
$3/4=.750000$	35.024272	35.026589	37.015691	42.483608	42.484072	44.475027
$7/10=.700000$	30.264875	30.266878	32.000603	36.711073	36.711473	38.446800
$2/3=.666667$	27.940315	27.942164	29.554931	33.894358	33.894728	35.508975
$1-1/e=.632121$	26.024776	26.026501	27.543222	31.575749	31.576094	33.094195
$3/5=.600000$	24.587821	24.589452	26.037878	29.838922	29.839248	31.288978
$1/2=.500000$	21.597063	21.598505	22.931569	26.240919	26.241207	27.575425
$2/5=.400000$	20.171993	20.173359	21.513729	24.563203	24.563476	25.904906
$1/e=.367879$	19.987028	19.988386	21.359202	24.361928	24.362199	25.734108
$1/3=.333333$	19.939395	19.940761	21.363129	24.334054	24.334327	25.758050
$3/10=.300000$	20.057060	20.058454	21.553464	24.513116	24.513395	26.009373
$1/4=.250000$	20.597325	20.598770	22.264902	25.236409	25.236688	26.908758
$1/5=.200000$	21.747677	21.749187	23.710148	26.747651	26.747953	28.710148
$1/6=.166667$	23.044737	23.046287	25.331750	28.456435	28.456773	30.721216
$1/7=.142858$	24.478707	24.480785	27.015855	30.269628	30.270044	32.805467
$1/8=.125000$	25.873694	25.875964	28.718282	31.944435	31.944782	34.910090
$1/9=.111111$	27.159658	27.161709	30.419694	33.615851	33.616149	37.012075
$1/10=.100000$	28.383163	28.384893	32.111111	35.410405	35.410772	39.100811
$10^{-2}=.010000$	110.606422	110.613354	152.010101	140.851648	140.852253	186.958601
$10^{-3}=.001000$	634.233080	634.305077	1,002.001001	999.243969	999.271416	1,234.991002
$10^{-4}=.000100$	952.635596	952.771074	7,502.000100	3,936.388179	3,936.345081	9,249.425111
$10^{-5}=.000010$	992.180156	996.031575	60,002.000010	4,877.714637	4,878.131331	73,981.400097
$10^{-6}=.000001$	964.821518	1,000.501666	500,002.000001	4,981.785555	4,988.528295	616,497.000724

Table 3: A comparison of our estimate $(C_p^*(n+1,n))$, the exact value $(C_p(n+1,n))$, and Pugh's bound $(C_p^+(n+1,n))$, of the expected search cost in searching for a value greater than the nth element in skip lists of size $n=1000$ and of size $n=5000$

[6] Pugh, W. "Whatever you might want to do using Balanced Trees, you can do it faster and more simply using Skip Lists". *Technical Report CS-TR-2286, Department of Computer Science, University of Maryland, College Park*, July 1989.

Lower Bounds for Monotonic List Labeling

Paul F. Dietz[1]
Ju Zhang

Department of Computer Science
University of Rochester
Rochester, NY 14627

We present optimal lower bound for special cases of the *list labeling problem*. This problem has diverse practical applications, including implementation of persistent data structures, in the incremental evaluation of computational circuits and in the maintenance of dense sequential files. We prove, under a reasonable restriction on the algorithms, that $\Omega(n \log^2 n)$ relabelings are necessary when inserting n items into list monotonically labeled from a label space of size $O(n)$. We also prove that $\Omega(n \log n)$ relabelings are required in the case of a label space of polynomial size.

1 Introduction

We consider the following problem, called the *list labeling problem*. We wish to maintain a list L containing at most n elements. Associated with each element x in the list is a label $0 \le lb(x) \le M$. In this paper, M is either $c_1 n$ for some constant $c_1 > 1$ or some polynomial $p(n)$. Initially, the list contains 1 element with label M (this will always be the final element in the list). Let $s(x)$ denotes the successor of x in L. Then, at all times the labeling must be *monotonic* and *well-spaced*: $lb(x) + 1 \le lb(s(x))$, where $x \in L$ and $s(x) \ne nil$.

We wish to execute a sequence of insertions on the list. An operation $Insert(x, y)$ causes a new element y not already in L to be made the predecessor of x. To maintain monotonic labels we may have to relabel some elements already in the list (it is assumed the label of the last element never changes). The work done in an insertion equals one plus the number of list elements whose labels have changed. The problem is to perform $n - 1$ insertions on L, relabeling in such a way as to minimize the total work.

A related problem, the *list integer labeling problem*, is of considerable practical importance. The integer list labeling problem arises in efficient algorithms for the *order maintenance problem* [3, 12, 5], which itself is important in algorithms for inheritance, incremental evaluation [1] and in making data structures fully persistent [6, 4]. Maintenance of padded sorted tables (which arise naturally in the implementation of databases of ordered records) [9, 8, 13, 14, 7] may be thought of as the integer list labeling problem with $O(n)$ labels. List labeling can be applied to the maintenance of priorities in randomized search trees [2], where it can be used to maintain random distinct priorities of tree elements.

---

[1]This work supported by NSF grant CCR–8909667

Upper bounds for integer list labeling were presented by Itai *et. al.* [8], who gave a $O(n \log^2 n)$ algorithm for the case of $O(n)$ labels, and Dietz [3], Tsakalidis [12] and Dietz and Sleator [5], who gave $O(n \log n)$ cost algorithms for polynomially bounded labels.

2 Notation

The ith insertion is denoted I_i. The list after I_i is L_i; the subscript will be omitted when the meaning is clear in context.

Let $L_k(0), \ldots, L_k(k)$ be the $k+1$ elements of L_k. The function s gives the next element in the list. In L_k, $s(L_k(i)) = L_k(i+1)$ $(0 \le i < k)$ and $s(L_k(k)) = nil$. The predecessor is defined similarly: in L_k, $p(L_k(i)) = L_k(i-1)$ if $0 < i \le k$; $p(L_k(0)) = nil$.

The label of list element x is denoted $lb(x)$. The label of x after I_i is $lb_i(x)$.

The *gap* of an element x is

$$g(x) = \begin{cases} lb(x) & \text{if } p(x) = nil \\ lb(x) - lb(p(x)) & \text{otherwise.} \end{cases} \tag{1}$$

The gap of x after the ith insertion is denoted $g_i(x)$.

3 Smoothness

If the sequence of insertions is known ahead of time, one can perform $n-1$ insertions with $n-1$ work simply by assigning to each element, as it is inserted, the label it will have in the final list. Relabeling is never necessary. To make the problem interesting we must either restrict the relabeling algorithm or consider an on-line adversary who decides where to perform insertions. We do the first, describing a special class of algorithms that satisfy a *smoothness condition*.

Definition 1 *A relabeling algorithm is* normalized *if, after I_i (for any $i > 0$), there are two integers a_i and b_i, $0 \le a_i < b_i \le i$, such that (1) the list element that was just inserted is one of $L_i(r_i)$ for some $a_i \le r_i < b_i$, and (2) $lb_{i-1}(L_i(j)) = lb_i(L_i(j))$ if and only if $0 \le j < a_i$ or $b_i \le j < i$.*

That is, the relabeled elements form a contiguous sublist containing the newly inserted element.

It is not hard to show that

Lemma 2 *For any relabeling algorithm, there is a normalized relabeling algorithm that does no more relabelings.*

Proof: We modify the normalized algorithm so that we only do the relabelings in a contiguous sublist containing the newly inserted element. We delay doing relabelings in other parts of the list until those elements are in or are next to such a sublist. The resulting algorithm does no more work than the original algorithm, yet satisfies the normalization condition. ∎

A relabeling algorithm for the list labeling problem satisfies the smoothness condition (or, it is a *smooth algorithm*) if it is normalized, and for each insertion I_i, $g_i(j) = g_i(j+1)$, $a_i \le j < b_i$.

There is a similar condition for the list integer labeling problem. In this case, the gaps must differ by at most one (some are rounded up, some rounded down). This is called the *approximate smoothness condition.*

We show that execution of any algorithm for the list (integer) labeling problem with $\Theta(n)$ labels that satisfies the smoothness condition has total work $\Omega(n \log^2 n)$ on a sequence of n insertions to the front of the list. The algorithm of Itai *et. al.* is smooth. We also prove an $\Omega(n \log n)$ lower bound for smooth algorithms with polynomially many labels.

4   Lower Bound for Linearly Bounded Labels

We first prove a nonoptimal lower bound for the case of smooth algorithms with real labels bounded by $M = c_1 n$. We consider $n - 1$ insertions to the front of the list. The lower bound is by means of a potential function [10, 11]. We will show that there is a function Φ on the labeling of the list that (1) is $O(n)$, (2) increases by $O(\log n)$ for at least $n/2$ of the insertions into the list, and (3) never decreases by more than the cost of an insertion. From this, we can deduce that the cost of $n - 1$ insertions is $\Omega(n \log n)$.

Lemma 3 *After every insertion, for any $x \in L$, if $s(x) \ne nil$ then $g(x) \le g(s(x))$.*

Proof: Immediate, by induction on the length of the list. ∎

Corollary 4 *For any $x \in L_i$, $lb_i(x) \le lb_{i+1}(x)$.*

For a sequence of n insertions at the front of the list, define the *history* of a smooth algorithm to be a sequence $[b_1, \ldots, b_n]$; that is, the list of the number of elements relabeled by each insertion (including the inserted element).

Lemma 5 *Let $H = [b_1, \ldots, b_n]$ be the history of a smooth algorithm for MLP. Then, any sequence of integers $H' = [b'_1, \ldots, b'_n]$ such that $h_i \le h'_i \le i$ for $i = 1, \ldots, n-1$ is also the history of some smooth algorithm for MLP, and for each list element, the labels assigned by the second algorithm are no less than the labels assigned by the first.*

Proof: Let $lb'(x)$ be the label assigned by the second algorithm. We proceed by induction on the insertion number i. For $i = 0$, $lb(L_0(0)) = lb'(L_0(0))$.

Now let $i > 0$. Assume the lemma holds for $i - 1$ insertions. Let $x = L_i(b_i)$ and $x' = L_i(b'_i)$. By lemma 3 and the induction hypothesis, for any $0 < j \leq b'_i$,

$$lb_{i-1}(L_i(j))/j \leq lb_{i-1}(x')/b'_i \leq lb'_{i-1}(x')/b'_i.$$

Therefore, the gaps left between renumbered elements by the second algorithm are at least as large as those left by the first, so they are well-spaced; additionally, $lb_i(L_i(0)) \leq lb'_i(L_i(0))$. Moreover, for $0 < j \leq b_i$,

$$lb_i(L_i(j)) = ilb_{i-1}(x)/b_i \leq ilb'_{i-1}(x')/b'_i = lb'_i(L_i(j)).$$

For $j > b_i$, let $y = L_i(j)$. Then,

$$lb_i(y) = lb_{i-1}(y) \leq lb'_{i-1}(y) \leq lb'_i(y).$$

∎

Lemma 6 *After the ith insertion,* $g(L(0)) = \cdots = g(L(b_i)) \leq M/i$.

Proof: If not, then, by lemma 3, $M = lb(L(i)) - lb(L(0)) = \sum_{i=0}^m g(L(i)) > M$ which is impossible. ∎

Theorem 7 *For any smooth algorithm for LLP, a sequence of $n-1$ insertions causes $\Omega(n \log n)$ relabelings.*

Proof: We define the potential after I_t to be

$$\Phi_t = \sum_{i=0}^t \log(M/lb(L_t(i))).$$

We show that many insertions (before relabeling) causes Φ to increase by $\Theta(\log n)$, while a relabeling of k items causes Φ to decrease by $O(k)$. Since, at the end, Φ is $O(n)$, this implies that relabelings with total cost $\Omega(n \log n)$ must occur.

We consider only the final $n/2$ insertions. We imagine that each insertion I_i is done in two phases. The new element, $L(0)$, is inserted before $L(1)$, and is assigned the label $lb_{i-1}(L_i(1))/2$. Then, the first b_i list elements are evenly relabeled. For each of such insertion, $1 \leq lb(L_{i-1}(1)) \leq 2M/n$ (lemma 6). Therefore, for the first stage, Φ increases by $\log(n/2) = \Theta(\log n)$.

In the second stage, we know that before relabeling occurs, for each j, $0 \leq j < b_i$, $lb_{i-1}(L_i(j)) \geq (j + 1)/2$. After the second stage, $lb_i(L_i(j)) \leq 2c_1(j + 1)$. Therefore, the difference in the potential between the value of Φ before and after the relabeling is $O(b_i)$. We conclude that the change in potential in an insertion (amongst the final $n/2$) that relabels b_i items is $O(\log n) - O(b_i)$ and that, therefore, $\Omega(n \log n)$ relabelings must occur. ∎

5 An Optimal Lower Bound for $O(n)$ Real Labels

We now prove a lower bound for the case of algorithms satisfying the smoothness condition with linear labels. The key observation is that the "bad" renumberings in the previous lower bound (that cause Φ to decrease by $\Theta(b_i)$) must be infrequent. Most relabelings will cause the potential to decrease by $O(b_i/\log n)$.

We need a lemma and some notation. A list element y is a *boundary element* if $s(y) = nil$ or there is some nonnegative integer i such that y is the first list element with $lb(y) \geq 2^i$. If $y_0, \ldots, y_p$ are boundary elements of a list (in order), define n_i to be the number of list elements with labels at most $lb(y_i)$. Let $l_i = n_i - n_{i-1}$, where $n_{-1} = 0$. For $i = 0, \ldots, p$, let $\rho_i = n_i/lb(y_i)$ (the densities of the initial segments ending at boundary elements) and $\delta_i = \rho_i/\rho_{i-1} - 1$ (the ratios of densities of two initial segments).

Lemma 8 *Let $x_1, \ldots, x_k$ be positive reals such that $\prod_{i=1}^{k}(1+x_i)$ is some constant $c > 1$. Then, $S = \sum_{i=1}^{k} x_i^{-1}$ is $\Omega(k^2/c)$.*

Proof: We prove that S is minimized when the x_i are equal. It suffices to prove this for $k = 2$. If $x_1, x_2 > 0$ and $(1+x_1)(1+x_2) = c > 1$ then elimination of x_2 and differentiation with respect to x_1 of $1/x_1 + 1/x_2$ shows that this function achieves a minimum at $x_1 = x_2 = c^{1/2} - 1$.

So, $S \geq k(c^{1/k} - 1)^{-1}$. Since $c^{1/k} - 1$ is $c/k + O(k^{-2})$, S is $\Omega(k^2/c)$. ∎

Theorem 9 *Any smooth algorithm performs $\Omega(n \log^2 n)$ relabelings on $n - 1$ insertions to the front of the list.*

Proof: We assume the list already has $n/2$ elements; we ignore the cost up to that point. For each of the last $n/2$ insertions, note that $\rho_0 \geq \ldots \geq \rho_p$ (so $\delta_i \geq 0$ for $i = 1, \ldots, p$) and that all ρ_i are $\Theta(1)$. We assume that for each of the last $n/2$ insertions, the algorithm behaves as follows: the new element x is inserted and assigned the (illegal) label $1/2$, then the algorithm evenly relabels the initial segment up to y_0, then the initial segment up to y_1, and so on up to some boundary element y_j. This multiplies the cost by a constant factor, since the ρ_j are all $\Theta(1)$.

Let $\epsilon_i = \rho_{i-1}/\rho_i$, and let $\delta_i = \epsilon_i - 1$. If we let $\Phi = \sum_{j=0}^{k} \log(M/lb_k(L_k(j)))$ as before, it is not hard to prove that the relabeling out to y_i (after the elements out to y_{i-1} have been relabeled) causes $\Theta(2^i)$ relabelings to occur, and decreases Φ by $\Theta(\delta_i 2^i)$. We want to show that for many such relabelings, δ_i is $O(\log^{-1} n)$. We do this by a charging scheme. After the previous relabeling on some previous insertion out beyond the label 2^i, $n_i - n_{i-1}$ and n_{i-1} differed by at most a constant. Therefore, $\Theta(\delta_i 2^i)$ insertions have been performed since then. For the *last half of these*, charge each insertion $\Theta(\delta_i^{-1})$ units of work.

Since relabelings must decrease Φ by $\Theta(n \log n)$ in the last $n/2$ insertions, and each insertion can be charged $\Theta(1)$ units of reduction of Φ at most $O(\log n)$ times, there are $\Theta(n)$ relabelings that are charged $\Theta(\log n)$ times. Consider one such insertion; let C be the set of boundary indices that charged the insertion. At the time that insertion occurs, δ_i has attained at least half the value it will have at the time the relabeling out to y_i charges the insertion. Therefore, the insertion is charged at least $\Theta(\sum_{i \in C} \delta_i^{-1})$. Since $|C|$ is $\Theta(\log n)$, $\delta_i \geq 0$, and $\prod(1 + \delta_i)$ is $\Theta(1)$, lemma 8 implies the work charged to the insertion is minimized when $\delta_s = \delta_t$ for all $s, t \in C$. In this case, the insertion is charged $O(\log^2 n)$ work. This proves the lower bound. ∎

The proof is essentially the same for integer labels.

6 Polynomially Bounded Labels

We now give a proof of a slightly nonoptimal lower bound for smooth algorithms with labels bounded by some polynomial $p(n)$.

Theorem 10 *Any smooth algorithm with polynomial labels requires $\Omega(n \log n / \log \log n)$ relabelings to perform $n - 1$ insertions to the front of the list.*

Proof: Define the boundary elements y_i as in the previous section. As before, we modify the algorithm so that it relabels out to some boundary element (although we do not require this relabeling to be in stages as in the previous proof). This at most doubles the work done by the algorithm.

We define the *complexity* of a particular labeling to be $C = \sum_{i=0}^{p} l_i \log l_i$ (the term being zero if l_i is zero). C is initially 0, and is $\Theta(n \log n)$ after $n - 1$ insertions. An insertion (with no relabeling) increases C by 1. Therefore, relabelings must increase C by $\Theta(n \log n)$.

Suppose the algorithm relabels out to y_{j-1}. The work performed is $S = l_0 + \cdots + l_{j-1}$. Let $l_i' = l_i / S$. Then, the change in C induced by the relabelings is at most

$$\Delta C \leq S \log S - \sum_{i=0}^{j-1} S l_i' \log(S l_i') = -S \sum_{i=0}^{j-1} l_i' \log l_i'$$

The ratio $\Delta C / S$ is maximized when each $l_i' = 1/j$, where ΔC is $S \log j = O(S \log \log n)$. Therefore, at least $\Omega(n \log n / \log \log n)$ relabelings must occur to increase C by $\Theta(n \log n)$. ∎

A more refined proof can eliminate the $\log \log n$ term.

Lemma 11 *Let $x_1, \ldots, x_k$ and $y_1, \ldots, y_k$ be real numbers, where $\sum_{i=1}^{k} x_i = \sum_{i=1}^{k} y_i = 1$ and all y_i are positive. If $I = \{i | x_i / y_i \geq 1/2\}$, then $\sum_{i \in I} x_i \geq 1/2$.*

Proof: By the definition of I, $\sum_{i \notin I} x_i < \frac{1}{2} \sum_{i \notin I} y_i \leq \frac{1}{2}$. The lemma follows. ∎

Theorem 12 *Any smooth algorithm with polynomial labels requires $\Omega(n \log n)$ relabelings to perform $n - 1$ insertions to the front of the list.*

Proof: The proof is similar to the proof of theorem 10. Define the function $M = \sum_{i=0}^{p} i l_i$. Like C, M grows by $\Theta(n \log n)$ during the $n - 1$ insertions. M also increases monotonically, since the label of a list element never decreases. Let ΔC_i and ΔM_i be the change in C and M over I_i. By lemma 11, there is some positive constant c_2 such that at least half of the increase in C occurs during insertions in which $\Delta C_i / \Delta M_i \geq c_2$.

We argue that during these insertions, ΔC_i is $O(S)$, and so the work done during these insertions is $\Omega(n \log n)$. To prove the bound on ΔC_i, we first note that ΔC_i is $O(S + \Delta M_i)$. This can be proved by maximizing $-\sum l'_i \log l'_i / \sum (i l'_i)$ subject to $\sum l'_i = 1$. Then, we note that for $\Delta C_i / \Delta M_i \geq c_2$, ΔM_i is maximized when $l'_{i+1} / l'_i = 2^{c_2}$. Since $c_2 > 0$, it immediately follows that ΔC_i is $O(S)$. ∎

We note that the algorithm of Dietz and Sleator [5] can be modified to be smooth (as stated, it solves the slightly different problem of cyclic list labeling).

7 Directions for Further Research

Ideally, we would like to be able to eliminate the need for the smoothness assumption, and prove a general lower bound. For this, we would need an interactive adversary that inserts at a "hot spot" in the labeling scheme.

8 Acknowledgements

We would like to thank Joel Seiferas and Danny Sleator for useful discussions and encouragement.

References

[1] Bowen Alpern, Roger Hoover, Barry K. Rosen, Peter F. Sweeney, and F. Kenneth Zadeck. Incremental evaluation of computational circuits. In *Proc. 1st Annual ACM–SIAM Symp. on Disc. Alg.*, pages 32–42, January 1990.

[2] Cecelia R. Aragon and Raimund G. Seidel. Randomized search trees. In *Proc. 30th Ann. IEEE Symp. on Foundations of Computer Science*, pages 540–545, October 1989.

[3] Paul F. Dietz. Maintaining order in a linked list. In *Proc. 14th ACM STOC*, pages 122–127, May 1982.

180

[4] Paul F. Dietz. Fully persistent arrays. In *Workshop on Algorithms and Data Structures*, pages 67–74, August 1989.

[5] Paul F. Dietz and Daniel D. Sleator. Two algorithms for maintaining order in a list. In *Proc. 19th ACM STOC*, pages 365–372, May 1987. A revised version of the paper is available that tightens up the analysis.

[6] James R. Driscoll, Neil Sarnak, Daniel D. Sleator, and Robert E. Tarjan. Making data structures persistent. *JCSS*, 38(1):86–124, April 1989.

[7] M. Hofri, A. G. Konheim, and A. G. Rodeh. Padded lists revisited. *SIAM J. On Computing*, 16(6):1073–1114, December 1987.

[8] A. Itai, A. G. Konheim, and M. Rodeh. A spare table implementation of sorted sets. Research Report RC 9146, IBM, November 1981.

[9] Robert Melville and David Gries. Sorting and searching using controlled density arrays. Tech Report 78-362, Dept. of Computer Science, Cornell U., 1978.

[10] Daniel D. Sleator and Robert E. Tarjan. Amortized efficiency of list update and paging rules. *Communications of the ACM*, 28(2):202–206, February 1985.

[11] Robert E. Tarjan. Amortized computational complexity. *SIAM J. on Alg. and Disc. Meth.*, 6(2):306–318, 1985.

[12] A. K. Tsakalidis. Maintaining order in a generalized linked list. *Acta. Info.*, 21(1):101–112, 1984.

[13] Dan E. Willard. Maintaining dense sequential files in a dynamic environment. In *Proc. 14th ACM STOC*, pages 114–121, May 1982.

[14] Dan E. Willard. Good worst–case algorithms for inserting and deleting records in dense sequential files. In *ACM SIGMOD 86*, pages 251–260, May 1986.

Sorting Shuffled Monotone Sequences

Christos Levcopoulos*   Ola Petersson*

Abstract

We present a new sorting algorithm which adapts to existing order within the input sequence. Let k be the smallest integer such that a sequence X of length n can be reduced to the empty sequence by the removal of k monotone, increasing or decreasing, subsequences. The algorithm, Slabsort, sorts X in $O(n \log k)$ time, without knowing k beforehand, which is optimal in a comparison-based model. In the worst case Slabsort degenerates to a hybrid of Melsort and Exact Quicksort and runs in $\Theta(n \log n)$ time. Further, k is shown to capture various kinds of existing order proposed in the literature.

Key words: sorting algorithm, presortedness, measures, shuffled sequences, optimality.

1   Introduction

It is well known that, in a comparison-based model of computation, $\Omega(n \log n)$ time is necessary to sort n elements in both the worst case and the average case [1]. This result relies on the assumption that the elements are randomly permuted. In many applications this is not the case, but the sequence to be sorted is already partially ordered (*presorted*). Most $O(n \log n)$ time algorithms, e.g., Heapsort and Mergesort, do not take existing order within their input into account.

A *measure of presortedness* is an integer function on a permutation π of a totally ordered set that reflects how much π differs from the total order. Examples of presortedness measures include the number of runs or inversions. The term presortedness was coined by Mehlhorn [13], who used the number of inversions in the sequence as a measure. Mehlhorn [13, 14] also gave an algorithm, A-Sort, which is adaptive with respect to this measure. Intuitively, an algorithm is *adaptive* with respect to a measure of presortedness if it sorts all permutations, but performs particularly well on permutations that have a high degree of presortedness (without knowing the value of the measure beforehand). Mannila [12] (see also [6]) formalized the concept of presortedness and gave algorithms that are adaptive with respect to several measures. Recently, Estivill-Castro and Wood [7], Carlsson, Levcopoulos, and Petersson [3], Levcopoulos and Petersson [10, 11], and Skiena [16] have considered other measures. Cook and Kim [4], Dromey [5], and Wainwright [18], among others, have studied the problem of designing adaptive sorting algorithms empirically.

This paper defines two new natural measures of presortedness, *SUS* (*Shuffled UpSequences*) and *SMS* (*Shuffled Monotone Sequences*), which generalize some known measures. *SUS* tells how many, not necessarily consecutive, upsequences that have to be removed from a sequence in order to reduce it to the empty sequence. Similarly, *SMS* tells how many monotone sequences, upsequences or downsequences, that have to be removed. In this way, a sequence X can be viewed as a shuffle of at least $SUS(X)$ upsequences or $SMS(X)$ monotone sequences. Clearly, $SMS(X) \leq SUS(X)$. *SUS* is a generalization of the measures *Runs*, the minimun number of

---

*Algorithm Theory Group, Department of Computer Science, Lund University, Box 118, S-221 00 Lund, Sweden.

consecutive upsequences, and *Max*, the maximum distance that an element is from its correct position, since a low *Runs* or *Max* measure implies a low *SUS* measure. The other way around is not true. In the same sense, *SMS* is a generalization of the measure *Enc*, i.e., the number of encroaching sequences (see Section 3).

The main result of the paper is the sorting algorithm Slabsort, which is optimal with respect to *SMS*, *Enc*, *SUS*, *Runs*, and *Max*. Here, optimality with respect to a measure means maximum adaptation (in an asymptotic sense). Slabsort sorts a sequence X of length n in time $O(n \log SMS(X))$. In order to understand the intuition behind Slabsort, let us make a geometric interpretation of a sequence X which contains $SMS(X) = k$ monotone subsequences. Map each element x_i in X onto the point (i, x_i) in the plane. Then, draw line segments between every pair of points corresponding to consecutive elements in the monotone subsequences. This creates k monotone polygonal chains. Now, the observation is that, if no ascending chain "intersects" a descending chain, then X can be sorted in $O(n \log k)$ time by applying Melsort [16] (see Section 3). In most cases, however, there are intersections. If there are, it is detected and we divide the plane into a number of horizontal "slabs", by drawing horizontal lines. All slabs contain the same number of points and the number of slabs is a function of k. The idea here is to introduce enough slabs to ensure that at most half of them contain no intersection. Note that this is possible, because there cannot be more than $(k/2)^2$ intersections since the polygonal chains are monotone. Now, at least half of the slabs, that is, half of the elements in X, can be sorted by Melsort. The remaining slabs are then sorted recursively. The division into slabs is implemented by recursively finding the median and, at each level, performing a stable partitioning around the median.

We also present tight bounds on our measures in terms of known measures of presortedness. These prove that the new measures capture various kinds of existing order. A consequence of these results is that Slabsort is optimal with respect to several known measures of presortedness.

The remainder of the paper is organized as follows. In Section 2 we state some preliminary results and definitions, present the new measures of presortedness, and provide lower bounds for comparison-based sorting algorithms with respect to the measures. In Section 3 we start our bottom-up description of Slabsort by examining Melsort's behaviour on shuffled sequences. In Section 4 Slabsort is presented and analyzed. In Section 5 we study the relationship between our measures and other known measures of presortedness. Finally, in Section 6, we make some concluding remarks.

2 The New Measures of Presortedness

First, in Section 2.1, we state some preliminary definitions. Second, in Section 2.2, the new measures of presortedness are introduced together with some of their basic properties. In Section 2.3 we establish lower bounds for comparison-based sorting algorithms with respect to the measures.

2.1 Preliminaries

Let $X = \langle x_1, \ldots, x_n \rangle$ be a *sequence* of n elements x_i from some totally ordered set. For simplicity it can be assumed that the elements are distinct. If $x_1 < x_2 < \cdots < x_n$, X is called an *upsequence*. Similarly, if $x_1 > x_2 > \cdots > x_n$, it is called a *downsequence*. A sequence is *monotone* if it is either an upsequence or a downsequence. For two sequences, $X = \langle x_1, \ldots, x_n \rangle$ and $Y = \langle y_1, \ldots, y_m \rangle$, their *catenation* XY is the sequence $\langle x_1, \ldots, x_n, y_1, \ldots, y_m \rangle$. Further, let $|X|$ denote the *length* of X and $\|S\|$ the *cardinality* of a set S. If $Z = \langle x_{f(1)}, x_{f(2)}, \ldots, x_{f(m)} \rangle$, $m \leq n$, and $f\{1, \ldots, m\} \to \{1, \ldots, n\}$ is injective and monotonically increasing, then Z is called a *subsequence* of X. In particular, Z is a *consecutive subsequence* of X if there exists an i, $1 \leq i \leq n - m + 1$, such that $Z = \langle x_i, \ldots, x_{i+m-1} \rangle$. $LUS(X)$ denotes the length of a longest upsequence in X and $LDS(X)$ is the length of a longest downsequence.[1] Let $X_1, \ldots, X_k$ be subsequences of X. If each element in X appears in exactly one of $X_1, \ldots, X_k$, then X is called a *shuffle* of $X_1, \ldots, X_k$.

2.2 Shuffled Sequences

Mannila [12] proposed some general conditions, which have recently been refined by Estivill-Castro, Mannila, and Wood [6], that any measure of presortedness should satisfy. Our approach will be a bit looser. Instead of demanding that some conditions (axioms) are satisfied, we say that a function $m: S^{<\mathbb{N}} \to \mathbb{N}$, where $\mathbb{N}$ denotes the set of natural numbers, and $S^{<\mathbb{N}}$, the set of all finite sequences of elements taken from some totally ordered set S, is a *measure of presortedness* if it in some intuitive way quantifies disorder among the input sequence to be sorted.

As mentioned in Section 1, the two new measures tell the smallest number of upsequences or monotone sequences that a sequence is a shuffle of. More formally, *SUS* (*Shuffled UpSequences*) is defined as

$$SUS(X) = \min\{k \mid X \text{ is a shuffle of } k \text{ upsequences}\}$$

and *SMS* (*Shuffled Monotone Sequences*) is defined as

$$SMS(X) = \min\{k \mid X \text{ is a shuffle of } k \text{ monotone sequences}\}.$$

We will also make use of the function *SDS* (*Shuffled DownSequences*) whose definition follows that of *SUS*. We note that Brandstädt and Kratsch [2] studied the problem of decomposing sequences into upsequences and downsequences. They also observed that the size of the decomposition can be viewed as a measure of presortedness; however, no effort was spent on seeking for an adaptive sorting algorithm.

We proceed by giving some basic properties of *SUS* and *SMS*, out of which some coincide with Mannila's axioms.

Properties *For any sequences* $X = \langle x_1, \ldots, x_n \rangle$ *and* $Y = \langle y_1, \ldots, y_m \rangle$,

1. $1 \leq SUS(X) \leq n$.

2. $SMS(X) \leq SUS(X)$ and $SMS(X) \leq SDS(X)$.

3. $SMS(X) = SMS(\langle x_n, x_{n-1}, \ldots, x_1 \rangle)$.

4. *If* Y *is a subsequence of* X, *then* $SUS(Y) \leq SUS(X)$ *and* $SMS(Y) \leq SMS(X)$.

5. *If every element in* X *is no greater than every element in* Y, $SMS(XY) \leq SMS(X) + SMS(Y)$ *and* $SUS(XY) = \max\{SUS(X), SUS(Y)\}$.

6. $SUS(X) = LDS(X)$ *and* $SDS(X) = LUS(X)$.

7. $1 \leq SMS(X) \leq \left\lfloor \sqrt{2n + 1/4} - 1/2 \right\rfloor$, *which is tight.*

Proof. The first four properties are obvious. Property 6 has been proved by both Brandstädt and Kratsch [2] and Fredman [8]. A proof of Property 7 can be found in [2]. We proceed by proving Property 5.

---

[1]The reader familiar with Young tableaux [9, 15] should recall that $LUS(X)$ is the number of columns and $LDS(X)$ is the number of rows in the tableau corresponding to X.

The *SMS*-part is obvious. (Equality holds if, for example, X is an upsequence and Y is a downsequence.) For the *SUS*-part, let X, $SUS(X) = k$, be a shuffle of $X_1, \ldots, X_k$, and Y, $SUS(Y) = \ell$, a shuffle of $Y_1, \ldots, Y_\ell$. If $k \leq \ell$, then XY is a shuffle of $X_1 Y_1, \ldots, X_k Y_k, Y_{k+1}, \ldots, Y_\ell$. Since $X \leq Y$, it follows that $X_i Y_i$, $1 \leq i \leq k$, are upsequences, and thus, $SUS(XY) \leq \ell$. Strict inequality would contradict Property 4, because Y is a subsequence of XY. Hence, $SUS(XY) = \ell$. If $k > \ell$, an analogous argument gives that $SUS(XY) = k$. ∎

2.3 Algorithmic Lower Bounds

The concept of an optimal algorithm with respect to a measure of presortedness was given in a general form by Mannila [12].

Definition 1 *Let M be a measure of presortedness, and A a sorting algorithm which uses $T_A(X)$ comparisons to sort a sequence X. We say that A is* M-optimal, *or optimal with respect to* M *if, for some $c > 0$, we have, for all $X = \langle x_1, \ldots, x_n \rangle$,*

$$T_A(X) \leq c \cdot \max\{|X|, \log(\|below(X, M)\|)\},$$

where $below(X, M) = \{\pi \mid \pi$ is a permutation of $\{1, \ldots, n\}$ and $M(\pi(X)) \leq M(X)\}$.

Giving a lower bound for algorithms with respect to a measure of presortedness thus means to bound the cardinality of the *below*-set from below.

A lower bound with respect to *SUS* is obtained from the lower bound with respect to the measure *Runs* [9, 12], that is,

$$Runs(X) = \min\{k \mid X \text{ is a catenation of } k \text{ upsequences}\}.$$

Mannila [12] proved that $\log(\|below(X, Runs)\|) = \Omega(n \log Runs(X))$. It is clear that $SUS(X) \leq Runs(X)$. For any sequence X, choose a sequence Y such that $SUS(X) = Runs(Y)$. We claim that

$$below(Y, Runs) \subseteq below(X, SUS). \tag{1}$$

Let $Z \in below(Y, Runs)$. Then

$$SUS(Z) \leq Runs(Z) \leq Runs(Y) = SUS(X),$$

proving that $Z \in below(X, SUS)$. Combining (1) and Mannila's lower bound gives

$$\log(\|below(X, SUS)\|) \geq \log(\|below(Y, Runs)\|) = \Omega(n \log Runs(Y)) = \Omega(n \log SUS(X)),$$

which by Definition 1 completes the proof of

Lemma 1 *Any SUS-optimal comparison-based sorting algorithm needs $\Omega(n \log SUS(X))$ comparisons to sort a sequence X of length n.*

As $SMS(X) \leq SUS(X)$, a similar argument gives

Lemma 2 *Any SMS-optimal comparison-based sorting algorithm needs $\Omega(n \log SMS(X))$ comparisons to sort a sequence X of length n.*

3 Encroaching Sequences and Melsort

Skiena [16] introduced a new measure of presortedness, called *Enc*, the number of *encroaching sequences*. The measure is easiest described by an algorithm that constructs the encroaching sequences. The algorithm scans the input sequence X from left to right while maintaining an ordered set of sorted sequences. Initially, there is just one sequence consisting of x_1. The insertions of the remaining elements follow the strategy: if an element can be placed on either end of an existing sequence without violating the sortedness, put it there, in the oldest such sequence; otherwise, create a new sequence, consisting of this element only. $Enc(X)$ is defined to be the number of (encroaching) sequences created. Note that, in most cases, the encroaching sequences are not subsequences of X.

Skiena also provided an algorithm, *Melsort*, which adapts to *Enc*. Melsort first constructs the encroaching sequences and then merges them to obtain the sorted output. Properly implemented, Melsort sorts a sequence X of length n in $O(n \log Enc(X))$ time.

Skiena proved that, for any sequence X, $Enc(X) \leq LDS(X)$, which by Property 6 is $SUS(X)$. Hence, Melsort runs in $O(n \log SUS(X))$ time, which matches the lower bound in Lemma 1, and we have proved

Theorem 3 *Melsort is SUS-optimal.*

We also take the opportunity to settle an open question of Skiena's of whether Melsort is *Enc*-optimal or not. By applying the same idea as in the proof of Lemma 1 it is easily verified that

$$\log(\|below(X, Enc)\|) = \Omega(n \log Enc(X)).$$

Since this matches the above upper bound we have

Theorem 4 *Melsort is Enc-optimal.*

It should be observed that Melsort is not optimal with respect to *SMS*: Given the input sequence $X = \langle 1, 2n, 2, 2n-1, \ldots, n, n+1 \rangle$, Melsort creates n encroaching sequences, and sorts in time $\Theta(n \log n)$. However, since $SMS(X) = 2$, Slabsort will complete the sorting in linear time as suggested by the lower bound in Lemma 2 (and shown in the next section).

4 Slabsort

We give a bottom-up description of Slabsort. Actually, the description started with Melsort in the last section. Melsort was shown to sort a shuffle of upsequences optimally. Next, in Section 4.1, we show that Melsort can be applied to sort certain kinds of shuffled sequences, called *zig-zag shuffles*. Then, in Section 4.2, Slabsort is presented and an analysis is given.

4.1 Melsort and Zig-Zag Shuffles

Let $X = X_1 X_2 \ldots X_\ell$, where

$$\sum_{i=1}^{\lceil \ell/2 \rceil} SUS(X_{2i-1}) = k_1 \quad \text{and} \quad \sum_{i=1}^{\lfloor \ell/2 \rfloor} SDS(X_{2i}) = k_2,$$

and $k_1 + k_2 = k$. Then X is called a *zig-zag shuffle of size k*.

To motivate the name zig-zag shuffle, consider a geometric interpretation of a sequence X with $SMS(X) = k$, which is a shuffle of k_1 upsequences and k_2 downsequences. Map each element x_i onto the point (i, x_i) in the plane. By joining all points that correspond to consecutive elements in the monotone sequences by line segments we end up with k_1 ascending and k_2 descending polygonal chains. X is a zig-zag shuffle of size k if there is a decomposition of the corresponding point set into k monotone polygonal chains such that no vertical line intersects both an ascending and a descending chain. (See Figure 1.)

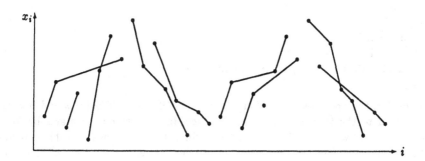

Figure 1: *A zig-zag shuffle of size 10*

How do we sort zig-zag shuffles of size k? Since these form a subset of all sequences with $SMS(X) = k$, Lemma 2 suggests that an $O(n \log k)$ time algorithm should be sought for. We know that $Enc(X) \leq SUS(X)$. Similarly, it can easily be proven that $Enc(X) \leq SDS(X)$. Consider running the encroaching sequences construction method on a sequence X which is a zig-zag shuffle of size k. Then each upsequence in X, and similarly each downsequence, cause at most one encroaching sequence. Hence, at most k encroaching sequences are created. Therefore,

Lemma 5 *If X is a zig-zag shuffle of size k, Melsort sorts X in $O(|X| \log k)$ time.*

4.2 The Algorithm

The idea in Slabsort is to exploit the fact that Melsort can be applied to sort zig-zag shuffles optimally. This is done by partitioning the input sequence into subsequences out of which most are zig-zag shuffles of small enough size.

Assume that $SMS(X) = k$ is known a priori. Again, consider a geometric interpretation of a sequence X whose corresponding point set is decomposed into k_1 ascending and k_2 descending polygonal chains, where $k_1 + k_2 = k$. Imagine dividing the point set into $k^2/2$ horizontal slabs, each containing $2n/k^2$ points. Now, consider any pair of an ascending polygonal chain and a descending polygonal chain. It is clear that no vertical line can intersect both of them in more than one single slab. Consequently, since there are $k_1 \cdot k_2 \leq k^2/4$ such pairs, in at least $k^2/4$ of the slabs no vertical line intersects both an ascending and a descending polygonal chain. Hence, at least half of the elements in X belong to zig-zag shuffles of size k. These slabs are sorted by Melsort in $O(n \log k)$ time, while the remaining are sorted recursively.

We next turn to the implementation of the algorithm. Let $Melsort(X, k)$ be a boolean function which runs Melsort on the sequence X, but stops and returns false if more than k encroaching sequences have been created, that is, if $Enc(X) > k$. Otherwise, it completes the sorting task and returns true. The division into slabs is implemented by first finding the median in X. Then, X is rearranged such that all elements that are less than the median appear in the first half of X and the other in the second half. Moreover, the rearrangment is done in a stable way, i.e., elements within each half obey the original order. Finally, we recur on each half until there are $k^2/2$ "parts". Let $Partition(X, k)$ be a procedure that performs the outlined partitioning of X. Denote by $X_1, X_2, \ldots, X_{k^2/2}$ the new parts of X produced by *Partition*. Note that what *Partition* carries out corresponds to a division of the plane into slabs in the geometric sense. Also, note that the stableness of *Partition* ensures that $X_1, X_2, \ldots, X_{k^2/2}$ are subsequences of X. We are now ready for a less informal description of Slabsort:

```
procedure Slabsort(var X: sequence; k: integer);
   var i: integer;
   begin
     Partition(X, k);
     for i: = 1 to k²/2 do
        if not Melsort(Xᵢ, k) then Slabsort(Xᵢ, k);
   end;
```

We claim that the presented algorithm runs in time $O(n \log k)$ if $|X| = n$ and $SMS(X) = k$. Let us first turn to the time consumed by the procedure *Partition*. It starts by finding the median of n elements which can be done in linear time [1]. The following stable rearrangement around the median can likewise be performed in linear time. After $O(\log(k^2/2)) = O(\log k)$ levels of recursion we are done, and therefore the time used by *Partition* is $O(n \log k)$.

The X_i's that are finished by *Melsort* contain at most k encroaching sequences and are thus sorted in $O(|X_i| \log k)$ time each. The sorting of those who contain more is interrupted when this is detected, and at that time we have spent at most $O(\log k)$ time on each element in finding its encroaching sequence. Hence, the total time consumed by *Melsort* is $O(n \log k)$ on each level of recursion.

Moreover, from the above discussion we know that at least half of the X_i's which *Melsort* is applied to sort are zig-zag shuffles of size k. By Lemma 5, all these will be sorted by *Melsort*. Hence, the number of recursive calls to *Slabsort* is at most $k^2/4$, each containing $2n/k^2$ elements. The time complexity of *Slabsort* is thus

$$T(n, k) \leq \begin{cases} O(n \log k) & \text{if } n \leq k^2/2, \\ k^2/4 \cdot T(2n/k^2, k) + O(n \log k) & \text{otherwise,} \end{cases}$$

which is $O(n \log k)$.

So far we have required $SMS(X) = k$ to be known beforehand. In most cases this is unrealistic. It is, however, no problem to slightly modify the algorithm to run within the same asymptotic time bound with no a priori knowledge about the value of the measure. The idea is simply to start by guessing that $k = 2$. Run the given algorithm, with the exception that all computations on one level of recursion are finished before making any new recursive calls. On each level we check that at least half of the calls to *Melsort* do not lead to further calls. (If the current guess is greater than or equal to $SMS(X)$, at least half of the X_i's will be sorted by *Melsort* on each level.) If this is the case, just continue with the next level. Otherwise, our guess was too small. Then, interrupt the algorithm and restart it with the new guess being the square of the last guess.

We claim that, if we repeatedly square our current guess and run the sorting as outlined, we are done after $O(n \log SMS(X))$ time. The largest possible guess for which we might not succeed finishing the sorting is $SMS(X) - 1$. It is clear that each time we try to run the algorithm $O(n \log k)$ time is spent, where k is our current guess. Therefore, the total time spent is

$$\begin{aligned} T(n) &\leq c \cdot (n \log 2 + n \log 4 + \cdots + n \log(SMS(X) - 1) + n \log(SMS(X) - 1)^2) \\ &\leq 4 c n \log SMS(X) = O(n \log SMS(X)). \end{aligned}$$

Since this matches the lower bound in Lemma 2, we have proved

Theorem 6 *Slabsort is SMS-optimal.*

5 A Comparison with Other Measures

We study the relationship between our new measures, SUS and SMS, and other measures of presortedness that appear in the literature. The approach is to take another measure m and establish an upper bound on $SMS(X)$ or $SUS(X)$ in terms of $m(X)$. The measures we consider are:

- $Inv(X) = \|\{(i,j) \mid 1 \leq i < j \leq n \text{ and } x_i > x_j\}\|$,

- $Osc(X) = \sum_{i=1}^{n} \|\{j \mid 1 \leq j < n \text{ and } \min\{x_j, x_{j+1}\} < x_i < \max\{x_j, x_{j+1}\}\}\|$,

- $Rem(X) = n - LUS(X)$,

- $Max(X) = \max_{1 \leq i \leq n} |i - \pi(i)|$, where π is the permutation such that $x_{\pi(1)} < x_{\pi(2)} < \cdots < x_{\pi(n)}$.

The most well-studied measure is Inv, the number of *inversions* within a sequence [9, 12, 13]. One of the reasons is its interpretation as the number of exchanges of adjacent elements needed in order to bring a sequence into sorted order (cf. Bubble Sort). Consider the following method for decomposing a sequence X, with k inversions, into monotone subsequences. As long as there is a downsequence of length at least $k^{1/3}$, remove a longest downsequence from X. If we at some point run out of such downsequences, then what remains in X can be decomposed into less than $k^{1/3}$ upsequences, by Property 6. Let $D(X)$ denote the number of downsequences we were able to remove. Then,

$$SMS(X) < D(X) + k^{1/3}. \tag{2}$$

Further, each downsequence of length at least $k^{1/3}$ must have contributed to $Inv(X)$ by $\Omega(k^{2/3})$. Hence, $D(X) \leq Inv(X)/\Omega(k^{2/3}) = O(k^{1/3})$, which, if inserted into (2) yields

$$SMS(X) = O(k^{1/3}) = O\left(\sqrt[3]{Inv(X)}\right).$$

To see that the bound is the best possible, let

$$Y_i = \langle m \cdot i, m \cdot i - 1, \ldots, m \cdot i - (m-1)\rangle,$$

for $1 \leq i \leq m$, and form the catenation of the Y_i's, $Y = Y_1 Y_2 \cdots Y_m$. Then, $SMS(Y) = m$ and $Inv(Y) = \Theta(m^3)$.

The measure Osc comes from a geometric interpretation of the input sequence, and intuitively tells how much it oscillates [10]. (See Figure 2.) For the sequence Y, which was used above to prove the tightness of the upper bound with respect to Inv, it holds that $Osc(Y) = \Theta(m^2)$. Consequently, $SMS(X) = \Omega(\sqrt{Osc(X)})$ in the worst case. Assume there is a sequence Z such that $SMS(Z) \neq O(\sqrt{Osc(Z)})$. Then $Osc(Z) = o(n)$, because $SMS(Z) = O(\sqrt{n})$, by Property 7. If $Osc(Z)$ is sub-linear, however, there is a monotone subsequence of Z that does not contribute to $Osc(Z)$ at all. If we remove this monotone subsequence, we are left with $\hat{Z}$, containing $O(Osc(Z))$ elements. Again, by Property 7,

$$SMS(Z) \leq 1 + SMS(\hat{Z}) \leq 1 + O\left(\sqrt{|\hat{Z}|}\right) = O\left(\sqrt{Osc(Z)}\right).$$

Hence, $SMS(X) = O(\sqrt{Osc(X)})$, which is tight in the worst case.

The measure Rem tells the smallest number of elements that have to be removed from a sequence to leave a sorted sequence [4, 12]. To get an upper bound on $SMS(X)$ in terms of $Rem(X)$, remove a longest upsequence from X. Then, we are left with $Rem(X)$ elements

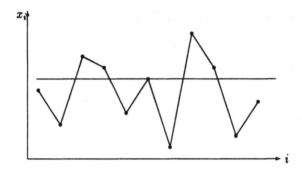

Figure 2: $Osc(X)$ *is the number of proper intersections between horizontal lines through the* x_i*'s and line segments defined by consecutive* x_i*'s. Here,* $Osc(X) = 35$

which by Property 7 can be decomposed into at most $\sqrt{2\,Rem(X)}$ monotone subsequences. Hence, $SMS(X) \leq 1 + \sqrt{2\,Rem(X)}$. The same family of sequences that proves Property 7 to be tight in [2] can be applied to show that the bound is the best possible in the worst case.

Max tells the maximum distance that an element is from its correct position. Let $SUS(X) = k$. Then, by Property 6, there is a downsequence $\langle x_{i_1}, x_{i_2}, \ldots, x_{i_k} \rangle$ of length k in X. It is clear that the correct position of x_{i_1}, $\pi(i_1)$, must be to the right of that of x_{i_k}, $\pi(i_k)$, that is $\pi(i_1) > \pi(i_k)$. Hence, $Max(X) \geq (i_k - i_1) \geq (k - 1)$, and thus, $SUS(X) \leq Max(X) + 1$. The reverse of a sorted sequence shows that the bound is the best possible.

We summarize the obtained bounds and related results in the following theorem.

Theorem 7 *For any sequence* X,

1. $SMS(X) = O\left(\sqrt[3]{Inv(X)}\right)$,

2. $SMS(X) = O\left(\sqrt{Osc(X)}\right)$,

3. $SMS(X) \leq 1 + \left\lfloor \sqrt{2\,Rem(X) + 1/4} - 1/2 \right\rfloor$,

4. $SUS(X) \leq Max(X) + 1$,

5. $Enc(X) \leq SUS(X) \leq Runs(X)$,

6. $SMS(X) \leq 2 \cdot Enc(X)$,

and the bounds are tight in the worst case.

Proof. The first four claims follow from the above discussion. The fifth was verified in Section 2.3 and Section 3. To prove the sixth, we note that, when constructing the encroaching sequences for a sequence X, all elements that are made the new head of a certain encroaching sequence form an upsequence in X. Similarly, all elements that are put in the lower end of the encroaching sequence form a downsequence in X. Hence, each encroaching sequence can be decomposed into one upsequence and one downsequence, which are subsequences of X. The result follows. ∎

At first, it might be easy to interpret Theorem 7 as saying that SMS is a better measure of presortedness than the others, in the sense that it "captures" sequences which are presorted according to any of the measures. That is, if a sequence X is considered to be presorted according to any of the measures, it is considered to be presorted according to *SMS* as well. To make such statement, we have to be very careful, however. For example, the ranges of the measures differ: $Inv: n \rightarrow \{0, \ldots, n(n-1)/2\}$ and $SMS: n \rightarrow \{1, \ldots, \lfloor \sqrt{2n + 1/4} - 1/2 \rfloor\}$. In order to say that a measure m_1 is (algorithmically) superior to another measure m_2 we require that any m_1-optimal sorting algorithm is also m_2-optimal. Then m_2 is in some sense included in m_1. Since a *SMS*-optimal algorithm runs in $O(n \log SMS(X))$ time we have,

Corollary 8 *Any SMS-optimal sorting algorithm is also optimal with respect to the measures Enc, SUS, Runs, and Max.*

Proof. The proofs for *Enc*, *SUS*, and *Runs* follow from Theorem 7 and their respective lower bounds presented in Section 3 and Section 2.3. For *Max*, Estivill-Castro and Wood [7] proved that $\Omega(n \log Max(X))$ is a lower bound. Again, the claim follows from Theorem 7. ∎

It is easy to find instances for which an m-optimal sorting algorithm, $m \in \{Enc, SUS, Runs, Max\}$, is not necessarily *SMS*-optimal. That is, Corollary 8 is *not* true the other way around.

By Theorem 6,

Corollary 9 *Slabsort is optimal with respect to the measures SMS, Enc, SUS, Runs, and Max.*

Now, consider the measures *Inv*, *Osc*, and *Rem*. A-Sort [13, 14] sorts a sequence X of length n in time $O(n \log(Inv(X)/n))$ which is *Inv*-optimal. We leave it as an exercise to find a sequence X such that $O(n \log(Inv(X)/n)) = o(n \log SMS(X))$, proving that Slabsort is not *Inv*-optimal. Analogous arguments show that Slabsort might consume more than $O(n \log(Osc(X)/n))$ time, which is too much for being *Osc*-optimal [10], and more than $O(n + Rem(X) \log Rem(X))$ which is required in order to be *Rem*-optimal [12].

Even though Slabsort is not m-optimal for any of the measures *Inv*, *Osc*, and *Rem*, it is adaptive with respect to them, by Theorem 7. In other words, *SMS* is sensitive to the kinds of existing order measured by these measures.

6   Concluding Remarks

A natural question to ask when a new measure of presortedness has been defined is how efficiently it can be computed, given a sequence. The measure *SUS* can be computed in $O(n \log SUS(X))$ time as shown by Brandstädt and Kratsch [2] and Fredman [8]. Moreover, Fredman proved that this is optimal.

Computing *SMS* turns out to be much more difficult: Wagner [17] proved that the decision problem, i.e., given a sequence X and an integer k, is $SMS(X) \leq k$, is NP-complete. However, some restricted variants of the problem are solvable in polynomial time [2, 17].

Acknowledgements

We would like to thank Svante Carlsson and Jyrki Katajainen for fruitful discussions concerning the combinatorial properties of the new measures. The observation that Slabsort is *Max*-optimal is due to Jingsen Chen, who also pointed out the work of Brandstädt and Kratsch.

References

[1] A.V. Aho, J.E. Hopcroft, and J.D. Ullman. *Data Structures and Algorithms*. Addison-Wesley, Reading, Mass., 1983.

[2] A. Brandstädt and D. Kratsch. On partitions of permutations into increasing and decreasing subsequences. *Journal of Information Processing and Cybernetics*, 22(5/6):263–273, 1986.

[3] S. Carlsson, C. Levcopoulos, and O. Petersson. Sublinear merging and Natural Merge Sort. In *Proc. SIGAL International Symposium on Algorithms*. LNCS, Springer-Verlag, 1990. To appear.

[4] C.R Cook and D.J. Kim. Best sorting algorithms for nearly sorted lists. *Communications of the ACM*, 23(11):620–624, 1980.

[5] G.R. Dromey. Exploiting partial order with Quicksort. *Software—Practice and Experience*, 14(6):509–518, 1984.

[6] V. Estivill-Castro, H. Mannila, and D. Wood. Right invariant metrics and measures of presortedness. Research Report CS–89–30, University of Waterloo, Department of Computer Science, Waterloo, Canada, 1989.

[7] V. Estivill-Castro and D. Wood. A new measure of presortedness. *Information and Computation*, 83(1):111–119, 1989.

[8] M.L. Fredman. On computing the length of longest increasing subsequences. *Discrete Mathematics*, 11:29–35, 1975.

[9] D.E. Knuth. *The Art of Computer Programming, Vol. 3: Sorting and Searching*. Addison-Wesley, Reading, Mass., 1973.

[10] C. Levcopoulos and O. Petersson. Heapsort—adapted for presorted files. In *Proc. 1989 Workshop on Algorithms and Data Structures*, pages 499–509. LNCS 382, Springer-Verlag, 1989.

[11] C. Levcopoulos and O. Petersson. Splitsort—an adaptive sorting algorithm. In *Proc. Fifteenth Symposium on Mathematical Foundations of Computer Science*. LNCS, Springer-Verlag, 1990. To appear.

[12] H. Mannila. Measures of presortedness and optimal sorting algorithms. *IEEE Transactions on Computers*, C-34(4):318–325, 1985.

[13] K. Mehlhorn. Sorting presorted files. In *Proc. 4th GI Conference on Theoretical Computer Science*, pages 199–212. LNCS 67, Springer-Verlag, 1979.

[14] K. Mehlhorn. *Data Structures and Algorithms, Vol 1: Sorting and Searching*. Springer-Verlag, Berlin/Heidelberg, F.R.Germany, 1984.

[15] C. Schensted. Longest increasing and decreasing subsequences. *Canadian Journal of Mathematics*, 13:179–191, 1961.

[16] S.S. Skiena. Encroaching lists as a measure of presortedness. *BIT*, 28(4):775–784, 1988.

[17] K. Wagner. Monotonic coverings of finite sets. *Journal of Information Processing and Cybernetics*, 20(12):633–639, 1984.

[18] R.L. Wainwright. A class of sorting algorithms based on Quicksort. *Communications of the ACM*, 28(4):396–402, 1985.

A Faster Parallel Algorithm for a Matrix Searching Problem

Mikhail J. Atallah*

Purdue University

Abstract

We give an improved parallel algorithm for the problem of computing the tube minima of a totally monotone $n \times n \times n$ matrix, an important matrix searching problem that was formalized by Aggarwal and Park and has many applications. Our algorithm runs in $O(\log \log n)$ time with $O(n^2/\log \log n)$ processors in the CRCW-PRAM model, whereas the previous best ran in $O((\log \log n)^2)$ time with $O(n^2/(\log \log n)^2)$ processors, also in the CRCW-PRAM model. Thus we improve the speed without any deterioration in the *time×processors* product. Our improved bound immediately translates into improved CRCW-PRAM bounds for the numerous applications of this problem, including string editing, construction of Huffmann codes and other coding trees, and many other combinatorial and geometric problems.

1 Introduction

First we briefly review the problem, which was formalized by Aggarwal and Park [1]. We use somewhat different conventions than Aggarwal and Park.

Suppose we have an $n_1 \times n_2 \times n_3$ matrix A and we wish to compute, for every $1 \leq i \leq n_1$ and $1 \leq j \leq n_3$, the $n_1 \times n_3$ matrix θ_A such that $\theta_A(i,j)$ is the smallest index k that minimizes $A(i,k,j)$ (that is, among all k's that minimize $A(i,k,j)$, $\theta_A(i,j)$ is the smallest). Assume that θ_A satisfies the following *sorted* property:

$$\theta_A(i,j) \leq \theta_A(i,j+1),$$

$$\theta_A(i,j) \leq \theta_A(i+1,j).$$

Furthermore, for any submatrix A' of A, $\theta_{A'}$ also satisfies the sorted property.

All the A matrices considered in this paper are assumed to satisfy the above two conditions (and these conditions are the only structure assumed by our algorithm).

Given A, the problem of computing the θ_A array is called by Aggarwal and Park [1] "computing the tube minima of a totally monotone matrix". Many aplications of this problem are mentioned in [1].

---

*Dept. of Computer Science, Purdue University, West Lafayette, IN 47907. This research was supported by the Office of Naval Research under Grants N00014-84-K-0502 and N00014-86-K-0689, the National Science Foundation under Grant DCR-8451393, and the National Library of Medicine under Grant R01-LM05118. Part of the research was done while the author was at Princeton University, visiting the DIMACS center.

The best CREW-PRAM algorithms for this problem run in $O(\log n)$ time and $O(n^2/\log n)$ processors [1, 3], and the best previous CRCW-PRAM algorithm ran in $O((\log \log n)^2)$ time and $O(n^2/(\log \log n)^2)$ processors [1] (where $n = n_1 = n_2 = n_3$).

The main result of this paper is a CRCW-PRAM algorithm of time complexity $O(\log \log n)$ and processor complexity $O(n^2/\log \log n)$. This improves on the speed of the previous $O((\log \log n)^2)$ bound, without any deterioration in the *time×processors* product.

There are applications to this problem in addition to those mentioned in [1]. Some of these are given in [5]: constructing Huffmann codes in parallel, and other tree-construction problems. In [5] the problem was given the name "multiplying two concave matrices": in that framework, we are given two $n \times n$ concave [5] matrices C and D and want to compute their $(\min, +)$ product $B = C * D$:

$$B(i, j) = \min_k \ (\ C(i, k) + D(k, j)\ ).$$

If we let $A(i, k, j) = C(i, k) + D(k, j)$, then this reduces the problem of $(\min, +)$-multiplying C and D to that of computing the matrix θ_A, where $\theta_A(i, j)$ is the smallest index k for which $A(i, k, j)$ is minimized. The concavity of C and D, and the fact that a submatrix of a concave matrix is itself concave, imply that A satisfies the two needed properties [1, 5].

Our improved CRCW-PRAM complexity bound for this problem immediately translates into improvements in the CRCW-PRAM complexities of its myriad applications, some of which were mentioned above (we refrain from tabulating these, and refer the reader to [1, 5] for some of the applications of this problem). We merely state our technical result:

Theorem 1 *The $n \times n$ matrix θ_A of an $n \times n \times n$ matrix A can be computed in $O(\log \log n)$ time with $O(n^2/\log \log n)$ processors in the CRCW-PRAM model.*

Before going into the details, we point out that the main new ingredient in our approach consists of the judicious use of the "aspect ratio condition" (described later). All the other ingredients of the recipe that follows can be found in one form or another in [3] or, independently, in [1]. These ingredients are put together here in a different way from either [3] or [1]. The way we have put the pieces together to achieve the better bound is made possible by the aspect ratio condition: intuitively, the idea is to allow the aspect ratio of the subproblems solved recursively to deteriorate, but not too much, and in a controlled fashion. This is made precise later in the paper.

Although aspect ratios play a crucial role in establishing the above theorem, once we have that theorem we can use it to solve problems of *arbitrary* aspect ratios. For example, it can easily be used to solve a problem where A is $\ell \times h \times \ell$ in $O(\log \log \ell + \log \log(h/\ell))$ time and with $O((\ell h + \ell^2)/T(\ell, h))$ processors.

It will be convenient to analyze our algorithms using the time and work (i.e., number of operations) complexities. The processor complexity is deduced from these using Brent's theorem

[6], which states that any synchronous parallel algorithm taking time T that consists of a total of W operations can be simulated by P processors in time $O((W/P) + T)$. There are actually two qualifications to Brent's theorem before one can apply it to a PRAM: (i) at the beginning of the i-th parallel step, we must be able to compute the amount of work W_i done by that step, in time $O(W_i/P)$ and with P processors, and (ii) we must know how to assign each processor to its task. Both qualifications (i) and (ii) to the theorem will be easily satisfied in our algorithms, therefore the main difficulty will be how to achieve W operations in time T.

In Section 2 we give a preliminary algorithm for the case where A is $\ell \times h \times \ell$ and $h \leq \ell^2$. That algorithm runs in $O(\log \log \ell)$ time and does $O((\ell h + \ell^2)(\log \ell)^2)$ work. Section 3 uses that preliminary algorithm to establish the above theorem. Section 4 makes further remarks, and Section 5 concludes.

Throughout, we refer to the first (resp., second, third) index of an entry of A as its *row* (resp., *column, height*) index. Thus an A that is $n_1 \times n_2 \times n_3$ has n_1 row indices, n_2 column indices, and n_3 height indices.

2   A Preliminary Algorithm

This section gives a preliminary algorithm that has the right time complexity, but does too much work (hence uses too many processors).

The procedure is recursive, and requires that A be $\ell \times h \times \ell$ with $h \leq \ell^2$. We call this last condition the *aspect ratio* requirement; we assume it to be true initially, and we maintain it through the recursion (doing so without damaging the time complexity or the work complexity is, in fact, the main difficulty in this preliminary algorithm). The preliminary CRCW-PRAM algorithm runs in $O(\log \log \ell)$ time and has work (that is, number of operations) complexity $O((\ell h + \ell^2)(\log \ell)^2)$.

Before describing the algorithm, we need a few definitions and a review of some properties.

Let X (resp., Y) be a subset of the row (resp., height) indices of A, and let Z be a contiguous interval of the column indices of A. The *problem induced by the triplet* (X, Z, Y) is that of finding $\theta_{A'}$ for the $|X| \times |Z| \times |Y|$ submatrix A' of A induced by X, Z and Y. That is, it consists of finding, for each pair u, v with $u \in X$ and $v \in Y$, the smallest index $k \in Z$ such that $A(u, k, v)$ is minimized. This k need not equal $\theta_A(u, v)$ since we are minimizing only over Z. However, the following property holds [1, 3]. Assume X (resp., Y) is a contiguous interval of row (resp., height) indices of A. Let x, z, y (resp., x', z', y') be the smallest (resp., largest) indices in X, Z, Y respectively. If $\theta_A(x, y) = z$ and $\theta_A(x', y') = z'$, then the solution to the triplet (X, Y, Z) gives the correct value of $\theta_A(u, v)$ for all $u \in X$ and $v \in Y$ (this follows from the sortedness of θ_A).

The first stage of the computation partitions the row indices of A into $\ell^{1/3}$ contiguous intervals $X_1, X_2, \ldots, X_{\ell^{1/3}}$ of size $\ell^{2/3}$ each. Similarly, the height indices of A are partitioned into $\ell^{1/3}$ contiguous intervals $Y_1, Y_2, \ldots, Y_{\ell^{1/3}}$ of size $\ell^{2/3}$ each. An *endpoint* of an interval X_i (resp., Y_i) is the largest or smallest index in it. For each pair v, w such that v is an endpoint of X_i and w is an endpoint of Y_j, we assign $h^{1+(1/6)}$ processors which compute, in constant time, the index $\theta_A(v, w)$. (Computing the minimum of h entries using $h^{1+(1/6)}$ processors is known to take constant time [9].) The total number of processors used in this step of the algorithm is $O(\ell^{1/3}\ell^{1/3}h^{1+(1/6)})$, which is $O(\ell h)$ because $h \leq \ell^2$.

Let x (resp., x') be the smallest (resp., largest) index in X_i, and let y (resp., y') be the smallest (resp., largest) index in Y_j. Let Z_{ij} be the interval $[a, b]$ where $a = \theta_A(x, y)$ and $b = \theta_A(x', y')$. In the future, when we want to define such a Z_{ij}, we shall simply say "let Z_{ij} denote the interval of column indices of A defined by the set $\theta_A(v, w)$ such that $v \in X_i$ and $w \in Y_j$"; we do so for simplicity of expression, although it is an abuse of language (because Z_{ij} might include an index k that is not the $\theta_A(u, v)$ of any pair $u \in X_i, v \in Y_j$).

After the above stage of the computation we know the beginning and end of each such interval Z_{ij}. As already observed, for any pair of indices u, v where $u \in X_i$ and $v \in Y_j$, we have $\theta_A(u, v) \in Z_{ij}$. Thus it suffices to solve all subproblems defined by triplets (X_i, Z_{ij}, Y_j). However, some of these subproblems violate the aspect ratio condition because their Z_{ij} is too large (larger than $|X_i|^2 = \ell^{4/3}$): each such troublesome subproblem (we call it a *bad* subproblem) will be further partitioned into $k_{ij} = \lceil |Z_{ij}|/\ell^{4/3} \rceil$ subsubproblems, by partitioning Z_{ij} into k_{ij} pieces of size $\ell^{4/3}$ each (except that possibly the k_{ij}-th piece might be smaller). Specifically, if $Z_{ij}^{(k)}$ denotes the k-th piece from this partition of Z_{ij}, then the k-th subsubproblem *spawned* by the bad subproblem (X_i, Z_{ij}, Y_j) is $(X_i, Z_{ij}^{(k)}, Y_j)$. Of course such a spawned subsubproblem $(X_i, Z_{ij}^{(k)}, Y_j)$ no longer has the property that $\theta_A(u, v) \in Z_{ij}^{(k)}$ for $u \in X_i$ and $v \in Y_j$. However, the answer returned by solving such an $(X_i, Z_{ij}^{(k)}, Y_j)$ is not meaningless: we can obtain $\theta_A(u, v)$ for $u \in X_i$ and $v \in Y_j$ by choosing the best among the k_{ij} candidates returned by the k_{ij} subsubproblems. We are now ready to give the details of the second stage of the computation.

The second stage of the computation "fills in the blanks" by doing one parallel recursive call on a number of problems, defined as follows. In what follows, we describe these problems one at a time, but one should keep in mind that they are all solved in parallel. The first class of problems to be solved recursively are the *good* ones, those defined by triplets (X_i, Z_{ij}, Y_j) where $|Z_{ij}| \leq \ell^{4/3}$. The Z_{ij} of such a good problem is not large enough to violate the aspect ratio constraint (because it satisfies $|Z_{ij}| \leq |X_i|^2$). The second class of problems to be solved recursively are those spawned by the *bad* triplets (X_i, Z_{ij}, Y_j). Using the terminology of the previous paragraph, these are the k_{ij} triplets $(X_i, Z_{ij}^{(k)}, Y_j)$, $1 \leq k \leq k_{ij}$. By definition, such a $(X_i, Z_{ij}^{(k)}, Y_j)$ satisfies the aspect ratio requirement. When these recursive calls return, we need not do further work for the good triplets, but for the bad ones we only have the answers for the k_{ij} subproblems they spawned. We can use these k_{ij} subanswers to get the correct answers in constant time, however. For each bad triplet (X_i, Z_{ij}, Y_j), we need to compute, for every pair $u \in X_i$ and $v \in Y_j$, the minimum among k_{ij} entries. We do so by using $k_{ij}h^{1/6}$ processors for each such pair $u \in X_i$ and $v \in Y_j$ (this is enough, since we are then computing the minimum of k_{ij} entries using $\geq k_{ij}^{1+(1/6)}$ processors). Since there are $\ell^{4/3}$ such u,v pairs per bad triplet, the total work done for this "bad triplet postprocessing" is upper-bounded by $\ell^{4/3} h^{1/6} \sum_{i,j} k_{ij}$; now, since

$$\sum_{i,j} k_{ij} = \sum_{\beta=-\ell^{1/3}+1}^{\ell^{1/3}-1} \sum_i k_{i,i+\beta} \leq \sum_{\beta=-\ell^{1/3}+1}^{\ell^{1/3}-1} (2h/\ell^{4/3}) = O(h/\ell),$$

this work is $O(\ell^{4/3} h^{1/6} h/\ell) = O(h\ell)$ (where the fact that $h \leq \ell^2$ was used).

The bottom of the recursion is as usual: we stop when ℓ is some small enough constant (note that, by the aspect ratio condition, a constant ℓ implies a constant h, since $h \leq \ell^2$).

Before we analyze the complexity of the above algorithm, we make a sraightforward observation that is needed in the analysis. Let δ_{ij} equal one if $|Z_{ij}| > \ell^{4/3}$, zero otherwise. Consider the sum:

$$H = \sum_{\delta_{ij}=0} |Z_{ij}| + \sum_{\delta_{ij}=1} \sum_{1 \leq k \leq k_{ij}} |Z_{ij}^{(k)}|.$$

This can be rewritten as follows, by changing the summation indices from i,j to i,β:

$$H = \sum_{\delta_{i,i+\beta}=0} |Z_{i,i+\beta}| + \sum_{\delta_{i,i+\beta}=1} \sum_{1 \leq k \leq k_{i,i+\beta}} |Z_{i,i+\beta}^{(k)}|.$$

Let H_β be the value of the above sum for a given value of β (that is, fixing β and summing over i). It is not hard to see that H_β is upper-bounded by $h + \ell^{1/3}$. Since there are $2\ell^{1/3} - 1$ possible choices for β, H is upper-bounded by $2h\ell^{1/3} + 2\ell^{2/3}$. This fact will be used in the analysis below.

The time and work complexities of the algorithm satisfy the recurrences:

$$T(\ell, h) \leq T(\ell^{2/3}, h') + c_1,$$

$$W(\ell, h) \leq \max\{c_2 \ell h, \sum_{\delta_{ij}=0} W(\ell^{2/3}, |Z_{ij}|)\} + \sum_{\delta_{ij}=1} \sum_{1 \leq k \leq k_{ij}} W(\ell^{2/3}, |Z_{ij}^{(k)}|)\},$$

where c_1 and c_2 are constants, and $h' \leq \ell^{4/3}$. The time recurrence implies that $T(\ell, h) = O(\log \log \ell)$. The work recurrence, together with the fact that $H \leq 2h\ell^{1/3} + 2\ell^{2/3}$, implies that $W(\ell, h) = O((\ell h + \ell^2)(\log \ell)^2)$.

3 Decreasing the Work Done

Let us go back to the original goal of computing the θ_A matrix for an $n \times n \times n$ matrix A, in $O(\log \log n)$ time and $O(n^2)$ work.

Let $ALGO_0$ denote the algorithm of the previous section (recall that it has the right time complexity but does a factor of $(\log n)^2$ too much work). There is more than one way to decrease the work done. The way we do it in this section has the advantage of being self-contained (in Section 4 we sketch another way, one that uses as a subroutine the CREW-PRAM algorithm of [1] or [3]).

Using algorithm $ALGO_0$, we create an algorithm $ALGO_1$ that runs in $O(\log \log n)$ time with $O(n^2(\log \log n)^2)$ work. Then, using $ALGO_1$, we create an algorithm $ALGO_2$ that runs in $O(\log \log n)$ time with $O(n^2(\log \log \log \log n)^2)$ work. Finally, using $ALGO_2$, we create an algorithm $ALGO_3$ that runs in $O(\log \log n)$ time with $O(n^2)$ work.

The method for getting $ALGO_k$ from $ALGO_{k-1}$ is somewhat similar for $k = 1, 2, 3$ (with some important differences, however). It follows the following **generic scheme**: the row indices of A get partitioned into n/s intervals $X_1, \ldots, X_{n/s}$ of length s each (the value of s will differ for each of the three usages we make of this scheme). The height indices of A get partitioned into n/s intervals $Y_1, \ldots, Y_{n/s}$ of length s each. The column indices of A get partitioned into s intervals $Z_1, \ldots, Z_s$ of length n/s each. Let E_X (resp., E_Y) be the set of $2(n/s)$ endpoints of the X_i's (resp., Y_i's). Then we do the following:

1. We run, in parallel, s copies of $ALGO_{k-1}$, one on each of the s triplets $(E_X, Z_1, E_Y), \ldots,$ (E_X, Z_s, E_Y).

2. For each $u \in E_X$ and $v \in E_Y$, we compute the correct $\theta_A(u, v)$ value by taking the best among the s answers for the pair u, v returned by the solutions to the s triplets of the previous stage. We do so in $O(\log \log s)$ time by assigning s processors to each such pair u, v. Since there are $O((n/s)^2)$ such pairs u, v and each pair gets s processors, the work done is $O(n^2/s) = O(n^2)$. If we let Z_{ij} denote the interval of column indices of A defined by the set $\theta_A(v, w)$ such that $v \in X_i$ and $w \in Y_j$, then after this stage of the computation we know the beginning and end of each such interval Z_{ij}.

3. For every Z_{ij} such that $|Z_{ij}| \leq s$, we solve the triplet (X_i, Z_{ij}, Y_j). Which method we use for this differs depending on whether we are designing $ALGO_1, ALGO_2$ or $ALGO_3$.

4. For every Z_{ij} such that $|Z_{ij}| > s$, we partition Z_{ij} into $k_{ij} = \lceil |Z_{ij}|/s \rceil$ intervals $Z_{ij}^{(1)}, Z_{ij}^{(2)},$ $\ldots, Z_{ij}^{(k_{ij})}$, of size s each (except that $Z_{ij}^{(k_{ij})}$ might be smaller). Then we solve each triplet $(X_i, Z_{ij}^{(k)}, Y_j)$. Again, which method we use for this differs depending on whether we are designing $ALGO_1$, $ALGO_2$ or $ALGO_3$.

5. For every Z_{ij} such that $|Z_{ij}| > s$, we compute the right answer for each pair $u \in X_i$ and $v \in Y_j$ from among the k_{ij} possibilities available from the previous stage. We do this in $O(\log \log n)$ time by assigning k_{ij} processors to each such pair u, v. There are s^2 such pairs u, v for each such Z_{ij}, and each pair gets k_{ij} processors and does $O(k_{ij})$ work. Thus the total work is:

$$s^2 \sum_{i,j} k_{ij} = s^2 \sum_{i,\beta} k_{i,i+\beta} \leq s^2 \sum_{\beta} 2n/s = O(s^2(n/s)(n/s)) = O(n^2).$$

Clearly, we need not worry about stages (2) and (5) of the generic scheme, since each has the right time and work complexities no matter what s we choose. The bottlenecks are stages (1),

(3) and (4).

We next explain in more detail how $ALGO_1$, $ALGO_2$, $ALGO_3$ are obtained.

3.1 Algorithm $ALGO_1$

We use the generic scheme with $s = (\log n)^2$.

In stage (1), each copy of $ALGO_0$ runs in $O(\log\log(n/s))$ time and does $O((n/s)^2(\log n)^2)$ work. Hence total work for stage (1) is $O(n^2)$.

In stage (3) we solve a triplet (X_i, Z_{ij}, Y_j) by using $ALGO_0$, which runs in $O(\log\log s)$ time and does $O(s|Z_{ij}|(\log s)^2)$ work. We are allowed to use $ALGO_0$ because, by definition of this stage, we have $|Z_{ij}| \le s < s^2 = |X_i|^2$ and hence the aspect ratio condition is not violated. We temporarily postpone the analysis of the total work done by this stage.

In stage (4) too we solve each triplet $(X_i, Z_{ij}^{(k)}, Y_j)$ by using $ALGO_0$. This takes $O(\log\log s)$ time and $O(|Z_{ij}^{(k)}|s(\log s)^2)$ work.

We now analyze the total work done. As in Section 2, we let δ_{ij} equal one if $|Z_{ij}| > s$, zero otherwise. Recall from stage (4) that if $\delta_{ij} = 1$ then $k_{ij} = \lceil |Z_{ij}|/s \rceil$. The total work done by stages (3) and (4) is $O(s(\log s)^2 H)$ where, as in Section 2, H is the sum:

$$H = \sum_{\delta_{ij}=0} |Z_{ij}| + \sum_{\delta_{ij}=1} \sum_{1 \le k \le k_{ij}} |Z_{ij}^{(k)}|.$$

An analysis like that of Section 2 shows that H is upper-bounded by $O(n^2/s)$. Thus the total work done by stages (3) and (4) is

$$O(s(\log s)^2 n^2/s) = O(n^2(\log\log n)^2).$$

3.2 Algorithm $ALGO_2$

We use the generic scheme with $s = (\log\log n)^2$.

In stage (1), each copy of $ALGO_1$ runs in $O(\log\log(n/s))$ time and has work complexity $O((n/s)^2(\log\log n)^2)$. Thus total work for stage (1) is $O(n^2)$.

In stage (3), we would like to solve a triplet (X_i, Z_{ij}, Y_j) by using algorithm $ALGO_1$, but that algorithm was designed for unit aspect ratio ("square" matrices) whereas here we might have $|Z_{ij}| < |X_i|$. We get around this problem simply by making the matrices square (padding with dummy $+\infty$ entries that cannot alter the correctness of the answer returned). Of course this means that we now do $O(s^2(\log\log s)^2)$ work for each such triplet. However, since there are $(n/s)^2$ such triplets, the total work for this stage is $O(n^2(\log\log s)^2) = O(n^2(\log\log\log\log n)^2)$.

In stage (4) too we solve each triplet $(X_i, Z_{ij}^{(k)}, Y_j)$ by using $ALGO_1$. This takes $O(\log\log s)$ time. The work for the k_{ij} triplets corresponding to a bad triplet (X_i, Z_{ij}, Y_j) is

$$O((|Z_{ij}|/s)s^2(\log\log s)^2) = O(|Z_{ij}|s(\log\log s)^2.$$

Since $\sum_{i,j} |Z_{ij}| = O(n^2/s)$, it follows that the total work for this stage is $O(n^2(\log\log s)^2) = O(n^2(\log\log\log\log n)^2)$.

3.3   Algorithm $ALGO_3$

We use the generic scheme with $s = (\log \log \log n)^2$.

In stage (1), each copy of $ALGO_2$ runs in $O(\log \log(n/s))$ time and has work complexity $O((n/s)^2 (\log \log \log \log n)^2)$. Hence total work for stage (1) is $O(n^2)$.

In stage (3) we solve each triplet (X_i, Z_{ij}, Y_j) sequentially, using only one processor. This is known to take $O(|X_i||Z_{ij}|)$ sequential time [3, 1], i.e., $O(s^2)$ time. Since there are $O((n/s)^2)$ such triplets, the total work for this stage is $O(n^2)$.

In stage (4) too we solve each triplet $(X_i, Z_{ij}^{(k)}, Y_j)$ sequentially, in $O(|X_i||Z_{ij}^{(k)}|) = O(s^2)$ time. The work for the k_{ij} triplets spawned by a bad triplet (X_i, Z_{ij}, Y_j) is $O((|Z_{ij}|/s)s^2) = O(|Z_{ij}|s)$. Since $\sum_{i,j} |Z_{ij}| = O(n^2/s)$, it follows that the total work for this stage is $O(n^2)$.

Using Brent's theorem [6] gives an $O(n^2/\log \log n)$ processor for $ALGO_3$, and completes the proof of the theorem.

4   Further Remarks

Using algorithm $ALGO_3$, we can tackle problems having different aspect ratios from those considered so far. By way of example, suppose A is $\ell \times h \times \ell$ where $h > \ell$ and the aspect ratio condition might be violated. For that case, we can get an $O(\log \log \ell + \log \log(h/\ell))$ time, $O(\ell h)$ work algorithm as follows. Let X (resp., Y) be the set of all row (resp., height) indices of A. Partition the column indices of A into $q = \lceil h/\ell \rceil$ intervals of size ℓ each (the q-th interval may be smaller). Let these q intervals be $Z_1, \ldots, Z_q$. Use q copies of $ALGO_3$ to solve in parallel all the (X, Z_i, Y) triplets. This takes $O(\log \log \ell)$ time and $O(q\ell^2) = O(\ell h)$ work. Then, for each pair $u \in X$ and $v \in Y$, we assign q processors to compute the correct $\theta_A(u, v)$ in $O(\log \log q)$ time (this involves taking the min of q quantities). The total work for this "postprocessing" is $(\ell^2 q) = O(\ell h)$, and the time is $O(\log \log(h/\ell))$.

We also note that, even if our $ALGO_0$ had done a polylog factor more work than the $ALGO_0$ we gave, we would still have been able to design an $ALGO_k$ that runs in $O(\log \log n)$ time and does only quadratic work. We would simply have had to use the generic scheme a few more times, and end up with an $ALGO_k$ with $k > 3$.

An alternative method of decreasing the work done was suggested to us by Professor L.L. Larmore [7], after he read a draft of this paper. The idea is to do everything as in the description of $ALGO_1$ except that, in stages (3) and (4), instead of using $ALGO_0$ we use any of the two best CREW-PRAM algorithms [1, 3]. An analysis similar to those done in Section 3 easily shows that this modification in fact yields an $ALGO_1$ that has the complexity bounds we seek (we omit the details).

5 Conclusion

We gave an $O(\log \log n)$ time, $O(n^2/\log \log n)$ processor for computing the tube minima of an $n \times n \times n$ totally monotone matrix in the CRCW-PRAM model. As observed by Professor L.L. Larmore [7], these bounds can easily be shown to be optimal among quadratic-work algorithms for this model (this follows from Valiant's $\Omega(\log \log n)$ lower bound for computing the minimum of n entries with $O(n)$ work [10]). Aggarwal and Park [1] introduced other versions of such parallel matrix searching problems. Optimal (or improved) algorithms for other such problems would be of great interest, especially in view of the wide applicability of matrix searching problems. In this respect, improved bounds for computing the *row minima* of an $n \times n$ totally monotone matrix were recently discovered [4], achieving $O(\log n)$ time and $O(n \log n)$ work in the EREW-PRAM model, as opposed to the previous $O(\log n \log \log n)$ time and $O(n \log n)$ work in the stronger CREW-PRAM model. Another interesting avenue of research is the parallel solution of these problems on more realistic models than the PRAM, such as hypercubes (see, e.g., [2]).

Acknowledgement. The author is grateful to Professor L.L. Larmore for his comments on a draft of this paper.

References

[1] A. Aggarwal and J. Park, "Parallel searching in multidimensional monotone arrays," to appear in *J. of Algorithms*. (A preliminary version appeared in *Proc. 29th Annual IEEE Symposium on Foundations of Computer Science*, 1988, pp. 497–512.)

[2] A. Aggarwal, D. Kravets, J. Park, and S. Sen, "Parallel searching in generalized Monge arrays with applications," *Proc. 2d Annual ACM Symposium on Parallel Algorithms and Architectures*, 1990 (to appear).

[3] A. Apostolico, M. J. Atallah, L. Larmore, and H. S. McFaddin, "Efficient Parallel Algorithms for String Editing and Related Problems," *Proc. 26th Annual Allerton Conf. on Communication, Control, and Computing*, Monticello, Illinois, 1988, pp. 253–263. To appear, *SIAM J. on Computing*.

[4] M. J. Atallah and S. R. Kosaraju, "An Efficient Parallel Algorithm for the Row Minima of a Totally Monotone Matrix," Purdue CS Tech. Rept. 959 (2/28/90).

[5] M. J. Atallah, G. L. Miller, S. R. Kosaraju, L. Larmore and S. Teng, "Constructing Trees in Parallel," *Proc. 1st Annual ACM Symp. on Parallel Algorithms and Architectures*, Santa Fe, New Mexico, 1989, pp. 421–431.

[6] R. P. Brent, "The Parallel Evaluation of General Arithmetic Expressions," *Journal of the ACM* 21, 2, pp.201–206 (1974).

[7] L. L. Larmore, personal communication.

[8] T. R. Mathies, "A Fast Parallel Algorithm to Determine Edit Distance," Tech. Rept. CMU-CS-88-130 (1988).

[9] Y. Shiloach and U. Vishkin, "Finding the Maximum, Merging and Sorting in a Parallel Model of Computation," *Journal of Algorithms* 2, pp.88–102 (1981).

[10] L. Valiant. Parallelism in Comparison Problems, *SIAM J. on Computing* 4, 3, pp.348–355 (1975).

A RECTILINEAR STEINER MINIMAL TREE ALGORITHM
FOR CONVEX POINT SETS

Dana S. Richards and Jeffrey S. Salowe

Department of Computer Science
University of Virginia
Charlottesville, Virginia 22903

ABSTRACT

A k-extremal point set is a point set on the boundary of a k-sided rectilinear convex hull. Given a k-extremal point set of size n, we present an algorithm that computes a rectilinear Steiner minimal tree in time $O(k^5 + k^4 n)$. For constant k, this algorithm runs in $O(n)$ time and is asymptotically optimal, and for arbitrary k, the algorithm is the fastest known for this problem.

1. Introduction

Given a set S of n terminals in the plane, the rectilinear Steiner minimal tree problem is to find a minimum-length interconnection for S using only vertical and horizontal edges. In finding this tree, new points called *Steiner points* can be created. The decision problem associated with the rectilinear Steiner minimal tree problem is NP-complete [6]. The problem is of practical importance, and substantial research activity has focused on heuristic and special-case algorithms, as well as discovering new properties of Steiner trees [8].

In this paper, we devise an algorithm to construct a rectilinear Steiner minimal tree when S lies on the boundary of its *rectilinear convex hull*. A rectilinear convex hull $Rconv(S)$ is a smallest-area simply-connected figure containing a shortest path between every pair of terminals in S, where paths consist of horizontal and vertical edges. Equivalently, if one places a coordinate axis at any point on the boundary of $Rconv(S)$, at least one of the quadrants is empty [3]. The boundary $B(S)$ of $Rconv(S)$ consists of vertical and horizontal edges.

Several researchers have looked at special cases of this problem. Aho, Garey, and Hwang [2] gave a linear-time algorithm when S lies on two parallel lines, aligned with one coordinate axis, and they gave a cubic-time algorithm when S lies on the boundary of a rectangle whose sides are aligned with the coordinate axes. The latter result was subsequently improved to linear time by Agarwal and Shing [1] and Cohoon, Richards, and Salowe [4]. Bern [3] and Provan [9] independently devised algorithms to find a rectilinear Steiner minimal tree when the input set S lies on $B = B(S)$. Such a point set is called *extremal*. If S is of size n, Bern's algorithm runs in $O(n^5)$ time and Provan's in $O(n^6)$ time. Our contribution is a Steiner tree algorithm for k-extremal point sets; a k-extremal point set is an extremal point set where B has exactly k sides. We give an algorithm that finds a rectilinear Steiner minimal tree for k-extremal point sets of size n which runs in $O(k^5 + k^4 n)$ time. If k is a constant, our algorithm runs in time linear in the number of input points. If k is near n, it is as efficient as Bern's. Furthermore, all of the algorithmic results above are subsumed or improved on by this paper.

Our algorithm differs substantially from the Bern and Provan algorithms. We use a general characterization theorem (Theorem 1), and several decomposition results (Lemmas 9, 10, and 12) to define a series of subproblems. These subproblems are limited in number by a second characterization result (Theorem 2) and combined using dynamic programming. The basis cases in the dynamic programming are the "simple Steiner trees" of Section 3. We present an algorithm to compute "minimal simple Steiner trees." It turns out that after preprocessing, the time complexity depends on the shape and number of sides of the subproblems. We use a series of additional results to reduce the time complexity of computing minimal simple Steiner trees and to carefully combine subproblems, proving our main result.

The remainder of the paper is divided up into four sections. In Section 2, the theoretical underpinnings of the algorithm are presented. The main result is Theorem 1, which describes a canonical form for rectilinear Steiner minimal trees. In Section 3, we give an algorithm to find minimal simple Steiner trees. In Section 4, the general algorithm is presented. The algorithm is based on a dynamic programming formulation whose subproblems are restricted by Theorem 2. There are a total of $O(k^4)$ subproblems, each taking $O(k^4 + n)$ time, giving an $O(k^8 + k^4 n)$ time algorithm. In Section 5, we reduce the time complexity to $O(k^5 + k^4 n)$.

2. A Canonical Form for Steiner Minimal Trees

In this section, we present several results to restrict our search for rectilinear Steiner minimal trees. Let the set S of input points be called *terminals*. Define the *grid graph* for S in the following manner. Draw vertical and horizontal lines through each terminal; the vertices of the grid graph are the intersection points, and the edges are line segments connecting two adjacent grid points. Hanan [7] proved that there is a rectilinear Steiner minimal tree which is a subset of the grid graph. Yang and Wing [10] further proved that there is a rectilinear Steiner minimal tree which is a subset of the grid graph within *Rconv (S)*.

We further restrict the search for rectilinear Steiner minimal trees by developing and applying tie-breaking rules. Among the set τ of rectilinear Steiner minimal trees for S lying on the grid graph within *Rconv (S)*, we prefer certain trees. Given τ, let $\tau_1 \subseteq \tau$ be the trees which maximize length along the boundary B. Let $\tau_2 \subseteq \tau_1$ be the trees where in addition the vertex degree of the terminals and the "boundary corners" of B is maximized. The terminals and the boundary corners of B will be called *nodes*. Finally, let $\tau_3 \subseteq \tau_2$ be trees which are *leftmost* in the sense described below. Trees in τ_3 are called *optimal Steiner trees*.

Three basic operations are used to prove our results: sliding, flipping, and transplanting. The sliding operation is performed on H-shaped subtrees as indicated in Figure 2.1. The crossbar edge can be *slid*, one way or the other. The *flipping* operation is performed on *corners* inside B. The two edges forming the corner can be *flipped* as depicted in Figure 2.1. Finally, in the *transplanting* operation, an edge e is added to a tree to form a single cycle, and a longer edge e' on the cycle is deleted.

We need to be more specific in defining slides and flips to say when trees are *leftmost*. There are four directions to slide in and four directions to flip in. The directions and two examples are shown in Figure 2.1. For a tree to be leftmost, it is not possible to make a sequence of N- and S-slides ending in a NW-flip, a SW-flip, or a W-slide.

An *interior edge* is an edge inside B that does not contain any terminals or Steiner points. An *interior line* consists of one or more adjacent, collinear interior edges. A *complete interior line* is an interior line which stretches completely across *Rconv (S)*.

A non-terminal of degree two is called a *corner-vertex*, a Steiner point of degree three is called a *T-vertex*, and a Steiner point of degree four is called a *cross-vertex*. Steiner points which are connected by a single edge are said to be *adjacent*. Two interior lines l_1 and l_2 are said to *alternate* off a third line l if l_1 and l_2 intersect

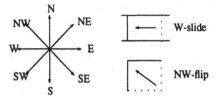

Figure 2.1 — Directions to Slide and Flip

the relative interior of l at two adjacent T-vertices, and l_1 and l_2 are on opposite sides of l. Edges are said to *alternate off of l* if every adjacent pair of lines incident to the relative interior of l alternates.

A corner consists of two lines incident to a corner-vertex. These lines are called the *legs* of the corner. A corner is a *complete corner* if the corner is inside B and both legs intersect B. A T consists of three lines intersecting at a T-vertex. The two collinear lines are called the *head* of the T, and the third line is called the *body*.

The boundary contains a series of *boundary edges*. Boundary edges can meet at terminals or *boundary corners*. An *inner boundary corner* bends away from the interior of $Rconv(S)$, and an *outer boundary corner* bends towards the interior of $Rconv(S)$. Note that all outer boundary corners are terminals. There are at most four special *boundary lines*, or series of collinear, adjacent boundary edges called *tabs*. Tabs connect two outer boundary corners and are at the positions as far to the north, south, east, and west as possible on B.

In general, there are only four extreme points, so tabs may be degenerate and contain only one point. There is another type of degeneracy where regions in $Rconv(S)$ of nonzero area are connected by rectilinear lines. We can remove these degeneracies in $O(n)$ time in a preprocessing step; details are in the full paper. We will therefore assume in the text below that no degeneracies are present.

In the remainder of this section, we present results to restrict the forms of optimal Steiner trees. The proofs are omitted.

> **Lemma 1:** Let p_1 and p_2 be two adjacent Steiner points lying on interior line l. In an optimal Steiner tree, the lines incident to p_1 and p_2 perpendicular to l must be on opposite sides of l.

> **Lemma 2:** In an optimal Steiner tree, edges properly incident to a leg of a corner must alternate. The edge closest to the corner-vertex must point in the direction opposite of the other leg of the corner.

> **Lemma 3:** In an optimal Steiner tree, the body of a T must hit the boundary. Furthermore, no interior edges are properly incident to the body of a T.

> **Lemma 4:** In an optimal Steiner tree, the body of a T intersects a node.

> **Lemma 5:** In an optimal Steiner tree, the legs of a corner must hit the boundary, that is, the corner is a complete corner. At most one of the legs can have more than one properly incident line.

> **Lemma 6:** In an optimal Steiner tree, the head of a T is either a leg of a complete corner or a complete line.

Let a *topology* be a connected portion of a Steiner tree in the relative interior of B. Since it is in the relative interior, a topology does not contain any terminals or Steiner points on the boundary. The main theorem limits the class of Steiner tree topologies for optimal Steiner trees.

> **Theorem 1:** A topology in an optimal Steiner tree satisfying the tie-breaking rules must be one of the following types: (1) Interior lines with no Steiner points, (2) two interior lines forming a cross with no other incident edges, (3) interior lines with alternating edges, (4) complete corners with no incident edges, (5) complete corners with alternating edges incident to one leg and no edges incident to the other leg, and (6) complete corners with one edge incident to one leg and alternating edges incident to the other leg. These topologies may occur several times, but they are disjoint in the relative interior of the convex hull. Furthermore, in each corner case, the edges closest to the corner-vertex must be directed opposite of the other leg of the corner.

3. Minimal Simple Steiner Trees

In a *simple* Steiner tree, only type 1 topologies from Theorem 1 can appear. Each interior line is an entire grid line, and there are no interior Steiner points.

We give a dynamic programming algorithm to find minimal simple Steiner trees. The basic step is of the following form. Consider any grid line ab between boundary points a and b. This defines a subproblem whose boundary B_{ab} is composed of the boundary clockwise from a to b combined with the edge ab itself. For the terminal set on B_{ab} we find several constrained simple Steiner trees such that *no interior line is incident to ab*; such a simple Steiner tree is an *ab-tree*. In particular we compute:

T_a — A minimum ab-tree with a regarded as a terminal (whether or not it actually is a terminal).

T_b — Analogous to T_a.

T_{ab} — A minimum ab-tree with both a and b regarded as terminals.

$T_{\overline{ab}}$ — A minimum ab-tree constrained to contain the edge ab.

A $T_{\overline{ab}}$ tree with the edge ab removed can be regarded as a "Steiner forest" with two components, separating a and b.

3.1. Overview

The algorithm generates each B_{ab} with four line sweeps beginning at each of the tabs of B and progressing across the interior to the opposite tab. The processing of all four sweeps is interleaved; we always work next on the B_{ab} that has the smallest area, and the four degenerate zero-area cases are processed first. Note that a simple case analysis shows that at most two tabs can have incident lines in any simple Steiner tree.

> **Lemma 7:** Let a and b be two endpoints connected by tab l_{ab} counterclockwise, and assume that there is a minimal simple Steiner tree where no interior line is incident to l_{ab}. Then a minimal simple Steiner tree is the shorter of $T_{ab} \cup k_{ab}$ and $T_{\overline{ab}}$, where k_{ab} is l_{ab} with its longest boundary edge deleted.

Proof: Omitted. □

The algorithm actually begins with a preprocessing step discussed below. Lemma 7 guarantees we will get a minimal simple Steiner tree if we correctly compute T_a, T_b, T_{ab}, and $T_{\overline{ab}}$. The remainder of this section describes how to construct these trees.

Note that the m interior edges of a simple Steiner tree divides the interior of B into $m + 1$ subregions, called *open regions* since they are free of internal lines. Some open regions are simple rectangles while others are more complex. Each open region is rectilinearly convex and so has four tabs. Its boundary consists of at most four tree edges along with edges from the original boundary.

A *canonical region* for ab is defined without reference to a tree. Formally, it is a rectilinearly convex region containing ab whose other three tabs contain complete grid lines; the remainder of the boundary is from B_{ab}. An analysis of a canonical region below will be over all ab-trees with no internal lines properly within the canonical region. Note that given an ab-tree the open region that includes ab is itself a canonical region. Therefore, if we exhaustively analyze all canonical regions then we will consider the open region of the optimal ab-tree.

We will see below that we can confine our search to a subset of all canonical regions. In particular, we define a *minimum canonical region* to be a nonempty canonical region not contained in another canonical region.

We give some definitions that allow us to collapse many cases into one analysis. Let d_{xy} be the (rectilinear) distance from x to y. Let b_{xy} be the length of the boundary B_{ab} (the portion not including ab) from x to y. Let g_{xy} be the length of the boundary from x to y with the longest inter-terminal gap removed (where x and y are regarded as terminals in this definition, even if they are not). Let f_{xy} be the length of the boundary from x to y with the gap from y to the terminal nearest y removed; it is zero if there are no terminals between x and y. Note that f_{xy} is not necessarily f_{yx}. Let L_{ab} be the length of T_{ab}, and so on.

3.2. The General Step

We begin by giving the details of how to compute the best possible T_{ab} etc. for a particular canonical region. In a subsection below we show how to speed up this procedure considerably. We will assume hereafter, without loss of generality, that ab is horizontal and B_{ab} surrounds a region above it.

To delimit a canonical region we need to specify its upper horizontal tab and the vertical tabs on the left and right. A generic instance is shown in Figure 3.1. Some, but not all, of these regions may be empty, thus creating simpler degenerate cases.

Note that we have already calculated $T_{\overline{cd}}$, T_{cd}, T_c, T_d, $T_{\overline{ef}}$, and so on, since these correspond to smaller-area subproblems. For $T_{\overline{ab}}$ if we remove ab there is an inter-terminal gap on B_{ab}, excluding ab, between the (clockwise) last terminal connected to a and the first terminal connected to b; call this gap the *cut* for $T_{\overline{ab}}$. To calculate $T_{\overline{ab}}$ we try all possible places where the cut can occur. There are several possibilities. If the cut occurs properly between h and e, for instance, then

$$L_{\overline{ab}} = d_{ab} + b_{ag} + L_{gh} + g_{he} + L_{ef} + b_{fc} + L_{cd} + b_{db}$$

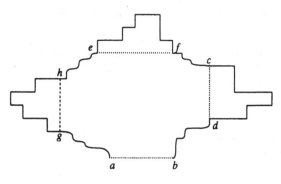

Figure 3.1

The other cases are analogous. $L_{\overline{ab}}$ for a particular canonical region is the minimum over all these choices. The various "inter-region" quantities, such as b_{ag} and f_{db}, are computed in a preprocessing step.

To compute T_{ab} for a particular open region we get this simple relation,

$$L_{ab} = \min \begin{cases} L_{\overline{ab}} \\ b_{ag} + L_{gh} + b_{he} + L_{ef} + b_{fc} + L_{cd} + b_{db} \end{cases}$$

The calculation of L_a depends on whether b is a terminal or not. If so we use L_{ab}, otherwise there may be terminals between d and b so we get

$$L_a = \min \begin{cases} b_{ag} + L_{gh} + b_{he} + L_{ef} + b_{fc} + L_{cd} + f_{db} \\ L_{ab} \end{cases}$$

The calculation of L_b is symmetrical. Note that in some of these cases it is possible for more than one tab to be "behind," say, gh; in that way Figure 3.1 is not completely general. However the algorithm is general enough to include all these cases, since degenerate cases make a zero contribution.

3.3. Using Minimal Canonical Regions

What is the time complexity of this approach, not counting the preprocessing step? Each sweep does $O(n)$ iterations of the general step. Each such step exhausts all $O(n^3)$ choices of tabs (defining a canonical region), leading to $O(n^4)$ total time. We can show how to implement this in $O(n + k^2)$ time, the chief result being the following lemma restricting our attention to minimal canonical regions, defined above; details are in the full paper.

> **Lemma 8:** If canonical region F is contained in canonical region E then the trees constructed for E will be no shorter than those for F.

3.4. The Three-sided Case

We can make an improvement for certain boundaries B that will occur in the next section. Let the portion of the boundary properly between two tabs be called a *staircase*. While every rectilinearly convex region has four tabs, some of its four staircases may be empty. We say a boundary is *three-sided* if at least one staircase is empty. In the three-sided case, we can restrict the number of placements even more by showing there are only $O(n+k)$ minimal canonical regions; in the full paper, we show this implies that a minimal simple Steiner tree can be constructed in $O(n)$ time plus the cost of preprocessing.

3.5. Preprocessing

It turns out that b_{xy} and f_{xy} can be computed in constant time after a sweep about the point set. The computation of g_{xy}, for any new x and y, is more complicated, but can be mapped to a solved problem [5]. We give details in the full paper. Each of the $O(k^2)$ quantities g_{xy} can be computed in constant time, leading to $O(n + k^2)$ time overall.

4. The Basic Algorithm

In this section, we present a polynomial-time algorithm to find a rectilinear Steiner minimal tree for k-extremal point set S by considering the remaining cases of Theorem 1, called nontrivial topologies. Subproblems will be divided along special grid lines called *blue lines*. We will show that there are $O(k)$ blue lines, and each subproblem is bounded by at most four blue lines, thereby restricting the number of subproblems to $O(k^4)$. To combine subproblem solutions, a method introduced by Aho, Garey, and Hwang [2] is described in Lemmas 9 and 10.

4.1. Preliminaries

Let l be a complete vertical interior line. Let $S_1 \subseteq S$ be the set of points to the left of and including l, and let $S_2 \subseteq S$ be the set of points to the right of and including l. Let B_1 be the boundary defined by l and the portion of B to the left of l, and let B_2 be the analogous boundary to the right.

> **Lemma 9:** Suppose some Steiner minimal tree for S contains a vertical line l. Then the union of the following two Steiner trees is a Steiner minimal tree T for S containing l: a Steiner tree T_1 for S_1 of least weight guaranteed to contain l, and a similar tree T_2 for S_2.

Proof: Choosing a tree of greater weight in place of T_1 or T_2 contradicts the minimality of T. □

> **Lemma 10:** A Steiner minimal tree for the point set consisting of S_1 and all Hanan grid points along l can be transformed to a Steiner tree T_1 for S_1 of least weight guaranteed to contain l.

Proof: Consider adjacent Hanan grid points x and y in B_1 along l. Any Steiner minimal tree T_1 either contains the edge between x and y, or there is some other path from x to y. This path must use an edge e between two grid points x' and y', where the distance from x' to y' equals the distance from x to y. If e is transplanted to the boundary edge between x and y, total length remains the same. Repeating this gives a minimum length rectilinear Steiner tree for S_1 containing l. □

A similar result can be developed to guarantee that there is a Steiner minimal tree with two boundary lines l and s forced to occur, provided l and s are orthogonal. Note that the number of points on the boundary is at most $2n$, since a rectilinearly convex boundary can intersect at most $2n$ grid points.

Our algorithm subdivides B by attempting to find the location of nontrivial topological *backbones*. For a complete line l with incident edges, the backbone is l, and for a complete corner, the backbone consists of the legs of the corner and the lines on each leg closest to the corner-vertex (if these lines exist). The backbones will be used to define the four or fewer recursive subproblems, each bounded by blue lines and portions of the original boundary.

Blue grid lines are colored by examining inner boundary corners. Consider an inner boundary corner b, as depicted in Figure 4.1. For each choice of b, at most ten Hanan grid lines are marked blue; these lines are the dotted lines in Figure 4.1. Reorient B by rotation so that it appears as in Figure 4.1. Point b_1 is the first terminal below b, b_2 is the first terminal to the right of b, d_1 is the first terminal on A below b, a_1 is the first terminal on A above b, e_1 is the second terminal on A above b, and so on. (In this specification, if there is a terminal at b, $b = b_1 = b_2$, and if there is a terminal a on A across from b, $a = a_1 = d_1$, and the same for C.) Note that $O(k)$ blue lines are colored for boundary B.

> **Theorem 2:** There is a Steiner minimal tree where all nontrivial topological backbones appear on blue lines.

Proof: Omitted.

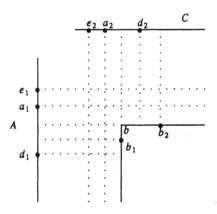

Figure 4.1 — Blue Lines

4.2. Description of Algorithm

We use dynamic programming to devise an efficient algorithm for k-extremal point sets. By Theorem 1, a Steiner minimal tree is either simple or contains one or more nontrivial topological instances. There are seven cases: case 0 is the simple Steiner tree, and cases 1-6 are depicted in Figure 4.2. In the recursive cases 1-6, subproblems are split at blue grid lines and are combined by the algorithm in the proof of Lemma 10. There are two important properties of the subproblems: blue lines only appear on tabs in recursive subproblems, and no new inner boundary corners are created, so the original set of $O(k)$ blue lines is sufficient. □

Case 0: A Steiner minimal tree may be a simple Steiner tree for which an algorithm was given in Section 3. Cases 1 and 2: (complete interior lines.) Theorem 2 implies that the complete line must lie on a blue grid line. There are a total of $O(k)$ candidate blue grid lines, each of which splits the problem involving boundary B into two convex subproblems of smaller area. Case 3: (even-even corner.) The placement of an even-even corner depends on the choice of a, b, and $e \cap B$. For each choice, there are three subproblems to resolve: Problem I is bounded by e and l_2, problem II is bounded by l_1 and l_2, and problem III is bounded by l_1, part of l_2, and e. Theorem 2 implies that e, l_1, and l_2 must lie on blue grid lines. For problems I and II, the blue lines are tabs. For problem III, we first flip the corner-vertex $l_1 \cap l_2$; in this subproblem, the blue lines are tabs and no new inner boundary corners are induced. This flip is justified in Lemma 12. Cases 4 and 5: (even-odd and odd-even corners.) Similar to Case 3. Case 6: (odd-odd corner.) Like cases 3, 4 and 5, but the placement of odd-odd corners depends on the choice of four parameters, namely a, b, c, and d, and there are four subproblems. Problem I is bounded by lines f and l_2, problem II is bounded by l_1 and l_2, problem III is bounded by l_1 and e, and problem IV is bounded by e, l_1, l_2, and f. In problem IV, the corner is flipped before the subproblem is solved.

4.3. Analysis of Algorithm

We now prove that the basic algorithm runs in $O(k^8 + k^4 n)$ time and is correct.

Lemma 11: No subproblem is bounded by more than four blue grid lines.

Corollary 1: There are a total of $O(k^4)$ subproblems.

In order to prove correctness, we must justify the corner flips in cases 3-6. Let e_1 and e_2 be two boundary edges

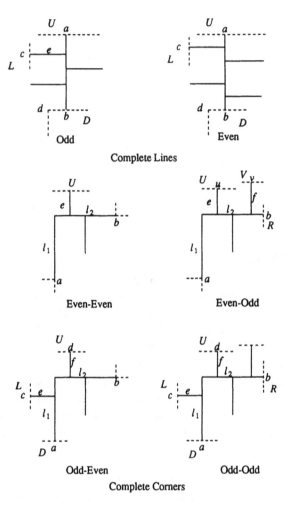

Figure 4.2 — Nontrivial Topologies

intersecting at inner boundary corner b. Assume that no original terminals lie on b or in the relative interior of e_1 or e_2. Let R be the rectangle bounded by e_1 and e_2, and let the other edges of R be e'_1 and e'_2, appearing clockwise as e_1, e_2, e'_2, and e'_1. Let S_3 be the set of points consisting of S and all the grid points along e_1 and e_2, and let S'_3 be the set of points consisting of S and all the grid points along e'_1 and e'_2.

> **Lemma 12:** Suppose there exists a Steiner minimal tree for S_3 containing e_1 and e_2 and having no interior line incident to b or properly incident to either e_1 or e_2. Then the length of a Steiner minimal tree for S_3 is the same as the length of a Steiner minimal tree for S'_3.

Proof: Let T be the asserted tree for S_3. Let T' be a Steiner minimal tree for S'_3. Note that by Lemma 10, we can assume that T' contains all of e'_1 and e'_2. Then *length* $(T') \geq$ *length* (T), since the lines e'_1 and e'_2 can be flipped to e_1 and e_2, giving a feasible Steiner tree for S_3 (it may be that some cycles may be induced as well, so certain edges would be removed). Second, as there is a Steiner minimal tree T where e_1 and e_2 appear with no properly incident lines, e_1 and e_2 can be flipped, giving a feasible solution for S'_3 containing e'_1 and e'_2, so *length* $(T) \geq length$ (T'), and the lengths of T and T' must be equal.

Further note that except at the intersection points of e_1 and e'_1 and e_2 and e'_2, no edges in T' are incident to the closed rectangle R. This is because such an edge implies *length* $(T') >$ *length* (T), a contradiction. $\square$

In the corner cases described in Theorem 2, the problem solved when the corner is flipped is a S'_3 problem; Lemma 12 states that the resulting Steiner minimal tree can be trivially transformed into a Steiner minimal tree for the S_3 problem which is needed to apply Lemma 10.

Theorem 3: A Steiner minimal tree for a k-extremal point set of size n can be found in $O(k^8 + k^4 n)$ time.

Proof: Theorem 1 implies that optimal Steiner trees contain only certain topologies. Corollary 1 implies that only $O(k^4)$ subproblems must be considered, each of which has at most $2n$ terminals. In each subproblem, there is a Steiner minimal tree computation. For each of the $O(k^4)$ subproblems considered in increasing area, a minimal simple Steiner tree is found in $O(k^2 + n)$ time, and the $O(k^4)$ decompositions described in cases 1-6 are considered. For each decomposition, the subproblems have smaller area, so a minimal tree is already stored. Lemmas 9, 10, and 12 imply that the subproblem solutions can be combined to give a Steiner minimal tree. The cost of this algorithm is therefore $O(k^8 + k^4 n)$. $\square$

5. The Improved Algorithm

The algorithm from the preceding section can be improved to $O(k^5 + k^4 n)$ time. The most time-consuming cases are the corner cases 3-6. In the even-even, even-odd, and odd-even corners, as many as three boundary intersections need to be selected, and in the odd-odd corners, as many as four boundary intersections must be chosen. It also appears that the algorithm to compute minimal simple Steiner trees is more efficient, after preprocessing, when restricted to the three-sided case. To get a more efficient algorithm, we define several classes of subproblems.

5.1. Preliminaries

Let e_1 and e_2 be two perpendicular edges intersecting at point c and intersecting B at points a and b, respectively. Let R be the closed rectangle determined by e_1 and e_2; that is, three of R's vertices are a, b, and c.

Lemma 13: R cannot contain an outer corner of B.

Proof: Omitted. $\square$

A simple corollary is used several times below: The boundary from a to b contains at least two edges, including a tab, in addition to the edges containing a and b.

Now consider a corner C with legs l_1 and l_2 and edges e and f incident to l_1 and l_2, respectively. Assume that e and f are the edges closest to the corner-vertex c. Define C to be a *solid corner* if e or f cannot be slid so that c becomes a T-vertex. Let R be the closed 6-gon bounded by l_1, l_2, e, f, a line segment perpendicular to e on the same side of e as c and intersecting e at its endpoint opposite l_1, and a line segment perpendicular to f on the same side of f as c and intersecting f at its endpoint opposite l_2.

Lemma 14: Let C be a solid corner as described above. Some portion of the boundary between e and f (i. e., on the same side of e as c and on the same side of f as c) must lie outside R.

Proof: Omitted. □

In the analysis in Section 4, solutions corresponding to non-solid corners would be discovered in the subproblems for cases 1 and 2. Therefore, only solid corners need be explicitly considered in cases 3-6. In the remainder of the section, corners are assumed to be solid.

Define a *paired tab* to be a structure where two tabs intersect at a single point, and define a *triple tab* to be a structure where three tabs intersect at two points.

> **Lemma 15:** If a boundary has a paired tab, the odd-odd corner cannot occur.

> **Lemma 16:** If a boundary has two paired tabs the odd-even and even-odd corners cannot occur.

> **Lemma 17:** If a boundary has a triple tab, no corner can occur.

Recall that when a corner is present, at most four subproblems are defined. Each subproblem is bounded by a paired tab, implying the following result:

> **Lemma 18:** At most one odd-odd corner can appear, and if one does appear, no other corners can appear. Further, at most two odd-even or even-odd corners can appear.

5.2. Description of the Improved Algorithm

In order to state the $O(k^5 + k^4 n)$ time algorithm, we define four new special-case algorithms: (A) Three-sided with no corners, (B) general with no corners, (C) zero-edge corners (the solution contains one or more even-odd, odd-even, or even-even corners), and (D) odd-odd corners. Note that the special-case algorithms B, C, and D partition the set of possible Steiner minimal trees.

A: If there are at most three staircases and no interior corners, a Steiner minimal tree can appear in one of two forms. First, it may be a minimal simple Steiner tree. Aside from preprocessing, this tree can be computed in $O(n)$ time. Otherwise, the Steiner minimal tree contains a complete line. There are $O(k)$ possible blue lines for the topological backbone position, and each subproblem is also of type A. Aside from preprocessing, the total time per subproblem is therefore $O(n + k)$, and since there may be as many as $O(k^4)$ subproblems, algorithm A takes $O(k^5 + k^4 n)$ time.

Regarding preprocessing for b_{xy}, f_{xy}, and g_{xy} queries, it can be shown that only $O(n)$ extra processing time is needed after an initial expenditure of $O(k^2 + n)$ time. The details are left to the reader, so the time bound stated above holds.

B: Omitted.

C: A zero-edge corner is one with at least one leg with no incident edges. Though odd-even and even-odd corners can only appear at most twice, even-even corners can appear many times. By a tab analysis, one can show that the zero-edge corners are arranged so that if corners are visited in a left-to-right sweep, some number of even-even corners are bracketed with at most one odd-even or even-odd corner on each.

In algorithm C, all subproblems save the original are bounded by exactly two blue lines forming a paired tab. Therefore, consider subproblems bounded by a paired tab in the upper left end; the other cases are symmetric. There are $O(k^3)$ locations for zero-edge corners in this subproblem, as there are as many as three edges to anchor on blue lines: two legs and the edge nearest the corner-vertex.

For one of these placements, no other corner appears between the paired tab at the upper left end and the two legs of the corner. If the corner is even-even, then it can be flipped so that it bends in the same way as the upper left paired tab; three subproblems are defined, two of which are of type A and one of which is a smaller instance of type C. If the corner is even-odd, then it also defines three subproblems except that by Lemma 16, all of the subproblems are instances of A. In summary, there are $O(k^2)$ subproblems, each solved in $O(k^3)$ time.

To process the original convex hull, $O(k^3)$ placements for a zero-edge corner are tried. Suppose the corner bends to the right with the vertical leg below the horizontal. Then two subproblems are of type A and the other is as described above. Algorithm C takes $O(k^5)$ time.

D: If an odd-odd corner is present, Lemma 18 implies that it is the only corner appearing in the Steiner tree. Therefore, starting with the original convex hull, the $O(k^4)$ placements of the odd-odd corner are checked. Each subproblem is of type A, so algorithm D takes $O(k^4)$ time.

This proves the main result of the paper:

Theorem 4: A Steiner minimal tree for a k-extremal point set of size n can be found in $O(k^5 + k^4 n)$ time.

6. References

1. P. K. Agarwal and M. T. Shing, Algorithms for the Special Cases of Rectilinear Steiner Trees: I. Points on the Boundary of a Rectilinear Rectangle, Tech. Report TRCS86-17, Univ. of California at Santa Barbara, 1986.

2. A. V. Aho, M. R. Garey and F. K. Hwang, Rectilinear Steiner Trees: Efficient Special-Case Algorithms, *Networks*, **7**, 1977, pp. 37-58.

3. M. W. Bern, *Network Design Problems: Steiner Trees and Spanning k-Trees*, PhD Thesis, Univerisity of California at Berkeley, 1987.

4. J. P. Cohoon, D. S. Richards and J. S. Salowe, A Linear-Time Steiner Tree Routing Algorithm for Terminals on the Boundary of a Rectangle, *Internation Conference on Computer-Aided Design*, 1988, pp. 402-405.

5. H. Gabow, J. L. Bentley and R. E. Tarjan, Scaling and Related Techniques for Geometry Problems, *16th Symposium on the Theory of Computing*, 1984, pp. 135-143.

6. M. R. Garey and D. S. Johnson, *Computers and Intractability*, Freeman, 1979.

7. M. Hanan, On Steiner's Problem with Rectilinear Distance, *SIAM J. of Appl. Math*, **14**, 1966, pp. 255-265.

8. F. K. Hwang and D. S. Richards, *Steiner Tree Problems*, University of Virginia, 1988.

9. J. S. Provan, Convexity and the Steiner Tree Problem, *Networks*, **18**, 1988.

10. Y. Y. Yang and O. Wing, Optimal and Suboptimal Solution Algorithms for the Wiring Problem, *Proc. IEEE Internation Symposium Circuit Theory*, 1972, pp. 154-158.

Finding Shortest Paths in the Presence of Orthogonal Obstacles Using a Combined L_1 and Link Metric*

Mark de Berg[†]   Marc van Kreveld[†]   Bengt J. Nilsson[‡]   Mark H. Overmars[†]
markdb@cs.ruu.nl   marc@cs.ruu.nl   bengt@dna.lu.se   markov@cs.ruu.nl

Abstract

The problem of computing shortest paths in obstacle environments has received considerable attention recently. We study this problem for a new metric that generalizes the L_1 metric and the link metric. In this combined metric, the length of a path is defined as its L_1 length plus some non-negative constant C times the number of turns the path makes.

Given an environment of n axis parallel line segments and a target point, we present a data structure in which an obstacle free shortest rectilinear path from a query point to the target can be computed efficiently. The data structure uses $O(n \log n)$ storage and its construction takes $O(n^2)$ time. Queries can be performed in $O(\log n)$ time, and the shortest path can be reported in additional time proportional to its size.

1   Introduction

In this paper we discuss a variation on the problem of finding the shortest path between two points in an environment consisting of obstacles that are not allowed to be crossed by the path. This problem has numerous applications, for example in robotics, circuit design and Geographical Information Systems (GIS).

Several researchers have studied instances of this problem. In [Sha78] the shortest path problem between two points inside a simple polygon is considered. In [WPW74, LPW79, SS86] the shortest path problem in the more general environment of polyhedral obstacles is discussed. In [LL81, CKV87] the problem is studied in the L_1 metric.

Most solutions to these problems first find a 'critical graph' which contains the shortest path between the source and the target. Dijkstra's algorithm [Dij59] or some other shortest path-finding algorithm is then applied to the graph.

In some cases more efficient solutions have been developed. Such methods solve the problem more directly instead of transforming it to a graph problem. In [LP84, dRLW89] it is shown that the shortest paths in environments of vertical line segment barriers or non-intersecting axis parallel rectangles are monotone with respect to one of the axes. This is used to give an efficient solution based on a plane sweep approach.

Mitchell [Mit89] has devised an optimal $O(n \log n)$ algorithm for the shortest path problem in the L_1 metric when the obstacles are disjoint polygons and the paths are rectilinear. Moreover, a data structure is given in which a shortest path from a query point to a fixed target point can

---

*This research was supported by the ESPRIT II Basic Research Actions Program of the EC under contract No. 3075 (project ALCOM). Work of the first author was supported by the Dutch Organization for Scientific Research (N.W.O.). The research was done while the third author was visiting Utrecht University.

[†]Department of Computer Science, Utrecht University, P.O.Box 80.089, 3508 TB Utrecht, the Netherlands.

[‡]Department of Computer Science, Lund University, Box 118, 221 00 Lund, Sweden.

be found in $O(\log n)$ time and the complete path can be reported in additional time proportional to its size.

Recently, shortest path problems in the *link metric* have gained considerable attention ([dB89, DLS89, Ke89, LPS*87, Su86, MRW90]). In this metric, the length of a path is equal to the number of turns in the path.

In this paper we study the shortest path problem for rectilinear paths in a new metric that generalizes the L_1 metric and the (rectilinear) link metric. In this *combined metric*, the length of a path is its length in the L_1 metric plus some non-negative constant C times the number of turns in the path. Hence, for $C = 0$ the combined metric is equal to the L_1 metric, whereas for $C = \infty$ the combined metric is equal to the link metric. Such a combined metric is realistic in many applications. For example, when a mobile robot that travels along a straight line wants to make a turn, it has to slow down. Hence, the time it takes to travel the path depends not only on the length of the path, but also on the number of turns the robot has to make.

We provide a data structure such that the shortest path from a query point to some fixed target point can be computed in $O(\log n + k)$ time. Here k is the size of the resulting path, i.e., the number of links of the path. The structure uses $O(n \log n)$ storage and can be built in $O(n^2)$ time. The obstacle environment in our problem consists of n axis parallel line segments. Unlike [CKV87, LP84, LL81, Mit89, dRLW89], we allow the segments to intersect.

The basic idea to solve the problem is the following. First we construct a weighted rectilinear visibility graph on the $2n$ end points of the segments and target point. (Here a point p 'sees' another point q if there is a rectilinear path from p to q, that does not cross an obstacle, consisting of at most two links.) The length of a path in this graph corresponds to the L_1 length of a path. Then the graph is altered such that it accounts for our combined metric. Dijkstra's algorithm [Dij59] is used to find the shortest path from every node in the graph that corresponds to an end point of a segment to the target node t.

Next we show that for any point p in the plane there is a shortest path to t which, with at most one turn, connects p to some end point of an obstacle segment. This induces a subdivision of the plane such that if two points are in the same region, then they have to be connected to the same end point. Because this subdivision can have quadratic size, we have to represent it implicitly. This is done by considering different starting directions separately. For a fixed starting direction, an efficient implicit representation of the corresponding subdivision can be given.

Once we have the structure, queries are performed as follows. Perform an (implicit) point location to locate the region in which the query point lies. Connect the query point to the end point corresponding to this region and follow the path in the shortest path graph.

The rest of this paper is organized in five sections. In section 2 we give a number of useful results in connection with shortest paths. In section 3 the construction of our modified visibility graph is described. Section 4 discusses the problem of finding the correct end point to go to from the source point. In section 5 we state our main result, and in the concluding section we give a brief overview of our results and discuss some open problems.

2 The Geometry of Shortest Paths

In this section we will do some preliminary work on which our solution is based.

We are given a set S of n axis parallel line segments (the *obstacles*), a *target point* t, and a non-negative real C, that is the cost of making one turn.

DEFINITION 2.1 A polygonal chain $s_1 s_2 \ldots s_m$ between two points p and q is called a *rectilinear*

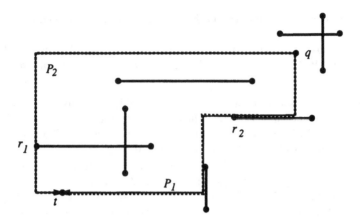

Figure 1: For $C = 0$, the shortest path from q to t is P_1, for $C = \infty$ it is P_2. P_1 first takes a simple step to r_1 and P_2 takes a simple step to r_2.

path between p and q if and only if each link s_i of the path is axis parallel.
A rectilinear path between p and q is called *legal* if and only if it does not intersect the interior of any of the line segments in S.
A rectilinear path between p and q is called a *simple step* between p and q if and only if it is legal, and consists of at most two links.
p and q are said to *see* each other if and only if there is a simple step between p and q.

In the following we only consider legal rectilinear paths, so we omit the words 'legal' and 'recti-linear'.

DEFINITION 2.2 A path $P = s_1 \ldots s_m$ between two points p and q has *length* $C \cdot (m-1) + \sum_{i=1}^{m} |s_i|$, where $|s_i|$ denotes the length of s_i. The *distance* between p and q is the minimum of the lengths over all paths between p and q.

Observe that the shortest path between two points is not necessarily unique. To see this, consider an obstacle-free environment. A shortest path is traced along either boundary segment pair of the rectangle defined by the two points.

Also notice that if $C = \infty$, then the shortest path in the combined metric will be the shortest path in the link metric, and if $C = 0$, then it is the shortest path in the L_1 metric. See Figure 1 for an illustration of these definitions.

Now we can give the basic lemma on which our solution is based.

LEMMA 2.1 *For any point p in the plane, there exists a shortest path $P = s_1 \ldots s_m$ between p and t such that any link s_i, with $1 < i < m$, of the path touches some end point of a line segment of S.*

PROOF: Let P' be a shortest path from p to t. We will transform P' into P, such that the cost will not increase, and without intersecting a line segment of S. Let $s_i = \overline{p_i p_{i+1}}$ $(1 < i < m)$ be the first segment of P' that does not contain any end point of a line segment in S. Assume without loss of generality that s_{i-1} is directed upward and s_i rightward.

If s_{i+1} is directed downward, then the segment s_i can be moved downward, thus shortening the segments s_{i-1} and s_{i+1}, until either s_i touches an end point of a line segment in S, or one of the segments s_{i-1} or s_{i+1} becomes redundant. In the latter case we have reduced the cost of the path, a contradiction with P' being a shortest path.

If s_{i+1} is directed upward, then the segment s_i can be moved upward, thus making s_{i-1} longer and s_{i+1} shorter with the same amount, until either s_i touches an end point, or s_{i+1} has length zero. In the latter case, s_{i+1} becomes redundant, contradicting the optimality of P'.

Thus we can move s_i such that it contains an end point without increasing the length of the path. This can be repeated with the rest of the path, until every segment s_i $(1 < i < m)$ contains an end point. $\qquad\square$

The lemma has the following useful consequence. Suppose we know for every end point of the line segments in S which is the best next end point to go to (including going to the target) on its way to the target. Suppose we also know for each end point the distance to the target. Then a shortest path from an arbitrary query point q to the target can be found in the following way. First take a simple step to that end point p of a line segment in S, such that the sum of the distance from q to p and the distance from p to t is minimized. Then follow the path from p to t.

There is one difficulty with this approach. It only works when the length function is additive in the sense that the length of a path from q to t via p is equal to the sum of the length of the path from q to p and the length of the path from p to t. Unfortunately, the link distance, and thus our length function, does not have this property. The problem is that when the length of the path between q and p is added to the length of the path between p and t, then the turn that might be made at p is not counted, whereas it is counted in the length of the path between q and t. This problem is circumvented as follows. Duplicate every end point p of the line segments in S, such that there is a horizontal version p_h, and a vertical version p_v. If a path arrives at p with a horizontal link, then it arrives at p_h, otherwise it arrives at p_v. It is not allowed to leave p_h vertically or p_v horizontally. But it is allowed to go from p_v to p_h and vice versa, at the cost of one turn. With this extension, the length function becomes additive.

So now we want to find for a query point q the end point p such that the sum of the distance from q to p plus the distance from p to t is minimized. The approach we use to solve this is the *locus approach*: subdivide the plane into regions such that if two points are in the same region then they have to be connected to the same end point. This subdivision, however, can have quadratic size (see Figure 2), so we have to store it implicitly. In order to be able to do this, we consider the eight different possibilities of leaving p with a simple step seperately. Suppose we are only allowed to leave p by a 'left–down' simple step, i.e., the path moves from the query point to the left in the plane, makes a turn and continues downward until it reaches an end point of an obstacle or the target. (For example, the simple step from q to r_1 in Figure 1 is a left–down simple step, whereas the step to r_2 is down–left.) Thus p can only see points that are to its left and below it. The crucial observation that makes an efficient implicit representation of the subdivision possible for a fixed starting step is the following:

LEMMA 2.2 *Suppose that only the left–down simple step is allowed, that p can see two end points e_1 and e_2, and that the path via e_1 has lower cost. Then for any point p' that can see e_1 and e_2, the path via e_1 has lower cost.*

PROOF: Follows from the additivity of the length function for paths. $\qquad\square$

This lemma does not imply that the 'left–down subdivision' has subquadratic size. In fact, it can still be quadratic. However, using this lemma we can give an implicit representation of the left–down subdivision that requires subquadratic storage. How this is done is described in section 4.

217

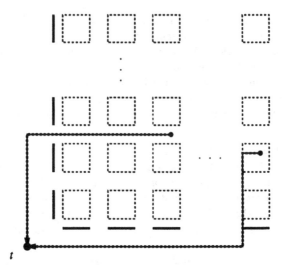

Figure 2: Example showing the quadratic worst case complexity of the subdivision. The bold segments are the obstacles and the dotted squares are regions of the subdivision.

3 Constructing the Shortest Path Graph

The (rectilinear) visibility graph of the set $S \cup \{t\}$ is defined as the graph $G = (V, E)$ where V is a set of $2n + 1$ nodes each corresponding to an end point of a segment in S or to t. The set E of edges is defined by the visibility property, that is, there exists an edge between nodes of the visibility graph if and only if the points corresponding to the nodes see each other. (Recall that two points see each other if and only if there is a simple step connecting the two points that does not cross an obstacle.) Furthermore, we assign to each edge a weight that is the length of the corresponding simple step in the L_1 metric. Thus, by Lemma 2.1, a shortest path between two nodes in G corresponds to a shortest path in the L_1 metric between the two points corresponding to these nodes.

We wish to modify the visibility graph G to take into account the cost for making turns. The modified visibility graph G^* is constructed from G by duplicating the nodes that correspond to the end point of an obstacle. We now have two nodes for each end point p of an obstacle, an h-node p_h and a v-node p_v. There can never be edges between two h-nodes or between two v-nodes, only between an h-node and a v-node (or between an h- or v-node and the target node t). Edges incident on an h-node represent horizontal path links to the corresponding end point. Edges incident on a v-node represent vertical path links. There is an edge between an h-node p_h and a v-node r_v if and only if there is a simple step between p and r that leaves p horizontally and r vertically. Observe that it is possible that there is an edge between p_h and r_v and also between p_v and r_h. We let the weight of the edge be the L_1 distance between the points plus C (the cost of making one turn). Between two nodes p_h and p_v corresponding to the same end point p of a segment there is an edge of weight C. This edge accounts for the turn that has to be made on a path that reaches p horizontally and leaves p vertically. See Figure 3 for an illustration of this modification of the visibility graph.

From the modified visibility graph we can easily construct the single target shortest path graph from the target t using Dijkstra's algorithm [Dij59]. This graph enables us to answer

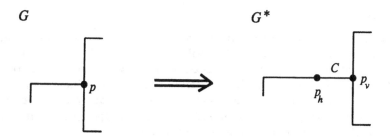

Figure 3: Modification of the Visibility Graph

shortest path queries where the source point is some end point of a segment. Such a query can be answered in time proportional to the number of links in the resulting path, and the graph uses linear storage in the number of nodes.

The construction of the modified visibility graph can be done in $O(n \log n + |E|) = O(n^2)$ time by a plane sweep algorithm. The application of Dijkstra's algorithm also requires $O(n^2)$ time. This leads to:

LEMMA 3.1 *The Single Target Shortest Path graph for the combined metric of a set of n axis parallel obstacle segments and a fixed target point can be computed in $O(n^2)$ time. The graph has size $O(n)$.*

4 Finding Initial Steps

Our next goal is to build a subdivision of the plane where each region is associated with an end point of a segment, namely the end point which yields the shortest path to the target for each query point q that lies in that region. As was noted before (refer to Figure 2), such a subdivision may require quadratic storage. However, by considering different ways of leaving q separately, we will be able to give an efficient implicit representation of this subdivision. Thus we have eight 'initial step finding' structures, one for each possible simple step with which we can leave q. Each structure again consists of two structures, one for finding the region in which to look for the appropriate connecting end point and the other for choosing the correct end point among a set of possible ones.

So, let us fix a simple step with which we will leave the query point, say the 'left–down' simple step. In the following subsections, we describe how the subdivision corresponding to a left–down initial step can be stored implicitly. The (seven) other initial steps can be handled similarly.

4.1 The Interval Finding Structure

When we want to start from the query point q with a left–down step, we need to be able to tell how far to the left the first link can go. In other words, we want to know the first obstacle that we hit when we move to the left from q. For this problem, a structure exists that uses linear storage and in which the first obstacle that is hit by a horizontal ray can be found in $O(\log n)$ time ([EOS84]). This structure can be built in $O(n \log n)$ time.

LEMMA 4.1 ([EOS84]) *The first obstacle to the left of the query point can be found in $O(\log n)$ with a structure that uses $O(n)$ storage. This structure can be built in $O(n \log n)$ time.*

4.2 The Slab Tree

The query with the source point in the structure of Lemma 4.1 results in a horizontal interval, which defines the region in which end points should be considered. This interval is used as the query object in the structure to be described next. This structure, which we call the *slab tree*, is an implicit representation of the subdivision corresponding to an initial left–down step. To simplify the description, we assign colours to each end point of an obstacle and we give the region from which we should go to a certain end point the corresponding colour. The special colour *white* is used when no end point can be reached at all.

The idea of the implicit colouring is as follows. Consider the intersection of a vertical line with the (coloured) subdivision. The colouring of this line tells us were to go when the source point lies on that line. But by Lemma 2.2, the colouring of this line gives us even more information: it tells us what the best end point to the left of the line is for points to the right of it. In other words, if the interval (that is found with the 'interval finding' structure) of a source point intersects this line in a red region, then the best end point to go to that lies to the left of the line is the red end point. Of course, the best end point need not lie to the left of this line. Therefore we do the following. We divide the plane in vertical *slabs* and we colour the right boundaries of these slabs with the colours of end points in that slab. (The fact that we only use colours of end points in the slab is crucial for the reduction in storage.) Then, for a query interval, we have to select a number of slabs that together cover the interval. The right boundaries of these slabs are all intersected by the interval and thus give us a number of colours. Of the corresponding end points, we have to select the best one.

We next give a more precise description of the slab tree.

Draw vertical lines through all end points of the line segments. This results in a number of *elementary slabs*. Associate the elementary slabs from left to right with the leaves of a balanced binary tree. Every internal node δ corresponds to a slab which is the union of the elementary slabs that are associated with the leaves of the subtree rooted at δ. We will paint the right bounding line of the slab of δ with the colours of end points that lie in the slab. (Points lying on the boundary between two slabs belong to the leftmost of these two slabs.) This results in a list ordered on y-coordinate, such that for any query interval that crosses the slab completely, the 'best' end point in this slab can be found by searching in this list. This list is called the *associated list* of δ. Although one colour can appear more than once, the size of an associated list is linear in the number of end points in the corresponding slab. This can be proved by showing that the enumeration of the m colours on a line forms an $(m, 2)$ Davenport-Schinzel sequence. It is well known that such a sequence has linear size (see e.g. [HS86]). Hence, the slab tree uses $O(n)$ space at every level of the tree which sums op to $O(n \log n)$ space in total. In Figure 4, an example of a slab tree is given.

A query in the slab tree is performed in the following way. The query object is a horizontal interval that defines the region in which the end point via which we have to go to the target must be contained. (This interval has been found with the 'interval finding' structure of Lemma 4.1.)

Move with the interval down the slab tree to the node δ where the path to its left end point and right end point split. From here follow the path to the left end point of the interval and for every node where we go left, search in the associated structure of its right son to find a colour. Similarly, follow the path from node δ to the right end point of the interval and for every node on the path where we go right, search in the associated structure of its left son to find a colour. Searching in the associated structures is done with the y-coordinate of the interval. In this way we search in the associated structures of $O(\log n)$ nodes whose slabs together cover the horizontal

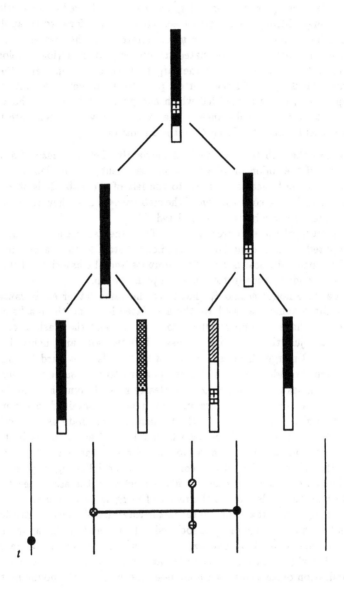

Figure 4: An example of a slab tree for the left–down simple step. The colour white corresponds to areas where no end point can be reached (with a left–down simple step).

query interval. Thus the query time is $O(\log^2 n)$, which can be reduced to $O(\log n)$ by applying fractional cascading ([CG86]) to the slab tree. (It is straightforward to apply this technique, because all associated lists are ordered on y-coordinate, and when we search, we always search with the same y-coordinate in the associated structures.) Of the $O(\log n)$ colours found, choose the one that gives the shortest path to the target. This colour represents the end point of an obstacle segment which will yield the shortest path to the target if the path must start with a left–down step. It can easily be decided which end point results in the shortest path, because for each end point we have the distance to the target, and we can calculate the distance from the source to any end point by adding C to the L_1 distance.

Next we show how the slab tree can be built efficiently. Let the *mask* of a node δ be defined as the projection of the union of all vertical line segments in the slab of δ. Thus, the mask of a slab represents the 'inaccessible area' to the left of the slab. It is stored in a linear list. Masks are only used for the construction of the slab tree. Hence, they can be removed when the construction of the slab tree has been completed.

First the skeleton of the slab tree is built. Then the associated list and the masks of the nodes are computed. This is done bottom up, i.e., we start at the leaves and work our way up to the root. The mask of a leaf is trivial to compute and the associated lists of the leaves can be computed in $O(n \log n)$ time in total using a plane sweep.

Now consider two nodes δ_l and δ_r that have the same father δ. Because we are working bottom up, we can assume that we have the associated lists and the masks of δ_l and δ_r. From this information we have to compute the associated list and the mask of δ. The mask of δ is easy to obtain: it is just the union of the masks of its two sons and, hence, it can be computed by a straightforward merge. The associated list of δ is also not hard to compute: consider a point x on the right boundary of the slab corresponding to δ. Assume that this point is coloured green in δ_r. If the mask of δ_r contains x then clearly no end point in δ_l can be reached from x. Hence, point x remains green. Now assume that x is not contained in the mask of δ_r and let the point on the right boundary of δ_l with the same y-coordinate as x be coloured red. Then x may have to be coloured red. Deciding whether the red end point is better than the green one can be done in constant time because we know for each end point its distance to the target. Summarizing, the associated list of δ is computed in the following way. Filter the associated list of δ_l by the mask of δ_r and merge this filtered list with the associated list of δ_r, using the shortest path graph to decide what is the result of merging two colours.

This building algorithm takes time linear in the total size of all associated lists and all masks. We already noted that the size of an associated list of a node is linear in the number of end points in the corresponding slab. The size of a mask is easily seen to be linear in this number as well. Hence, the algorithm takes $O(n)$ time at every level of the tree and $O(n \log n)$ time in total. The application of fractional cascading does not influence the bound on the preprocessing. This leads to:

LEMMA 4.2 *The slab tree can be built in $O(n \log n)$ time, it uses $O(n \log n)$ storage and queries in the slab tree take time $O(\log n)$.*

5 The Main Result

In this section we give an overview of the preprocessing and the query algorithm, and we state our main result.

The preprocessing of our algorithm runs as shown in the following description.

Algorithm	Preprocess(S, t, C)
Step 1	Build the modified visibility graph of $S \cup \{t\}$ for the constant C
Step 2	Compute the Single Target Shortest Path graph of the visibility graph
Step 3	for each initial simple step do
Step 3.1	Build the interval finding structure
Step 3.2	Build the slab tree
	endfor
end Preprocess	

By Lemma 3.1, steps 1 and 2 of the preprocessing algorithm take $O(n^2)$ time. Steps 3.1 and 3.2 take $O(n \log n)$ time by Lemmas 4.1 and 4.2 and, because there are eight different simple steps, step 3 takes $O(n \log n)$ time in total. The Single Target Shortest Path Graph, as well as the interval finding structures, use $O(n)$ storage and the slab trees use $O(n \log n)$ storage. Hence, the total amount of storage used is $O(n \log n)$.

Queries are performed according to the following description.

Algorithm	Query(s)
Step 1	for each initial simple step do
Step 1.1	Do an interval finding query with s in the appropriate structure
Step 1.2	Do a query with the interval thus found in the appropriate slab tree
	endfor
Step 2	Of the eight end points found, select the best one
Step 3	Follow the path from s via this end point in the STSP graph to the target t
end Query	

Interval finding queries take $O(\log n)$ time, see Lemma 4.1. Queries in the slab tree also take $O(\log n)$ time by Lemma 4.2. Since there are eight different initial steps for which the queries have to be done, steps 1 and 2 of the query algorithm take $O(\log n)$ time. Note that the distance of the query point to t (i.e. the length of a shortest path) can be reported now. Reporting the path itself is done in step 3 and takes time proportional to the number of links of the path.

We can now state the main theorem of our paper.

THEOREM 1 *For a set of n axis parallel obstacle segments, there exists a data structure of size $O(n \log n)$, such that the distance in the combined metric from a query point to a fixed target point can be found in $O(\log n)$ time. A shortest path can be reported in $O(k)$ additional time, where k is the size of the resulting path. The data structure can be built in $O(n^2)$ time and it uses $O(n \log n)$ storage.*

6 Conclusion

We have provided a data structure to compute the shortest path to a fixed target in an obstacle environment consisting of n possibly intersecting axis parallel line segments. The length of a path was measured in a new metric that generalizes the L_1 and the link metric. The query time is $O(\log n + k)$, and the structure uses $O(n \log n)$ storage. Here n is the number of obstacle segments and k is the number of links in a shortest path.

A number of open problems do remain. A first interesting problem is the construction of a data structure that allows both the source and the target to be query points. It would also be nice if the cost of a turn could be specified in the query. Finally, in this paper we studied axis parallel obstacles and paths. It would be interesting to consider arbitrary polygonal obstacles and paths. Here the length of a path could be defined as its Euclidean length plus some extra term which is dependent on the number of turns as well as on the angle of the turns. No results for this general model are known to date.

References

[CG86]   B. CHAZELLE, L.J. GUIBAS. Fractional Cascading: I. A data structuring technique. *Algorithmica*, 1:133–162, 1986.

[CKV87]   K. CLARKSON, S. KAPOOR, P. VAIDYA. Rectilinear Shortest Paths through Polygonal Obstacles in $O(n \log^2 n)$ Time. In *Proc. 3rd ACM Symp. on Computational Geometry*, pages 251–257, 1987.

[dB89]   M. DE BERG. *On Rectilinear Link Distance*. Technical Report RUU-CS-89-13, Department of Computer Science, Utrecht University, 1989.

[Dij59]   E.W. DIJKSTRA. A Note on Two Problems in Connection with Graphs. *Numer. Math.*, 1:269–271, 1959.

[DLS89]   H.N. DJIDJEV, A. LINGAS AND J. SACK. An $O(n \log n)$ time Algorithm for Computing the Link Center in a Simple Polygon. In B. Monien and R. Cori (Eds.) *Proceedings STACS '89*, Lect. Notes in Comp. Science 349, Springer Verlag, pages 96–107, 1989.

[EOS84]   H. EDELSBRUNNER, M.H. OVERMARS, R. SEIDEL. Some Methods of Computational geometry Applied to Computer Graphics. *Computer Vision, Graphics, and Image Processing*, 28:92–108, 1984.

[dRLW89] P.J. DE REZENDE, D.T. LEE, Y.F. WU. Rectilinear Shortest Paths with Rectangular Barriers. *Journal of Discrete and Computational Geometry*, 4:41–53, 1989.

[HS86]   S. HART AND M. SHARIR. Nonlinearity of Davenport-Schinzel Sequences and of Generalized Path Compression Schemes. *Combinatorica* , 6:151–177, 1986.

[Ke89]   Y. KE. An Efficient Algorithm for Link Distance Problems. In *Proc. 5th ACM Symp. on Computational Geometry*, pages 69–78, 1989.

[LL81]   R.C. LARSON, V.O. LI. Finding Minimum Rectilinear Distance Paths in the Presence of Barriers. *Networks*, 11:285–304, 1981.

[LP84]   D.T. LEE, F.P. PREPARATA. Euclidean Shortest Paths in the Presence of Rectilinear Barriers. *Networks*, 14:393–410, 1984.

[LPS*87]   W. LENHART, R. POLLACK, J. SACK, R. SEIDEL, M. SHARIR, S. SURI, G. TOUSSAINT, S. WHITESIDES AND C. YAP. Computing the Link Center of a Simple Polygon. In *Proc. 3rd ACM Symp. on Computational Geometry*, pages 1–10, 1987.

[LPW79]   T. LOZANO-PEREZ, M.A. WESLEY. An Algorithm for Planning Collision-free Paths among Polyhedral Obstacles. *Communications of the ACM*, 22:560–570, 1979.

[MRW90]   J.S.B. MITCHELL, G. ROTE AND G. WÖGINGER. Computing the Minimum Link Path Among a Set of Obstacles in the Plane. To appear in *Proc. 6th ACM Symp. on Computational Geometry*, 1990.

[Mit89]   J.S.B. MITCHELL. An Optimal Algorithm for Shortest Rectilinear Paths Among Obstacles in the Plane. In *Abstracts of the First Canadian Conference on Computational Geometry*, page 22, 1989.

224

[Sha78]   M.I. SHAMOS. *Computational Geometry*. PhD thesis, Yale University, New Haven, CN, 1978.

[SS86]   M. SHARIR, A. SCHORR. On Shortest Paths in Polyhedral Spaces. *SIAM Journal of Computing*, 15(1):193–215, 1986.

[Su86]   S. SURI. *Minimum Link Paths in Polygons and Related Problems*. Ph. D. thesis, Johns Hopkins University, 1986.

[WPW74] G.E. WANGDAHL, S.M. POLLACK, J.B. WOODWARD. Minimum-Trajectory Pipe Routing. *Journal of Ship Research*, 18:46–49, 1974.

Input-Sensitive Compliant Motion in the Plane

Joseph Friedman*   John Hershberger   Jack Snoeyink*

Stanford University   DEC Systems Research Center   Stanford University

Abstract

In 1989, Briggs [1] gave an $O(n^2 \log n)$ algorithm for single-step planar compliant motion planning among polygonal obstacles. At the same time, Friedman, Hershberger, and Snoeyink [8] gave an $O(n \log n)$ preprocessing, $O(n)$ space, $O(\log n)$ query time algorithm for the same problem when the environment is the interior of a simple polygon. In this paper, we bridge the complexity gap between these two results. For a polygonal environment with n vertices and k connected components on its boundary, our algorithm requires $O(kn \log n)$ preprocessing time, $O(kn \log n)$ space (the preprocessing phase requires $O(kn \log n)$ space to run, and produces a data structure of size $O(kn)$), and $O(k \log n)$ query time.

1 Introduction

Research in robot motion planning provides many problems that are interesting both practically and theoretically. Given a robot's initial and goal position in some environment, there are typically many paths connecting the initial position and the goal position. It is important, therefore, to specify which of these paths is most desirable (e.g., a shortest path [13] or a high-clearance path [12]). The choice of path may also depend on the control and sensing abilities of the robot.

In this paper, we investigate a motion paradigm called *compliant motion*. Compliant motion has the property that the paths the robot follows are "stable" in the sense that even if the robot diverges slightly from the commanded path, it still reaches the goal. A robot performing compliant motion is equipped with a compass and a force sensor. We instruct it to go in a certain direction α, and when it hits an obstacle, it slides along its

Figure 1

boundary so long as α points into the obstacle. When the robot reaches the "end" of the obstacle (see Figure 1), it resumes the direction α through free space. Because of friction, the robot may be unable to slide along the boundaries of some obstacles. If the robot hits one of these, it stops moving. This type of motion enables the robot to "grope" its

*Research supported in part by Digital Equipment Corporation.

way towards the goal. In the basic model, the only way the robot can stop is by getting stuck [3, 4, 7]. In less restricted cases, some forms of goal sensing are possible [10].

Sliding on a boundary wall is "stable" in the sense mentioned above, since there is a whole range of directions that force the robot to slide in the same way along that wall. Given a robot's control uncertainty, that is, the amount its actual direction of motion in free space may differ from the commanded direction, it is possible to ensure that the robot will always slide left (or right) along any particular wall it encounters. It follows that when the goal lies on a boundary of the environment, it is sometimes possible to find a stable commanded direction that will get the robot all the way from the starting point to the goal. In general, however, there may not be a stable direction via which the robot can reach the goal. In this case we need to outline a *plan*, which specifies a number of subgoals, each one directly attainable from the previous one, that eventually lead to the desired goal. In this case, it is desirable to find a plan that contains as few subgoals as possible.

Compliant motion planning in 3-space is *PSPACE*-hard and *NEXPTIME*-hard [2, 11]; however, the planar problem is more tractable [1, 3, 4, 6, 8]. For a polygonal environment with n walls and a start region and goal region of constant size, Donald [4] presented an $O(n^4 \log n)$ algorithm for the single-step case, and an $O(n^{r^{O(1)}})$ algorithm for the r-step case. Briggs [1] improved the single-step bound to $O(n^2 \log n)$.

Friedman, Hershberger, and Snoeyink [8] consider the special case in which the environment is the interior of a simple polygon. They separate the planning process into preprocessing and query phases: given the polygon and a goal vertex or edge, the preprocessing phase builds a linear-size data structure in $O(n \log n)$ time; using the data structure, the query algorithm can find a single-step plan for a query initial position in $O(\log n)$ time. They also give an $O(n^2 \log n)$ preprocessing, $O(n^2)$ space, and $O(n \log n)$ query time algorithm for a restricted version of the multistep sensorless case, improving the previous exponential bound.

In this paper, we bridge the complexity gap between the two previous single-step algorithms [1, 8]. We extend the simple polygon algorithm [8] to work in the general case of a polygonal environment with k polygonal obstacles. For the preprocessing phase, we require $O(kn \log n)$ time and space; we produce a data structure of size $O(kn)$ that supports queries in $O(k \log n)$ time. The complexity of our new algorithm, therefore, is the same as that of the old algorithm [8] for the simple polygon case and the same as that of Briggs' algorithm [1] for the worst case of $\Theta(n)$ obstacles of average constant size. For repeated queries with a fixed environment and the same goal, our new algorithm improves Briggs' algorithm even in the case of $\Theta(n)$ obstacles. In practice, a typical environment contains a small number of obstacles, which nonetheless require many vertices for detailed description. Our technique allows precise specification of the environment without incurring a quadratic blowup in time complexity.

Since our new result is a direct extension of the simple polygon algorithm, we give a summary of that algorithm in Section 3. In Section 4 we describe the modifications to the simple polygon algorithm in the perfect control case, and prove their correctness. In Section 5 we address the issues of imperfect control and position sensing. We conclude with an open problem.

2 The Motion Model

In this section we introduce some notation, give a formal definition of compliant motion, and state a fundamental property of paths followed by a robot doing compliant motion.

We first define notation. If a and b are points, we denote the direction from a to b by $\vec{ab}$. If α is a direction, then $-\alpha$ is the reverse direction. If α and β are directions, then the interval $[\alpha, \beta]$ denotes the range of directions from α counterclockwise to β. We later define a special direction 0 such that $\alpha < \beta$ means that $[\alpha, \beta]$ does not contain direction 0. If ϵ is an angle, then $\alpha + \epsilon$ is the direction ϵ counterclockwise from α. We use α^+ to represent a direction that is infinitesimally counterclockwise from α, so that if $\beta \geq \alpha^+$ then $\beta > \alpha$. Similarly, α^- denotes a direction that is infinitesimally clockwise from α.

The environment, which we call P, is described by a set of k disjoint simple polygons that have a total of n vertices. Without loss of generality, we assume that one of the polygons, called the *outside polygon*, contains all the other polygons, called the *islands*. This assumption is valid since we can always place a bounding polygon of constant complexity around an environment that has only islands. None of the islands contains any other island inside it. The *interior* of the environment is the interior of the outside polygon excluding the islands. The robot always moves through the interior of the environment, which we refer to as *free space*, or slides along a boundary edge (or simply an *edge*) of the environment.

Each edge of the environment has an associated *friction cone* that defines the robot's behavior when it hits the edge. Let $\vec{ab}$ be an edge of the environment, and assume that the interior of the environment is locally to the left of $\vec{ab}$. The friction cone of $\vec{ab}$ divides the direction interval $[\vec{ba}, \vec{ab}]$ into three zones: *zone a*, *zone b*, and the *stop zone*. See Figure 2. Suppose a robot moving in direction α hits $\vec{ab}$. If α is in zone a, then the robot slides toward a; if α is in zone b, the robot slides toward b; and if α is in the stop zone, then the robot is stopped by friction when it hits $\vec{ab}$. In our model, the stop zone is a closed interval. We do not require the friction cones to be symmetric, or even require the stop zone to include the normal to $\vec{ab}$.

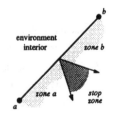

Figure 2

A robot that slides along an edge to an endpoint will do one of three things: leave the boundary and resume travel through free space, slide along the next edge in the same direction, or get stuck. The robot gets stuck on a vertex whenever the sliding directions on the adjacent edges are inconsistent. This includes the case in which the robot reaches such a vertex from free space.

In the next two sections, we assume perfect control, so a starting point p and a commanded direction α define a unique path that the robot takes (a *trail*), which we denote by $T_\alpha(p)$. When there is no ambiguity, we may use T_α instead of $T_\alpha(p)$.

Let us now give the key property of compliant motion paths. Let p be a point in the environment, and let α and β be arbitrary directions. The following theorem states that although trails T_α and T_β start together at p, and may touch at other points, they do not cross. This property holds whether or not the environment has islands, and for arbitrary friction cones.

Theorem 2.1 *Let p be a point in the environment, and let α be a direction. Then T_α has no loops. Furthermore, if β is some other direction, then it is possible to partition the plane into three disjoint pieces, A, B, and Γ, such that*

- *A and B are simply-connected open sets with Γ as their common boundary, and*
- *T_α is completely contained in $A \cup \Gamma$, and T_β is completely contained in $B \cup \Gamma$.*

It is beyond the scope of this paper to give a full proof of Theorem 2.1, but here is the main idea of the proof of the second part of the theorem. First we define an *infinite extension* of a trail as follows: whenever the trail gets stuck on an island, we extend the trail through the interior of the island to the extreme point of the island in the commanded direction, then we continue the trail through free space from that point. Similarly, for the outside polygon we continue the trail in a straight line through the outside of the environment. Next, we show that the infinite extensions of T_α and $T_{-\alpha}$ never cross and use them to divide the plane into two simply-connected regions. Finally, we show by case analysis that T_β never crosses this boundary.

An important corollary to Theorem 2.1 is that if q is any common point of T_α and T_β other than p, then when we go from p to q along T_α and then come back to p along T_β, we trace the boundary of a simply-connected polygonal region. We denote the region of the plane to the left of this boundary, including the boundary, by $R_{\alpha,\beta,q}$. (Note that $R_{\alpha,\beta,q}$ may be either the interior or the exterior of the polygonal region.)

The preprocessing phase of our algorithm works with a fixed goal on the boundary of the environment. Throughout the rest of this abstract, we assume that the goal is a vertex g; however, our results also hold for a goal that is an edge. When $T_\alpha(p)$ contains g, we say that "α is a *good* direction for p." When α and β are both good directions for some point p, we simply write $R_{\alpha,\beta}$ instead of $R_{\alpha,\beta,g}$.

To conclude this section, let us remark on the choice of direction 0, mentioned earlier. The direction 0 is one that is not good for any point in the environment other than the goal itself. (An example of such a direction is the one that points into the environment along the bisector of the two edges incident to the goal. This direction does not point into either edge incident to g, so a robot following commanded direction 0 cannot slide along them. Furthermore, in free space locally around g, it points away from g. Therefore, no compliant motion path with a programmed direction 0 can end at g.) The usefulness of this definition of direction 0 will become apparent in the next section.

3 Summary of the Simple Polygon Algorithm

In this section we sketch the simple polygon algorithm [8]. We describe only the case of single-step motion, perfect control, perfect position sensing, and a vertex goal. The first step is to preprocess the environment, which is a simple polygon P with n vertices, and the goal vertex g. The result of the preprocessing phase is a linear-size data structure that subsequently enables us to produce a "plan" (i.e., report all the good directions), once the starting point p is given.

For any direction α, the α-backprojection, denoted by B_α, is the set of all the points p for which α is a good direction, that is, such that a robot starting at p with a commanded direction α reaches the goal. The basic idea of the simple polygon algorithm is to rotate the direction α a full turn counterclockwise, keeping track of how B_α changes and recording the changes in a data structure.

The α-backprojection is a subpolygon of P. For a given α, B_α has three kinds of edges on its boundary: *bottom edges*, which are edges of P ("P-edges") that the robot slides along; *free-space edges*, which are edges based at vertices of P and extending in direction $-\alpha$ through free space; and the remaining *top edges*, which are portions of edges of P. If we partition P into trapezoids using the α-visibility graph, we can see that B_α is a union of trapezoids, and we can identify these trapezoids by working our way backwards from the trapezoid(s) incident to g.

The following theorem provides motivation for restricting the environment to be a simple polygon:

Theorem 3.1 *For each point p inside P, if $\alpha < \beta$ are good directions for p then every $\gamma \in [\alpha, \beta]$ is a good direction for p.*
(Note: for this theorem to hold, we need direction 0 to have the property stated at the end of section 2.)

This theorem implies that as we rotate α a full turn, points enter and leave B_α at most once. For a given point p, the good directions are all the directions between *start*(p), the direction at which p entered B_α, and *stop*(p), the direction at which p left B_α.

Most of the time, the combinatorial structure of B_α does not change much as we rotate α; only the free-space edges turn, following α. However, for some directions, which we call *events*, there are structural changes. There are three types of events:

Sliding event: a bottom edge of B_α ceases to be slidable towards g, or a P-edge next to the base of a free-space edge becomes slidable towards g;

Visibility event: a free-space edge encounters a P-vertex;

Turnover: a top edge becomes a bottom edge, or a bottom edge becomes a top edge.

At each of the first two types of events, B_α gains or loses triangular sectors and zero or more trapezoids. We collect the regions that B_α gains in one pile, and the regions it loses in another. Since points enter B_α and leave it at most once, these two piles have the property that no two regions in the same pile intersect, and hence each pile forms a *subdivision*. By the same argument (applied to vertices), there are only $O(n)$ events and only $O(n)$ regions in each pile.

At query time, we locate the region that contains the query point p in each subdivision. For the start subdivision, if p is in a trapezoid with direction α, then $start(p) = \alpha^+$; if p is in a sector and sees the base of the sector in direction α, then $start(p) = \alpha^+$. For the stop subdivision, the corresponding direction is α^-. Figures 3 and 4 below illustrate the query processing.

Figure 3   Figure 4

We discover the events and compute the trapezoids "on the fly". We start out with a triangulation of the polygon. During the rotation of α, we maintain, for each free-space edge f of B_α, a data structure called a *path hull*, which stores all the P-vertices that belong to triangles through which f extends and lie ahead of f in its motion; the path hull represents the convex hull of these vertices. The path hulls enable us to determine quickly the next vertex that f is going to hit. We maintain a priority queue in order to determine the next global event. Whenever a free-space edge moves, we update its path hull and generate new events for it in the priority queue. The total number of path hull operations is proportional to the number of regions in the start and stop piles.

In spite of the fact that there can be $O(n)$ free-space edges at any given time, and the size of the path-hull for a given free-space edge can be $O(n)$, we can show that the total size of all the path-hulls is $O(n)$. Each path hull operation is carried out in amortized time proportional to the logarithm of the size of the path hull(s) involved, and since the total size of the path hulls is $O(n)$, processing the path hulls takes $O(n \log n)$ time. The same bound applies for maintaining the event queue and for constructing the subdivisions.

4   Polygons with Islands

Now suppose the environment P has $k - 1$ islands ($k \geq 1$), as described in Section 2. For any given α, the α-backprojection has the same structure as in the simple polygon case, namely, it has bottom, free-space, and top edges, and it can be computed from the trapezoids of the α-visibility graph. The problem is that Theorem 3.1 does not hold when the polygon has islands, and hence as we rotate α, points may come and go more than once. This means that the piles of Section 3 are not subdivisions any more. In order to make the simple polygon algorithm work for the case of polygons with islands, we need a way of separating each pile into layers so that every layer is indeed a subdivision. For the analysis of the complexity of this amended algorithm, we need to bound the number of events that take place as we rotate α, the total size of the piles, and the storage requirements of the path hulls used during the rotation.

The following lemmas bound the number of times a point can enter and leave the α-backprojection:

Lemma 4.1 *If $\alpha < \beta$ are good directions for p, and $R_{\alpha,\beta}$ does not contain any islands, then every $\gamma \in [\alpha, \beta]$ is a good direction for p.*

Proof: We can reduce this scenario to the simple polygon case. We turn the islands into peninsulas by connecting them to the outside polygon by disjoint paths that don't touch $R_{\alpha,\beta}$. Since trails do not cross (Theorem 2.1), all the trails $T_\gamma(p)$ must stay inside $R_{\alpha,\beta}$ and therefore the new edges we introduce cannot affect them. Now the lemma follows from Theorem 3.1. ∎

Lemma 4.2 *For a given point p, there are at most k ranges that define all the good directions for p.*

Proof: Let the smallest good direction for p be α, and the largest good direction for p be ω. If the good directions for p form l ranges, then there are $l-1$ "bad" ranges in between α and ω. Every such bad range must contain an island by Lemma 4.1. An island can appear in at most one bad range, since it can never be split by a trail. Therefore, $l-1$ cannot exceed the number of islands, which is $k-1$. ∎

We proceed as follows: in Section 4.1 we describe the layering scheme of the preprocessing phase and prove its correctness. In Section 4.2 we describe how to answer queries for the single-step, perfect control, perfect position sensing case using the data structures produced by the preprocessing phase. (Section 5 shows how to answer more realistic queries.) Finally, we analyze the complexity of the new scheme in Section 4.3.

4.1 Preprocessing

Let us limit our attention to the pile that contains trapezoids and triangles that are *added* to B_α, known as the "start pile" (the treatment of the stop pile is symmetric). We need to specify the layering scheme that we use to partition this pile into subdivisions.

We use k layers, one for each of the k boundary components of the environment. Recall that a (non-turnover) event involves adding a triangular sector and zero or more trapezoids. The base vertex of a triangular sector belongs to a particular boundary component; we place the sector in the layer of that component. All the trapezoids go to the layer of the component where the *event* took place, as follows:[1]

---

[1] We must treat sectors and trapezoids differently because some sectors in the start pile can be generated by stop events. See [8] for more details.

- for a sliding event, when an edge next to B_α becomes slidable, we use the boundary component that contains that edge;
- for a visibility event, when a leading free-space edge encounters a vertex, we use the boundary component that contains the base vertex of the free-space edge.

Figure 5 illustrates the layering scheme. In this example we use idealized friction cones that are just the normals to the edges. The goal is the vertex at the bottom of the environment. On the left is the layer that corresponds to the outside polygon, and on the right is the layer that corresponds to the island. Note that some trapezoids that are based on the island go into the layer of the outside polygon, because the event in which they were created took place on the outside polygon. In order to prove the correctness of this scheme, we need to show that all the regions in any layer are disjoint. First we need the following lemmas:

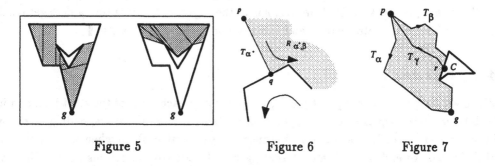

Figure 5   Figure 6   Figure 7

Lemma 4.3 *Suppose p enters the α-backprojection at α^+ during an event associated with boundary component C, and suppose $\beta > \alpha^+$ is also a good direction for p. Then C is outside $R_{\alpha^+,\beta}$.*

Proof: We know that $p \notin B_\alpha$, $p \in B_{\alpha^+}$ so T_{α^+} has almost the same prefix as T_α, but T_α either gets stuck or takes a wrong turn at a point q on C. The point q is arbitrarily close to a point r on C where T_{α^+} goes through. See Figure 6.

Now, since we know that T_{α^+} goes counterclockwise around $R_{\alpha^+,\beta}$, the point q cannot be inside $R_{\alpha^+,\beta}$. Hence the whole interior of C must lie outside $R_{\alpha^+,\beta}$. ∎

Lemma 4.4 *Let p be a point in the environment, let $\alpha < \beta$ be good directions for p, and let C be a boundary component such that $R_{\alpha,\beta}$ does not contain C. Then for any $\alpha < \gamma < \beta$, the trail T_γ cannot get stuck on C.*

Proof: By Theorem 2.1, T_γ stays inside $R_{\alpha,\beta}$. The only way T_γ can get stuck on C is if C touches $R_{\alpha,\beta}$ where T_γ gets stuck. Suppose T_γ gets stuck at a point r on the boundary of $R_{\alpha,\beta}$. This means that T_γ cuts $R_{\alpha,\beta}$ into two pieces: piece A is bounded by T_α and T_γ near p, and piece B is bounded by T_γ and T_β near p. Only one of these two pieces contains the goal g. See Figure 7.

Suppose A contains g. Again by Theorem 2.1, T_β stays in B up to r. But T_β does not get stuck at r, so it crosses to the "left" (to A) at that point. Because T_γ always slides left more easily than T_β, it cannot get stuck at r.

The other case is symmetric. ∎

Using the above lemmas, we can now prove the correctness of the layering scheme:

Theorem 4.5 *For any layer, any two regions are disjoint.*

Proof: We prove the theorem by contradiction. Suppose that in the layer of boundary component C there are two regions that have a nonempty intersection, and suppose p is in their intersection. One region implies α^+ as a start direction for p, and the other region implies β^+, where $\beta > \alpha$.

By Lemma 4.3, C is not contained in R_{α^+,β^+}. Since $\alpha^+ < \beta < \beta^+$, by Lemma 4.4 T_β cannot get stuck on C. However, p is in the region of β^+ because T_β *does* get stuck on C. This is a contradiction, and therefore no such point p exists. ∎

4.2 Queries

When we are given a query point $p \in P$, we locate it in each layer of the start pile, and in each layer of the stop pile. Every region that contains p implies a corresponding direction in which p enters or leaves the α-backprojection (see Section 3), so when we sort the list of directions we end up with an alternating list of start-stop directions for p. This list specifies the good ranges for p, and this is the answer to the query.

We can use the same data structure to answer queries in the imperfect control case, or when we don't know the exact starting point of the robot. We give more details about these cases in Section 5 below.

4.3 Analysis

There are two quantities to consider, space and time. To help bound working storage, we have the following lemma:

Lemma 4.6 *For a fixed α, every triangulation edge intersects B_α in at most k segments. Consequently, the total size of all the path hulls is $O(kn)$.*

The total preprocessing time is the total number of trapezoids and sectors plus the number of turnover events (which do not contribute any regions to the pile, but have to be processed nevertheless), multiplied by the time it takes to process each one.

Lemma 4.7 *The total number of sectors and trapezoids in the start and stop piles is $O(nk)$.*

Proof: Each vertex of P enters B_α at most k times as α rotates. Every trapezoid or sector added to the start pile can be associated with one of these vertex entrances, and each vertex entrance has at most one trapezoid and one sector associated with it. A symmetric argument applies to the stop pile. ∎

Every P-edge can turn over at most twice, so there are $O(n)$ turnovers. Processing a trapezoid, a sector, or a turnover involves a path hull operation, and since the size of each path hull is $O(n)$, each operation takes $O(\log n)$ time. Therefore, the total preprocessing time is $O((kn + n)\log n) = O(kn \log n)$.

It is easy to keep track of the adjacency of the regions we generate in each layer (since α rotates monotonically). All the regions in each layer are connected by adjacency to the island that is associated with that layer, so if we triangulate the island and add the triangles to the layer, we get a connected subdivision at no extra cost. We can preprocess this subdivision for point location using time and space linear in the size of the layer [5, 9].

Query answering involves $O(k)$ point location queries, which take $O(k \log n)$ total time, and sorting at most $2k$ numbers, which takes $O(k \log k) = O(k \log n)$ time.

For memory management during preprocessing, we store each path hull in a chunk of memory no more than four times the size of the path hull. When a path hull gets too large or too small, we copy it to a new chunk. (This is the second allocation scheme described in our previous paper [8].) We preallocate enough buffers to satisfy the demand at any given time: for $0 \le i \le \log n$, we allocate $O(kn/2^i)$ buffers of size $O(2^i)$ each. The total storage used is $O(kn \log n)$, and this suffices: at any moment, a path hull is of size at most $O(n)$, and the total size of all path hulls is $O(kn)$, so there will always be enough buffers. The total cost of copying path hulls is $O(kn \log n)$.

5 Imperfect Control

The preprocessing of Section 4.1 is aimed toward answering queries for the perfect control, perfect position sensing case. However, once we have completed the preprocessing, we can easily answer more realistic queries, allowing imperfect control and imperfect position sensing.

The control uncertainty of a robot is an angular constant ϵ in the range $0 \le \epsilon < 90°$. If the robot is programmed to move in direction α, its actual direction of motion (in free space) at any given instant lies in the range $[\alpha - \epsilon, \alpha + \epsilon]$. The path the robot follows now is not unique. We say that a direction α is *good* for a starting point p if every path the robot may take, when commanded to move from p in direction α, is guaranteed to get the robot to the goal. We relate perfect and imperfect control in Lemma 5.2 below, but first we need another property of paths under perfect control:

Lemma 5.1 *If α and β are good directions for p, and $R_{\alpha,\beta}$ contains an island inside it, then there exists some $\gamma \in [\alpha, \beta]$ that is not a good direction for p.*

Lemma 5.2 *Let p be in P, and let (α, β) be one of the (at most k) ranges of good directions for p for a robot with perfect control. Let the control uncertainty be ϵ.*

- *if (α, β) is wider than 2ϵ, then any $\gamma \in (\alpha + \epsilon, \beta - \epsilon)$ is a good direction for the imperfect robot;*
- *if (α, β) is narrower than 2ϵ, then no $\gamma \in (\alpha, \beta)$ is a good direction for the imperfect robot.*

Using Lemma 5.2, it is clear how we can use the structure from Section 4.1 to answer queries under the imperfect control model. First, we find the good ranges under the perfect control model, as described in Section 4.2. Next, we increase every start direction by ϵ and decrease every stop direction by ϵ. Every range that stays nonempty is a good range under the imperfect control model. Note that we use ϵ only at query time, so the preprocessing is independent of the control uncertainty. This means that the same data structures can support several robots, with different capabilities, that operate in the same environment.

The *nondirectional backprojection* is the set of all points that can reach the goal in a single compliant motion. When we assume perfect control, this set is the union of all the regions in the start pile (which is identical to the union of all the regions in the stop pile). When the control uncertainty ϵ is greater than 0, consider ν, the locus of all the points p such that p has at least one good range of size exactly 2ϵ. It is not hard to see that ν is a non-simple curve made up from line segments and circular arcs. Furthermore, the interior of ν, as defined by its winding number, is the nondirectional backprojection. We compute ν in $O(kn)$ steps by marching through the layer data structure.

We have so far implicitly assumed that the robot can detect the goal when it reaches it. If this is not the case (the robot is *sensorless*), we can still use our data structures to answer queries. We compute good ranges as above, then intersect them with the range of directions guaranteed to stick at the goal [8].

When the robot has imperfect position sensing, we may know its starting position only approximately. Formally, we know that the robot is somewhere in a (typically small) subset S of the environment, known as the *start region*. In this case, we need to find directions that are good for all the points in S simultaneously. It is not hard to see that it suffices to find directions that are good for the boundary of S. We can modify the technique discussed in our simple polygon paper so that we obtain these good directions by $O(k)$ walks along the boundary of S through the layer data structure, where we never visit the same triangle or trapezoid more than the number of times the boundary of S crosses it. For many "standard" starting regions, such as a disc (intersected with the environment) or a polygon with a constant number of sides, this technique yields a bound of $O(t + k \log n)$, where t is $O(kn)$ in the worst case, but is typically much less, depending on the number of triangles and trapezoids that intersect S.

Finally, using our data structures, we can solve a restricted version of the multi-step problem for sensorless robots in $O(kn^2 \log n)$ preprocessing time, $O(kn^2)$ space, and $O(kn \log n)$ query time, again by a slight modification of our previous algorithm [8, Section 7]. For the sake of brevity we omit the details.

6 An Open Problem

In the standard model of compliant motion [1, 2, 3, 4, 6, 10], a vertex of the environment can be "sticky" when the robot encounters it coming from free space. In reality, however, if the robot "shakes" a little, it may get unstuck and reach the goal. In other words, it is possible that this vertex separates two good ranges for the starting point of the robot. Under the model in which the robot is able to "shake loose" when it gets stuck, it may be possible to unite these two ranges. This can produce a good range for the imperfect control case, even when no good range exists under the standard model. We conjecture that it is possible to unite ranges, given the control uncertainty of the robot, in $O(kn \log n)$ preprocessing and $O(k \log n)$ query time.

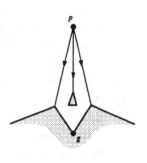

Figure 8

Acknowledgement

We would like to thank Leo Guibas for reading an early draft of this paper and for making useful comments about Theorem 2.1.

References

[1] A. J. Briggs. An efficient algorithm for one-step planar compliant motion planning with uncertainty. In *Proceedings of the 5th ACM Symposium on Computational Geometry*, pages 187–196, 1989.

[2] J. F. Canny and J. Reif. New lower bound techniques for robot motion planning problems. In *Proceedings of the 28th IEEE Symposium on Foundations of Computer Science*, pages 49–60, 1987.

[3] B. R. Donald. Error detection and recovery for robot motion planning with uncertainty. Technical Report MIT-AI-TR 982, Artificial Intelligence Laboratory, Massachusetts Institute of Technology, 1987.

[4] B. R. Donald. The complexity of planar compliant motion planning under uncertainty. In *Proceedings of the 4th ACM Symposium on Computational Geometry*, pages 309–318, 1988.

[5] H. Edelsbrunner, L. J. Guibas, and J. Stolfi. Optimal point location in a monotone subdivision. *SIAM Journal on Computing*, 15:317–340, 1986.

[6] M. A. Erdmann. Using backprojections for fine motion planning with uncertainty. *International Journal of Robotics Research*, 5(1), 1986.

[7] M. A. Erdmann and M. Mason. An exploration of sensorless manipulation. In *IEEE International Conference on Robotics*, pages 1569–1574, 1986.

[8] J. Friedman, J. Hershberger, and J. Snoeyink. Compliant motion in a simple polygon. In *Proceedings of the 5th ACM Symposium on Computational Geometry*, pages 175–186, 1989.

[9] D. Kirkpatrick. Optimal search in planar subdivisions. *SIAM Journal on Computing*, 12:28–35, 1983.

[10] J.-C. Latombe. Motion planning with uncertainty: The preimage backchaining approach. Technical Report STAN-CS-88-1196, Department of Compuuter Science, Stanford University, Stanford, California 94305, March 1988.

[11] B. K. Natarajan. *On Moving and Orienting Objects*. PhD thesis, Cornell University Department of Computer Science, Ithaca, N.Y., 1986.

[12] C. Ó'Dúnlaing, M. Sharir, and C. K. Yap. Retraction: A new approach to motion-planning. In *Proceedings of the 15th Annual ACM Symposium on Theory of Computing*, pages 207–220. ACM, 1983.

[13] M. Sharir and A. Schorr. On shortest paths in polyhedral spaces. In *Proceedings of the 16th Annual ACM Symposium on Theory of Computing*, pages 144–153, 1984.

FAST ALGORITHMS FOR GREEDY TRIANGULATION

Christos Levcopoulos and Andrzej Lingas
Department of Computer Science
Lund University, Box 118, 22100 Lund, Sweden

Abstract: We present the first quadratic-time algorithm for the greedy triangulation of a finite planar point set, and the first linear-time algorithm for the greedy triangulation of a convex polygon.

1. Introduction

Planar point sets and simple polygons are special cases of the so called planar straight-line graphs (PSLG for short) [PS]. A PSLG G is a pair (V, E) such that V is a set of points in the plane and E is a set of non-intersecting, open straight-line segments whose endpoints are in V. The points in V are called vertices of G, whereas the segments in E are called edges of G. If G is a simple cycle, it is called a (simple) *polygon*. If G has no edges, it is a *planar point set*. A *diagonal* of G is an open straight-line segment that neither intersects any edge of G nor includes any vertex of G and that has its endpoints in V. A *triangulation* of G is a maximal set of non-intersecting diagonals of G.

A minimum weight triangulation of G is a triangulation of G that minimizes the total diagonal length. The problem of finding a minimum weight triangulation of a planar point set was raised in numerical analysis several years ago but its complexity status has not been resolved yet (see [A,Ll,PS]).

Among the polynomial-time heuristics for minimum weight triangulation of a planar point set, one of the most known and simplest is the so called *greedy triangulation*. It is known that it does not approximate the optimum [Le,MZ]. The greedy triangulation method can be generalized to include any PSLG G. It inserts a diagonal d of G into the plane if d is the shortest among all diagonals of G that neither intersect nor overlap with those already in the plane.

For a long time, the most efficient known algorithm for computing the greedy triangulation of a planar point set was that due to Gilbert [Gi], taking $O(n^2 \log n)$ time and $O(n^2)$ space. In [Li87,Li89], Lingas showed that by using the so called *Voronoi diagrams with barriers*, the greedy triangulation of a PSLG, in particular, a planar point set, can be constructed in time $O(n^2 \log n)$ and only linear space. Independently, an analogous result only for planar point sets was obtained by Goldman [Go]. Interestingly, if the input planar point set is uniformly distributed then by using Voronoi diagrams with barriers, the greedy triangulation can be computed in $O(n \log n)$ expected time [Li88]. In this paper, we combine the concept of Voronoi diagrams with barriers with a generalization of the linear-time algorithm for Voronoi diagrams of convex polygons due to Aggarwal, Guibas, Saxe and Shor [AGSS] to obtain the first $O(n^2)$-time (and linear-space) algorithm for the greedy triangulation of a planar point set, and more generally, a PSLG. In this way, we make progress on one of the open problems in computational geometry listed by Aggarwal (see P188 in [A]).

In [LL87], Levcopoulos and Lingas showed that the greedy triangulation for convex polygons approximates the minimum weight triangulation. They also presented a quadratic-time algorithm for the greedy triangulation of convex polygons [LL87], and together with Sack, a linear-time algorithm for the greedy triangulation for a special class of convex polygons called semi-circular polygons [LLS88]. Our second main result in this paper is a linear-time algorithm for the greedy triangulation of convex polygons which affirmatively resolves another open problem in Aggarwal's list (See P188 in [A]).

The structure of the paper is as follows. In Section 2 we introduce Voronoi diagrams with barriers. In Section 3 we present the quadratic-time algorithm for the greedy triangulation of a PSLG. The linear-time algorithm for the greedy triangulation of convex polygons is presented in two sections. In Section 5 we consider the special case where the input polygon resembles a long and thin rod whereas in Section 4 we consider the more natural, other case.

2. Voronoi diagrams with barriers

We start with a more formal definition of Voronoi diagrams with barriers.

Definition 2.1: Let $G = (V, E)$ be a PSLG. For $v \in V$, the region $P(v)$ consists of all points p in the plane for which the shortest, open straight-line segment between p and a vertex of G that does not intersect any edge of G is (p, v). The minimal set of straight-line segments and half-lines that complements G to the partition of the plane into the regions $P(v)$, $v \in V$, is called the *Voronoi diagram with barriers* of G ($Vorb(G)$ for short). The maximal straight-line segments or half-lines on the boundaries of the regions $P(v)$, $v \in V$, that do not overlap with edges of G, are called edges of $Vorb(G)$. The endpoints of edges of $Vorb(G)$ are called vertices of $Vorb(G)$.

$Vorb(G)$ has several properties analogous to those of classical Voronoi diagrams (comp. [Li89] with [PS]). Every edge of $Vorb(G)$ is a continuous part of the perpendicular bisector of a pair of vertices of G from which the edge is visible. Next, every vertex of $Vorb(G)$ is the common intersection of at least three edges of $Vorb(G)$ or residues inside an edge of G. For all $v \in V$, $P(v)$ is a collection of polygonal sub-regions separated by the edges incident to v. There are at most $3n - 6$ edges in $Vorb(G)$ (see [Li89]). Wang and Schubert have presented an asymptotically optimal algorithm for constructing the Voronoi diagram with barriers of G (called the bounded diagram of G in [WS]).

Fact 2.1 [WS]: Let G be a PSLG with n vertices. The Voronoi diagram with barriers of G can be constructed in time $O(n \log n)$ and linear space.

3. Greedy triangulation in quadratic time

In [Li89], the following characterization of shortest diagonals of planar straight-line graphs has been proved:

Fact 3.1 [Li89]: A shortest diagonal of a PSLG $G = (V, E)$ either lies inside the union of the regions of its endpoints in $Vorb(G)$ or cuts off an empty triangular face from G and lies within at most four different regions in $Vorb(G)$.

Remark 3.1: It follows also from the proof of Fact 3.1 that a shortest diagonal of PSLG can intersect properly at most three different regions of $Vorb(G)$ (see also Fig. 3 in

[Li89]). Moreover, if it properly intersects three different regions of $Vorb(G)$ then it cuts off an empty triangle and the middle region intersected belongs to the apex of the triangle different from the endpoints of the diagonal (see Fig. 1).

By Fact 3.1, we can use the following algorithm to construct a greedy triangulation.

Algorithm 1

Input: a PSLG G

Output: a greedy triangulation T of G

1. $T \leftarrow \emptyset$;
2. Compute $Vorb(G)$;
3. <u>while</u> T is not a complete triangulation <u>do</u>
   <u>begin</u>
      a. Using $Vorb(G \cup T)$ find a shortest diagonal d of $G \cup T$;
      b. $T \leftarrow T \cup \{d\}$
      c. Update $Vorb(G \cup T)$ (i.e. compute $Vorb(G \cup T)$ from $Vorb(G \cup T - \{d\})$)
   <u>end</u>
4. Output T

By Fact 2.1, Step 1 can be implemented in time $O(n \log n)$ and space $O(n)$. Thus, to implement Algorithm 1 in quadratic time and linear space it remains to show how to implement Step 3.a and Step 3.c in linear time.

Theorem 3.1: Let H be a PSLG. Given H and $Vorb(H)$, we can find a shortest diagonal of H in linear time.

Proof: Let D be the set of diagonals d of H such that the regions of the endpoints of d share a continuous piece of boundary in $Vorb(H)$. The set D is sometimes called a generalized Delaunay triangulation of H (see [C,Li89,LL86,WS]), and can be constructed in linear time, given a DCEL representation (see [PS]) of $Vorb(H)$. Next, let E be the set of all diagonals that cut off empty triangular faces from H and intersect (not necessarily properly) three regions in $Vorb(H)$. By Fact 3.1 and Remark 3.1, any shortest diagonal of H is in the union of D and E. Thus, it suffices to show how to construct E in linear time.

To begin, we find all pairs e_1, e_2 of edges in H such that e_1, e_2 share an endpoint and form an acute angle which does not include any other edge incident to this endpoint. Given H, we can easily find the above pairs in linear time. Next, for each such pair e_1, e_2, we perform the following test:

Let w be the common endpoint of e_1, e_2. For $i = 1, 2$ find the intersection point n_i of e_i with the bisector b_i (see [PS]) of w and the other endpoint of e_i. For $i = 1, 2$, check whether the intersection points n_i lie on the boundary of the region of w and the region of the other endpoint of e_i in $Vorb(H)$.

By the convexity of the sector of the region of w between e_1 and e_2, the segment connecting the other endpoints of e_1, e_2 is in E iff both answers are positive. Given $Vorb(H)$, we can perform the test in constant time following the boundary of the region of w from the intersection of e_i with b_i, for $i = 1, 2$.

Thus, we can construct also E in linear time. Finally, we can find a shortest diagonal in $D \cup E$ in linear time, which completes the proof. Q.E.D.

In [AGSS], Aggarwal *et al.* sketch a linear-time method of updating a classical Voronoi diagram of planar point sets after deleting a point from the set, using a generalization of their linear-time algorithm for Voronoi diagrams of convex polygons. In the proof of the following theorem we show that the problem of implementing Step 3.c in linear time, i.e. updating $Vorb(G \cup T)$ after inserting a shortest diagonal into T, is similar to that of updating a classical Voronoi diagram after a point deletion.

Theorem 3.2: Let d be a shortest diagonal of a PSLG H. Given $Vorb(H)$, we can compute $Vorb(H \cup \{d\})$ in linear time.

Proof: By Fact 3.1 and remarks following it, we need to update $Vorb(H)$ only if d properly intersects three regions in $Vorb(H)$ cutting off an empty triangle. Then, as in Fig. 1, the middle of the three regions belongs to the apex v_1 of the triangle different from the endpoints of d. Let P be the part of the region of v_1 in $Vorb(H)$ cut off from v_1 by d. Next, let S be the set of vertices in H different from v_1 whose regions in $Vorb(H)$ touch the boundary of P. It is easy to see that $Vorb(H \cup \{d\})$ overlaps with $Vorb(S)$ within P and with $Vorb(H)$ outside P. Thus, the problem of updating $Vorb(H)$ to $Vorb(H \cup \{d\})$ reduces to that of computing the intersection of the classical Voronoi diagram of S with the inside of P.

On the other hand, consider the Delaunay triangulation D of $S \cup \{v_1\}$. Note that all vertices in S are Delaunay neighbours of v_1 in D, i.e. are adjacent to v_1 in D. Apply the standard lifting map $\mu(x, y) \rightarrow (x, y, x^2 + y^2)$ to $S \cup \{v\}$. It follows from the known relationship between the Delaunay triangulation of a planar point set W and the convex hull of $\mu(W)$ that for each $v \in S$, $(\mu(v), \mu(v_1))$ is an edge of the convex hull of $\mu(S \cup \{v_1\})$ (see [E]).

Hence, the points in $\mu(S)$ lie on a common convex cone whose apex is $\mu(v_1)$. Therefore, we can compute the convex hull of $\mu(S)$ using a generalization of the algorithm for Voronoi diagrams of convex polygons given in [AGSS] (see p.45) in linear time. In turn, given the convex hull of $\mu(S)$, we can easily compute the Voronoi diagram of S in linear time by the above relationship [E].

Now, it remains to glue the Voronoi diagram of S inside P with $Vorb(P)$ outside P to obtain $Vorb(H \cup \{d\})$. Given $Vorb(H)$, d and v_1, we can construct P, S and cut off P from $Vorb(H)$ in time proportional to the size of P. Given the Voronoi diagram of S and P, we can find the intersection of the diagram with P in linear time by the convexity of the regions and P (see [PS] pp. 263-269). We conclude that the whole updating of $Vorb(H)$ to $Vorb(H \cup \{d\})$ can be done in linear time. Q.E.D.

Putting everything together, we obtain our first main result.

Theorem 3.3: The greedy triangulation of a PSLG on n vertices can be computed in time $O(n^2)$ and space $O(n)$.

4. The linear-time algorithm for non-rod convex polygons

Our algorithm for greedy triangulation of convex polygons makes distinction between two cases of the input convex polygon P. In one of the cases, P has roughly a shape of a long and thin rod. In this section, we present a linear-time algorithm for the greedy triangulation of convex polygons in the other, much more natural case.

Formally, we can distinguish the two cases as follows. Turn the input convex polygon P to obtain its maximum length perpendicular

projection on the X axis. Compute the ratio $rod(P)$ between the length of perpendicular projection of P on the X axis and the length of perpendicular projection of P on the Y axis. Now, we can choose an arbitrary positive constant c which is greater than 10, and call P a rod if $rod(P) > c$. Clearly, the choice of c influences the other constants in the algorithms for the greedy triangulation in both cases.

Our algorithm for the non-rod case (Algorithm 2) is slightly simpler than that presented in the next section for the rod case. The reason is that if P is not a rod then there is always a continuous piece of the perimeter of P with a constant number of vertices which certifies that some diagonal can be inserted in the greedy triangulation. Thus, Algorithm 2 walks around P in order to test such perimeter pieces, and possibly produces new greedy diagonals. If it performs a non-constant number of unsuccessful tests it will eventually walk backwards by producing greedy diagonals. Algorithm 2 uses a counter which tells how long is the chain of unsuccessful tests. The counter causes Algorithm 2 to jump slightly backwards whenever a new greedy diagonal is produced. Additionally, in some special case, in order to avoid rotating, Algorithm 2 treats a chosen vertex as the so called *guard vertex*, impossible to pass. The use of the counter and the guard vertex make it impossible for Algorithm 2 to rotate more than once without producing $\Omega(n)$ edges.

In our specification of Algorithm 2, we assume that no two diagonals of P have the same length. If it is not so then we can impose a total order which is the extension of the diagonal length partial order by using for instance their lexicographic order. Note that the above assumption implies that the greedy triangulation is unique. To specify Algorithm 2, we need the following notation.

Definition 4.1: Let P be a convex polygon, and let p be a point on its perimeter. The set of all vertices of P that are reachable from p by following at most k edges of P clockwise or counterclockwise is denoted by $V_k(p, P)$ (When P follows from the context, we shall simply write $V_k(p)$ instead). A vertex w of P is said to be in the V-distance i, where i is a natural number, from p if $w \in V_i(p)$ and $w \notin V_{i-1}(p)$.

Algorithm 2

Input: a non-rod convex polygon P given as a circular doubly-linked list of vertices of P.
Output: the greedy triangulation T of P.

Step a. Set v to an arbitrary vertex of P, and C to 0.

Step b. If P has $\leq 8k$ vertices then find the greedy triangulation of P by using a naive cubic-time algorithm, extend T by the diagonals from the above triangulation, output T and stop.

comment: k denotes here any sufficiently large positive constant whose minimum value follows from the algorithm analysis.

Step c. Set V'_{2k} to V_{2k} if the guard vertex is not in $V_{2k}(v)$. Otherwise set V'_{2k} to the set of all vertices in $V_{2k}(v)$ on the same side of the guard vertex as v. Analogously define $V'_k(v)$. Find the shortest diagonal s with both endpoints in $V'_{2k}(v)$. If s has both endpoints in $V'_k(v)$ then go to Step e, otherwise go to Step d.

Step d. If s cuts off at least k vertices from V'_{2k} then set the guard vertex to a vertex in V'_{2k} cut off by s whose shortest V-distance to an endpoint of s is maximum. Let v' be the last endpoint of s clockwise. If v' is after v clockwise then set v to v' and increase C by one. Otherwise set v to the other endpoint of s and decrease C by one. Go to Step c.

Step e. Insert s into T. Find the greedy triangulation of the subpolygon cut off by s from $V_{k'}(v)$ by applying a naive cubic-time algorithm. Insert the diagonals comprising the above greedy triangulation into T. If $C \geq 0$ then set v to the last vertex in $V'_{2k}(v)$ counterclockwise. Otherwise set v to the last vertex clockwise. If C is nonnegative set it to $\max\{0, C - 2k\}$. Otherwise, set C to $\min\{0, C + 2k\}$. Go to Step b.

To prove the correctness of Algorithm 2, we need introduce the concept of a *locally shortest* diagonal.

Definition 4.2: A diagonal d of a simple polygon P is said to be locally shortest if there is no shorter diagonal of P that intersects d.

By the definition of the greedy triangulation and Definition 4.2, the following lemma implies that all edges produced by Algorithm 2 belong to the greedy triangulation.

Lemma 4.1: If the diagonal d in Step e is inserted in T then it is locally shortest.

Sketch: Since nor two of k vertices preceding d clockwise neither two vertices of k vertices following d clockwise are endpoints of a diagonal of P shorter than d, there are two continuous pieces L and R of the perimeter of P such that:

1. L immediately preceds d in clockwise order and R immediately follows d in clockwise order.

2. L and R are of length $\geq \frac{1}{2}kd$.

3. there is no diagonal intersecting d that both endpoints in $L \cup R$.

Suppose now that d is not locally shortest. Then, d is intersected by a diagonal e shorter than d with both endpoints outside L and R. Hence, it follows by (2) that the ratio $rod(P)$ is at least $\frac{k-2}{4}$. We obtain a contradiction. Q.E.D.

In order to complete the proof of correctness of Algorithm 2, and analyse its time complexity, we distinguish a special case in Step d called a *balloon case*. It means the situation where the shortest diagonal s with both endpoints in $V'_{2k}(v)$ cuts off at least k vertices from $V'_{2k}(v)$. The following observation will be helpful in showing that the balloon case can happen only once.

Remark 4.1: Let d be a diagonal of a convex polygon P. Next, let P_1, P_2 be the two subpolygons of P cut off by d. There is $i \in \{0, 1\}$ such that the length of perimeter of P_i is $O(|d|)$.

Lemma 4.2: A balloon case occurs at most once during a performance of Algorithm 2.

Proof: Suppose that a balloon case occurs. Let $s = (v_1, v_2)$ be the diagonal with both endpoints in $V'_{2k}(v)$ found in Step c that satisfies the requirements of the balloon case according to Definition 4.3. Next, let P_1 be the subpolygon cut off by s from P that includes the at least k vertices from $V_{2k}(v)$ and does not contain vertices outside $V'_{2k}(v)$. Since s is the shortest diagonal of P_1, the subpolygon P_1 satisfies the following property: For each vertex w of P_1 in V-distance from $\{v_1, v_2\}$ equal to i, $i > 1$, w is in the Euclidean distance $\Omega(i \mid s \mid)$ from s.

By the above property, P_1 has perimeter $\Omega(k \mid s \mid)$. It follows also that any diagonal of P shorter than s which intersects s is incident to a vertex of P_1 in $O(1)$ V-distance from s. We shall call the vertices of P_1 in $O(1)$ V-distance from s border vertices of P_1. Let P_2 be the other subpolygon cut off from P by s. By Remark 4.1 and the $\Omega(k \mid s \mid)$ bound on the perimeter length of P_1, the perimeter of P_2 has length $O(\mid s \mid)$. Delete s from P_2 and extend the remaining convex polygonal chain to the minimal polygonal convex chain C_2 that includes the $O(1)$ border vertices of P_1. Clearly, C_2 has also length $O(\mid s \mid)$. Hence, Algorithm 2 always finds a diagonal of C_2 shorter than $\mid s \mid$ until C_2 is peeled to a chain with $O(1)$ vertices. Note that then P is peeled to a subpolygon with $\leq 4k$ vertices which is triangulated in Step b. Thus, a new ballon case cannot occur so late. We shall show that neither any new balloon case can occur in C_2 earlier. Suppose otherwise. Let s' be a diagonal of P'_2 found in Step c that satisfies the requirements of the balloon case. Similarly, as s, it splits the current subpolygon P' into a small perimeter subpolygon P'_1, and a large perimeter subpolygon P'_2 on $\theta(k)$ vertices. After v became the guard vertex no later test done in Step c can cover all vertices in $P_1 - C_2$. For this reason, the large perimeter subpolygon P'_2 consists of a continuous piece of C_2 and s, and the small perimeter subpolygon P'_1 includes almost whole P_1 possibly but for the $O(1)$ border vertices of P_1. By the bounds on the length of perimeters of P_1 and P_2 previously derived, and convexity of all considered subpolygons, we obtain a contradiction. Q.E.D.

The following lemma completes the proof of correctness of Algorithm 2 and implies its linear-time performance.

Lemma 4.3: The absolute value of the counter C during performance of Algorithm 2 is always $O(n)$.

Proof: Suppose first that the balloon case does not occur. Then, the only way for Algorithm 2 to increase the value of the counter C to a value greater than $2k$ is to perform at least $2k$ consecutive, unsuccessful tests which result in $\geq 2k$ jumps in the clockwise direction. Let v_1 be the first vertex set to v in the above sequence of unsuccessful tests. Until the counter is set to zero, Algorithm 2 will operate at the end of a chain of vertices and shorter and shorter diagonals, starting from v_1 clockwise. The chain corresponds to a sequence of unsuccessful tests done in Step c, where consecutive tests overlap on at least $2k$ vertices. Whenever Algorithm 2 inserts diagonals satisfying the tests into T in Step c, it jumps $2k$ vertices backwards counterclockwise. If unsuccessful tests occur again, Algorithm 2 proceeds clockwise again since the shortest diagonal s intersecting the previous shortest one cannot come from the already tested neighbourhood. The only way for the counter C to get a value close to n would be to reach v_1 from the other side. In the latter case, Algorithm 2 would certainly find a locally shortest diagonal which would force it to jump back. We conclude that in the balloon-free case the counter C is always $O(n)$. If the balloon case occurs, let c be the value of the counter C in Step d at that moment. By repeating the argumentation from the previous case, we know that the counter cannot exceed $\max(c, 0) + O(n)$ which is $O(n)$.

Symmetrically, we can prove that the counter C has always value greater than $-an$, for some positive constant a. Q.E.D.

By Lemma 4.3, Algorithm 2 always produces a complete triangulation of P. Otherwise, it would rotate clockwise or counterclockwise such that the absolute value of C would be unbounded. This combined with Lemma 4.1 proves the correctness of Algorithm 2. It remains to estimate the time complexity of Algorithm 2.

Lemma 4.4. Algorithm 2 takes $O(k^2 \cdot k \cdot n)$ time.

Proof: The time complexity of Algorithm 2 can be estimated from above by the total number q of times C is decreased or increased by one times the cost of Steps c,d and e. It is easy to see that Steps c, d take $O(k^2)$ time, whereas Step e takes $O(k^3)$ time. On the other hand, by the rules of changing the value of C, and by Lemma 4.3, q can be estimated by $n(2k+1) + O(kn)$ from above. It follows that Algorithm 2 takes $O(k^3 \cdot k \cdot n)$ time. We can improve the above estimation to $O(k^2 \cdot k \cdot n)$ by charging the edges of the greedy triangulations of the local subpolygons with all the vertices in $V_k(v)$ with the cost of the cubic-time algorithm applied locally in Step e. The charge is then $O(k^2)$ per such an edge. The remaining work in Step e takes time $O(k^2)$ which implies that Algorithm 2 takes $O(k^2 \cdot q) + O(k^2 \cdot n) = O(k^2 \cdot k \cdot n)$ time. Q.E.D.

Putting everything together, we obtain the following theorem.

Theorem 4.1: Algorithm 2 constructs the greedy triangulation of a non-rod convex polygon in linear time.

5. The rod case

The rod case of convex polygons, although very peculiar if the rod ratio is large, is slightly more complicated than the non-rod one. We assume that the input polygon P is so placed that it maximizes the length of its perpendicular projection on the X-axis. If not, then we can easily turn P to such a position in linear time. The main difference from the non-rod case is that now we have to take into account both the upper side and the lower side of P, while searching for a shortest diagonal in a constant neighbourhood of v and trying to certify that the diagonal found is locally shortest. For this reason, instead of the neighbourhoods $V_k(v)$, $V_{2k}(v)$, we use the neighbourhoods $X_k(v)$, $X_{2k}(v)$ according to the following definition.

Definition 5.1: Let P be a convex polygon, and let v be its vertex. Next, let $p(v)$ be the vertical projection of v on the opposite (horizontal) side. Draw vertical straight-lines through the extreme leftmost and rightmost vertices in $V_k(v)$ and $V_k(p(v))$. The set of all vertices of P between the rightmost of the two lines passing through the leftmost vertices and the leftmost of the two lines passing through the rightmost vertices is denoted by $X_k(v)$.

In order to access $X_k(v)$ easily, we preprocess P by sorting its vertices by their X-coordinates, and forming a doubly linked list of so sorted vertices.

Our recursive procedure for the greedy triangulation in the rod case can be seen as a generalization of Algorithm 2 in part. It searches for the shortest diagonal s in $X_{2k}(v)$. If s does not have both endpoints in $X_k(v)$ it sets v to an endpoint of s and repeats the search. If s has both endpoints in $X_k(v)$ then the procedure inserts s into the triangulation constructed, and splits the current subpolygon of P into two subpolygons P_1 and P_2. Next, it calls recursively itself two times, first with P set to P_1 and v to an endpoint of s, then with P set to P_2 and v to the endpoint of s.

Unfortunately, the above generalization of Algorithm 2 to include both sides of P by considering the neighborhoods $X_{2k}(v)$ instead of $V_{2k}(v)$ may lead to endless oscillations at the horizontal extremes of P or production of a non locally shortest diagonal. For example, consider the situation shown in Fig. 5.1.

Fig. 5.1

Suppose that the current vertex v is v_1. In $X_{2k}(v_1)$, (v_{2k-2}, v_{2k}) is the shortest diagonal and there is no diagonal with both endpoints in $X_k(v_1)$ that is not intersected by a shorter diagonal with both endpoints in $X_{2k}(v_1)$. Therefore, v is set to v_{2k-2} or v_{2k}. In $X_k(v_{2k-2})$ and $X_k(v_{2k})$, (v_{2k}, v_{2k+2}) is the shortest diagonal which is not intersected by any shorter diagonal with both endpoints in $X_{2k}(v_{2k-2})$ or $X_{2k}(v_{2k})$ respectively. Thus, we would produce (v_{2k}, v_{2k+2}) as a greedy edge inspite of the fact that it intersects (v_1, v_{2k+1}) and $|(v_1, v_{2k+1})| < |(v_{2k}, v_{2k+2})|$.

For this reason, the recursive procedure Rod switches to a subroutine $Rod - end$ when the current vertex v defining the local neighbourhood is enough close to a horizontal extreme of P. $Rod - end$ uses the neighbourhood $V_k(v)$ to search for locally shortest diagonals in a similar fashion as Algorithm 2. The procedure Rod together with preceeding preprocessing will be termed as Algorithm 3.

Algorithm 3

Step a. For each vertex v of P compute the length $L(v)$ of the intersection of the vertical straight-line passing through v with the inside of P.

Step b. Set T to $\emptyset$.

Step c. Set v to the middle vertex of P with respect to the X-coordinate.

Step d. Call $Rod(P, v)$

procedure $Rod(P, v)$
begin

Step R1. If v is in the distance $\leq 2L(v)$ from the leftmost or the rightmost vertex of P then switch to $Rod - end$.

Step R2. The same as Step b in Algorithm 2.

Step R3. Find the shortest diagonal s with both endpoints in $X_{2k}(v)$. If it does not have both endpoints in $X_k(v)$ then go to $R2$.

Step R4. Insert s into T, and split P into the two subpolygons along s. Next, if one of the two subpolygons has all vertices in $X_k(v)$ then find its greedy triangulation using a cubic-time algorithm, insert the edges of the triangulation found into T, set P to the other subpolygon and v to an arbitrary endpoint of s, and go to R1. Otherwise set P_1 to the leftmost of the two subpolygons and P_2 to the rightmost one. Set v_1 to the rightmost endpoint of s and v_2 to the leftmost one, call $Rod(P_1, v_1)$, $Rod(P_2, v_2)$, and stop.

Step R5. Set v to any endpoint of s and go to R2.

end

To prove the correctness of Algorithm 3, first, we prove the following lemma.

Lemma 5.1: If *Rod* inserts a diagonal in the greedy triangulation under construction, outside the subroutine *Rod – end*, then the diagonal is locally shortest.

Sketch: Let CH be the convex hull of $X_{2k}(v)$. The parts of CH that do not overlap with the perimeter of P are in the distance $\Omega(k \mid s \mid)$ from s. This put together with (iii) forces any diagonal intersecting s to have length greater than $\mid s \mid$. We conclude that s is locally shortest. Q.E.D.

The next crucial step in proving correctness of Algorithm 3 is the following lemma.

Lemma 5.2: Assuming that *Rod – end* produces only locally shortest diagonals and each call of *Rod – end* results in at least one new greedy diagonal, Algorithm 3 constructs the greedy triangulation of P.

Proof: First, we observe that after an unsuccessful test of s which follows such a previous unsuccessful test of, say s', the procedure *Rod* moves the current v in the same direction as previously, i.e. it keeps moving to the left or right, respectively. It follows by $\mid s \mid < \mid s' \mid$ and the fact that we cannot find a shorter diagonal than s in the area from which we came, and where s' was the shortest one.

Thus, we can imagine the procedure as operating at the end of a chain of unsuccessful tests moving to the left or right all the time, which eventually is stopped either by a diagonal insertion or reaching one of horizontal extremes of P, and switching to *Rod – end*. In the latter case, a locally shortest diagonal is found by our assumptions. Thus, *Rod* eventually constructs the greedy triangulation of P. Q.E.D.

To estimate the time performance of Algorithm 3 in the next lemma, we assume the following definition.

Definition 5.2: The X-distance between a pair of vertices of P is the number of other vertices of P whose X-coordinates lie between these of the pair of vertices.

Lemma 5.3: Under the assumption that *Rod – end* never switches to *Rod* when the X-distance between the current vertex v and the entry vertex v is greater than $3k$, Algorithm 3 outside *Rod – end* takes $O(nk^3)$ time.

Proof: Clearly, the steps a,b,c in Algorithm 3 can be done in linear time. It remains to estimate the time taken by $Rod(P, v)$ outside *Rod – end*.

Let w_l be the rightmost vertex of P such that its X-coordinate is not greater than that of the leftmost vertex of P plus $2L(w_l)$. Next, let L_l be the vertical straight-line passing through the vertex of P that is in the X-distance $3k$ to the right of w_l. Symmetrically, define the vertical straight-line L_r with respect to the rightmost vertex of P. Next, let S be the vertical strip, possibly empty, bounded by L_l on the left side and by L_r on the right side. Note that during the performance of *Rod*, the strip S may shrink.

First, let us estimate the time used by *Rod* outside *Rod – end* when the current vertex v is outside the current strip S. In this case, *Rod* uses $O(k^3)$ time until it either moves the current vertex v to S or switches to *Rod – end*. In the former subcase, it has to produce a greedy diagonal to come back to S. On the other hand, by Lemma 5.1, *Rod* can switch only $O(n)$ times to $Rod - end$. Thus, the total time spent by Rod outside $Rod - end$, with the current vertex v outside S is $O(nk^3)$.

In turn, let us estimate the total time spent by Rod with the current vertex v within S. Note that when Rod within S leaves some local neighbourhood moving a non-constant number of times to the left, or to the right, then it can pass over this neighbourhood again only by performing diagonal insertions all the way back. As each diagonal inserted is in $X_k(v)$, for some v, in the worst case it has to pay for the tests done for $O(k)$ vertices in a local neighbourhood (one can say, in $X_{2k}(v)$). As each test costs $O(k^2)$, we conclude that the procedure takes $O(k^2 n + kk^2 n)$ time. In the above, the first subterm stands for the cost of initial passing. Q.E.D.

To finish the proof of correctness and complexity analysis of Algorithm 3, it remains to prove the correctness and estimate time performance of the subroutine $Rod - end$ which is as follows.

Subroutine $Rod - end(v)$
begin
Step A. Set the entry vertex to v.
Step B. Set L to the vertical straight line passing in the distance $2L(v)$ from the extreme vertex of P closest to v.
Step C. Set CH to the convex hull of the union of $V_k(v)$ with the set of vertical projections of vertices in $V_k(v)$ on the opposite side of P and the doubleton consisting of the intersection points of L with the perimeter of P.
Step D. Find the set D of all diagonals d of P with both endpoints in $V_k(v)$ such that:
(i) no straight line segment with one endpoint in $V_k(v)$ and the other on the part of the perimeter of CH outside the minimum fragment of the perimeter including $V_k(v)$ can both be shorter than d and intersect d,
(ii) no diagonal with both endpoints in $V_k(v)$ can both be shorter than d and intersect d.
If $D \neq \emptyset$ then go to F;
Step E. Set s to the shortest diagonal in the set of diagonals of P that have both endpoints in $V_k(v)$ among which at least one is on the same side of L as v lies on L. Set v to an endpoint of s. If the entry vertex is on the upper side of P, then set the counter C to the V-distance between v and the entry vertex times -1. Otherwise, set C to the V-distance between these two vertices. Go to C.
Step F. Suppose first that v_e is closer to the left extreme of P than to the right one. Set v to the vertex of P such that
(i) the X-coordinate of w is not less than the X-coordinate t of the leftmost vertex of P plus $2L(v_e)$ and not greater than $t + 10L(v_e)$, and
(ii) the number of vertices of P whose X-coordinates lie between $t + 2L(v)$ and the X-coordinate of w is the largest possible bounded by $3k$.
Go to Step R1 in Rod. If v_e lies closer to the rightmost vertex of P proceed symmetrically.
end

It follows immediately from the definition of the subroutine $Rod-end$ that it can augment the current partial triangulation only by locally shortest diagonals. This proves the correctness of $Rod - end$. The following lemma estimates time performance of $Rod - end$.

Lemma 5.4: Each call of the subroutine $Rod - end$ augments the current partial triangulation by at least one diagonal and takes $O(k^4 q)$ time where q is the number of greedy edges produced during this call.

Proof: We may assume without loss of generality that $Rod - end$ is called for a vertex v on the lower side of P closer to the leftmost vertex of P, say v_l, then to the rightmost one. Let L be the vertical straight-line passing in the distance $2L(v)$ to the right of v_l.

First, we shall prove that the call of $Rod - end$ results in at least one greedy diagonal produced. By the definition of the procedure Rod, v is an endpoint of a shortest diagonal s with both endpoints in $X_{2k}(w)$ where w is a vertex of P to the right of L. If $Rod - end$ does not produce any greedy diagonal with both endpoints in $V_k(v)$ then the new vertex v follows v clockwise by D. Note that the new v is an endpoint of a diagonal shorter than d since d has one endpoint on the left side of L and either d has both endpoints in $V_k(v)$ or there is a shorter diagonal with both endpoints in $V_k(v)$ at least one of them on the left side of L. Next, $Rod - end$ moves the current vertex v clockwise as long as the sets D are empty, following shorter and shorter diagonals of P, by argumentation analogous to that in the proof of correctness of Algorithm 2.

Suppose now that the current vertex v to the left of L is in a V-distance $\leq k$ from some vertices on the upper side of P to the right of L. By the definition of d, there is a vertical strip S with the left boundary equal to L and breadth $\Omega(k|d|)$ such that no diagonal s with both endpoints in S, shorter than d, lies in S. On the other hand, the current vertex v is an endpoint of a diagonal shorter than d with at least one vertex to the left of L. Thus, no diagonal shorter than s with both endpoints in S can intersect s. It follows that eventually $Rod - end$ produces a greedy diagonal with at least one endpoint not to the right of L.

After each greedy diagonal production, $Rod - end$ backs with the current vertex v by $\Omega(k)$ V-distance but for some very special cases (see step E) when the counter is $O(k)$. The only way for $Rod - end$ to return to the procedure Rod is to decrease the counter to $O(k)$. The maximum value of the counter during the call of $Rod - end$ is clearly the maximum V-distance m from the entry vertex v reached by a current vertex v. It follows that $Rod - end$ has to produce $\Omega(m/k)$ greedy diagonals in order to switch to the procedure Rod. On the other hand, it takes $O(k^2)$ time to determine the set D in $V_{2k}(v)$. Therefore, arguing analogously as in Lemma 4.4, we can prove that the call takes $O(mk^3)$ steps. Q.E.D.

Combining Lemmas 5.2, 5.3 and 5.4, we obtain the following theorem.

Theorem 5.1: Algorithm 3 constructs the greedy triangulation of a rod convex polygon in linear time.

Theorem 5.1 combined with linear-time solution to the non-rod case given in the previous section yields our second main result.

Theorem 5.2: A greedy triangulation of convex polygon can be constructed in linear time.

250

References

[A] A. Aggarwal, *Research topics in computational geometry*, Bulletin of EATCS 39 (October 1989), pp. 388-408.

[AGSS] A. Aggarwal, L. Guibas, J. Saxe, P. Shor, *A Linear Time Algorithm for Computing the Voronoi Diagram of a Convex Polygon*, in Proc. of the 19th ACM Symposium on Theory of Computing, New York, pp. 39-45,1987.

[C] L.P. Chew, *Constrained Delaunay triangulations*, in: Proc. of the 3rd ACM Symposium on Computational Geometry, Waterloo, pp. 215-222, 1987.

[Gi] P.D. Gilbert, *New Results in Planar Triangulations*, M.S. Thesis, Coordinated Science Laboratory, University of Illinois, Urbana, Illinois, 1979.

[E] H. Edelsbrunner, *Algorithms in Combinatorial Geometry*, Texts and Monographs in Computer Science, Springer Verlag, Berlin, 1987.

[Go] S.A. Goldman, *A Space Efficient Greedy Triangulation Algorithm*, Information Processing Letters 31 (1989) pp. 191-196.

[Le] C. Levcopoulos, *An $\Omega(\sqrt{n})$ lower bound for non-optimality of the greedy triangulation*, Information Processing Letters 25 (1987), pp. 247-251.

[LL86] D.T. Lee and A. Lin, *Generalized Delaunay Triangulation for Planar Graphs*, Discrete and Computational Geometry 1(1986), Springer Verlag, pp. 201-217.

[LL87] C. Levcopoulos, A. Lingas, *On Approximation Behavior of the Greedy Triangulation for Convex Polygons*, Algorithmica (1987) 2, pp. 175-193.

[LLS] C. Levcopoulos, A. Lingas and J. Sack, *Heuristics for Optimum Binary Search Trees and Minimum Weight Triangulation Problems*, Theoretical Computer Science 66 (1989), pp. 181-203.

[Li87] A. Lingas, *A space efficient algorithm for the greedy triangulation*, in : M. Thoma and A. Wyner, eds., Proc. 13th IFIP Conf. on System Modelling and Optimization, Tokyo (August 1987), Lecture Notes in Control and Information Sciences, Vol. 113 (Springer, Berlin, 1987), pp. 359-364.

[Li88] A. Lingas, *Greedy Triangulation Can be Efficiently Implemented in the Average Case*, in : J. van Leeuwen, ed., Proc. International Workshop WG'88, Amsterdam (June 1988), Lecture Notes in Computer Science, Vol. 344 (Springer, Berlin, 1988), pp. 253-261.

[Li89] A. Lingas, *Voronoi Diagrams with Barriers and their Applications*, Information Processing Letters 32(1989), pp. 191-198.

[Ll] E.L. Lloyd, *On Triangulations of a Set of Points in the Plane*, Proc. of the 18th Annual IEEE Conference on the Foundations of Computer Science, Providence, 1977.

[MZ] G. Manacher and A. Zorbrist, *Neither the greedy nor the Delaunay triangulation of a planar point set approximates the optimal triangulation*, Information Processing Letters 9 (1979).

[PS] F.P. Preparata and M.I. Shamos, *Computational Geometry, An Introduction*, Texts and Monographs in Computer Science, Springer Verlag, New York, 1985.

[WS] C. Wang and L. Schubert, *An optimal algorithm for constructing the Delaunay triangulation of a set of line segments*, in: Proc of the 3rd ACM Symposium on Computational Geometry, Waterloo, pp. 223-232, 1987.

Star Unfolding of a Polytope with Applications[*]

(Extended Abstract)

Pankaj K. Agarwal,[†] Boris Aronov,[‡] Joseph O'Rourke,[§]
and
Catherine A. Schevon[¶]

Abstract

We define the notion of a "star unfolding" of the surface $\mathcal{P}$ of a convex polytope with n vertices and use it to construct an algorithm for computing a small superset of the set of all sequences of edges traversed by shortest paths on $\mathcal{P}$. It requires $O(n^6)$ time and produces $O(n^6)$ sequences, thereby improving an earlier algorithm of Sharir that in $O(n^8 \log n)$ time produces $O(n^7)$ sequences. A variant of our algorithm runs in $O(n^5 \log n)$ time and produces a more compact representation of size $O(n^5)$ for the same set of $O(n^6)$ sequences. In addition, we describe an $O(n^{10})$ time procedure for computing the geodesic diameter of $\mathcal{P}$, which is the maximum possible separation of two points on $\mathcal{P}$, with the distance measured along $\mathcal{P}$, improving an earlier $O(n^{14} \log n)$ algorithm of O'Rourke and Schevon.

1 Introduction

In this paper we consider a number of problems related to shortest paths on the surface $\mathcal{P}$ of a convex polytope in $\mathbb{R}^3$. A shortest path on $\mathcal{P}$ is identified uniquely by its endpoints and the sequence of edges that it encounters. Sharir [Sha87] proved that no more than $O(n^7)$ distinct sequences of edges are actually traversed by the shortest paths on $\mathcal{P}$, and gave an $O(n^8 \log n)$ algorithm to compute an $O(n^7)$ superset of such sequences. The bound on the maximum possible number of shortest-path edge sequences was later improved to $\Theta(n^4)$ [Mou85, SO88]. Recently Schevon and O'Rourke [SO89] presented an algorithm that computes the exact set of all shortest-path edge sequences and also identifies, in logarithmic time, the edge sequences traversed by all shortest paths connecting a given pair of query points lying on edges of $\mathcal{P}$. The sequences can be explicitly

---

[*]Part of the work was carried out when the first two authors were at Courant Institute of Mathematical Sciences, New York University and the fourth author was at the Department of Computer Science, The Johns Hopkins University. The work of the second author was partially supported by an AT&T Bell Laboratories Ph.D. Scholarship. The work of the first two authors has also been partially supported by DIMACS (Center for Discrete Mathematics and Theoretical Computer Science), a National Science Foundation Science and Technology Center — NSF-STC88-09648. The third author's work is supported by NSF grant CCR-882194.

[†]DIMACS Center, Rutgers University, Piscataway, NJ 08855, USA and Computer Science Department, Duke University, Durham, NC 27706, USA

[‡]DIMACS Center, Rutgers University, Piscataway, NJ 08855, USA

[§]Department of Computer Science, Smith College, Northampton, MA 01063, USA

[¶]AT&T Bell Laboratories, Holmdel, NJ 07733, USA

generated, if necessary, in time proportional to their length. Their algorithm, however, requires $O(n^9 \log n)$ time and $O(n^8)$ space. Hwang et al. [HCT89] claimed to have a more efficient procedure for solving the same problem, but their argument, as stated, is flawed, and no corrected version appeared in the literature. In this paper we propose a significantly faster and simpler algorithm to compute a superset of shortest-path edge sequences of size $O(n^6)$, which runs in time $O(n^6)$, thus improving the result of [Sha87]. The computation of the collection of all shortest-path edge sequences on a polytope is an intermediate step of several algorithms [Sha87, OS89], and is of interest in its own right.

The second problem studied in this paper is that of computing the *geodesic diameter* of $\mathcal{P}$, i.e. the maximum distance along $\mathcal{P}$ between any two points on $\mathcal{P}$. O'Rourke and Schevon [OS89] gave an $O(n^{14} \log n)$ time procedure for determining the geodesic diameter of $\mathcal{P}$. We present a simpler and faster algorithm whose running time is $O(n^{10})$.

Our algorithms are based on a common geometric construction, the *star unfolding*. Intuitively, the star unfolding of $\mathcal{P}$ with respect to a point $x \in \mathcal{P}$ can be viewed as follows. Suppose there is unique shortest path from x to every vertex of $\mathcal{P}$. The object obtained after removing these n paths from $\mathcal{P}$ is the star unfolding of $\mathcal{P}$. We show that the combinatorial complexity of a star unfolding is $\Theta(n^2)$, and its combinatorial structure remains the same for all points in a sufficiently small neighborhood of x, as long as the shortest path from x to each corner of $\mathcal{P}$ is unique. We prove a number of other properties of star unfoldings, which facilitate the construction of efficient algorithms for the above two problems.

Chen and Han [CH90a] have independently discovered the star unfolding and used it for computing the shortest-path information from a single fixed point on the surface of a polytope.

The paper is organized as follows. In Section 2, we formalize our terminology and mention some basic properties of shortest paths. Section 3 defines star unfolding and proves some of its properties. Section 4 describes an efficient algorithm to compute a superset of all possible shortest-path edge sequences, based on the star unfolding, and in Section 5 we use the notion of star unfolding to obtain a faster algorithm for determining the geodesic diameter of a polytope. Section 6 contains some concluding remarks. Due to lack of space most of the proofs are omitted ¿from this version.

2 Geometric Preliminaries

We begin by reviewing the geometry and combinatorics of shortest paths on convex polytopes that we need to define the star unfolding in the next section.

Let $\mathcal{P}$ be the surface of a polytope with n vertices. We refer to vertices of $\mathcal{P}$ as *corners*; the unqualified terms *face* and *edge* are reserved for faces and edges of $\mathcal{P}$. We assume that $\mathcal{P}$ is triangulated. This does not change the number of faces and edges of $\mathcal{P}$ by more than a multiplicative constant, but simplifies the description of our algorithms.

2.1 Geodesics and Shortest Paths

A path π on $\mathcal{P}$ that cannot be shortened by a local change at any point in its interior is referred to as a *geodesic*. Equivalently, a geodesic on the surface of a convex polytope is either a subsegment of an edge or a path that (1) does not pass through corners, though may possibly terminate at them, (2) is straight near any point in the interior of a face and (3) is transverse to every edge it meets in such a fashion that it would appear straight if one were to "unfold" the two faces incident on this edge until they lie in a common plane [SS86]. In the following discussion we disregard the geodesics lying completely within a single edge of $\mathcal{P}$.

The *(geodesic) distance* $d(x,y)$ between two points $x, y \in \mathcal{P}$ is the Euclidean length of a shortest path connecting them and confined to $\mathcal{P}$. Trivially every shortest path is a geodesic and no shortest path meets a face or an edge more than once. The following additional property of shortest paths is crucial for our analysis.[1]

Lemma 2.1 (Sharir and Schorr [SS86]) *Let π_1 and π_2 be distinct shortest paths emanating from x. Let $y \in \pi_1 \cap \pi_2$ be a point distinct from x. Then either one of the paths is a subpath of the other, or neither π_1 nor π_2 can be extended past y while remaining a shortest path.* □

The *geodesic diameter* of $\mathcal{P}$, $D(\mathcal{P})$, is the maximum geodesic distance between any two points on $\mathcal{P}$, i.e. $D(\mathcal{P}) = \max_{x,y \in \mathcal{P}} d(x,y)$; (x,y) is a *diametral pair* if $d(x,y) = D(\mathcal{P})$.

2.2 Edge Sequences and Sequence Trees

A *shortest-path edge sequence* is the sequence of edges intersected by some shortest path π connecting two points on $\mathcal{P}$, in the order met by π. Such a sequence is *maximal* if it cannot be extended in either direction while remaining a shortest-path edge sequence; it is *half-maximal* if no extension is possible at one of the two ends. It has been shown by Schevon and O'Rourke[SO88] that the maximum total number of half-maximal sequences is $\Theta(n^3)$.

Observe that every shortest-path edge sequence σ is a prefix of some half-maximal sequence, namely the one obtained by extending σ maximally at one end. Thus an exhaustive list of $O(n^3)$ half-maximal sequences contains, in a sense, all the shortest-path edge sequence information of $\mathcal{P}$. More formally, given an arbitrary collection of edge sequences emanating from a fixed edge e, let the *sequence tree* Σ_e of this set be the tree with all possible non-empty prefixes of the given sequences as vertices, the trivial sequence consisting solely of e as the root, and such that σ is an ancestor of σ' in the tree if and only if σ is prefix of σ' [HCT89]. The $\Theta(n^3)$ bound on the number of half-maximal sequences implies that the collection of $O(n)$ sequence trees obtained by considering all shortest-path edge sequences from each edge of $\mathcal{P}$ in turn, has a total of $\Theta(n^3)$ leaves and $\Theta(n^4)$ nodes.

2.3 Ridge Trees

Given a point x on $\mathcal{P}$, $y \in \mathcal{P}$ is a *ridge point* with respect to x if there are two or more distinct shortest paths between x and y. Ridge points with respect to x form a *ridge tree* T_x embedded on $\mathcal{P}$,[2] whose leaves are the corners of $\mathcal{P}$, and whose internal vertices have degree at least three and correspond to points of $\mathcal{P}$ with three or more distinct shortest paths to x. The edges of T_x, referred to as *ridges*, are (open) geodesics consisting of

---

[1] This lemma is known to hold in a much wider context. In particular, if neither path is a subpath of the other, then y is known as a *conjugate* point of x, and the lemma is equivalent to Theorem 4.1 (p. 112) of Kobayashi [Kob67].

[2] For more general surfaces, the ridge tree is known as the *cut locus* [Kob67].

points with exactly two distinct shortest paths to x [SS86]. In the full version of this paper, we establish a stronger characterization of ridges:

Lemma 2.2 *Every edge of the ridge tree of any point on $\mathcal{P}$ is a shortest path.*   □

The worst-case combinatorial size of T_x jumps from $\Theta(n)$ to $\Theta(n^2)$ if one takes into account the fact that a ridge is a shortest path comprised of $\Theta(n)$ line segments in the worst case—and it is possible to exhibit a ridge tree for which the number of ridge-edge incidences is indeed $\Omega(n^2)$ [Mou85]. Henceforth we will use the unqualified term "ridge point" to refer to a ridge point with respect to some corner of $\mathcal{P}$. For simplicity we assume that ridges intersect each edge of $\mathcal{P}$ transversally.

3   Star Unfolding

In this section we introduce the notion of the *star unfolding* of $\mathcal{P}$ and describe its geometric and combinatorial properties. Working independently, Chen and Han have used the same notion in their algorithms for computing all shortest paths from a fixed point of $\mathcal{P}$ [CH90a] and for storing the induced subdivision of $\mathcal{P}$ [CH90b].

3.1   Geometry of the Star Unfolding

Let $x \in \mathcal{P}$ be a non-ridge point, so that there is a unique shortest path connecting x to each corner of $\mathcal{P}$. These paths are called *cuts* and are comprised of *cut points* (see Fig. 1(b)). The cuts together with edges induce a convex decomposition of $\mathcal{P}$, which we will treat as a surface $\mathcal{P}_x$ of a polytope. It is geometrically identical to $\mathcal{P}$, but combinatorially different.

Now form a two-dimensional complex from the faces of $\mathcal{P}_x$ as follows. The cells or *plates* of the complex are the faces of $\mathcal{P}_x$, each a compact convex polygon. For each pair of adjacent plates of $\mathcal{P}_x$ sharing an edge of $\mathcal{P}_x$, which is a portion of an edge of $\mathcal{P}$, topologically identify the two plates along that edge. We define the *star unfolding* S_x as the resulting two-dimensional complex. We next define the *pasting tree* Π_x as a graph whose nodes are the plates of S_x, and two nodes are connected by an arc if the corresponding plates share a topologically identified edge. It is easy to show that Π_x is a tree, which implies that S_x is topologically equivalent to a closed disk. Its polygonal boundary ∂S_x consists entirely of edges originating from cuts. Define a *segment* in S_x as a path $\pi \in S_x$ that unfolds straight: laying the plates that π crosses end-to-end in a common plane results in a straight line segment. It can be shown that there is at most one segment connecting any two points of S_x. Every segment in S_x is a shortest path in the intrinsic metric of the complex.

We now describe two intuitive "models" for S_x, both of which are flawed, but are nevertheless considerable aids to intuition.

1. Let $\mathcal{P}_x^-$ be $\mathcal{P}$ minus all cuts from x. $\mathcal{P}_x^-$ naturally corresponds to the interior of S_x: every geodesic in $\mathcal{P}_x^-$ maps to a segment completely interior to S_x. But any segment in S_x that includes points of ∂S_x has no associated geodesic on $\mathcal{P}_x^-$.

2. If S_x is unfolded into a common plane, it is conceivable that the plates might overlap. However, we have not found an example where overlap actually occurs,

and we conjecture that overlap is impossible. But in the absence of a proof of this conjecture, we must imagine the unfolding as lying on different "levels" in order for each point of S_x to have a unique image in $\mathbb{R}^2$. Fig. 1(c) shows an example of S_x unfolded.

Although occasionally we will use these models for ease of presentation, the definition of S_x as a two-dimensional complex should be considered primary.

Notice that $x \in \mathcal{P}$ corresponds to n distinct points (*images*) in S_x, a non-corner point $y \in \mathcal{P}$ distinct from x and lying on a cut has two exactly images in S_x, and all other points of $\mathcal{P}$ have unique counterparts in S_x. In particular, we have:

Lemma 3.1 (Sharir and Schorr [SS86]) *For a point* $y \in \mathcal{P}$, *any shortest path* π *from* x *to* y *maps to a segment* $\pi^* \subseteq S_x$ *connecting some image of* y *in* S_x *to some image of* x. □

In Section 5 we will need the concept of the "kernel" of a star unfolding. S_x is a topological disk bounded by the polygonal cycle $x_1 p_1 x_2 \ldots p_n x_1$ comprised of $2n$ segments, where $x_1, \ldots, x_n$ are the n images of x and $p_1, \ldots, p_n$ are the images of the corners of $\mathcal{P}$. It can be shown that triangles $\triangle p_{i-1} x_i p_i$, for $i = 1, \ldots, n$, are fully contained in S_x and meet only at vertices.[3] Their removal yields a two-dimensional complex K_x which we call the *kernel* of $\mathcal{P}$, bounded by the cycle of segments $\partial K_x = p_1 p_2 \ldots p_n p_1$ and topologically equivalent to a closed disk. Fig. 2 illustrates the kernel of the star unfoldings for several randomly generated polytopes.[4] The main property of the kernel that we will need later is:

Lemma 3.2 *The image of the ridge tree is completely contained within the kernel, which is itself a subset of the star unfolding:* $\text{Image}(T_x) \subset K_x \subset S_x$. □

3.2 Combinatorics of the Star Unfolding

We now describe the combinatorial structure of S_x. A *vertex* of S_x is an image of x, of a corner of $\mathcal{P}$, or of an intersection of an edge of $\mathcal{P}$ with a cut. An *edge* of S_x is a maximal portion of an image of a cut or an edge of $\mathcal{P}$ delimited by vertices of S_x. It is easy to see that S_x consists of $\Theta(n^2)$ convex plates in the worst case, even though its boundary is formed by only $2n$ "segments", the images of the cuts. We define the *combinatorial structure* of S_x as the 1-skeleton of S_x, i.e. the graph whose nodes and arcs are the vertices and edges of S_x. Intuitively, the combinatorial structure of S_x describes the ordering of edges of $\mathcal{P}$ along each cut, and also the ordering of cuts along each edge of $\mathcal{P}$. The combinatorial structure of a star unfolding has the following crucial property:

Lemma 3.3 *Let* x *and* y *be two non-corner points lying in the same face or on the same edge of* $\mathcal{P}$, *so that* xy *contains no ridge points. Then* S_x *and* S_y *have the same combinatorial structure.* □

This lemma holds under more general conditions. Namely, instead of requiring that xy be free of ridge points, it is sufficient to assume that the number of distinct shortest paths connecting z to any corner does not change as z varies along xy.

---

[3] More precisely, each such set is a subset of S_x bounded by the three segments $p_{i-1} x_i$, $x_i p_i$, $p_i p_{i-1}$ and isometric to a planar triangle.

[4] The unfoldings were produced with code written by Julie DiBiase and Stacia Wyman of Smith College.

4 Shortest-Path Edge Sequences

In this section we offer a new algorithm for constructing a superset of the shortest-path edge sequences, which is both more efficient than previously suggested procedures, and conceptually simpler. However, reducing the size of this set by deleting unrealizable sequences seems to be as difficult as the original problem. We also provide a mechanism for supporting queries of the form "Given two points x, y on the edges of $\mathcal{P}$, determine $d(x, y)$, or more generally, find a shortest path between x and y." But unlike [SO89], we can only guarantee $O(n)$ worst-case performance.

We propose a two-phase algorithm based on the star unfolding introduced in the previous section. In the first phase, we construct pasting trees, which form a compact representation of a superset of all shortest-path edge sequences. In the second phase, these trees are traversed to construct $O(n)$ sequence trees for this set, thereby obtaining $O(n^6)$ sequences.

4.1 Construction of Pasting Trees

4.1.1 One Pasting Tree

We first discuss the geometric and combinatorial structure of a single such tree, and then move to consideration of all relevant pasting trees.

Recall that the *pasting tree* Π_x is a tree whose nodes are the $O(n^2)$ plates of S_x, and two nodes are connected by an arc if the corresponding plates share a topologically identified edge. Π_x has only n leaves, corresponding to the triangular plates incident to the n images of x in S_x or, equivalently, to x in $\mathcal{P}_x$. By Lemma 3.1, any shortest path from x to $y \in \mathcal{P}$ maps to a segment in S_x connecting some image of x to an image of y. This segment in turn corresponds to a simple path in Π_x, originating at one of the leaves. Hence from each image x_i of x there are at most $O(n^2)$ edge sequences (one for each node in Π_x, i.e. each plate in S_x), for a total of $O(n^3)$ sequences from x. In fact, there are at most $\binom{n}{2}$ maximal (with respect to Π_x) sequences in this collection, since Π_x has n leaves and every maximal simple path in a tree always connects one leaf to another. Thus Π_x already constitutes a compact representation of a superset (of size $O(n^3)$) of the edge sequences traversed by shortest paths emanating from x; it can be computed in $O(n^2 \log n)$ time by an easy modification of the algorithm of Mount [Mou85].

4.1.2 $O(n^3)$ Pasting Trees

Observe that all shortest-path edge sequences are realized by pairs of points lying on edges of $\mathcal{P}$—any other shortest path can be contracted without affecting its edge sequence so that its endpoints lie on edges of $\mathcal{P}$. We now consider the set of all distinct pasting trees arising from non-ridge points on edges of $\mathcal{P}$.

Since the ridge tree of a corner intersects a fixed edge of $\mathcal{P}$ in $O(n)$ points and there are n corners and $O(n)$ edges in $\mathcal{P}$, the edges of $\mathcal{P}$ can be partitioned into a total of at most $O(n^3)$ open segments, each of which is free of ridge points. Call each such segment an *edgelet*. On each edgelet ε, the combinatorial structure of the star unfolding S_x of $\mathcal{P}$ with respect to a point $x \in \varepsilon$ is independent of the choice of x by Lemma 3.3. In particular, the pasting tree Π_x does not change as x ranges over ε. Thus the set of $O(n^3)$ pasting trees, each of size $O(n^2)$, contains all possible shortest-path edge sequences, with possible exception of the ones that emanate from ridge points—they can be computed by brute force. It can be produced in time $O(n^5 \log n)$ by first employing, for each corner, Mount's ridge tree algorithm [Mou85], obtaining the above partitioning of each edge into edgelets, and then constructing the $O(n^3)$ star unfoldings by executing Mount's algorithm from an arbitrary interior point of each edgelet.

In pseudocode, the first phase of the algorithm is:

Algorithm 1a: PASTING TREES
for each corner c of $\mathcal{P}$ do
        Construct the ridge tree T_c with respect to c.
Form edgelets.
for each edgelet ε **do**
        Construct unfolding S_x from a point $x \in \varepsilon$.
        From S_x, build pasting tree Π_x.

Thus we have presented an algorithm that, in $O(n^5 \log n)$ time, produces a data structure of size $O(n^5)$ containing an implicit representation of a set of $O(n^6)$ sequences ($O(n^5)$ of which are maximal in this set), which includes all shortest-path edge sequences. We now turn to the second phase of the algorithm, which makes these sequences explicit.

4.2 Constructing Sequence Trees

We now show how to produce, for each edge e, the sequence tree representation (cf. Section 2.2) of those edge sequences computed above that emanate from e, in overall time $O(n^6)$. Since each face of $\mathcal{P}$ is a triangle, any sequence tree node has at most two children. The sequence tree for a fixed edge e is constructed incrementally. Given the current tree Σ_e, a new maximal sequence σ (potentially not in Σ_e) is selected from the collection of pasting trees $\{\Pi_x \mid x \in e\}$ computed by Algorithm 1a. The sequence σ is traced in Σ_e from the root e, repeatedly advancing to that child of the current node which corresponds to the next edge in σ. If σ diverges from the paths in Σ_e, then Σ_e is augmented at this node by the untraversed portion of σ.

Furthermore, the following test is easily incorporated in the above procedure at no additional asymptotic cost: Every path is truncated at the first point where a face already visited by it is encountered for the second time. As shortest-path edge sequences never revisit a face or an edge, they are not affected by this trimming. On the other hand, this guarantees that the length of each sequence produced, and thus the depth of the sequence tree, is bounded by $O(n)$, so that having inserted $O(n^5)$ sequences, maximal in their respective pasting trees and not visiting any edge twice, we obtain a collection of trees with $O(n^5)$ leaves and $O(n^6)$ nodes altogether. Having spent $O(1)$ time per edge of a maximal sequence, the algorithm runs in $O(n^6)$ time.

The following is a pseudocode summary of the second phase of the algorithm.

Algorithm 1b: Sequence Trees
for each edge e of $\mathcal{P}$ do
    Initialize sequence tree $\Sigma_e = \{e\}$.
    for each pasting tree Π_x for $x \in \varepsilon \subseteq e$ do
        for each sequence σ maximal in Π_x do
            Traverse σ, augmenting Σ_e.

Thus we have improved the algorithm of Sharir [Sha87] which requires $O(n^8 \log n)$ time to compute the sequence tree representation for a superset of size $O(n^7)$ of all shortest-path sequences.

Theorem 4.1 *Algorithm 1 (1a and 1b together) takes $O(n^6)$ time to produce a set of $O(n^6)$ edge sequences (represented as sequence trees) that includes all shortest-path edge sequences.* □

Note that our algorithm uses nothing more complex than Mount's peel unfolding algorithm, and tree traversals. It achieves an improvement over previous algorithms mainly by reorganizing the computation around the star unfolding.

As mentioned above, the sequence tree representation for just the shortest-path edge sequences is smaller by a factor of n^2 than our estimate on the size of the set produced by Algorithm 1, but computing it efficiently seems difficult. In addition, it is not clear how far the output of our algorithm is from the set of all shortest-path edge sequences. We have a sketch of a construction for a class of polytopes that force our algorithm to produce at least cn^5 *non*-shortest-path edge sequences, for some constant $c > 0$.

4.3 Query Problem

We have also promised an $O(n)$ query-time procedure for determining the distance between two given points lying on edges of $\mathcal{P}$. It proceeds as follows. First run Algorithm 1a. Given a point $x \in e$, use binary search on e to identify the edgelet containing x. This gives us the combinatorial structure of S_x. It remains to identify the subset of images of x in S_x that can "see" the image y^* of y and find the one(s) among them closest to y (as will be detailed in the following section). This gives the distance from x to y. It will be shown in Section 5.2 that all this can be accomplished in $O(n)$ time. The actual paths and/or edge sequences can be generated by tracing their images in S_x in time proportional to the number of edges traversed by the paths.

5 Computing the Geodesic Diameter

In this section we present an algorithm to compute the geodesic diameter of $\mathcal{P}$. As mentioned in the introduction, this question was first investigated by O'Rourke and Schevon [OS89] who presented an $O(n^{14} \log n)$ time algorithm for computing it. Their algorithm relies on the following observation:

Lemma 5.1 (O'Rourke and Schevon [OS89]) *If a pair of points $x, y \in \mathcal{P}$ realizes the diameter of $\mathcal{P}$, then either p or q is a corner of $\mathcal{P}$, or there are at least five distinct shortest paths between p and q.* □

Lemma 5.1 suggests the following strategy for locating all diametral pairs. First dispose of the possibility that either x or y is a corner in $n \times O(n^2 \log n) = O(n^3 \log n)$ time just as in [OS89]. Next locate all pairs of points on $\mathcal{P}$ which have five or more distinct shortest paths between them. In implementing the latter procedure, which dominates the running time of the algorithm, we take an approach somewhat different from that used in [OS89]. Specifically, we use two facts: (a) for a point $x \in \mathcal{P}$, all shortest paths from x are contained in the star unfolding S_x as segments, and (b) the combinatorial structure of S_x does not change as long as x is confined to a ridge-free region (cf. Lemma 3.1). For a fixed ridge-free region R, we locate all pairs (x, y) of points with $x \in R$ and connected to y by five or more shortest paths. The diameter is obtained by identifying the pair(s) with largest separation.

Thus the question of locating all diametral pairs can be reduced to the following two problems: (1) identifying all ridge-free regions and (2) for each region, determining all potential diametral pairs with one point confined to it.

5.1 Computing Ridge-Free Regions

In this subsection we bound the number of ridge-free regions on $\mathcal{P}$ and describe a procedure for computing them.

Lemma 5.2 *In the worst case, there are $\Theta(n^4)$ ridge-free regions on $\mathcal{P}$.*
Proof: The overlay of n ridge trees, one from each corner of $\mathcal{P}$, produces a subdivision of $\mathcal{P}$ in which every region is bounded by at least three edges. Thus, by Euler's formula, the number of regions in this subdivision is proportional to the number of its vertices, which we proceed to estimate.

By Lemma 2.2 ridges are shortest paths and therefore two of them intersect in at most two points (cf. Lemma 2.1) or overlap. In the latter case no new vertex of the ridge-partition of $\mathcal{P}$ is created, so we restrict our attention to the former. In particular, as there are $n \times O(n) = O(n^2)$ ridges, the total number of their intersection points is $O(n^4)$. Refining this partition further by adding the edges of $\mathcal{P}$ does not affect the asymptotic complexity of the partition, as ridges intersect edges in $O(n^3)$ points altogether. This establishes the upper bound.

It is easily checked that in Mount's example of a polytope with $\Omega(n^4)$ shortest-path edge sequences [Mou87], there are $\Omega(n^4)$ ridge-free regions. Hence there are $\Theta(n^4)$ ridge-free regions on $\mathcal{P}$ in the worst case. □

The ridge-free regions can be computed by first calculating the ridge tree for every corner, and then overlaying the trees in each face of $\mathcal{P}$. The first step takes $n \times O(n^2 \log n) = O(n^3 \log n)$ time (cf. [Mou85]), while the second step can be accomplished in time $O((r + n^3) \log n) = O(n^4 \log n)$, where r is the number of ridge-free regions in $\mathcal{P}$, using the line-sweep algorithm of Bentley and Ottmann [BO79].

5.2 Locating Diametral Pairs in a Ridge-Free Region

Next we describe how to find diametral pairs with one point in a fixed ridge-free region. For such a region R, we aim to locate all pairs of points (x, y) such that $x \in R$ and has at least five distinct shortest paths connecting it to y. All of these paths correspond to segments in S_x. Points that lie on cuts have exactly one shortest path to x, so y is not a cut point and thus has a unique image $y^* \in S_x$.

We now introduce the notion of *visibility* in S_x. For $a, b \in S_x$, we say that a is *visible* from b if there is segment in S_x connecting a to b (cf. Section 3.1).

The distance between x and y on $\mathcal{P}$ is the minimum distance from y^* to an image of x visible from y^* in the above sense—this follows from Lemma 3.1 and the fact that any segment in S_x, not passing through the image of a corner, is an unfolded geodesic.

For convenience, we view S_x as unfolded in the plane (as in the second model of S_x; see Section 3.1), positioning it so that the images of the corners of $\mathcal{P}$ are fixed and the images of x move as x moves. Recall that S_x is not known to be a simple polygon and may overlap itself when laid out in the plane. Let $(x_1, \ldots, x_5)$ be a quintuple of images of x in S_x corresponding to the five shortest paths. In view of the above discussion, the 5-shortest-paths condition reduces to the following:

1. $|y^*x_1| = |y^*x_2| = |y^*x_3| = |y^*x_4| = |y^*x_5|$, where $|\cdots|$ denotes the Euclidean distance in the plane,

2. every x_i, $i = 1 \ldots 5$, is visible from y, and

3. no image of x visible from y^* is closer to y^* than x_1.

To simplify the presentation we initially assume that the planar drawing of S_x in $\mathbb{R}^2$ does not self-overlap, and then we describe the general case. Observe that, for a given quintuple of images of x, the coordinates of x_i in the plane are known linear functions of the position of x in R. Thus condition (1) is a system of four quadratic equations in four unknowns, namely the coordinates of x and y^*. Any solutions to (1) other than isolated points can be disregarded, as such solutions can never correspond to diametral pairs (cf. [OS89]). Thus we obtain $O(1)$ tuples of points $(x_1, x_2, x_3, x_4, x_5, y^*)$ in the plane satisfying the first condition. We will assume that these tuples can be obtained in constant time. At this point, we also check whether the actual point x that gives rise to the five images in the tuple lies in R. This can be easily done in logarithmic time, since R is convex. Next, we compute the set of images of x visible from y^* and verify conditions (2) and (3).

Recall from Section 3 the definition of the kernel K_x bounded by the cycle $\partial K_x = p_1 p_2 \ldots p_n p_1$. With the points p_i "anchored" in the plane, K_x is the same for all points in $x \in R$. By Lemma 3.2, the kernel fully contains (the image of) the ridge tree T_x with respect to x. Thus if there indeed are 5 shortest paths between x and y, y^* is a vertex of T_x of degree at least 5 and must lie in K_x. If an image x_i is visible from a point $y^* \in K_x$, the segment $x_i y^*$ intersects ∂K_x at a unique point $z_i \in p_{i-1} p_i$. Given a position of $x \in R$ and an image x_i, z_i can be computed in $O(1)$ time. Therefore it suffices to check whether z_i is visible from y^*. Since K_x is a simple n-gon,[5] the portion of ∂K_x visible from y^* can be computed in $O(n)$ time, using the algorithm of El Gindy and Avis [EA81]. Thus given a pair of points $(x, y) \in \mathcal{P}$, the set of images visible from y^* can be computed in time $O(n)$. Once we have the set of visible images, (2) and (3) can be easily verified and $d(x, y)$ determined in $O(n)$ time. Repeating the above process for each quintuple and then for every ridge-free region, one can determine all diametral pairs in $O(n^4 \times n^5 \times n) = O(n^{10})$ time.

The following pseudocode summarizes the algorithm.

---

[5] Simple under our (temporary) assumption of non-overlap.

Algorithm 2: GEODESIC DIAMETER

Find a furthest neighbor $\hat{c}$ for each corner c of $\mathcal{P}$; $d(c, \hat{c})$ is a diameter candidate.

for each corner c of $\mathcal{P}$ do

    Construct the ridge tree T_c with respect to c.

Overlay all T_c to construct ridge-free regions.

for each ridge-free region R do

    Compute S_x for some $x \in R$.

    for each quintuple of images of x do

        Compute the set of solutions (x, y^*) to (1).

        for each (x, y^*) do

            Compute the portion of ∂K_x visible from y^*.

            Determine which images of x are visible from y^*.

            if conditions (2) and (3) are satisfied

                then $d(x, y)$ is a diameter candidate.

As mentioned in Section 3, we do not know whether S_x can in general be isometrically embedded in the plane without overlapping. Overlapping of S_x creates two complications: (i) K_x is no longer necessarily a simple polygon, so we cannot use the algorithm of El Gindy and Avis directly, and (ii) the solution of condition (1) does not give the actual point $y \in \mathcal{P}$ (or $y^* \in S_x$), but a point $\bar{y}$ in the planar drawing of S_x. If S_x overlaps, there may be as many as $O(n)$ points y^* on S_x mapped to the point $\bar{y} \in \mathbb{R}^2$. Both of these problems can be handled without increasing the worst case running time, by modifying the visibility algorithm, in the following manner.

Although Π_x has $\Theta(n^2)$ nodes in the worst case, all but $O(n)$ of these nodes have degree two: they lie on paths created when sequences of faces of $\mathcal{P}$ are sliced by adjacent cuts. Merging such paths of degree-two nodes will reduce the combinatorial complexity of the tree to $O(n)$. Applying this idea to the kernel K_x yields a two-dimensional complex of only $O(n)$ polygonal cells. Triangulating each cell, we obtain a decomposition of K_x into $O(n)$ triangles. Notice that the same kernel will be used to process $\binom{n}{5}$ visibility queries, so that a slow triangulation procedure does not affect the asymptotic performance of the overall algorithm. Using this triangulation as a guide, visibility calculations can be performed in K_x within the same time bounds as in the nonoverlapping case. Details are provided in the full version of this paper. Hence we have

Theorem 5.3 *Given a convex polytope in $\mathbb{R}^3$ with n vertices, its geodesic diameter and the set of all diametral pairs can be computed in time $O(n^{10})$.*

6 Discussion and Open Problems

We have shown that use of the star unfolding of a polytope leads to substantial improvements in the time complexity of two shortest-path algorithms: finding edge sequences and computing the geodesic diameter. Moreover, the algorithms are not only theoretical improvements, but also, we believe, significant conceptual simplifications. This demonstrates the utility of the star unfolding.

It would be pleasing to establish that S_x always unfolds in the plane without overlap. This would further simplify our algorithms. In the full version of the paper we argue that showing that S_x is unfoldable without self-overlap is equivalent to proving that the kernel K_x or the ridge tree T_x are unfoldable. Namely, given an example of S_x that overlaps when unfolded in the plane, we can construct a polytope in which K_x (resp. T_x) would overlap as well. On the other hand, should S_x be shown to always unfold without self-overlap, the same would surely hold for $T_x, K_x \subset S_x$.

References

[BO79] J.L. Bentley and T.A. Ottmann, "Algorithms for reporting and counting geometric intersections," *IEEE Transactions on Computers* C-28 (1979), 643–647.

[CH90a] J. Chen and Y. Han, "Shortest paths on a polyhedron," to appear in *Proc. of* 6th *ACM Symp. on Comput. Geom.*, 1990.

[CH90b] J. Chen and Y. Han, "Shortest paths on a polyhedron, part II: Storing shortest paths," Tech. Report 161-90, Dept. of Comp. Sci., Univ. of Kentucky, Lexington, February, 1990.

[EGS86] H. Edelsbrunner, L. Guibas and J. Stolfi, "Optimal point location in a monotone planar subdivision," *SIAM J. Computing* 15 (1986), 317–340.

[EA81] H. El Gindy and D. Avis, "A linear algorithm for computing the visibility polygon from a point," *J. Algorithms* 2 (1981), 186–197.

[HCT89] Y. Hwang, C. Chang and H. Tu, "Finding all shortest path edge sequences on a convex polyhedron," in *Workshop on Algorithms and Data Structures, Lectures Notes in Computer Science*, 382 (1989), Springer-Verlag, pp. 251-266.

[Kob67] S. Kobayashi, "On conjugate and cut loci," in *Studies in Global Geometry and Analysis*, S. S. Chern, Ed., Mathematical Association of America, 1967, pp. 96-122.

[Mou85] D. Mount, "On finding shortest paths on convex polyhedra," Tech. Report 1495, University of Maryland, 1985.

[Mou87] D. Mount, "The number of shortest paths on the surface of a polyhedron," Tech. Report (July 9, 1987), University of Maryland, 1987.

[OS89] J. O'Rourke and C. Schevon, "Computing the geodesic diameter of a 3-polytope," In *Proc. 5th ACM Symp. Computational Geometry*, June 1989, pp. 370-379.

[SO88] C. Schevon and J. O'Rourke, "The number of maximal edge sequences on a convex polytope" In *Proc. 26th Annual Allerton Conference on Communication, Controls, and Computing*, University of Illinois at Urbana-Champaign, October 1988, pp. 49-57.

[SO89] C. Schevon and J. O'Rourke, "An algorithm to compute edge sequences on a convex polytope," Tech. Report JHU-89-3, Dept. of Computer Science, The Johns Hopkins University, Baltimore, Maryland, 1989.

[SS86]   M. Sharir and A. Schorr, "On shortest paths in polyhedral spaces," *SIAM J. Comput.* **15** (1986), 193–215.

[Sha87] M. Sharir, "On shortest paths amidst convex polyhedra," *SIAM J. Comput.* **16** (1987), 561–572.

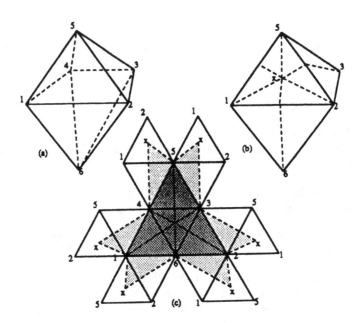

Figure 1: The star unfolding of an octahedron: (a) An octahedron with vertices labeled. (b) Shortest paths from the source x to the vertices are shown dashed. (c) The star-unfolding S_x. The lightly-shaded region bound by dashed lines is the star-unfolding; the dashed lines are the images of the cuts. The dark region is the kernel. The ridge tree is shown dotted (inside the kernel).

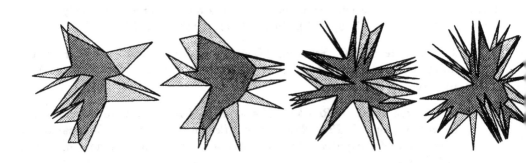

Figure 2: Four star unfoldings: $n = 13, 13, 36, 42$ vertices, left to right. The kernel is shaded darker in each figure.

Space-Sweep Algorithms for Parametric Optimization

(Extended Abstract)

DAVID FERNÁNDEZ-BACA*

Department of Computer Science, Iowa State University Ames, IA 50011

1 Introduction

Combinatorial optimization is essentially the problem of minimizing $\mathbf{c}^\mathsf{T}\mathbf{x} = \sum_{j=1}^{n} c_j x_j$, subject to $\mathbf{x} \in D$, where D is a nonempty finite set of feasible solutions. Indeed, both polynomially-solvable problems, such as finding the minimum spanning tree of a graph, and NP-hard ones, such as the traveling salesman problem, can be formulated in this way [PaSt82]. *Sensitivity analysis* studies the effect that altering the parameters of a problem will have on the optimum solution to the problem. Such an analysis might be motivated by, among other reasons, uncertainty about the parameters (which, e.g, may depend on fluctuating prices) or because the problem occurs in a dynamic context, and the solution must be obtained for different parameter values. One problem that has received much attention, for example, has been that of determining the effect of processor load variation on the optimal assignment of tasks to processors in a distributed system [Bok87]. An up-to-date bibliography of the literature on sensitivity analysis has been compiled by van Hoesel *et al.* [vHKRW89].

The sensitivity analysis problem of interest to us here is the case where the coefficients of the objective function depend linearly on one or more parameters. Formally, the *k-parameter problem* is to compute $P(\lambda_1, \ldots, \lambda_k) = \min\{\mathbf{c}(\lambda_1, \ldots, \lambda_k)^\mathsf{T}\mathbf{x} = \sum_{j=1}^{n} c_j(\lambda_1, \ldots, \lambda_k)x_j \mid \mathbf{x} \in D\}$, where $c_j(\lambda_1, \ldots, \lambda_k) = a_j + \sum_{i=1}^{k} \lambda_i b_{ji}$. Parametric optimization is, by now, a classical topic in linear programming (see, e.g. [Chv83]), where the computational procedure for the one-parameter case, known as the *parametric simplex method*, has received extensive treatment. In contrast, the multiparameter case has not been studied as closely, with some exceptions. Among the latter, Klein Haneveld *et al.* [KHvdMP79] showed how extend the parametric simplex method to handle two-parameter linear programming problems. Their method has the disadvantages that it works only for linear programming problems, and that, since it derives from simplex, in the worst case it may require time proportional to the number of feasible solutions. The techniques we use here derive in part from two other works, [Gus83] and [FeSr90], that deal with the two-parameter problem. We will discuss the relevant aspects of those papers later.

---

*Supported in part by the National Science Foundation under grant No. CCR-8909626.

The Minimization Diagram. To each $\mathbf{x}^* \in D$ there corresponds a hyperplane in R^{k+1}, defined as the set of all points $(\lambda_1, \ldots, \lambda_k, z)$ such that $z = \mathbf{c}(\lambda_1, \ldots, \lambda_k)^{\mathsf{T}} \mathbf{x}^*$. $P(\lambda_1, \ldots, \lambda_k)$ is the lower envelope of these hyperplanes. Thus, $P(\lambda_1, \ldots, \lambda_k)$ is a well-defined, piecewise-linear concave function from R^k, the *parameter space*, to the real numbers. $P(\lambda_1, \ldots, \lambda_k)$ induces a subdivision of the parameter space into convex polyhedral regions, called *cells*, where the relative interior of each cell is the maximal connected set of points at which $P(\lambda_1, \ldots, \lambda_k)$ is attained by a particular $\mathbf{x}^* \in D$. This subdivision, or *cell complex*, together with the optimum solution vector for each region, is called the *minimization diagram* of the problem [ScSh87] and is denoted by M_P (if more than one solution is optimum for a given region, only one is stored). M_P provides a complete combinatorial description of P. The *complexity* of P, denoted $\kappa(P)$ is the geometric complexity of M_P, namely, the number of faces of dimension 0 through k of M_P.

Given the minimization diagram, the problem of determining the value of P for a point v in the parameter space is reduced to a point location problem in M_P: finding the cell of M_P that contains v. In one-parameter problems $P(\lambda_1)$ subdivides R into a sequence of intervals. Thus, if b denotes the number of intervals, $\kappa(P)$ is $O(b)$ and binary search allows one to find the optimum solution for any value of λ_1 in $O(\log b)$ time. For two-parameter problems, M_P is a subdivision of the plane. If f denotes the number of faces of M_P, then $\kappa(P)$ is $O(f)$ [PrSh85]. It is well-known (see, e.g., [PrSh85,SaTa86]) that, in $O(f \log f)$ time, one can construct a data structure that solves the planar point location problem in $O(\log f)$ time. For $k = 3$, M_P is a subdivision of three-dimensional space. If N denotes the number of facets of M_P, then $\kappa(P)$ is $O(N)$. In $O(N \log^2 N)$ time, it is possible to construct a search structure that answers point location queries in $O(\log^2 N)$ time [PrTa89].

Our Results. Constructing M_P basically amounts to constructing the lower envelope of a set of hyperplanes. While the latter problem has been addressed in the past (e.g., [ScSh87]), the parametric optimization problem differs in at least one important respect. In most cases it has been assumed that the hyperplanes are known in advance. In our problem, however, even though all hyperplanes could, in principle, be obtained by enumerating the feasible solutions, this would be impractical, since, in general $|D| \gg \kappa(P)$. Our goal is, instead, to obtain algorithms whose running time depends only on the input size, n, and on $\kappa(P)$. For this, we rely on a technique due to Gusfield [Gus83] (based on ideas by Megiddo [Meg79]), which provides an answer to the following online parametric problem (see section 2 for details). We are given a point p in the parameter space and a ray emanating from p. Suppose $\mathbf{x}_p$ denotes the optimum solution at p. The problem is to return the point q on the ray that is closest to p such that $\mathbf{x}_p$ is optimal at q but not at any point beyond q. We must also obtain a solution $\mathbf{x}_q$ that is optimal at q and also at some point beyond q. Thus, the ray out of p is like a probe into the parameter space, and Gusfield's method allows us to obtain the "next" optimum solution along the ray.

The algorithm for the above problem is derived from an algorithm A for the non-parametric version of the problem. If the running time of algorithm A is $T_A(n)$, then the running time $T_X(n)$ of Gusfield's technique depends only on n and on $T_A(n)$. Gusfield showed that, in general, $T_X(n) = O(T_A^2(n))$. This bound can sometimes be reduced considerably using ideas due to Megiddo [Meg83] and Cole [Col84]. Further improvements are possible in certain special cases [GGT89].

Gusfied's method provides an immediate $O(b \cdot T_X(n))$ solution to the one-parameter problem: simply obtain the intervals of M_P from left to right. Gusfield also indicated that his method may be extended to two-parameter problems, but the analysis of the algorithm he proposed has an error, which makes it unclear whether it can be implemented in $O(f \cdot T_X(n))$ time as implied in [Gus83].

In this paper, we present a $O(f \cdot T_X(n) + f \log f)$ plane sweep algorithm for the two-parameter problem. We then use the ideas developed for that procedure in order to obtain a $O(N \cdot T_X(n) + N^2)$ time space sweep algorithm for the three-parameter problem. Thus, our algorithms are *output-sensitive* in the sense that their running times depend on n and $\kappa(P)$, not on the number of feasible solutions. A different $O(f \cdot T_A(n) + f^2)$ algorithm for the two-parameter problem was presented in [FeSr90]. Which two-parameter algorithm to choose depends on the number of faces, f, and on the relationship between T_X and T_A. In the worst case, the dominant factor tends to be f, since it can be very large, and the method presented here would be the algorithm of choice. Furthermore, in practice, a space sweep algorithm could be preferable, since it builds M_P incrementally, from top to bottom, thus allowing one to stop the construction at any point. To our knowledge, the three-parameter algorithm presented here is the first of its kind.

Outline of the Paper. Our notation and the basic primitive required by our algorithms are described in section 2. Algorithms for constructing the minimization diagram for two- and three-parameter problems are presented in sections 3 and 4, respectively. Section 5 summarizes our results and discusses possible extensions.

2 Preliminaries

2.1 Definitions and Notation

A point in R^k, the *parameter space*, is denoted by the k-tuple $(\lambda_1, \ldots, \lambda_k)$. A point in R^{k+1} is denoted by the $(k+1)$-tuple $(\lambda_1, \lambda_2, \ldots, \lambda_k, z)$. We denote the λ_i-coordinate of v in R^k or R^{k+1} by $\lambda_i(v)$. The z-coordinate of $v \in R^{k+1}$ is $z(p)$.

Definition 1 For $p, q \in E^k$, $p \prec q$ if

1. $\lambda_k(p) < \lambda_k(q)$ or

2. there is a j, $1 < j < k$ such that $\lambda_j(p) < \lambda_j(q)$ and $\lambda_i(p) = \lambda_i(q)$ for $j < i \leq k$.

Clearly, for $p \neq q$, either $p \prec q$ or $q \prec p$.

Definition 2 Let x_0 be a optimal solution at a point v in the parameter space and let $r(\lambda_1, \ldots, \lambda_k) = c(\lambda_1, \ldots, \lambda_k)^T x_0$. The set of all points $p \in R^{k+1}$ such that $z(p) = r(\lambda_1(p), \ldots, \lambda_k(p))$ is an *optimal plane at* v.

Definition 3 Suppose $S \subseteq R^{k+1}$. Then, orth(S) denotes the orthogonal projection of S onto the parameter space.

For instance, in the two-parameter case, every edge of M_P is contained in orth($\pi_1 \cap \pi_2$) for some optimal planes π_1 and π_2.

Definition 4 Let π_1 and π_2 denote two nonparallel optimal hyperplanes, and let $L = $ orth($\pi_1 \cap \pi_2$). Then, if $v \in L$, ray(π_1, π_2, v) $= \{u : u \in L$ and $v \prec p\}$.

2.2 Procedure NextOpt

Our algorithms rely on the following procedure:

NextOpt(L): Given a ray L in the parameter space emanating from p, return a pair $[q, \pi_q]$ satisfying the following conditions. Let $\mathbf{x}_p$ denote an optimal solution at p. Then, if $\mathbf{x}_p$ is optimal for *all* $q' \in L$, q and π_q are undefined. Otherwise q and π_q are chosen so that:

1. $\mathbf{x}_p$ is optimal for all $q' \in L$ such that $\| q' - p \| \leq \| q - p \|$, and

2. $\mathbf{x}_p$ is not optimal for any $q' \in L$ such that $\| q' - p \| > \| q - p \|$,

3. π_q is an optimal hyperplane at q that is also optimal at some point $q' \in L$ such that $\| q' - p \| > \| q - p \|$.

We assume that NextOpt has a worst-case running time of $T_X(n)$. As we mentioned in the Introduction, such a procedure can be derived from an algorithm for the non-parametric version of the problem.

3 The Two-Parameter Problem

Our approach is to view the λ_2-coordinate as a time parameter and to reconstruct M_P from a sequence of snapshots at different times. The resulting algorithm is another example of *plane sweep*, a paradigm that has found many applications in computational geometry [PrSh85].

More formally, the intersection of the horizontal line $\lambda_2 = t$ with M_P defines a one-parameter problem where only λ_1 varies. The minimization diagram for this problem, which we denote by $M_P(t)$, is a subdivision of that line into a sequence of intervals, ordered from left to right, where for each interval there is an associated optimal plane (see Figure 1). To construct M_P, we vary t from $-\infty$ to $+\infty$, while recording the changes in $M_P(t)$. This is equivalent to sweeping a horizontal line, the *sweepline*, from bottom to top of M_P. The efficiency of this procedure depends on the fact that, while $M_P(t)$ evolves continuously, changes in its topology only occur at instants of time where vertices of M_P are encountered.

We first describe the sequence of objects that are encountered during the sweep. Edge (u, v) of M_P is assumed to be directed from u to v if $u \prec v$ and from v to u otherwise. If $e = (u, v)$ is directed from u to v, then tail(e) $= u$ and head(e) $= v$. Edges that go to $-\infty$ ($+\infty$) in the λ_2-coordinate are assumed to have tail(e) (head(e)) at $-\infty$ ($+\infty$). We partition the edges incident on a vertex v of M_P into two sets: IN(v) $= \{(w, v) \mid w \prec v\}$ and and OUT(v) $= \{(v, w) \mid v \prec w\}$.

Let $v_1 \prec v_2 \prec \cdots \prec v_V$ denote the set of vertices of M_P. Let t_j denote the λ_2-coordinate of v_j, $1 \leq j \leq V$, and, for convenience, put $t_0 = -\infty$ and $t_{V+1} = +\infty$. Throughout the remainder of this extended abstract, we will assume that all t_i's are distinct. It is not difficult to eliminate this assumption, but we will not discuss the topic further here.

Let $G(t)$ be the graph obtained by taking the line $\lambda_2 = t$ and putting vertices at the points of intersection between edges of M_P and the sweep line. This graph has two infinite edges, corresponding to the first and last intervals. $G(t)$ gives the topology of

$M_P(t)$. We say that two one-parameter minimization diagrams are *isomorphic* if we can establish a one-to-one correspondence between the respective vertices and edges of the corresponding graphs. The following is an immediate observation.

Lemma 3.1 *For all t', t'' such that $t_i < t', t'' < t_{i+1}$, $0 \le i \le V$, $M_P(t')$ is isomorphic to $M_P(t'')$.*

An *event* occurs whenever the sweep line goes through a vertex v, such that $\lambda_2(v) = t$. Let t_- and t_+ denote the times immediately before and immediately after the sweep line goes through v. Clearly, $G(t)$ can be obtained from $G(t_-)$ by contracting certain edges in the latter graph into a single vertex, v. $G(t_+)$ can be obtained from $G(t)$ by, perhaps, expanding v into a set of intervals (see, e.g., [SaTa86]).

We must now show how vertices, edges and faces of M_P are identified during the sweep. For this, we must explain how to process an event.

3.1 Event Processing

Assume that the sweep line is at some instant t_- between the last and next events, and that, at this point, we have determined all faces, edges, and vertices of M_P that lie below the sweep line. The status of the sweep line $\lambda_2 = \tau$ is determined by $G(\tau)$, a list of the optimal planes associated with the edges of $G(\tau)$, and a priority queue Q containing one entry per vertex of $G(\tau)$. Thus, for all vertices u such that $\lambda_2(u) < t_-$, we have determined both IN(u) and OUT(u). Each vertex $u \in G(t_-)$ is the intersection of the sweep line with an edge e of M_P whose head and tail are known. Q contains all these edges. Within Q, e_1 has the same priority as e_2 if head(e_1) = head(e_2). Otherwise, e_1 has higher priority than e_2 if and only if head(e_1) $\prec$ head(e_2).

Edges are processed in *batches*, where each batch corresponds to the set of edges in Q of highest priority. Let e be an edge of highest priority in Q. If head(e) is at $+\infty$, then all edges in Q are rays, and we are done: there are no further vertices of M_P beyond t_-. Otherwise, all edges in the batch meet at vertex $v =$ head(e). These edges are precisely the edges in IN(v). Let $t = \lambda_2(v)$, denote the time of this next event. The first part of our task is to advance the sweep line to t. This is accomplished by applying the following lemma.

Lemma 3.2 *Let $e_1, e_2, \ldots, e_b$ denote the edges in a batch and let $v_1, v_2, \ldots, v_b$ denote the corresponding vertices of $G(t_-)$. Then, the subgraph J of $G(t_-)$ induced by $\{v_1, \ldots, v_b\}$ is connected. Furthermore, $G(t)$ is obtained from $G(t_-)$ by contracting the edges of $E(J)$.*

We must now advance the sweep line to t_+, in order to construct $M(t_+)$. Our goal is to obtain all planes π satisfying:

(P1) π is optimal at v and at some point p above the sweep line.

Two such planes are already known: π_A and π_B, the optimal planes associated with the intervals (edges) of $M_P(t)$ to the right and left of v. Both planes are clearly optimal at v and, since there are no other vertices of M_P with the same λ_2-coordinate as v, they must also be optimal at some point above the sweep line. We now use NextOpt to probe the parameter space in order to find all faces incident on v that lie above the sweep line. Each probe is carried out along a ray emanating from v and directed toward the region above the sweep line. We shall show that, even though there are infinitely many rays emanating from v, the number of rays that must be considered is proportional to the total number of faces incident on v that lie above the sweep line. The probing process can be viewed as a simulation of an algorithm for the one-parameter problem due to Eisner and Severance [EiSe76].

At any stage in our algorithm, we will have a set S of planes such that each $\pi \in S$ satisfies P1. Initially, $S = \{\pi_A, \pi_B\}$. At the end of the algorithm, S will contain all planes satisfying P1 that are associated with faces of M_P lying above the sweep line at t. We shall need some more notation. Let R denote a set of planes. $L_R(\lambda_1, \lambda_2)$ denotes their lower envelope, M_R denotes the minimization diagram of L_R, and $M_R(\tau)$ denotes the intersection of the line $\lambda_2 = \tau$ with M_R.

Let T be the set of all optimal cost planes satisfying P1 that are associated with faces of M_P. Each edge in $\mathrm{OUT}(v)$ is contained in the orthogonal projection of the intersection of two faces of T. Suppose D is the set of all optimal planes associated with the intervals of $M_P(t)$. All planes in D must also be optimal at some point above the sweep line, although not necessarily at v. Thus, $M_P(t_+) = M_{T \cup D}(t_+)$, Planes π_A and π_B are known to be in T and all the intervals in $M_P(t_+)$ that are not in $M_P(t)$ are sandwiched between the intervals associated with π_A and π_B. Therefore, we can focus our attention only on π_A and π_B and the intervals in between. Our algorithm starts with $S = \{\pi_A, \pi_B\}$, and keeps adding planes to S until $L_S(\lambda_1, t_+) = L_T(\lambda_1, t_+)$, for all λ_1. At that point, we will have determined all the edges in $\mathrm{OUT}(v)$. We need not worry about the precise choice of t_+ because of the following lemma, which states that the topology of $M_S(t')$ is the same for all $t' > t$. The proof follows from the fact that all planes in S are optimal at v.

Lemma 3.3 For all $t', t'' > t = \lambda_2(v)$, $M_S(t')$ is isomorphic to $M_S(t'')$.

Consider now $M_S(t_+)$. Each vertex u of this minimization diagram is the intersection of a ray emanating from v with the sweep line at t_+. This ray is contained in the projection of the intersection of the two optimal planes associated with the edges to the left and right of u. We shall say that u is verified if the corresponding ray emanating from v is known to contain an edge of M_P. If this is the case, we shall also refer to the ray as being verified. Otherwise, the vertex (or ray) is said to be unverified. Initially, $M_S(t_+)$ contains only one vertex, which is the intersection of $L = \mathrm{ray}(\pi_A, \pi_B, v)$ with the line $\lambda_2 = t_+$. This vertex is assumed to be unverified. To determine whether to declare a vertex verified or not, we use the following fact.

Lemma 3.4 Let π_1 and π_2 be two distinct optimal planes and let $L = \mathrm{orth}(\pi_1, \pi_2)$. Consider any segment $\overline{pq} \subset L$, $p \neq q$. Then $\overline{pq}$ is contained in an edge of M_P if and only if both π_1 and π_2 are optimal at all points in $\overline{pq}$.

The algorithm repeats the following procedure until all vertices of $M_S(t_+)$ are verified. First, pick any unverified vertex u of $M_S(t_+)$. Let L denote the corresponding ray emanating from v. Compute $[p, \pi_C] = \mathrm{NextOpt}(L)$. There are two possibilities: $v \prec p$ or $v = p$. In the first case, because of Lemma 3.4, (v, p) is an edge of M_P (if p is undefined, L is an edge of M_P). Vertex u becomes verified. If $v = p$, L does not contain an edge of $\mathrm{OUT}(v)$, and we have obtained a new plane π_C satisfying P1. The plane π_C intersects with at most two faces of $L_S(\lambda_1, t_+)$. We add π_C to S and, in the process, create new unverified vertices in $M_S(t_+)$. It is possible that adding π_C to S may yield a ray identical to one that has already been verified. If this occurs, the resulting ray will also be assumed to be verified.

Assume, as before, that D is the set of all optimal planes associated with edges of $M_P(t)$.

Lemma 3.5 *If $M_S(t_+)$ has no unverified vertices, then $M_{S \cup D}(t_+) = M_P(t_+)$.*

PROOF SKETCH: (By contradiction.) Assume that all vertices of $M_S(t_+)$ are verified. We have $P(\lambda_1, t_+) \leq L_{S \cup D}(\lambda_1, t_+)$, for all λ_1. If the two sides are not equal, there must be a value of λ_1 such that $P(\lambda_1, t_+) < L_{S \cup D}(\lambda_1, t_+)$. Therefore, there must exist a vertex in $L_{S \cup D}$ lying above P. This vertex must be unverified, a contradiction. $\square$

The number of probes required is bounded by the lemma below.

Lemma 3.6 *$M_P(t_+)$ is computed using $O(|OUT(v)|)$ calls to NextOpt.*

PROOF SKETCH: We have seen that the optimal planes associated with intervals of $M_P(t)$ are a subset of those of $M_P(t_+)$ and that the underlying change from one instant to the next is local in the sense that $M_P(t_+)$ is obtained by inserting new intervals into $M_P(t)$. This enables us to restrict our attention only to the part of $M_P(t)$ that corresponds to the faces satisfying P1. This portion is essentially $M_T(t_+)$. $M_T(t_+)$ consists of consecutive intervals, the first and last corresponding to the regions where π_A and π_B are optimal and the middle ones to faces of M_P whose interior lies entirely above the horizontal line passing through v. Now consider the sweep line at t_+. For each point in the sweep line, there is a ray emanating from v and going through that point. Each such ray could be used during a call to NextOpt. We can partition these rays into distinct classes, depending on whether they go through a vertex of $M_T(t_+)$ or whether they go through the interior of an edge. The proof consists of showing that for any edge in $M_T(t_+)$, NextOpt will be called at most once to probe a ray that goes through the interior of that edge. For any vertex in $M_T(t_+)$, NextOpt will be called at most twice to probe the ray that goes through that vertex. $\square$

Theorem 3.1 *M_P can be constructed in $O(f \cdot T_X(n) + f \log f)$ time.*

PROOF: $M_P(t_+)$ is computed from $M_P(t_-)$ in $O(|OUT(v)| \cdot T_X(n) + (|IN(v)| + |OUT(v)|) \cdot \log f)$ time. To see this, note that there are $|IN(v)|$ edges in a batch, each of which is extracted from Q in $O(\log f)$ time. This is followed by a contraction of $O(|IN(v)|)$ edges of $G(t_-)$. The time spent on probing during an expansion is $O(|OUT(v)| \cdot T_X(n))$, by Lemma 3.6. Each time a plane π is added to S because of a probe along a ray L, we must update $M_S(t_+)$. This can be done in $O(1)$ time, since π will only intersect the planes associated with the intervals associated with faces to the left and right of L. There is an additional $O(|OUT(v)| \log f)$ time to update Q. The theorem is proved by adding up the work per vertex over all vertices of M_P and using Euler's relation. $\square$

As the sweep is carried out, one can simultaneously construct an efficient data structure to support point location queries in M_P. Sarnak and Tarjan's method [SaTa86] is well-suited for this purpose, since it can build the required data structure in $O(f \log f)$ time, as M_P is swept.

One important detail we have omitted is how to start the sweep. One possibility is to choose a time τ such that, for all $t', t'' < \tau$, $M_P(t')$ is isomorphic to $M_P(t'')$. This is the basically the *steady state problem* discussed in section 5. It is more common, however, to be interested only in constructing a certain region in M_P. Thus, if this region starts at time τ, we build $M_P(\tau)$ and start the sweep from there.

4 The Three-Parameter Case

We shall now sketch how to extend the techniques developed in the previous section to three-parameter problems. In this case, we shall view λ_3 as our time parameter. $M_P(t)$ is the minimization diagram for the two-parameter problem obtained by intersecting M_P with the horizontal *sweep plane* $\lambda_3 = t$. As usual, $v_1 \prec v_2 \prec \cdots \prec v_V$, where $\lambda_3(v_i) = t_i$, and $t_0 = -\infty$, $t_{V+1} = +\infty$. We assume that all t_i's are distinct. $M_P(t)$ consists of a planar graph $G(t)$, together a list that gives the optimal hyperplane associated with each face of $G(t)$. A lemma similar to the following one was proved in [PrTa89].

Lemma 4.1 *For all t', t'' such that $t_j < t', t'' < t_{j+1}$, $M_P(t')$ and $M_P(t'')$ are isomorphic.*

As before, events are associated with vertices. Consider what happens as the sweep plane goes through vertex v. Let $t = \lambda_3(v)$ and let t_- and t_+ denote instants immediately before and after t. The change from $M_P(t_-)$ to $M_P(t)$ amounts to contracting a subgraph J of $G(t_-)$ into vertex v. The change from $M_P(t)$ to $M_P(t_+)$ amounts to expanding v into a subgraph H of $G(t_+)$, together with finding the optimal hyperplanes associated with the new faces. Thus, our problem is to determine J and H.

We first show how to determine J. The vertices of $M_P(t_-)$ are the intersections of edges of M_P with the sweep plane at t_-. These edges are maintained in a priority queue Q, ordered by the "$\prec$" relation on their heads, as in the two-parameter case. Edges in Q are processed in batches, consisting of edges with the same head.

Lemma 4.2 *Let $\{f_1, \ldots, f_r\}$ be the set of edges in a batch, and let $U = \{u_1, \ldots, u_r\}$ denote the corresponding vertices of $G(t_-)$. Then J is the subgraph induced by U.*

To find the edges of J to be contracted, we simply conduct a search in $G(t_-)$, which can be done in $O(|E(J)|)$ time, where $E(J)$ is the set of vertices of J.

The second problem, finding H, is harder than its counterpart in the two-parameter case. The strategy, as before, is to probe the parameter space using rays emanating from v and directed towards the region above the sweep plane. The technique we use for this purpose is based upon a generalization of the Eisner-Severance method presented in [FeSr90].

Since every plane that is optimal at t is also optimal at t_+, the optimal hyperplanes associated with the faces of $M_P(t)$ are a subset of those associated with faces of $M_P(t_+)$. To construct $M_P(t_+)$, we must find all additional hyperplanes that are optimal at v, and at some point above the sweep plane at t. At this point, we know at least three such hyperplanes. These are the hyperplanes associated with faces of $M_P(t)$ that are incident on v. Now, consider H in isolation, as a graph embedded in the plane. We shall refer to all faces of H, except for the unbounded face as *interior faces*. Each interior face is associated with a hyperplane that is optimal at v and some instant after t, but that is not optimal at any instant prior to t. Call this set of hyperplanes T.

As in section 3, let M_R denote the minimization diagram of a set of hyperplanes R. If D denotes the set of all hyperplanes associated with faces of $M_P(t)$, we have $M_P(t_+) = M_{D \cup T}(t_+)$. As is the two-parameter case, we concentrate on obtaining $M_T(t_+)$. All unbounded faces of $M_T(t_+)$ correspond to the faces that surrounded v in $M_P(t)$. The hyperplanes associated with these faces remain optimal at instant t_+. The bounded faces are the interior faces of H which we are looking for.

Our initial set S consists of all hyperplanes associated with faces surrounding v in $M_P(t)$. We keep adding hyperplanes optimal at v and at some instant after t to this set until $M_S(t_+) = M_T(t_+)$. The precise choice of t_+ does not matter, as long as $t_+ > t$, because of the following lemma.

Lemma 4.3 *For all $t, t'' > t$, $M_S(t')$ is isomorphic to $M_S(t'')$.*

We first construct $M_S(t_+)$. While in the two-parameter case, $M_S(t_+)$ would have initially contained a single vertex (corresponding to the intersection of two planes), in the current case $M_S(t_+)$ might have more than one vertex. Indeed, in general, $M_S(t_+)$ will be the result of expanding S into a tree (see Figure 2). Every vertex of $M_S(t_+)$ is the intersection of a ray originating at v with the sweep plane at t_+. Each ray provides a direction that may, at some time, be probed. Vertices of $M_S(t_+)$ are classified as being either verified or unverified, depending on whether or not they contain an edge of M_P. Testing whether a ray L contains an edge of M_P is done by computing $[p, \pi] = $ NextOpt(L). If $v = p$, the answer is no, and a new face, π is added to S, creating new unverified vertices. If $v \neq p$, the vertex becomes verified. The probing process continues until all vertices of $M_S(t_+)$ are verified. At this point, it can be shown that $M_S(t_+) = M_T(t_+)$. Therefore, all interior faces of H have been found.

The number of probes can be shown to be $O(F_H + V_H + E_H)$, where F_H, V_H, and E_H are the number of interior faces, vertices, and edges of H [FeSr90]. The process of updating $M_S(t_+)$ after adding π to S takes $O(E_H)$ time. Thus, the total time spent in updates is (ignoring certain other steps) $O(E_H^2)$, the time required to build $M_P(t)$ from $M_P(t_-)$ is $O(|E(J)|)$ and the total time spent probing is $O(E_H \cdot T_X(n))$. Since every edge of a plane section of M_P is the intersection of a facet of M_P with a plane, the total time spent probing over all vertices of M_P is $O(N \cdot T_X(n))$ and the total time spent in expansions and contractions is $O(N^2)$, where N is the number of facets. Thus, we have:

Theorem 4.1 *The minimization diagram for a three-parameter problem can be constructed in $O(N \cdot T_X(n) + N^2)$ time.*

Preparata and Tamassia's method [PrTa89] can be used to build a data structure that supports $O(\log^2 N)$ time point location in M_P. This structure can be built in $O(N \log^2 N)$ time, as M_P is swept.

5 Discussion

We have presented space-sweep algorithms to construct the minimization diagram for two- and three-parameter problems. The approach is general, in the sense that the same method can be used for different optimization problems by simply changing the NextOpt subroutine. We have assumed that NextOpt is implemented using Gusfield's approach mainly because the method has polynomial worst-case performance. In fact any other algorithm that would allow us to implement the probing process could be used for this purpose. This includes the parametric simplex method.

There are several issues that must be addressed in order to generalize our approach to higher dimensions. These include the problem of handling the probing process once a vertex is identified and the problem of building a data structure suitable for point location in dimensions higher than three.

In addition to the algorithms presented here, we have also used space-sweep to solve a related problem, which we call finding the *steady state* [Meg86]. We are given a point $p \in R^k$ and we are asked to find an interval $I = \{q \mid \lambda_i^{\min}(q) \leq \lambda_i(q) \leq \lambda_i^{\max}, i = 1, 2, \ldots, k\}$, such that $p \in I$ and so are all the vertices of M_P. For lack of space, those results are not discussed here.

Acknowledgement

I thank Roberto Tamassia for useful discussions.

References

[Bok87]   S.H. Bokhari. *Assignment Problems in Parallel and Distributed Computation*. Kluwer, Boston, 1987.

[Chv83]   V. Chvátal. *Linear Programming*. W.H. Freeman and Co., San Francisco (1983).

[Col84]   R. Cole. Slowing down sorting networks to obtain faster sorting algorithms. *25th Annual Symposium on Foundations of Computer Science*, Singer Island (October 1984), 255–259.

[EiSe76]   M.J. Eisner and D.G. Severance. Mathematical techniques for efficient record segmentation in large shared databases. *J. Assoc. Comput. Mach.*, 23:619–635, 1976.

[FeSr90]   D. Fernández-Baca and S. Srinivasan. *Constructing the minimization diagram of a two-parameter problem*. Tech. Report 90-01, Department of Computer Science, Iowa State University, January 1990. To appear in *Operations Research Letters*.

[GaJo79]   M. Garey and D. Johnson. *Computers and Intractability: A Guide to the theory of NP-Completeness*. Freeman, San Francisco, 1979.

[GGT89]   G. Gallo, M.D. Grigoriades, and R.E. Tarjan. A fast parametric maximum flow algorithm and its applications. *SIAM J. Computing*, 18(1):30–55, 1989.

[Gus83]   D. Gusfield. Parametric combinatorial computing and a problem in program module allocation. *J. Assoc. Comput. Mach.*, 30(3):551–563, July 1983.

[vHKRW89]   C.P.M. van Hoesel, A.W.J. Kolen, A.H.G. Rinooy Kan, and A.P.M. Wagelmans, *Sensitivity analysis in combinatorial optimization: a bibliography*. Report 8944/A, Econometric Institute, Erasmus University Rotterdam, 1989.

[KHvdMP79]   W.K. Klein Haneveld, C.L.J. van der Meer, and R.J. Peters. A construction method in parametric programming. *Mathematical Programming*, 16:21–36, 1979.

[Meg79]   N. Megiddo. Combinatorial optimization with rational objective functions. *Math. Oper. Res.*, 4:414–424, 1979.

[Meg83]   N. Megiddo. Applying parallel computation algorithms in the design of serial algorithms. *J. Assoc. Comput. Mach.*, 30(4):852–865, 1983.

[Meg86]   N. Megiddo. Dynamic location problems. IBM Almaden Research Center, Tech. Report RJ 4983, 1986.

[PaSt82] C.H. Papadimitriou and K. Steiglitz. *Combinatorial Optimization: Algorithms and Complexity*. Prentice-Hall, Englewood Cliffs, NJ, 1982.

[PrSh85] F.P. Preparata and M.I. Shamos. *Computational Geometry: An Introduction*. Springer-Verlag, New York, 1985.

[PrTa89] F.P. Preparata and R. Tamassia. *Efficient point location in a convex spatial cell complex*. Technical Report No. CS-89-47, Department of Computer Science, Brown University, December 1989. A preliminary version appeared in *Algorithms and Data Structures*, F. Dehne *et al.*, eds, Lecture Notes in Computer Science 382, Springer-Verlag, Berlin, 1989, pp. 3–11.

[ScSh87] J. Schwartz and M. Sharir. *On the two-dimensional Davenport-Schinzel problem*. Technical Report 193, NYU Courant Institute of Mathematical Sciences, August 1987.

[SaTa86] N. Sarnak and R.E. Tarjan. Planar point location using persistent search trees. *Communications ACM*, 29:669–679, 1986.

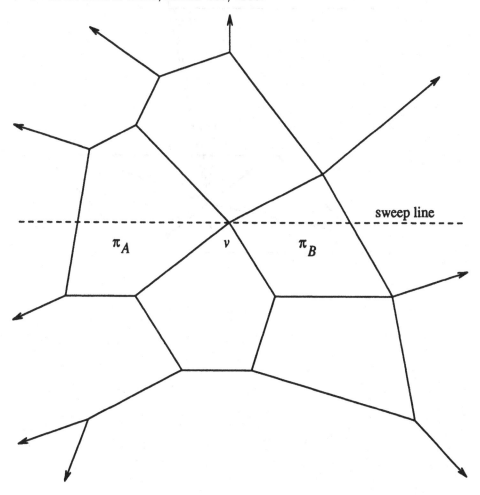

Figure 1.

275

$M_S(t_+)$

$M_S(t)$

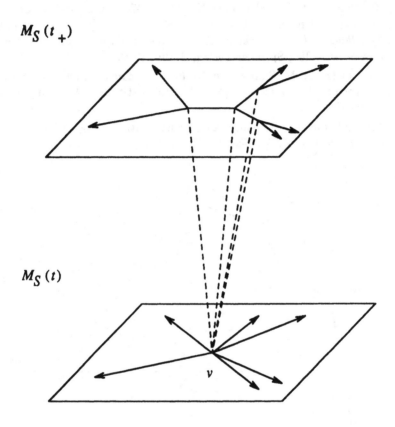

Figure 2.

APPROXIMATING FINITE WEIGHTED POINT SETS BY HYPERPLANES

N. M. Korneenko[x] and H. Martini[xx]

[x] Institute of Mathematics, BSSR Academy of Sciences, Surganova 11, Minsk, 220604, USSR

[xx] Sektion Mathematik, PH Dresden, Wigardstr. 17, 8060 Dresden, DDR

We shall discuss the complexity of the following geometric location problem: Given n fixed points p_i in Euclidean d-space $(d \geq 2)$ with corresponding positive weights w_i, i=1,...,n, find the hyperplanes H minimizing

$$S(H) = \sum_{i=1}^{n} w_i \, \text{dist} \, (H,p_i) , \qquad (1)$$

where dist $(\cdot , \cdot)$ is the usual Euclidean distance. The one-dimensional version of this problem and the determination of (d-1)-subspaces maximizing (1) are considered, too.

§ 1 NOTATION AND BASIC FACTS

As usual in combinatorial computational geometry, we adopt the infinite precision real RAM with unit-cost operations +, -, /, *, "square root" as a modell of computation.

Let $P = \{p_1, ..., p_n\}$ be an arbitrary point set in Euclidean d-space $(d \geq 2)$ with corresponding positive weights w_i (i=1,...,n) and

$W = \sum_{i=1}^{n} w_i$. Excluding unimportant cases, we assume *dim aff* $P = d$.

For a hyperplane H define H^+, H^-, H^0 to be the sets of indices of given points lying above, below and in H, respectively, where as W^+, W^-, W^0 denote the corresponding sums of weights.

A hyperplane minimizing (1) is called a *minsum hyperplane* with respect to P.

Further on we suppose the dimension d as well as P to be fixed, and the notational dependence on P shall be suppressed.

For a set X, $|X|$ gives its cardinality.

For a convenient approach to necessary criteria with respect to minsum hyperplanes, we now introduce the geometric machinery of zonotopes (cf. [M 71, M 87, SW 83]). These considerations refer to Euclidean vector space $\mathbb{R}^d$ $(d \geq 2)$ with inner product $< \cdot , \cdot >$ and unit sphere $S^{d-1} = \{u \in \mathbb{R}^d \mid <u, u> = 1\}$. The position vector of $x \in \mathbb{E}^d$ is given by $x \in \mathbb{R}^d$, in particular o denotes the origin.

The vector sum

$$Z = \left\{ \sum_{i=1}^{n} \lambda_i \, (w_i \, p_i) \mid -1 \leq \lambda_i \leq 1 \right\}$$

of the line segments $S_i = \mathrm{conv}\,(-w_i p_i, w_i p_i)$, $i=1,...,n$, is named a *d-zonotope*, i.e., Z is a convex d-polytope all faces of which are centrally symmetric. By

$$h\,(Z, u) = \max\,\{<u, x> \mid x \in Z\}\ , u \in S^{d-1}\ ,$$

the *support function* of Z is defined. Since the support function of the vector sum of line segments is the same as the sum of their single support functions, we obtain

$$h\,(Z, u) \;=\; \sum_{i=1}^{n} w_i \, |<p_i, u>|\,,\ u \in S^{d-1}\ , \tag{2}$$

Obviously, if a direction $u_{min} \in S^{d-1}$ yields the *minimum of the support function* of a convex d-polytope Q with $o \in \mathrm{int}\,Q$, then u_{min} is the outer normal direction of some (d-1)-face of Q.

In general, a face of Z has the representation

$$S_{i_1} +...+ S_{i_k} + \delta_{i_{k+1}} (w_i p_i)_{k+1} +...+ \delta_{i_n} (w_i p_i)_n$$

with $\{i_1,..., i_n\} = \{1,...,n\}$ and $\delta_{i_j} \in \{-1;1\}$, $j \in \{k+1,..,n\}$.

Without loss of generality, let

$$\overline{F} = S_1 +...+ S_m + \delta_{m+1} (w_{m+1} p_{m+1}) +...+ \delta_n (w_n p_n)$$

be a (d-1)-face of Z normal to u_{min}. Hence the translate

$$F = S_1 + \dots + S_m = \sum_{i=1}^{m} conv\, (-w_i p_i,\ w_i p_i)$$

of F is a zonotope of dimension d-1, i.e. we have

$$<p_i, u_{min}> = 0,\ i=1,\dots,m,\ \text{and}\ \text{dim lin}\ \{p_1,\dots,p_m\} = d-1. \tag{3}$$

As we shall see, this is a useful base for the following necessary criterion on a minsum hyperplane.

THEOREM:

Let $P = \{p_1,\dots,p_n\}$ be a set of arbitrary points in $\mathbb{E}^d$ ($d \geq 2$) with *dim aff* $P = d$ and corresponding positive weights w_i, $i=1,\dots,n$. The minsum hyperplane with respect to P is the affine hull of d affinely independent points from P.

Proof:

We identify an arbitrary point $x \in \mathbb{R}^d$ with the origin and consider all (d-1)-subspaces (through x). (Thus, the set of all hyperplanes in $\mathbb{R}^d$ is taken into consideration.)

The sum of weighted Euclidean distances from P to such a (d-1)-subspace H normal to $u \in S^{d-1}$ then reads

$$S(H) = \sum_{i=1}^{n} w_i\ |<p_i, u>|\ ,$$

which is precisely the support function of

$$Z = \sum_{i=1}^{n} conv\, (-\, w_i p_i,\ w_i p_i).$$

By (3) the minsum (d-1)-subspace must contain d-1 linearly independent position vectors belonging to P. Now we identify one of these d-1 position vectors with the "new" origin and apply repeatedly (3), getting d affinely independent points necessarily contained in each minsum hyperplane.

q.e.d.

We shall say that a hyperplane passing through d affinely independent points of a weighted set P is a *blocked hyperplane* .

§ 2 ALGORITHMS

The immediate consequence of our Theorem is that the minsum hyperplane can be found in $O(n^{d+1})$ *time:* Enumerate all $C_n^d = O(n^d)$ candidate hyperplanes and compute the corresponding weighted distance sums. (Constant time per candidate-k-subset enumeration algorithms can be found e.g. in [RND 77] , sect. 5.2.2.).

Now we shall describe some improvements being the extensions of the two-dimensional algorithms in [DKM 84, DKM 85, MT 83] and [LC 85] , respectively.

2.1. $O(n^d \log n)$ time, $O(n)$ space

The restriction of our considerations to a set of hyperplanes carrying one fixed point (cf. the first part of the proof in § 1) shall be called the *anchored minsum hyperplane problem* , because this carrier point obviously can be named the *anchor* of the hyperplane set. We first reformulate briefly the planar $O(n^2 \log n)$ algorithms from [DKM 84, DKM 85, MT 83] by solving the anchored version of the main problem. By the Theorem, the minsum line must pass through some given point distinct from the anchor. This suggests a straightforward $O(n^2)$ algorithm. But we can trade one n for $\log n$ via some sorting.

Algorithm 1. (input: pointcount n, pount set P, weight set, anchor q; output: anchored minsum line)

Step 1.   Sort all candidate lines $H_i = \text{aff}(q, p_i)$, i=1,...,n, with $q \neq p_i$ by their slopes. Without loss of generality assume that the indices give us this order.

Setp 2.   For a line H: (ax+by+c = 0) the weighted distance sum (1) may be rewritten as

$$S(H) = (a^2+b^2)^{1/2} \left[\sum_{i \in H^+} w_i (ax_i+by_i+c) - \sum_{j \in H^-} w_j(ax_j+by_j+c) \right]. \tag{4}$$

Starting now from the horizontal position, rotate the anchored line stopping it where it hids given points and updating S(H) and its current minimum. Once the given points are sorted by slope and initial $S(H_{\text{horiz}})$ is computed, one may easily see that the i-th update of (4) can be done in time $O(|H_i^o| + |H_{i+1}^o|)$ by keeping track of sets H^+, H^-, H^o and sigmas in (4) during the rotation.

Step 1 takes $O(n \log n)$ time , Step 2 needs $O(\sum |H_i^o|) = O(n)$ time .

As for the original problem in the plane, take as anchors all given points in turn spending in total $O(n^2 \log n)$ time and *linear space* .

Now it is easy to go into d dimensions: We anchor our candidate hyperplanes by all (d-1)-point sets in turn leaving them with the single rotational degree of freedom to sort by and proceed as above arriving at the claimed $O(n^d \log n)$ time algorithm. The laziest description is as follows.

Algorithm 2. (input: d,n,P, set of weights; output: minsum hyperplane)
begin

for ind: =1 to $\binom{n}{d-1}$ do

begin

Generate the next (d-1)-point subset P_{ind} [RND 77] . It determines a (d-2)-flat F_{ind}. Take its orthogonal complement C_{ind} (being a 2-flat) and project P orthogonally onto it with P_{ind} mapping onto a point q_{ind};

Solve the instance of the anchored minsum line problem in the plane C_{ind} with anchor q_{ind};
Update the current minimum of (4);
    end
end.

2.2 $O(n^d)$ time and space

Here again we sketch the planar case [LC 85]. In the algorithms from sect. 2.1 the bottleneck was angular sorting of P n times. It turns out that we can trade *log n time* for *n space* by means of the geometric duality transform [LC 85]. One possible kind of such a transform is (we prefer it for its symmetry with its inverse)

object			T			parameter
	(x,y)-plane	$\leftrightarrow$		(k,b)-plane		
space	$L: (y+kx+b=0)$	$\leftrightarrow$		$L_T=(k,b)$		space
	$p=(x,y)$	$\leftrightarrow$		$p_T:(b+xk+y=0)$		

mapping points into lines and vice versa. Assume that no two given points share their x-coordinate (it can be provided in $O(n \log n)$ *time* by an appropriate rotation). Then it is readily seen that lines sharing a common point p and sorted by slope are mapped via T onto points lying on and sorted along the line p_T (cf. [E 87]).

Now let T transform our given point set P onto some line arrangement P_T. The full description of P_T can be constructed in $O(n^2)$ *time and space* (see [CGL 85, EOS 86, E 87]). Moreover, for any given anchor p from P we may spend *time $O(n)$* to extract from this description the relevant line p_T together with the point set sorted along p_T and corresponding with the pair-of-point line set H_i, i=1,...,n (see sect. 2.1), sorted by slope [LC 85]. Having got the sorted line set we proceed as in sect. 2.1 spending in total $O(n^2)$ *time and space*.

Now get back to d dimensions. The extended form of the duality transform will be

$$T$$
$$H:(y+a_1x_1+...+a_{d-1}x_{d-1}+a_d=0) \quad \rightarrow \quad H_T=(a_1,...,a_d)$$
$$p=(x_1,...,x_{d-1},y) \quad \rightarrow \quad p_T:(a_d+x_1a_1+..+x_{d-1}a_{d-1}+y=0,$$

which also preserves various positional relations: incidence, sideness, order, etc. We may extend it for all non-vertical flats in $I\!E^d$.

$$F_{d-k} = \begin{matrix} H_1: (y+...=0) \\ \cdot \\ \cdot \\ \cdot \\ H_k:(y+...=0) \\ \text{(k simultaneous} \\ \text{linear equations)} \end{matrix} \quad \overset{T}{\underset{\leftarrow}{\rightarrow}} \quad F_{k-1} = aff~((H_1)_T,...,(H_k)_T)$$

Here again T casts our given point set P into the hyperplane arrangement P_T, and full description of the related cell complex may be constructed in $O(n^d)$ time and space (cf. [EOS 86, EM 87]). Moreover, as one may already expect, the object space hyperplanes anchored by a (d-1)-point anchor Q (which may be regarded as a (d-2)-flat) and ordered by turn angle are mapped via T onto a point set lying on and ordered along the 1-flat Q_T. Along the lines of [LC 85] we may extract the desired ordered point set, back-transform it into the anchored hyperplane set sorted by turn angle and make a rotate-and-update trick as in sect. 2.1 in total *time and space $O(n^d)$.*

2.3 $O(n^d)$ time and $O(n^2)$ space, d > 2

To explain getting the idea of the space improvements we may look in history. In [MS 83] it was proposed to sort all pair-of-point lines by slope spending $O(n^2)$ *space.* In [DKM 85, LC 85] it was noted that $O(n)$ *space* is enough for such an algorithm, if we sort only locally around each fixed given point in turn. Further, [LC 85] blows up space requirements back, to save *log n time.* Our straightforward extension (sect. 2.2) needs all $O(n^d)$ *space* for all-simultaneous sort. Now we try and localize this sorting by dealing not with all $O(n^{d-1})$ (d-1)-point anchored hyperplane bundles but rather grouping them into thicker bundles treated separatedly.

Intuitively, we make anchored hyperplane bundles thicker by adding one degree of freedom and sorting them simultaneously by the first d.o.f. for all possible discrete values of the added d.o.f. just as in [MS 85].

More formally, take as an anchor a (d-2)-point set P_{d2} defining a (d-3)-flat $Q = aff (P_{d2})$. (Obviously, one may regard Q as the anchor.) Then the Q-anchored hyperplanes are mapped via T into points on the 2-flat Q_T. A point $p_i \in P \setminus Q$ is mapped via T onto the line L_i on Q_T being the intersection of Q_T and the hyperplane $(p_i)_T$.

If we now fix some additional $p_i \in P \setminus Q$ to obtain a (d-1)-point anchor $P_{d1} = P_{d2} \cup \{p_i\}$ for use in Algorithm 2, then the P_{d1}-anchored hyperplanes are mapped via T onto the points along the line L_i in the plane Q_T. Therefore we arrive at a situation amenable to the simultaneous sorting of [MS 85].

We apply this $\binom{n}{d-2}$ times. Each such major step takes $O(n)$ *time* to map $P \setminus Q$ onto a line set $\{L_i\}$ in the plane Q_T plus $O(n^2)$ *time and space* to construct the arrangement in Q_T and to extract the desired ordered point sets for Algorithm 2 just as in [LC 85].

Hence we have the claimed complexity.

2.4 $O(n^d)$ time and linear space

Yamamoto et al. [YKII 88] used only linear space in the planar case to enumerate all weighted minsum lines by traversing the weighted minsum belt in the dual plane using the topological sweep of a 2-dimensional arrangement [EG 86].

Note now that the topological sweep [EG 86] gives the sorted orders of intersection points of the planar line arrangement along all arrangement lines "in parallel" during the pseudoline sweep, e.g. from the left to the right. Remember that if we work in the image plane then a line of the arrangement corresponds with a given point, and the intersection points correspond with pair-of-point lines in the object space. Therefore, here is one additional planar minsum line algorithm:

Intuitively, we interleave n "processes" of Algorithm 1 in accordance with the transitions of the pseudoline sweeping the arrangement as in [EG 86]. We keep track of the pair-of-point lines in sorted order around each of the given points "in parallel" during the pseudoline sweep [EG 86] of the image-space arrangement by first initializing n candidate minsum lines through each given point and the updating them in turn switching from one to another in accordance with sweeping pseudoline transitions in [EG 86] and recomputing (4) as in Algorithm 1. Obviously, we shall need no more than in [EG 86].

Now we replace the [LC 85]-based step in the Algorithm from sect. 2.3 by the [EG 86]-based step to obtain finally the *space-optimal* $O(n^d)$ *time* algorithm.

2.5 Rudimentary parallelism

Clearly, each anchored bundle in sect. 2.4 may be processed by a seperate processor with memory $O(n)$; so for p processors, $1 < r < n^{d-2}$, we obtain the optimal speedup for $d > 2$.

In the planar case n processors trivially cope with the problem in *time $O(n \log n)$*.

2.6 Remarks to the unweighted case

Looking at § 1, one might miss an obvious additional necessary criterion for minsum hyperplanes, namely: For each minsum hyperplane with respect to n arbitrary weighted points in Euclidean d-space (here with $d \geq 1$) we have

$$W^+ < \frac{W}{2} \quad \text{and} \quad W^- < \frac{W}{2} . \tag{5}$$

(Note that a blocked hyperplane satisfying (5) is called a *halving hyperplane.*) This condition strengthens somewhat statements from [MN 80, DKM 84, DKM 85, MT 83, YKII 88] and was given in the complete sense already in [NM 80]. In [DKM 84, YKII 88] it was shown that (5) in the planar weighted case cannot improve the worst-case complexity. But in the planar unweighted version (i.e. $w_i = 1$ and $p_i \neq p_j$ for $i,j=1,...,n$ with $i \neq j$) this criterion allows some progress, using suitable upper bounds on the number of halving lines in terms of n. In general, the unweighted planar case waits its best because of the mismatch between the *Omega (n log n)* lower bound, proved in [YKII 88], and the complexity of existing algorithms. For results in this direction (as well as a short proof of (5)) we refer to the survey [KM 90].

§ 3 THE ONE-DIMENSIONAL WEIGHTED CASE

Of course, the one-dimensional version of our problem does not depend on the Theorem, but on (5). It is almost trivially solved by sorting points followed by a single scan. However, we can do better in *optimal O(n) time*. Namely, we see that here (1) is a piecewise-linear unimodal function, in principle amenable to binary search. To accomplish this, we need to repeat recursively three main steps.

1. Divide the current domain into roughly equal parts;
2. Compute the middle value of (1);
3. Throw away half of the domain.

If we perform them in time proportional to the size of the current domain then we would have the recurrence relation for the time complexity $T(n)$: $T(k) = T(\frac{k}{2}) + O(k)$ with the solution $T(n) = O(n)$.

Step 1 can be accomplished by a well-known linear-time median finding procedure. Step 3 causes no troubles, too. However, Step 2 needs some care because direct computation of (1) needs all its $O(n)$ *time* on all levels of recursion.

Therefore we rearrange (1). For a point q let q^- denote the set of (indices of) given points not greater than (that of) q. Let further $H > p_t$ for some t. Then (1) may be rewritten as

$$S(H) = \sum_{i \in p_t^-} w_i(p_t - p_i + H - p_t) + \sum_{i \in P \backslash p_t^-} w_i \, |H\text{-}p_i| =$$

$$= \sum_{i \in p_t^-} w_i(p_t\text{-}p_i) + \left[\sum_{i \in P_t^-} w_i \right] (H\text{-}p_t) + \sum_{i \in P \backslash p_t^-} w_i \, |H\text{-}p_i| =$$

$$= S1 + SW \, (H - p_t) + S3 ,$$

where S1 is constant for $H > P_t$, S3 is an ordinary weight distance sum, and the second sum is the weighted distance from H to p_t, where p_t is given the "lumped weight" SW. Now we are ready for implicit binary search.

Algorithm 3.
```
    begin
        Initialize S1, SW, t in (6) to 0.
        while n > 3 do
        begin
            1. Find a median point pₘ and compute S(pₘ) by (6);
            2. Find the nearest right neighbour pᵣ for pₘ, compute S(pᵣ)
               by (6);
```

3. If $S(p_r) < S(p_m)$ then

    minimum of S(H) lies to the right of p_m^- and we throw away points of p^- and set

    t: = m and corresponding SW, S1

    else do symmetric actions on the right hand side;

  4. Decrease n by suitable half

end of while

for remaining 3 or less points compute minimum directly

end.

Thus, the weighted minsum point can be found in *optimal linear time*.

§ 4 CONCLUDING REMARKS

From the historical point of view, some papers shall be mentioned, partially already given in the text before.

Problem (1) was considered in [NM 80], and its two-dimensional version gained attention both in Operations Research community [MT 82, W 75] and in Computational Geometry community [D 84, DKM 84, DKM 85, K 88, K 89, LC 85, MT 82], and a number of algorithms is known for d= 2 [DKM 84, DKM 85, MT 83, MN 80], exploiting basically the same fact that a minsum line has to be blocked.

The forthcoming survey [KM 90] shall give a wider view on our problem and related ones.

A final remark shall be devoted to a natural inverse question to a subproblem in this paper, namely the

Anchored maxsum hyperplane problem: Given a set P of n positively weighted points in Euclidean d-space (d ≥ 2), find the hyperplane H through an anchor q maximizing S(H) (cf. (1)). Identifying q with the origin o of $\mathbb{R}^d$, this leads one directly (see [M 87]) to the Boolean optimization problem

$$\left|\left|\sum_{i=1}^{n} \rho_i (w_i p_i)\right|\right|^2 = \rho^T C \rho \quad \rightarrow \text{Max!}$$
$$\rho_i \in \{-1;1\}$$

Here C stands for the positively semidefinite matrix with elements $<p_h, p_k>$. Special techniques for solving such problems by means of "branch and bound" algorithms are given in [MY 80].

REFERENCES

[D 84]   N. N. Doroshko, The solution of two mainline problems in the plane. Manuscript Deposited in VINITI USSR (14.06.1984), Reg. no. 39614 (1984), 15 pp.

[DKM 84]   N. N. Doroshko, N. M. Korneenko, N. N. Metel'skij, Optimizing line location with respect to a planar point system. Preprint Inst. Math. Acad. Sci. BSSR, Minsk, USSR, no. 23 (1984), 19 pp.

[DKM 85]   N. N. Doroshko, N. M. Korneenko, N. N. Metel´skij, Mainline problems in the plane. Doklady AN BSSR 29 (1985), 872 - 873.

[CGL 85]   B. Chazelle, L. Guibas, D. T. Lee, The power of geometric duality. Proc. 24th Ann. IEEE Symp. Found. Computer Sci., 217-225 (1983), also in BIT 25 (1985), 76 - 90.

[EOS 86]   H. Edelsbrunner, J. O´Rourke, R. Seidel, Constructing arrangements of lines and hyperplanes with applications.Proc. 24th Ann. IEEE Symp. Found. Computer Sci., 83 - 91, also in SIAM J. Comput. 15 (1986), 341 - 363.

[EG 86]   H. Edelsbrunner, L. Guibas, Topologically sweeping the arrangement. Proc. 18th ACM Symp. Theory of Computing, 389 - 403 (1986).

[EM 87]   H. Edelsbrunner, E. Mücke, Simulation of simplicity: A technique to cope with degenerate cases in geometric algorithms. Proc. 4th ACM SYmp. Computational Geometry, Baltimore, USA, 118 - 133 (1987).

[E 87]   H. Edelsbrunner, Algorithms in Combinatorial Geometry. Springer-Verlag, New York, 1987.

[K 88]   N. M. Korneenko, Optimal straight lines for planar point sets. 33rd Internat. Wiss. Koll. TH Ilmenau, DDR, Posterbeitr. sekt. C1, 15 - 18 (1988).

[K 89]   N. M. Korneenko, Optimal lines in the plane: a survey. In: Geobild ´89, Proc. 4th Workshop Geom. Problems of Image Processing, Georgenthal, DDR, 43 - 52, Akademie-Verlag, Berlin, 1989.

[KM 90]   N. M. Korneenko, H. Martini, Least-distances lines and flats and related topics. Submitted.

[LC 85]   D. T. Lee, Y. T. Ching, The power of geometric duality revisited. Inform Process. Lett. 21 (1985), 117 - 122.

[M 87]   H. Martini, Some results and problems around zonotopes. Coll. Math. Soc. J. Bolyai 48 (Intuitive Geometry, Siofok 1985), 383 - 418. Eds. K. Böröczky and G. Fejes Toth, North-Holland, 1987.

[MY 80]   R. D. McBride, J. S. Yormark, An implicit enumeration algorithm for quadratic integer programming. Management Sci. 26 (1980), 282 - 296.

[M 71]   P. McMullen, On zonotopes. Trans. Amer. Math. Soc. 159 (1971), 91 - 110.

[MT 82]   N. Megiddo, A. Tamir, On the Complexity of locating linear facilities in the plane. Operations Research Lett. 1 (1982) 5, 194 - 197.

[MT 83]   N. Megiddo, A. Tamir, Finding least-distance lines. SIAM J. Algebr. Discrete Meth. 4 (1983), 207 - 211.

[MN 80]   J. G. Morris, J. P. Norback, A simple approach to linear facilities location. Transp. Sci. 14 (1980), 1 - 8.

[NM 80]   J. P. Norback, J. G. Morris, Fitting hyperplanes by minimizing orthogonal deviations. Math. Progr. 19 (1980), 102 - 105.

[RND 77]   E. M. Reingold, J. Nievergelt, N. Deo, Combinatorial Algorithms: Theory and Practice. Prentice-Hall, Englewood Cliffs, NJ, 1977.

[SW 83]   R. Schneider, W. Weil, Zonoids and related topics.Convexity and Its Applications, Eds. P. Gruber and J. M. Wills, Birkhäuser, Basel 1983, 296 - 317.

[W 75]   G. Wesolowski, Location of the median line for weighted points. Environ. Planning A 7 (1975), 163 - 170.

[YKII 88]   P. Yamamoto, K. Kato, K. Imai, H. Imai, Algorithms for vertical and orthogonal L_1 approximation of points. Proc. 4th ACM Symp. Computational Geometry, 1988, 352 - 361.

Data Structures for Traveling Salesmen

David S. Johnson

AT&T Bell Laboratories
Murray Hill, NJ 07974

Many of the best standard heuristics for the Traveling Salesman Problem (TSP), as described in computer science and operations research textbooks, are too inefficient to be usable for the large-scale instances (10,000 cities, 100,000 cities, and more) that arise in current VLSI applications. Fortunately, this obstacle can often be surmounted by the use of appropriate data structures and algorithmic adaptations.

In this talk, we first discuss techniques designed to take advantage of the geometric structure of such instances. These can reduce naive running times of n^2 and worse to what in practice looks more like $O(n \log n)$. We then turn to techniques, independent of geometry, that greatly speed up such local improvement algorithms as 2-Opt, 3-Opt, and the Lin-Kernighan algorithm. A simple algorithmic observation about the search space, included in the original Lin-Kernighan paper but often missed by would-be implementers, offers across-the-board improvements here, and we discuss alternative ways of exploiting it. In addition, we discuss the issue of how the current tour should be represented. For the largest instances, the standard linked-list or array-based representations can become major bottlenecks, and we evaluate several alternatives from both a theoretical and an empirical point of view.

This talk reports on work done by the author and, in various combinations, David Applegate, Jon Bentley, Vasek Chvatal, Bill Cook, Lyle McGeoch, Gretchen Ostheimer, and Ed Rothberg.

Efficient Parallel Algorithms for Shortest Paths in Planar Graphs*

Grammati E. Pantziou [1]   Paul G. Spirakis [1,2]   Christos D. Zaroliagis [1]

(1) Computer Technology Institute, P.O. Box 1122, 26110 Patras, Greece
Computer Sc and Eng Dept, University of Patras, Greece

(2) Courant Institute of Math. Sciences, NYU, NY 10012, USA

Abstract

Efficient parallel algorithms are presented, on the CREW PRAM model, for generating a succinct encoding of all pairs shortest path information in a directed planar graph G with real-valued edge costs but no negative cycles. We assume that a planar embedding of G is given, together with a set of q faces that cover all the vertices. Then our algorithm runs in $O(\log^2 n + \log^3 q)$ time and employs $O(nq)$ processors. $O(\log^2 n)$ time, n-processor algorithms are presented for various subproblems, including that of generating all pairs shortest path information in a directed *outerplanar* graph. Our work is based on the fundamental hammock-decomposition technique of G. Frederickson. We achieve this decomposition in $O(\log^2 n)$ parallel time by using $O(n)$ processors. The hammock-decomposition seems to be a fundamental operation that may help in improving efficiency of many parallel (and sequential) graph algorithms. Our algorithms avoid the matrix powering (sometimes called the *transitive closure bottleneck*) thus lead to a considerably smaller number of processors, and tighter processor-time products.

1   Introduction

Computing shortest path information is one of the fundamental problems in digraphs. Efficient *sequential* algorithms for this problem have been proposed (see e.g. [4, 7, 8, 9, 12, 19]. In some of the more recent works, the idea of exploiting topological features of the input graph (such as edge sparsity, planarity) has been emphasized. The very recent work of [10] deals with n-vertex weighted planar digraphs with real weights but no negative cycles and produces a *succinct encoding of all pairs shortest paths* in time ranging from $O(n)$ up to $O(n^2)$, depending on the topological properties of the input graph. This fundamental technique is based on a decomposition of the digraph into special subgraphs, called *hammocks*. The decomposition is defined relative to a given face-on-vertex covering of a planar graph. [10] manages to approximate (by a constant factor) the *minimum* cardinality covering over *all* embeddings of the graph, in $O(n)$ time. If this cardinality is

---

*This work was partially supported by the EEC ESPRIT Basic Research Action No. 3075 (ALCOM) and by the Ministry of Industry, Energy and Technology of Greece.

p, then the shortest path information is computed in $O(np)$ time, by Frederickson's techniques. This decomposition seems to be a fundamental operation that may help in improving efficiency of many sequential and parallel graph algorithms.

The hammock decomposition idea together with the use of succinct encoding (compact routing tables) of shortest paths, not only manages to avoid the $\Omega(n^2)$ sequential time lower bound, but also it does not use matrix multiplication. Thus, an efficient parallelization of these techniques might avoid the use of matrix powering (also called the *transitive closure bottleneck* in [16]). This, in turn, might lead to a number of processors much less than $M(n)$ (i.e. the number required for matrix multiplication) and, thus tighten the processor-time tradeoff. Up to now, parallel algorithms for all pairs shortest paths have only been considered in e.g. [5] and led to an $O(\log n \log d)$ CREW PRAM algorithm (where d is the *depth* of the digraph), by using $M(n)$ processors. Let us note that d (the maximum of the diameters of the connected components) may be $O(n)$.

Our work efficiently parallelizes the hammock decomposition technique of Frederickson. A combination of algorithmic techniques (such as shunt/rake tree operations, parallel connected components, etc.) and of graph properties of the hammock decomposition are essential in our method. The heart of our technique is a *reducing procedure* which produces a smaller instance of the same problem, in parallel. The smaller problem is solved recursively until this brings us below some threshold for the size of the problem. An alternative procedure is then used to complete the parallel algorithm. (This procedure can even be matrix multiplication, since its processor demands can now be efficiently handled.) Such reducing schemes were first considered in [3].

Our methods manage to solve the all pairs shortest paths problem in planar weighted digraphs (with real edge weights but no negative cycles) in parallel time $O(\log^2 n + \log^3 q)$ and $O(nq)$ processors in the CREW PRAM. Here q is the cardinality of a face-on-vertex covering $\mathcal{C}$ for a plane embedding $\hat{G}$ of the input graph G. We assume that both $\mathcal{C}$ and $\hat{G}$ are provided by the input to our techniques. (Plane embeddings of a planar n-vertex graph can be constructed in $O(\log^2 n)$ time by using $O(n)$ processors in the CREW PRAM model (see [15]). The problem of finding a minimum face-on-vertex covering was shown to be NP-complete in [2]. For a given embedding, we might approximate this problem by constructing the dual of the planar graph and by finding there a maximal matching. The resulted vertex cover (of the dual graph) can be used as a face-on-vertex covering of $\hat{G}$, but we can not claim that this face-on-vertex covering is at most a constant multiple from the minimum possible. The resource bounds for maximal matching on a CREW PRAM are the same as the above ones for constructing the embedding (see [13]); also, see [18] for an optimal implementation.)

The algorithms presented here avoid the transitive closure bottleneck and thus drop the processor complexity to $O(nq)$ and the processor time product to $O(nq(\log^2 n + \log^3 q))$.

Efficient parallel algorithms for several important subproblems are also presented here, to be used as intermediate steps in our main scheme. These include: (a) the finding of all pairs shortest paths information in *outerplanar* weighted digraphs (with real edge costs but no negative cycles) in parallel time $O(\log^2 n)$ and $O(n)$ processors in the CREW PRAM, (b) finding the all pairs shortest paths information between vertices of a hammock of h vertices, (even when the shortest paths have to leave and reenter the hammock) in time $O(\log^2 h)$ using $O(h)$ processors in the CREW PRAM, and (c) decomposing a planar digraph into hammocks in $O(\log^2 n)$ parallel time using $O(n)$ processors.

2 Definitions and Notation

A graph is called *outerplanar* if it can be embedded in the plane so that all vertices are on one face. If G is a planar graph and $\hat{G}$ an embedding of G in the plane, let a *face-on-vertex covering* $\mathcal{C}$ be a set of faces of $\hat{G}$ that together cover all vertices. (See figure 1.)

Let $G = (V, E)$ be a planar digraph with real edge costs but no negative cycles. For each edge $\langle v, w \rangle$ incident from any given vertex v, let $S(v, w)$ be the set of vertices such that there is a shortest path from v to each vertex in $S(v, w)$ with the first edge on this path being $\langle v, w \rangle$. A *tie* might occur if there is a vertex u with a shortest path from v to u starting with $\langle v, w \rangle$ and also a shortest path from v to u starting with $\langle v, w' \rangle$ where $w' \neq w$. In such a case, we employ an appropriate tie-breaking rule so that $\forall u, v \ (u \neq v)$, u is *in just one set* $S(v, w)$ for some w. Let each $S(v, w)$ be described as a union of a minimum number of subintervals of $[1, n]$, assuming that $V = \{1, 2, ..., n\}$. Here we allow a subinterval to wrap around from n to 1. We call the set $S(v, w)$ (described as above) the *label* of edge $\langle v, w \rangle$.

A *compact routing table* T for G will have an entry for each of the subintervals in an edge label at v ($\forall v \in V$). This entry is an initial value of the subinterval together with a pointer to the corresponding edge. It can be shown that the total size of T can be made $O(qn)$ by constructing T through a decomposition of the planar digraph into outerplanar subgraphs based on a face-on-vertex covering of cardinality q (see e.g. [6]).

We now use graph embeddings to describe a decomposition of a planar digraph into subgraphs, called *hammocks*, each of which is outerplanar and each of which shares at most four vertices with all other subgraphs in the decomposition. To generate the decomposition, we first convert the embedding $\hat{G}$ of G into an embedded undirected planar graph $\hat{G}_1$. $\hat{G}_1$ has no parallel edges. If $\mathcal{C}_1$ is the face-on-vertex covering of $\hat{G}_1$ that results out of the face-on-vertex covering $\mathcal{C}$ of $\hat{G}$, then all faces in $\hat{G}_1$ (except $\mathcal{C}_1$) are triangulated. Also, the faces in $\mathcal{C}_1$ do not share vertices. Furthermore, the boundary of each face is a simple cycle. Such a graph $\hat{G}_1$ is called *neatly prepared*. (See figure 2.)

We now group the faces of $\hat{G}_1$ together by using two operations: *absorption* and *sequencing* (see details in [10] and [11]). Initially mark all edges bordering the unbounded face. Let f_1, f_2 be two faces that share an edge. If f_1 contains two marked edges then absorb f_1 into f_2 as follows: first contract one edge that f_1 shares with the unbounded face. The first face then becomes a face bounded by two parallel edges, one of which is shared with the second face. Delete this edge, thus merging f_1 and f_2. Repeat this operation until it can be no longer applied.

Once absorption is no longer possible, identify maximal sequences of faces such that each face in the sequence has a marked edge and each pair of consecutive faces share an edge in common. Expand the faces that were absorbed into faces in the sequence. Each subgraph resulting from expanding such a sequence is called a *(major) hammock*. The two vertices at each end of the hammock are called *vertices of attachment*. Any edge not included in a major hammock induces a *(minor) hammock*. The set of all hammocks is a *hammock decomposition* of $\hat{G}$. In the sequel, $d(v, w)$ will denote the shortest distance from v to w.

3 Shortest Paths in Outerplanar Graphs

Let us assume that $G = (V, E)$ is an outerplanar weighted digraph with real edge costs but no negative cycles. Let us furthermore assume that a plane embedding $\hat{G}$ of G is provided. We will show how to efficiently compute (in parallel) the labels $S(v, w)$ for each edge $\langle v, w \rangle$. To simplify the discussion we assume that G is *nice*, i.e.

- (i) $\forall \langle v, w \rangle \in E, \quad \exists \langle w, v \rangle \in E$

- (ii) the undirected version of G is biconnected

- (iii) edge costs follow the generalized triangle inequality, i.e. $\langle v, w \rangle$ is the shortest path from v to w

- (iv) vertices are named in clockwise order around the unbounded face.

(These assumptions can be easily removed – see full paper [17].)

Remark 1 [6]: Each $S(v, w)$ is a single interval, if G is outerplanar.

Remark 2: A convergent (to the root) or divergent tree of shortest paths rooted at some vertex x of G can be constructed (given that edge labels are known to the algorithm) in $O(\log n)$ time using $n/\log n$ processors on an EREW PRAM.

Proof sketch: For the convergent tree case first execute in $O(1)$ time with n processors the following: "for each vertex $v \neq x$ such that $x \in S(v, w)$, add edge $\langle v, w \rangle$ to the tree". Then, a parallel prefix computation completes the construction. Similar for the other case. ∎

The algorithm which constructs the labels proceeds as follows:

(1) By using the adjacency lists of each vertex we construct a counterclockwise list of edges of each face, in $O(1)$ time.

(2) By pointer doubling, we assign a unique name to each face (just assign a processor to each edge and take the smallest processor id).

(2.1) For each face, let x be the minimum numbered vertex and x' the maximum numbered one. Then $\langle x, x' \rangle$ and $\langle x', x \rangle$ (which are called the associated pair of edges of that face) can be found in $O(\log n)$ time by n processors (using pointer doubling).

(2.2) The clockwise and counterclockwise weighted distances from x to every vertex in the same face can also be found with pointer doubling.

(3) For each face f_i, create a node. For each associated edge $\langle x, x' \rangle$ create a pointer to the face f_j containing $\langle x', x \rangle$. A *tree* is thus constructed.

(4) We order the tree nodes (by sorting) and convert it to a binary tree if necessary (see e.g. [1]).

(5) For each face, compute the edge labelling (inside the face) as follows: for each vertex v, find a vertex z such that the counterclockwise distance from v to z (in the face) is less than or equal to the clockwise distance from v to z. (Use binary search.) Then, $S(v, w_1) = \{u : u \in [z, w_1]\}$ (where w_1 is the counterclockwise neighbour of v). Similarly, for the clockwise neighbour w_2, we have $S(v, w_2) = \{u : u \in [w_2, z)\}$ (see figure 3). This step takes $O(\log n)$ time with n processors on a CREW PRAM.

(6) Suppose that we are given two outerplanar graphs G_1 and G_2 that share an edge. We will find shortest path information from every vertex of each G_i to the vertices of the other. We only describe how to find edge labelling information from each vertex of G_1 to vertices of G_2. Let v a vertex of G_1 and u a vertex of G_2. Let $\langle x, x' \rangle$ the common edge between G_1 and G_2 (see figure 4). The shortest path from v to u is through x iff either

$$d(v, x) + d(x, u) \leq d(v, x') + d(x', u) \quad or \quad d(v, x) - d(v, x') \leq d(x', u) - d(x, u) \quad (*)$$

These distances can be found by constructing (for x and x') a convergent tree for G_1 and a divergent tree for G_2 and then by using Euler tour or tree contraction in parallel (see e.g. [16]). Note that edge labelling information for edges incident on x or x' consist of conceentive vertices (in G_2). Thus, each $v \in G_1$ now uses (*) and a binary search on the vertices of G_2 in order to determine how $S(v, w)$ will be augmented. The binary search can be done because,

Lemma 1 *Let $\hat{G}$ be an embedded outerplanar graph. Let x, y be endpoints of an edge. Then the function $h_{xy}(v) = d(v, x) - d(v, y)$ is nondecreasing as v moves clockwise from x to y.*

Proof: See full paper [17]. Note that the above lemma extends the monotonicity property of [10]. ∎
From the above we conclude that,

Lemma 2 *Shortest path information between two outerplanar graphs G_1 and G_2, of n_1 and n_2 vertices accordingly, that share an edge, can be found in $O(\log(n_1 + n_2))$ time by using $n_1 + n_2$ processors on a CREW PRAM, provided that the shortest path information within each G_i is given.*

(7) The algorithm now is completed by a tree-contraction technique, using the RAKE operation as follows:
  The RAKE operation. (See figure 5.) Each l_x or f_x is either a simple face or an outerplanar graph (provided by unioning simple faces). Suppose l_j is performing a RAKE. Then using lemma 2 we can find edge labelling information between l_j, f_j and then with f_k, and substitute these three nodes of the tree with one (f_{jk}) containing edge labelling information among vertices in l_j, f_j, f_k.
From steps (1) through (7) we get:

Theorem 1 *All pairs shortest path information in an outerplanar graph of n vertices can be found in $O(\log^2 n)$ time, using $O(n)$ processors on a CREW PRAM.*

Remark 3: It is obvious that the above theorem also holds in the case where the names of vertices in clockwise order comprise a constant number of consequtive sequences.

4 The general case of a planar digraph

4.1 The basic steps of the algorithm

Our technique to determine the labels of all the edges of a planar digraph (whose an embedding and a face-on-vertex covering are given) consists of the following major steps:

ALGORITHM *LABEL*
BEGIN
(1) $i := 1$; $C(\hat{G}_1) := \hat{G}$; $q_1 := q$;
  repeat
    (1.a) Decompose the embedded digraph $C(\hat{G}_i)$ into $O(q_i)$ hammocks, where q_i is
      the cardinality of the face-on-vertex covering by first converting $C(\hat{G}_i)$ into
      a neatly prepared undirected graph and then by replacing the operations
      of absorption and sequencing with their fast parallel versions.
    (1.b) Run the outerplanar algorithm in each hammock and then compress
      it into an $O(1)$-sized outerplanar graph.
    (1.c) Compress the embedded digraph $C(\hat{G}_i)$ into a planar digraph $C(\hat{G}_{i+1})$ of

$O(q_i)$ nodes, with a face-on-vertex covering of q_{i+1} faces where q_{i+1}
is a *constant fraction* of q_i.
(1.d) $i := i + 1$
until the problem size is such that it can be solved by
matrix powering with $O(nq)$ processors at most.
(2) Run the matrix powering algorithm for all pairs shortest paths in $C(\hat{G}_i)$
(3) **for** $j := i - 1$ **downto** 2 **do**
(3.a) find edge labelling information in $C(\hat{G}_j)$
(3.b) find all pairs shortest distances in $C(\hat{G}_j)$
od
(4) find edge labelling information in the initial graph $\hat{G}$
END

We will analyze the steps of the above algorithm and will show that: steps (1.a), (1.b) can be done (by a CREW PRAM) in $O(\log^2 n)$ time with $O(n)$ processors, where $n = |C(\hat{G}_i)|$. Step (1.c) takes $O(\log q_i)$ time with $O(q_i)$ processors, where q_i is the size of the face-on-vertex covering of $C(\hat{G}_i)$. The recursion depth is at most $\log q$, and since the first iteration requires $O(\log^2 n)$ time, the second $O(\log^2 q)$ time, the third $O(\log^2 q/2)$ etc, step (1) needs $O(\log^2 n + \log^3 q)$ time with $O(n)$ processors. Steps (3.a) and (3.b) need $O(\log^2 q_i)$ time and $O(q_i^2)$ processors. Finally, step (4) needs $O(\log^2 n)$ time using $O(nq)$ processors. Hence, our method needs (in total) a parallel time of $O(\log^2 n + \log^3 q)$ and $O(nq)$ processors.

4.2 Parallel decomposition of a planar digraph into hammocks

4.2.1 Making G neatly prepared

Let $\hat{G}$ be an embedded planar digraph with set of faces F and a face-on-vertex covering $\mathcal{C}$, with $|\mathcal{C}| = q$, $(q > 2)$. (The case where $|\mathcal{C}| = 2$ can be easily handled.) Following the definitions of Section 2, we first have to produce the *neatly prepared graph* $\hat{G}_1$ with face-on-vertex covering $\mathcal{C}'$ with the following properties:

- it has no parallel edges;

- the boundary of each face is a simple cycle;

- no pair of faces of $\mathcal{C}'$ share a vertex;

- all faces of $\hat{G}_1$, except $\mathcal{C}'$, are triangulated.

We assume that each face of $\mathcal{C}'$ is given as a list of the vertices of the face in a counterclockwise ordering around it. By assigning one processor to each edge of the faces and by doing pointer doubling in the list of edges of each face (which can be easily derived), we can assign a unique name to each face (e.g. the highest id of the processors associated to it). Then, each face of $\mathcal{C}'$ can be given a distinct number in $\{1, ..., q\}$ by sorting. These operations take time $O(\log n)$ and require $O(n)$ processors. In the same time-processor bounds we can uniquely name the rest of the faces of F, count the vertex number in each $f \in F$ and number the vertices of f accordingly. We then remove directions and duplicate edges in parallel. The detection of a vertex v that appears more than once in a clockwise walk around a face f is done as follows. We again use pointer doubling to find the rank of each vertex in the clockwise list of f. Obviously, the vertices we are looking for are the vertices that have more than one ranks. Then, for all but the first one occurences of v in the clockwise list of f do the following: if v_1, v_2 are the predecessor and successor of v respectively (in the clockwise walk around f), add the edge $\{v_1, v_2\}$ to the graph.

Shared vertices among at least two faces of C', produced from C in this way, are assigned to just one face (that with the smallest id among those sharing the vertex). For each other face f_i (which shares the same vertex v) do the following. If v has even rank (in the list of vertices of f_i) then add the edge $\{v_1, v_2\}$ to the graph and delete v from f_i, where v_1, v_2 are the preceding and succeeding vertices of v in f_i. Do the same for each shared vertex v with odd rank. If f_i contains more than four vertices repeat the above steps. Clearly, after $O(\log n)$ steps each face will either have no shared vertices (associated with it), or have only four vertices some of which are shared with other faces. In the latter case, we choose any shared vertex v, add $\{v_1, v_2\}$ to the graph and delete v from the list of f_i. Now, if there is still a vertex shared by two faces f_i, f_j, replace v by v_i, v_j in the lists of f_i, f_j respectively.

Finally, the triangulation of faces not in C' can be easily achieved by using the vertex numbers of the face-on-vertex covering C' as follows:

(i) If all the vertices of the face belong to the same face of C' then just add an edge from each vertex of the face to the vertex with the smallest id.

(ii) If the vertices of a face f are split between two faces f_j, $f_{j'}$ in C', then we must add edges in stages to preserve planarity: in the first stage we add edges from all the vertices of $f \cap f_j$ to the vertex of $f \cap f_{j'}$ with the smallest id. The second stage takes care of all the vertices of $f_{j'} \cap f$ and completes the triangulation. Thus,

Lemma 3 *The digraph can be neatly prepared in $O(\log^2 n)$ time with $O(n)$ processors in a CREW PRAM.*

4.2.2 Decomposing $\hat{G}_1$ into hammocks

We now show how to efficiently parallelize the absorption and sequencing procedures. We start by renaming the vertices of the faces of the face-on-vertex covering of $\hat{G}_1$ by assigning to each vertex $v \in f_i$ ($f_i \in C'$) a number according to its clockwise order around f_i (we must do this since we have no shared vertices in C'). As in [10] we associate with each face f_1 of the set of faces F_1 of $\hat{G}_1$ (except for the faces in C') a data structure consisting of two fields. The *first field* is a 4-tuple (i, j, k, r). The values (i, j, k) are the indices of the faces in C' that contain the three vertices of f_1 (ordered in such a way that $k = j$ implies $i = j$). The value r is the number of edges of f_1 in common with faces of C'. We call this first field the *label of the face*. The *second field* is a triple (w_i, w_j, w_k) where w_i, w_j, w_k are the names of the vertices of f_1 belonging to faces i, j, k, respectively.

4.2.2.1 Absorption procedure

During this procedure, we construct a *forest of faces* of $F_1 - C'$ as follows: we create a node for each $f_k \in F_1 - C'$ with first field value either (i, i, i, r) with $r = 0, 1, 2$, or $(i, i, j, 0)$ with $j \neq i$ for some faces f_i, f_j in C'. For each $f_i \in C'$, if an f_k in $F_1 - C'$ (with first field (i, i, i, r) or $(i, i, j, 0)$ with $j \neq i$) has an edge in common with a face f_l in $F_1 - C'$ (with first field (i, i, i, r) or $(i, i, j', 0)$ for any $f_{j'}$ with $j' \neq i$) then we add the edge $\{f_k, f_l\}$ to the forest. Some trees are thus constructed. They are binary, unrooted and unordered. We then run a parallel connected components algorithm (see [14]) to identify each tree in the forest. By sorting the array of face numbers by component name, we get all the faces belonging to the same tree to be consequtive in the array.

We next select a face in each tree to be the root of that tree (the particular selection is explained in the full paper [17]). Once each tree is rooted, it can be ordered through parallel sorting. Then, by using the Euler Tour technique (see e.g. [16]) we can establish parent-child relationships between all faces (nodes) of the tree, efficiently in parallel. The trees help us to specify (i) the groups of faces that will be absorbed one into the other and (ii) the groups of faces in $F_1 - C'$ that constitute some of the hammocks. Each such group of faces is associated with a 4-tuple which determines a hammock. This 4-tuple consists of the names of four vertices of the hammock that may be shared by one or more other hammocks. These vertices are called attachment vertices. Since hammocks must be outerplanar subgraphs, we have to prevent the violation of the outerplanarity condition. So, when one attachment vertex is shared by two (or more) hammocks it must be splitted into two (or more) vertices.

Several cases have to be considered for grouping of faces according to the values of the first field of the root of the tree and the corresponding values in the other nodes. See the full paper [17] for details.

4.2.2.2 Hammock construction

This second procedure efficiently parallelizes the sequencing technique and constructs hammocks that were not yet constructed. We create a node for each face with label $(i, i, j, 1)$ or $(j, j, i, 1)$ (for some indices i, j corresponding to faces f_i, f_j in C'). We also create a node for each new face resulting from the absorption operation. If two faces f_1, f_2 have one edge in common, we add the edge $\{f_1, f_2\}$. The connected components of this new graph are:

1. a sequence of faces in $F_1 - C'$ between two faces f_i, f_j of C', or

2. a sequence of faces in $F_1 - C'$ around a face $f_i \in C'$, or

3. an isolated face in $F_1 - C'$ between two faces of C' or a face f around a face of C' (the face f may be the result of an absorption operation).

Each of these components include the faces that constitute a hammock. The attachment vertices of these hammocks are easily detected (see full paper [17]). Finally, we return all the faces that were absorbed back to each hammock. Thus,

Lemma 4 *Let $\hat{G}_1$ be an embedded, neatly prepared planar digraph with n vertices and a face-on-vertex covering of q faces. Then, $\hat{G}_1$ can be decomposed into $O(q)$ hammocks in $O(\log n \log^* n)$ time using $O(n)$ processors on a CREW PRAM.*

4.2.3 The compressed version of a hammock

We now discuss, how a compressed version of a hammock is generated using the idea of [10]. Let a_i, $i = 1, 2, 3, 4$ be the attachment vertices of a hammock H with $a_1, a_2 \in f$ and $a_3, a_4 \in f'$ (f, f' the two faces bounding H), and a_1 is adjacent to a_4 and a_2 adjacent to a_3. Let T_i denotes the following tree

$$T_i = SP(a_i, a_j) \bigcup SP(a_i, a_k), \quad i = 1, 2, 3, 4, \quad j, k \neq i, 5 - i, \quad j \neq k$$

where $SP(u, w)$ denotes the shortest path from u to w (if exists). Note that each T_i can be easily constructed as it was explained before.

Let $B(H)$ be the graph,

$$B(H) = \left(\bigcup_{i=1}^{4} T_i \right) \bigcup A$$

where $A = \{\langle a_1, a_4 \rangle, \langle a_4, a_1 \rangle, \langle a_2, a_3 \rangle, \langle a_3, a_2 \rangle\}$ is a set of pseudo-edges with costs equal to their corresponding shortest distances in H. The construction of $B(H)$ can be done as follows. First, form the graphs $T_1 \cup T_4$ and $T_2 \cup T_3$; also associate each edge with a color. If $e \in T_1 \cup T_4$ then $color(e) =$"blue", and if $e \in T_2 \cup T_3$ then $color(e) =$"red". Second, form the graph $(T_1 \cup T_4) \cup (T_2 \cup T_3) \cup A$. Moreover if $e \in A$ then $color(e) =$"black". Each union operation can be obviously done in $O(\log h)$ time using h processors on a CREW PRAM, where $h = |H|$. If there are edges $e, e' \in B(H)$ such that $e \in \cup_{i=1}^{4} T_i$ and $e' \in A$ (but $color(e) \neq color(e')$) and e, e' have the same endpoints, then delete e from $B(H)$.

Now we are able to compress $B(H)$ as much as possible by contraction of blue or red paths in $B(H)$. We perform two contractions, one for blue and one for red edges. We shall describe the former one, while the latter is similar. Suppose that the only two blue edges incident on a vertex v are $\langle u, v \rangle$ and $\langle v, w \rangle$. Then delete v and replace them with the blue edge $\langle u, w \rangle$ with $cost(u, w) = cost(u, v) + cost(v, w)$. The above operation is called *shortcutting*. It is similar to pointer doubling, thus after $O(\log h)$ steps all the blue or red paths will have been contracted. If there is a vertex of degree 0 after the contraction, then we delete that vertex. Let us call $C(H)$ the resulting graph. We then have,

Lemma 5 ([10]) $C(H)$ *is outerplanar and its size is fixed. For every pair a_i, a_j in H, if a shortest path from a_i to a_j exists in H then there is also one in $C(H)$ of equal cost.*

Lemma 6 $C(H)$ *can be constructed in $O(\log h)$ time by $O(h)$ processors on a CREW PRAM, where $h = |H|$.*

Once shortest paths between attachment vertices are known, then shortest path information among other vertices of the same hammock (or between two hammocks) can be easily recovered. The following two lemmas are proved in the full paper [17].

Lemma 7 *Let H be a hammock in an embedded planar graph $\hat{G}$, and let H have n_1 vertices. Correct shortest path (edge labelling) information for the edges of each vertex of H (even when the shortest paths leave and reenter the hammock) can be generated in $O(\log^2 n_1)$ time, using $O(n_1)$ processors on a CREW PRAM, provided that the shortest path distances among attachment vertices of H are given.*

Lemma 8 *Let H_1, H_2 be two hammocks of n_1, n_2 vertices respectively. If the shortest distances between their attachment vertices are known, then edge labelling information among vertices in H_1, H_2 can be generated in $O(\log(n_1 + n_2))$ time by using $O(n_1 + n_2)$ processors on a CREW PRAM.*

4.2.4 Compressing the graph (the recursive step)

By using the compressed versions of the hammocks (and by adding in the minor hammocks) we get a compressed version $C(\hat{G})$ of the graph $\hat{G}$. It is enough to find all pairs shortest paths information in $C(\hat{G})$ because then we shall have the shortest distances between all attachment vertices in $\hat{G}$ and this is enough to find everything, by lemmas 7 and 8. $C(\hat{G})$ is a planar graph of size $O(q)$. The face-on-vertex covering $\mathcal{C}$ of $\hat{G}$ appears as a "new" covering $C(\mathcal{C})$ in $C(\hat{G})$, but $C(\mathcal{C})$ is still of size q.

However, we can efficiently reduce the number of faces on the face-on-vertex covering by joining *adjacent* faces (i.e. faces that share at least one hammock) into one face. Cycles can be avoided if each face is joined with the adjacent face of the largest index (the index is an integer between 1 and q). This joining operation produces a face-on-vertex covering $\mathcal{C}_1$ (of $C(\hat{G})$) whose size is less than $q/2$.

Lemma 9 *The face-on-vertex covering $\mathcal{C}_1$ of the compressed graph $C(\hat{G})$ can be produced from $C(\hat{G})$ and $C(\mathcal{C})$ in $O(\log q)$ time by $O(q)$ processors in the CREW PRAM. Furthermore, the size of $\mathcal{C}_1$ is at most half the size of $C(\mathcal{C})$.*

Proof (sketch): Suppose that each vertex has a field that contains the index of the face (of $C(\mathcal{C})$) to which the vertex belongs. For each vertex v update the above field with the highest indexed face to which, some vertex v' adjacent to v belongs (if the face to which v' belongs has higher index than the face to which v belongs). Using this information, the vertices belonging to a face f_i, can find (by pointer doubling) the highest indexed face f_i' adjacent to f_i. The face f_i' (if it exists) must have an index higher than that of f_i.

For each face f_i assign one processor to each edge that connects a vertex v of f_i with a vertex v' of the highest indexed adjacent face f_i' and v, v' are not attachment vertices. Using prefix computation find the edge which is associated with the processor with the smallest id. If there is no edge $\{v_1, v_2\}$ between vertices of f_i, f_i' such that v_1, v_2 are not attachment vertices, do the following. From the structure of the hammock that connects f_i, f_i' we can find a vertex w that belongs to one of f_i, f_i' and w is not attachment. Suppose that a_1, a_2 are the attachments of f_i, a_3, a_4 the attachments of f_i' and w belongs to f_i' (see fig.6). Consider u a vertex between a_1, a_2 and draw the edges $\{a_1, u\}, \{u, a_2\}, \{u, w\}$, with costs $d(a_1, u) = d(u, a_2) = d(a_1, a_2)/2$ and $d(u, w) = \infty$ ($\{u, w\}$ is the edge we are looking for). Split each one of the two endpoints of the edge into two vertices. Reconnect the vertices, so that the faces f_i, f_i' are merged and the resulting graph is planar with the new face in the face-on-vertex covering $\mathcal{C}_1$. ∎

5 Conclusions and Further Research

The procedures explained in sections 4.2.1 through 4.2.4 prove our claim that algorithm *LABEL* constructs all pairs shortest path information in $O(\log^2 n + \log^3 q)$ time using $O(nq)$ processors in the CREW PRAM. Since the finding of an embedding of a planar graph G with a face-on-vertex covering of cardinality as small as possible is NP-complete, we pose as an open problem the efficient parallel construction of a face-on-vertex covering of cardinality at most a constant multiple of the minimum possible. (Such coverings can be constructed in polynomial sequential time, as is shown in [10].)

References

[1] K. Abrahamson, N. Dadoun, D. Kirkpatrick, T. Przytycka, "A Simple Parallel Tree Contraction Algorithm", *J. of Algorithms*, 10(1989), pp.287-302.

[2] D. Beinstock, C.L. Monma, "On the complexity of covering faces by vertices in a planar graph", *SIAM J. Comp.*, Vol.17, No.1, Feb. 1988, pp.53-76.

[3] R. Cole, U. Vishkin, "Deterministic Coin Tossing with Applications to Optimal Parallel List Ranking", *Inform. and Control*, Vol.70, No.1, pp.32-53, July 1986.

[4] E.W. Dijkstra, "A note on two problems in connexion with graphs", *Numerische Mathematik*, 1(1959), pp.275-323.

[5] E. Dekel, D. Nassimi, S. Sahni, "Parallel Matrix and Graph Algorithms", *SIAM J. Comp.*, Vol.10, No.4, Nov.1981, pp.657-675.

[6] G.N. Frederickson, R. Janardan, "Designing Networks with Compact Routing Tables", *Algorithmica*, 3(1988), pp.171-190.

[7] R.W. Floyd, Algorithm 97: shortest path, *Comm. ACM*, 5(1962), pp.345.

[8] M.L. Fredman, "New bounds on the complexity of the shortest path problem", *SIAM J. Comp.*, 5(1976), pp.83-89.

[9] G.N. Frederickson, "A new approach to all pairs shortest paths in planar graphs", *Proc. 19th ACM STOC*, New York City, May 1987, pp.19-28.

[10] G.N. Frederickson, "Planar Graph Decomposition and All Pairs Shortest Paths", TR–89-015, ICSI, Berkeley, March 1989.

[11] G.N. Frederickson, "Using Cellular Graph Embeddings in Solving All Pairs Shortest Path Problems", *Proc. 30th Annual IEEE Symp. on FOCS*, 1989, pp.448-453; also CSD–TR-897, Purdue University, August 1989.

[12] M.L. Fredman, R.E. Tarjan, "Fibonacci heaps and their uses in improved network optimization algorithms", *JACM*, 34(1987), pp. 596-615.

[13] A. Goldberg, S. Plotkin, G. Shannon, "Parallel Symmetry-Breaking in Sparse Graphs", *Proc. of the 19th ACM STOC*, 1987, pp.315-324.

[14] T. Hagerup, "Optimal Parallel Algorithms for Planar Graphs", *Inform. and Computation*, to appear.

[15] P. Klein, J.H. Reif, "An Efficient Parallel Algorithm for Planarity", *Proc. 27th Annual IEEE Symp. on FOCS*, 1986, pp.465-477.

[16] R.M. Karp, V. Ramachandran, "A Survey of Parallel Algorithms for Shared-Memory Machines", Rep.No. UCB/CSD 88/804, University of California, Berkeley, 1989.

[17] G. Pantziou, P. Spirakis, C. Zaroligis, "Efficient Parallel Algorithms for Shortest Paths in Planar Graphs", TR-90.01.02, Computer Technology Institute, Patras, January 1990.

[18] G. Pantziou, P. Spirakis, C. Zaroliagis, "Optimal Parallel Algorithms for Sparse Graphs", TR-90.04.08, Computer Technology Institute, Patras, April 1990 (revised version).

[19] S. Warshall, "A theorem on Boolean matrices", *JACM*, 9(1962), pp.11-12.

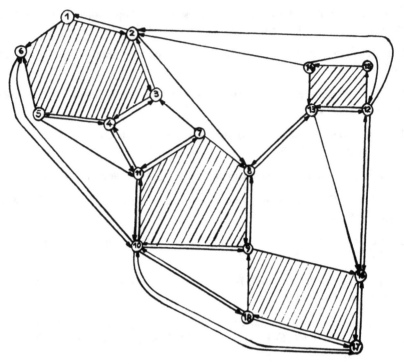

Fig.1 An embedding of a planar graph that has a face-on-vertex covering
of cardinality 4 (shaded faces)

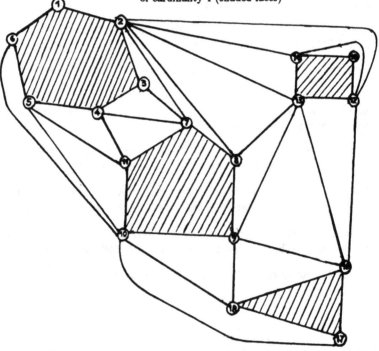

Fig.2 A neatly prepared planar graph generated from the embedded
planar graph of fig.1

300

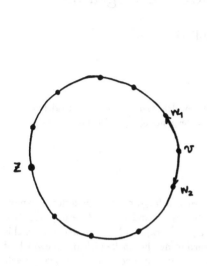

Fig.3

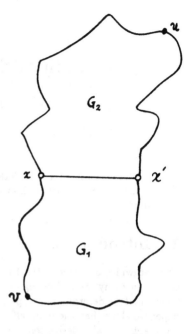

Fig.4

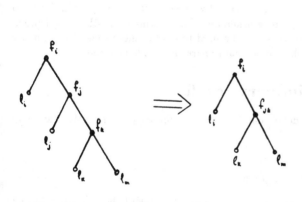

Fig.5 ℓ_j performs a RAKE

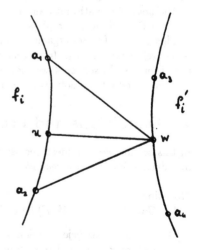

Fig.6

The Pathwidth and Treewidth of Cographs

Hans L. Bodlaender[*]   Rolf H. Möhring[†]

Abstract

It is shown that the pathwidth of a cograph equals its treewidth, and a linear time algorithm to determine the pathwidth of a cograph and build a corresponding path-decomposition is given.

1 Introduction

The pathwidth and treewidth of a graph are two notions with a large number of different applications in many areas, like algorithmic graph theory, VLSI-design and others (see e.g. [1, 13]). Unfortunately, determining the pathwidth or treewidth of a given graph is NP-complete [2]. In this paper we show that there are efficient algorithms for determining the pathwidth or treewidth of a cograph. We also derive some technical lemmas, which are not only necessary to prove correctness of the algorithms, but are also interesting in their own right. For instance, we show that the pathwidth of a cograph equals its treewidth.

The complexity of the problems to determine the pathwidth and treewidth of graphs has also been studied for other interesting classes of graphs. Gustedt [10] showed that the pathwidth problem stays NP-complete when restricted to chordal graphs. For fixed k, the problem of determining whether the pathwidth or treewidth of a given graph is at most k can be solved in polynomial time with dynamic programming [2, 8], and in $O(n^2)$ time with graph minor theory [5, 16].

The notions of pathwidth and treewidth have several equivalent characterizations (see e.g. [1, 13, 18].) For instance, a graph is a partial k-tree, if and only if its treewidth is at most k.

This paper is further organized as follows. In section 2 we give most necessary definitions and some preliminary results. In section 3 we prove a number of interesting graph-theoretic lemmas and theorems. In section 4 we show how these can be used to obtain linear time algorithms for pathwidth and related notions on cographs. Some final remarks are made in section 5.

2 Definitions and preliminary results

In this section we give most necessary definitions and some preliminary results. We start with introducing the notion of *cographs*.

Notation

Let $G = (V, E), H = (W, F)$ be undirected graphs.

(i) We denote the disjoint union of G and H by $G \dot{\cup} H = (V \dot{\cup} W, E \dot{\cup} F)$ (where $\dot{\cup}$ is the disjoint union on graphs, and sets, respectively.)

---

[*]Department of Computer Science, Utrecht University, P.O. Box 80.089, 3508 TB Utrecht, the Netherlands. The research of this author was partially supported by the ESPRIT II Basic Research Actions Program of the EC under Contract No. 3075 (project ALCOM).

[†]Department of Mathematics, Technical University of Berlin, Straße des 17. Juni 136, 1000 Berlin 12, West Germany. The research of this author was partially supported by the Deutsche Forschungsgemeinschaft under Grant No. Mo446/1-1.

(ii) With $G \times H$ we denote the following type of "product" of G and H: $G \times H = (V \dot\cup W, E \dot\cup F \cup \{(v, w) \mid v \in V, w \in W\})$

(iii) The complement of G is denoted by $G^c = \{V, E^c\}$, with $E^c = \{(v, w) \mid v, w \in V, v \neq w, (v, w) \notin E\}$.

Proposition 2.1 $\dot\cup, \times$ *are commutative and transitive.* $G \times H = (G^c \dot\cup H^c)^c$.

Definition 2.1 *A graph* $G = (V, E)$ *is a cograph, iff one of the following conditions holds:*

1. $|V| = 1$

2. There are cographs $G_1, \ldots, G_k$ *and* $G = G_1 \dot\cup G_2 \dot\cup \ldots \dot\cup G_k$

3. There are cographs $G_1, \ldots, G_k$ *and* $G = G_1 \times G_2 \times \ldots \times G_k$

There are other, equivalent characterizations of the class of cographs. Rule 3 can be replaced by:

3′. There is a cograph H *and* $G = H^c$.

Also, one can restrict k in rule 2 and 3 to be 2. Alternatively, one can define the class of cographs as the graphs that do not contain P_4, a path with 4 vertices, as an induced subgraph. (See e.g. [6]).

With each cograph $G = (V, E)$, one can associate a labeled tree, called the *cotree* T_G of G. Each vertex of G corresponds to a unique leaf in T_G. Internal vertices of T_G have a label $\in \{0, 1\}$. To each vertex in T_G one can associate a cotree in the following manner: a leaf corresponds to a cotree, consisting of one vertex. The cograph corresponding to a 0-labeled vertex v in T_G is the disjoint union of the cographs, corresponding to the sons of v in T_G. The cograph corresponding to a 1-labeled vertex v in T_G is the product ("$\times$") of the cographs, corresponding to the sons of v in T_G. Note that $(v, w) \in E$, if and only if the lowest common ancestor of v and w in T_G is labeled with a 1. (There are other very similar notions of "cotree").

Corneil, Perl and Stewart [7] gave an $O(n + e)$ algorithm for determining whether a given graph $G = (V, E)$ is a cograph, and if so, building the corresponding cotree.

A cotree T_G can easily be transformed to an equivalent cotree T'_G such that every internal vertex in T'_G has exactly 2 sons. (Note that $G_1 \dot\cup \ldots \dot\cup G_k = (G_1 \dot\cup \ldots \dot\cup G_{k-1}) \dot\cup G_k$, and $G_1 \times \ldots \times G_k = (G_1 \times \ldots G_{k-1}) \times G_k$. The resulting operation on trees is illustrated in Figure 1).

So, in the remainder of this paper we assume that cographs G are given together with a binary cotree T_G.

Next we give the definitions of pathwidth and treewidth, introduced by Robertson and Seymour [15, 16].

Definition 2.2 *A tree-decomposition of a graph* $G = (V, E)$ *is a pair* $(\{X_i \mid i \in I\}, T = (I, F))$ *with* $\{X_i \mid i \in I\}$ *a family of subsets of* V, *and* T *a tree, such that*

- $\bigcup_{i \in I} X_i = V$

- $\forall (v, w) \in E : \exists i \in I : v \in X_i \wedge w \in X_i$

- $\forall v \in V : \{i \in I \mid v \in X_i\}$ *forms a subtree of* T

The treewidth of a tree-decomposition $(\{X_i \mid i \in I\}, T = (I, F))$ *is* $\max_{i \in I} |X_i| - 1$. *The treewidth of* G *is the minimum treewidth over all possible tree-decompositions of* G.

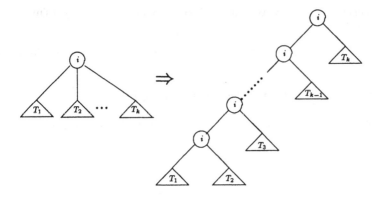

Figure 1: Transformation to a binary co-tree $i \in \{0, 1\}$

Note that the third condition can be replaced by:

$$\forall \, i, j, k \in I : \text{ if } j \text{ is on the path from } i \text{ to } k \text{ in } T, \text{ then } X_i \cap X_k \subseteq X_j$$

There are several other notions that are equivalent to the notion of treewidth, e.g. a graph G is a partial k-tree, if and only if $treewidth(G) \leq k$. (See [1, 18]).

The notion of pathwidth is obtained from the notion of treewidth by requiring that the tree T in the tree-decompositions is a path.

Definition 2.3 *A path-decomposition of a graph* $G = (V, E)$ *is a pair* $(\{X_i \mid i \in I\}, I)$, *with* $\{X_i \mid i \in I\}$ *a family of subsets, and* $\exists \, r \in \mathbb{N} : I = \{1, 2 \ldots, r\}$, *such that*

- $\bigcup_{i \in I} X_i = V$

- $\forall (v, w) \in E : \exists i \in I : v \in X_i \wedge w \in X_i$

- $\forall v \in V : \exists b_v, e_v \in I : \forall i \in I : v \in X_i \Leftrightarrow b_v \leq i \leq e_v$

The pathwidth of $(\{X_i \mid i \in I\}, I)$ *is* $\max_{i \in I} |X_i|$. *The pathwidth of* G *is the minimum pathwidth over all possible path-decompositions of* G.

The third condition states that for all $v \in V$ $\{i \in I \mid v \in X_i\}$ forms an interval in I, and is equivalent to " $\forall \, i, j, k : i < j < k \Rightarrow X_i \cap X_k \subseteq X_j$ ".

The notion of pathwidth is closely related to several other notions, including node search number and interval thickness.

Definition 2.4 *The node search number of a graph* $G = (V, E)$ *is the minimum number of searchers needed to clear all edges of* G, *under the following rules.*

- *Initially all edges are contaminated.*

- *A move can consist of*

    1. *Putting a searcher on a vertex,*

    2. *Removing a searcher from a vertex,*

    3. *Moving a searcher over an edge from a vertex to an adjacent vertex.*

- *A contaminated edge becomes cleared when there is a searcher on both ends of the edge.*

- *A cleared edge becomes recontaminated when there is a path from the edge to a contaminated edge that does not pass through a vertex with a searcher on it.*

Definition 2.5 *A graph $G = (V, E)$ is an interval graph if to each $v \in V$ an interval $[b_v, e_v] \subseteq \mathbb{R}$ can be associated such that $\forall\, v, w \in V : (v, w) \in E \Leftrightarrow [b_v, e_v] \cap [b_w, e_w] \neq \phi$.*

Lemma 2.1 *(See [4, 9])*
Let $G = (V, E)$ be an interval graph. Let the chromatic number of G be $\chi(G)$, let the maximum size of a clique in G be $\omega(G)$. Then $\chi(G) = \omega(G) = \text{treewidth}(G) + 1 = \text{pathwidth}(G) + 1$.

Definition 2.6 *The interval thickness of a graph $G = (V, E)$ is the minimum chromatic number of an interval graph H that contains G as a subgraph.*

Theorem 2.1 *[11, 13]: For every graph $G = (V, E)$, the following three numbers are equal:*

- *the pathwidth of $G + 1$*

- *the node search number of G*

- *the interval thickness of G.*

3 Graph-theoretic results

In this section we derive some new and interesting graph theoretic results which are needed to derive the algorithm, but have also interest on their own. We start with a very short proof of a known result.

Definition 3.1 *A family $\{T_i \mid i \in I\}$ of subsets of a set T is said to satisfy the Helly-property, if for all $J \subseteq I$ with for all $i, j \in J : T_i \cap T_j \neq \emptyset$ it holds that $\cap_{j \in J} T_j \neq \emptyset$.*

Theorem 3.1 *(See [9], p.92): A family of induced subtrees of a tree satisfies the Helly property.*

Lemma 3.1 *("Clique containment lemma")*
Let $(\{X_i \mid i \in I\}, T = (I, F))$ be a tree-decomposition of $G = (V, E)$, and let $W \subseteq V$ be a clique in G. Then $\exists\, i \in I : W \subseteq X_i$.

Proof
Let $T_v = \{i \in I \mid v \in X_i\}$. $\{T_v \mid v \in W\}$ is a family of subtrees of T. By theorem 3.1: $\exists\, i \in I : \forall u \in W : i \in T_v$. Hence $\exists \in I : W \subseteq X_i$. □

Older and longer proofs of lemma 3.1 can be found in [4, 18]. With the help of the Helly-property for trees we can also prove a variant of the clique-containment lemma for bipartite subgraphs.

Lemma 3.2 *"Complete bipartite subgraph containment lemma"*
Let $(\{X_i \mid i \in I\}, T = (I, F))$ be a tree-decomposition of $G = (V, E)$. Let $A, B \subseteq V$, and suppose $\{(v, w) \mid v \in A, w \in B\} \subseteq E, A \cap B = \emptyset$. Then $\exists i \in I : A \subseteq X_i$ or $B \subseteq X_i$.

Proof
Let $(\{X_i \mid i \in I\}, T = (I, F)), G = (V, E)$, and A and B be given. Suppose that for all $i \in I : B \not\subseteq X_i$. Let $T_v = \{i \in I \mid v \in X_i\}$. Consider the family $\{T_v \mid v \in B\}$ of subtrees of T. As $\cap_{v \in B} T_v = \emptyset$, it follows from theorem 3.1 that there are $b_1, b_2 \in B$ such that T_{b_1} and T_{b_2} are vertex disjoint. Consider the unique path of T connecting T_{b_1} and T_{b_2}, and let k and ℓ be the border vertices of this path. (See Figure 2).

Each $a \in A$ must be contained in a set X_i with $i \in T_{b_1}$ and in a set X_j with $j \in T_{b_2}$. Hence $a \in X_k$. (Use definition of tree-decomposition). So $A \subseteq X_k$. □

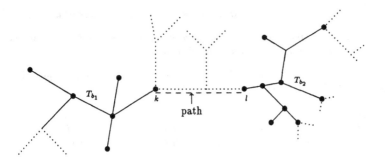

Figure 2. An illustration of the proof of Lemma 3.2

Lemma 3.3 *Let* $(\{X_i \mid i \in I\}, T = (I, F))$ *be a tree decomposition of* $G = (V, E)$, *and let* $A, B \subseteq V$ *and suppose* $\{(v, w) \mid v \in A, w \in B\} \subseteq E, A \cap B = \emptyset$. *Suppose* $\exists i \in I : A \subseteq X_i$. *Then there exists an induced subtree* $T' = (I', F')$ *of* T, *such that*

(i) $\forall i \in I' : A \subseteq X_i$

(ii) $B \subseteq \cup_{i \in I'} X_i$

(iii) $(\{X_i' \mid i \in I'\}, T' = (I', F'))$ *with* $X_i' = X_i \cap (A \cup B)$ *is a tree-decomposition of the subgraph of* G *induced by* $A \cup B$.

Proof

Let $I' = \{i \in I \mid A \subseteq X_i\}$. Take $T' = (I', F')$ to be the subgraph of T induced by I'. By definition of tree-decomposition, T' is again a tree. Clearly, condition (i) is fulfilled.

Because $\exists i : A \subseteq X_i, (\{X_i \mid i \in I\}, T)$ is a tree-decomposition of $G' = (V, E')$ with $E' = E \cup \{(v, w) \mid v, w \in A, v \neq w\}$. For all $b \in B, A \cup \{b\}$ forms a clique in G', and hence by the clique-containment lemma: $\exists i \in I : A \cup \{b\} \subseteq X_i \Rightarrow \exists i \in I' : b \in X_i$. So condition (ii) is fulfilled. Consider an edge $(b, c) \in E, b, c \in B$. $A \cup \{b, c\}$ forms a clique in G', and hence by the clique-containment lemma $\exists i \in I : A \cup \{b, c\} \subseteq X_i \Rightarrow \exists i \in I' : \{b, c\} \subseteq X_i$. It now easily follows that condition (iii) is fulfilled. □

Lemma 3.4 *Let* $G = (V, E), H = (W, F)$ *be graphs.*

(i) $treewidth(G \dot\cup H) = \max(treewidth(G), treewidth(H))$.

(ii) $pathwidth(G \dot\cup H) = \max(pathwidth(G), pathwidth(H))$.

(iii) $treewidth(G \times H) = \min(treewidth(G) + |W|, treewidth(H) + |V|)$.

(iv) $pathwidth(G \times H) = \min(pathwidth(G) + |W|, pathwidth(H) + |V|)$.

Proof

(i), (ii) Trivial.

(iii) First we show that $treewidth(G \times H) \leq treewidth(G) + |W|$. Take a tree-decomposition $(\{X_i \mid i \in I\}, T = (I, F))$ of G with treewidth $treewidth(G)$. Then $(\{X_i \cup W \mid i \in I\}, T = (I, F))$ is a tree-decomposition of $G \times H$ with treewidth $treewidth(G) + |W|$. So $treewidth(G \times H) \leq treewidth(G) + |W|$. Similarly one can show $treewidth(G \times H) \leq treewidth(H) + |V|$.

Next we show that $treewidth(G \times H) \geq \min(treewidth(G) + |W|, treewidth(H) + |V|)$. Consider a tree-decomposition $(\{X_i \mid i \in T\}, T = (I, F))$ of $G \times H$. From the complete bipartite subgraph containment lemma it follows that $\exists i : V \subseteq X_i$ or $\exists i : W \subseteq X_i$.

Suppose $\exists i : V \subseteq X_i$. Let $T = (I', F')$ be a subtree of T such that $\forall i \in I' : V \subseteq X_i, W \subseteq \cup_{i \in I'} X_i$ and $(\{X_i \mid i \in I'\}, T' = (I+, F'))$ is a tree-decomposition of $G \times H$. T' exists by lemma 3.3. Note that $(\{X_i \cap W \mid i \in I'\}, T' = (I', F))$ is a tree-decomposition of H, so $\exists i \in I' : |X_i \cap W| \geq treewidth(H) + 1 \Rightarrow \exists i \in I', |X_i| \geq |V| + treewidth(H) + 1$. So the treewidth of the tree-decomposition $(\{X_i \mid i \in I\}, T = (I, F))$ is at least $|V| + treewidth(H)$.

Similarly, if $\exists i : W \subseteq X_i$, one can show that the treewidth of $(\{X_i \mid i \in I\}, T = (I, F))$ is at least $|W| + treewidth(G)$. Hence $treewidth(G \times H) \geq \min(|V| + treewidth(H), |W| + treewidth(G))$.

(iv) Similar to (iii). □

Theorem 3.2 *For every cograph* $G = (V, E)$: $treewidth(G) = pathwidth(G)$.

Proof

Use induction on $|V|$.

If G consists of a single vertex, then $treewidth(G) = 0 = pathwidth(G)$. If $G = G_1 \dot\cup G_2$, then $treewidth(G) = \max(treewidth(G_1), treewidth(G_2)) = (i.h.) \max(pathwidth(G_1), pathwidth(G_2)) = pathwidth(G)$. If $G = G_1 \times G_2$, with $G_1 = (V_1, E_1)$, $G_2 = (V_2, E_2)$, then $treewidth(G) = \min(treewidth(G_1) + |V_2|, treewidth(G_2) + |V_1|) = (i.h.) \min(pathwidth(G_1) + |V_2|, pathwidth(G_2) + |V_1|) = pathwidth(G)$. □

4 Algorithms for pathwidth and related notions on cographs

In this section we give linear algorithms for determining treewidth, pathwidth, path-decompositions, optimal node search strategies, and interval graph augmentations with minimum clique size of cographs.

In Figure 3 we give two recursive procedures. COMPUTE-SIZE computes for every vertex in a binary cotree the number of vertices of the corresponding cograph. COMPUTE-PATHWIDTH computes for every vertex in a binary cotree the pathwidth of the cograph corresponding to that vertex. To compute the pathwidth of a cograph G, let r be the root of the binary cotree corresponding to G. Now first call COMPUTE-SIZE(r), and then COMPUTE-PATHWIDTH(r). As per vertex in the cotree a constant number of operations are performed, this costs $O(n)$ time in total. Correctness follows from lemma 3.4.

Theorem 4.1 *The pathwidth and treewidth of a cograph given with a corresponding cotree, can be computed in $O(n)$ time.*

It is not hard to construct corresponding path-decompositions in time, linear in the output, i.e. linear in $\sum_{i \in I} |X_i|$. However, in some cases this may be quadratic in n. (Consider a cograph $G = G_1 \times G_2$, where G_1 is a clique with $n/2$ vertices, and G_2 consists of $n/2$ isolated vertices. The optimal tree-decomposition of G will consist of $n/2$ sets X_i, each containing each vertex of G_1 and one vertex of G_2).

Thus, we are looking for a more compact representation of path-decompositions. We solve this in the following way: for each $v \in V$ we compute numbers $first(v) = \min\{i \in I \mid v \in X_i\}$ and $last(v) = \max\{i \in I \mid v \in X_i\}$. These numbers fix the path-decomposition, because for all $v \in V$, $i \in I : v \in X_i \Leftrightarrow first(v) \leq i \leq last(v)$.

Note that this representation corresponds by assigning to each $v \in V$ an interval such that the corresponding interval graph contains G; the chromatic number = maximum clique size of this interval graph equals the pathwidth of G plus 1. Thus we also find a representation of G corresponding to its interval thickness.

```
procedure  COMPUTE-SIZE (v: vertex);
     begin if v is a leaf of T_G
               then size (v) := 1
               else begin COMPUTE-SIZE (left son of v)
                          COMPUTE-SIZE (right son of v);
                          size (v) := size (left son of v) + size (right son of v)
                    end
     end


procedure  COMPUTE-PATHWIDTH (v: vertex);
     begin if v is a leaf of T_G
               then pathwidth(v) := 0
               else begin COMPUTE-PATHWIDTH (left son of v);
                          COMPUTE-PATHWIDTH (right son of v);
                          if label(v)=0
                            then pathwidth(v) := max(pathwidth (left son of v),
                                 pathwidth (right son of v))
                            else pathwidth(v) := min(size (left son of v) +
                                 pathwidth (right son of v),
                                 pathwidth (left son of v) + size (right son of v))
                    end
     end
```

Figure 3

The numbers $first(v)$ and $last(v)$ for all $v \in V$ are computed in the procedure MAKE-INTERVALS of Figure 4, which is called with MAKE-INTERVALS $(r, 1, m)$, where r is the root of the binary cotree of G, and m is an integer variable. In the procedure, $start$ always denotes the smallest value that can be used for $first(w)$ with w a leaf in the subtree of the cotree rooted at v, and $finish$ will yield the largest value used for $last(w)$, with w again leaf in the subtree rooted at v. Correctness of the procedure easily follows. Clearly, the procedure is linear in the size of the cotree $= O(n)$.

Theorem 4.2 *A representation of a path-decomposition with optimal pathwidth of a cograph, given with a corresponding cotree, can be computed in $O(n)$ time.*

Theorem 4.3 *The pathwidth and treewidth of cographs and corresponding path-decompositions or tree-decompositions can be computed in $O(n + e)$ time.*

Proof
Recall that the cotree of a cograph can be found in $O(n + e)$ time (see section 2). We now use the fact that optimal path-decompositions of cotrees fulfill $\sum_{i \in I} |X_i| = O(n + e)$. ☐

Theorem 4.4 *There exists an algorithm that, given a cograph G and a corresponding cotree of G, determines in $O(n)$ time an interval graph H that contains G as a subgraph and has chromatic number equal to the interval thickness of G.*

Theorem 4.5 *There exists an algorithm that, given a cograph G and a corresponding cotree of G, determines the node search number of G and corresponding search strategy in $O(n)$ time.*

procedure MAKE-INTERVALS (v: **vertex**, *start*: **integer**, *finish*: **var integer**);
  var *help*: **integer**:
  begin if v is a leaf of T_G
        then begin *first*(v) := *start*; *last*(v) := *start*; *finish* := *start* **end**
        else if label(v) = 0
            then begin MAKE-INTERVALS (left son of v, *start*,*help*);
                          MAKE-INTERVALS (right son of v, *help* +1,*finish*)
                end
            else ($*$ label(v) = 1)
            if size (left son of v) + *pathwidth*(right son of v) $>$
                  size (right son of v) + *pathwidth*(left son of v)
                then begin MAKE-INTERVALS (left son of v, *start*, *finish*);
                              for each $w \in V$ that is a leaf-descendant of the right son of v
                              do begin *first*(w) := *start*; *last*(w) := *finish* **end**
                    end
                else begin MAKE-INTERVALS (right son of v, *start*, *finish*)
                              for each $w \in V$ that is a leaf-descendant of the left son of v
                              do begin *first*(w) := *start*; *last*(w) := *finish* **end**
                    end

  end

Figure 4

Proof
Compute *first*(v) and *last*(v) for all $v \in V$ as described above. Now use the following search strategy:

put a searcher on each vertex v with *first*(v) = 1
for $i :=$ 1 **to** max$\{last(v) \mid v \in V\} - 1$
do begin for all $v \in V$ with *last*(v) = i: remove searcher from v;
          for all $v \in V$ with *first*(v) = $i + 1$: put a searcher on v
      end

With this search strategy, all edges will be cleared, no recontamination can take place, and the optimal number of searchers *(pathwidth(G)+1)* is used. Determining the sets $\{v \mid first(v) = i\}$, and $\{v \mid last(v) = i\}$ can be done with bucket sort in $O(n)$ time in total.
□

5 Final Remarks

In this paper we gave a linear time algorithm to determine the treewidth and pathwidth of cographs. Currently, we are investigating how to extend the results of this paper to larger classes of graphs, e.g., graphs that are built with modular composition with small neighborhood modules (see [14]). Another interesting problem is whether these results can be extended to distance-hereditary graphs.

References

[1] S. Arnborg. Efficient algorithms for combinatorial problems on graphs with bounded decomposability - A Survey. *BIT*, 25:2-23, 1985.

[2] S. Arnborg, D.G. Corneil and A. Proskurowski. Complexity of finding embeddings in a k-tree. *SIAM J. Alg. Disc. Meth.*, 8:277-284, 1987.

[3] H.L. Bodlaender. Classes of graphs with bounded treewidth. Technical Report RUU-CS-86-22, Dept. of Computer Science, University of Utrecht, Utrecht, 1986.

[4] H.L. Bodlaender. Dynamic programming algorithms on graphs with bounded tree-width. Techn. rep., Lab. for Computer Science, M.I.T., 1987. Ext. abstract in proceedings ICALP 88.

[5] H.L. Bodlaender. Improved self-reduction algorithms for graphs with bounded treewidth. Technical Report RUU-CS-88-29, Dept. of Computer Science, Univ. of Utrecht, 1988. To appear in: Proc. Workshop on Graph Theoretic Concepts in Comp. Sc. '89.

[6] D.G. Corneil, H. Lerchs, and L. Stewart Burlingham. Complement reducible graphs, Disc. Appl. Math. 3: 163-174, 1981.

[7] D.G Corneil, Y. Perl and L. Stewart. A linear recognition algorithm for cographs. *SIAM J. Comput.*, 4:926-934, 1985.

[8] J. Ellis, I.H. Sudborough, and J. Turner. Graph separation and search number. Report DCS-66-IR, University of Victoria, 1987.

[9] M.C. Golumbic. *Algorithmic Graph Theory and Perfect Graphs.* Academic Press, New York, 1980.

[10] J. Gustedt. Path width for chordal graphs is NP-complete. Preprint TU Berlin, 1989.

[11] L.M. Kirousis and C.H. Papadimitriou. Interval graphs and searching. *Disc. Math.*, 55:181-184, 1985.

[12] L.M. Kirousis and C.H. Papadimitriou. Searching and pebbling. *Theor. Comp. Sc.*, 47:205-218, 1986.

[13] R.H. Möhring. Graph problems related to gate matrix layout and PLA folding. Technical Report 223/1989, Department of Mathematics, Technical University of Berlin, 1989.

[14] M.B. Novick. Fast parallel algorithms for the modular decomposition. Technical Report TR 89-1016, Department of Computer Science, Cornell University, 1989.

[15] N. Robertson and P. Seymour. Graph minors. I. Excluding a forest. *J. Comb. Theory Series B*, 35:39-61, 1983.

[16] N. Robertson and P. Seymour. Graph minors. II. Algorithmic aspects of tree-width. *J. Algorithms*, 7:309-322, 1986.

[17] N. Robertson and P. Seymour. Graph minors. XIII. The disjoint paths problem. Manuscript, 1986.

[18] P. Scheffler. *Die Baumweite von Graphen als ein Maß für die Kompliziertheit algorithmischer Probleme.* PhD thesis, Akademie der Wissenschaften der DDR, Berlin, 1989.

Canonical representations of partial 2- and 3-trees

Stefan Arnborg*
NADA, KTH
S-100 44 Stockholm, Sweden

Andrzej Proskurowski[†]
Department of Computer and Information Science
University of Oregon, Eugene, Oregon 97403, USA

Abstract

We give linear time algorithms constructing canonical representations of partial 2-trees (series parallel graphs) and partial 3-trees. These algorithms directly give a linear time isomorphism algorithm for partial 3-trees.

1 Introduction

A canonical representation of a family of graphs assigns to each member of the family a label that is independent of any arbitrary vertex numbering: two graphs have the same canonical representation if and only if they are isomorphic. Thus, the graph isomorphism problem can be solved using canonical representations and solved efficiently if such representations can be efficiently computed and compared. Other uses of canonical representations are to investigate the structure of the automorphism group of a graph and to generate random graphs with some distribution over isomorphism classes.

Most graph representations are not canonical since vertices are arbitrarily numbered. But if we consider all possible vertex permutations, compute the corresponding representations, and select the lexicographically smallest, then we get a canonical representation. The set of vertex permutations yielding the lexicographically smallest representation is a coset of the automorphism group for the graph, regarded as a subgroup of the symmetric group on the vertex set.

A straightforward application of the above procedure has exponential (in the graph size) cost since there are exponentially many vertex permutations to minimize over. But in some cases it is possible to constrain the set of explicitly considered permutations in such a way that the whole procedure can be performed in polynomial time. We need only consider a set guaranteed to contain at least one coset of the automorphism group of the given graph. In this paper we show how these ideas yield an algorithm which produces a canonical representation for partial 3-trees in linear time, and thus also solves the isomorphism problem for partial 3-trees in linear time. Previously, the graph isomorphism problems for graphs of bounded valence (Luks [15]), genus (Filotti and Mayer [11], Miller [16], [17]), and tree-width (Bodlaender [7]) (none of which is a subfamily of another) have been shown solvable in polynomial time. Linear time algorithms for isomorphism of planar graphs (and thus also for partial 2-trees, which are planar) are already known (Fontet [12]; Hopcroft and Wong [14]; Colbourn and Booth [10]).

Bodlaender proposed an algorithm for deciding isomorphism of partial k-trees [7]. His method is based on the brute force k-tree embedding method of Arnborg, Corneil and Proskurowski [3], where all k-vertex separators of the given partial k-tree are tested for suitability as separators in a k-tree embedding. This algorithm requires solving bipartite matching problems and takes $O(n^{k+4.5})$ time. The procedure we use is based on a canonical reduction sequence obtained from the safe reduction rules reported in Arnborg and Proskurowski [5]. For any given graph, the set of vertices reducible according to a given rule is fixed by the automorphism group of the graph[1]. Each reduction involves a separator of the graph with one, two or three vertices. Whether two reduced vertices are automorphic depends on symmetries between the corresponding separators. Our method keeps a record of symmetries of the reduced parts of the graph through a sequence of labels and orientations attached to the separators used in the reduction process. Two reduced subgraphs cut off by such separators are isomorphic (the isomorphism mapping

---

*Supported in part by a grant from NFR.
[†]Research supported in part by the Office of Naval Research Contract N00014-86-K-0419.

[1]This means that any automorphism of the graph permutes these vertices among themselves.

one separator to the other) if and only if the labels of the two separators are equal, and their orientations indicate the correspondence between the separator vertex sets.

A reducible vertex represents a k-leaf in an embedding k-tree. Thus, adjacent vertices cannot be reduced in parallel. To deal with this, we refine the reduction rules to handle overlapping reduction instances. These refined reduction rules allow us to construct a *parse tree*, where each node is associated with a reduction instance and two nodes are adjacent if the reduction instance corresponding to one of the nodes creates a (hyper-) edge involved in the reduction corresponding to the other node. This tree is used to implement efficiently the algorithms computing canonical representations. In order to get linear time performance we use a technique of label numbering and bucket sorting described in Aho, Hopcroft and Ullman [1].

Our paper is organized as follows. After defining the necessary terminology in section 2 we introduce the method in section 3 by applying it to partial 2-trees. This special case is much simpler. We then derive the additional reduction instances necessary for partial 3-trees in section 4. The algorithms used to give linear time performance are presented and analyzed in section 5.

2 Definitions and terminology

We will use standard graph theory terminology, as found, for instance, in Bondy and Murty [8]. We will also make use of concepts from the realm of hypergraphs, but will introduce them first in section 4. Some elementary and completely standard group theory is also used, see *e.g.*, Rotman [21]. We will now define some basic concepts.

A *walk* is a sequence of vertices such that every two consecutive vertices are adjacent. If all the vertices are different, we have a *path*. A walk forms a *cycle* if only its first and last vertices are identical. A set of k vertices, every two of which are adjacent, is called a k-clique. The graph on k vertices whose vertex set is a k-clique will be denoted K_k. A (minimal) subset of vertices of a graph such that their removal disconnects the graph is a (minimal) *separator*. A k-tree is a connected graph with no K_{k+2} subgraph such that every minimal separator is a k-clique. Equivalently, the complete graph on k vertices (K_k) is a k-tree, and any k-tree with $n > k$ vertices can be constructed from a k-tree with $n - 1$ vertices by adding a new vertex adjacent to all vertices of a k-clique of that graph. In this new graph, the added vertex is a k-*leaf*. A *partial k-tree* is any subgraph of a k-tree.

While partial k-trees are undirected, simple graphs (without multiple edges or self-loops), in the course of our presentation we will allow both undirected *edges* and directed *arcs* (ordered pairs of vertices), as well as parallel edges and arcs. Those mixed graphs will be intermediate results of applying *graph rewriting rules*, consisting of replacing a subgraph isomorphic to the lefthand side of such a rule by the righthand side subgraph. In our case, the latter has always fewer vertices than the former and thus a set of such rules defines possible *reduction sequences*. Given a class of graphs, a rewriting rule such that its application preserves membership in both the class and its complement is called a *safe* rule. A set of reduction rules such that any non-trivial graph in the class contains as a subgraph the lefthand side of some rule is called a *complete* set of rules.

Complete sets of safe and confluent reduction rules for partial k-trees are known for $k \leq 3$ (Arnborg and Proskurowski [5]). Intuitively, they correspond to *pruning* of k-leaves in an embedding k-tree. For partial 1-trees (forests), the set of reduction rules consists of the removal of pendant vertices (of degree 1) and of isolated vertices. For partial 2-trees (series-parallel graphs) we have additional *series* and *parallel* rules, in which a path of degree 2 vertices is replaced by an edge, and multiple edges are replaced by one edge, respectively. For partial 3-trees, the additional rules deal with three cases of degree 3 vertex reductions: the *triangle*, the *buddy*, and the *cube* rules (*cf.* Figure 1). The cube-like configuration with the 'hub' (vertex x in Figure 1) of degree greater than 3 can be safely reduced, as well.

A reduction sequence can be used to obtain a canonical representation under certain circumstances, namely that there is a unique way to choose the next rule in every situation. A difficulty arises when applicable rules can make adjacent vertices into k-leaves; obviously, for a k-tree embedding, only one of two such vertices is a k-leaf. This situation has to be dealt with separately. We note that there is a simple solution if the conflicting rules are different, *e.g.* a vertex reducible according to the triangle rule is adjacent to a vertex reducible according to the buddy rule. We simply order the rules and say

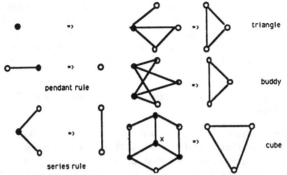

Figure 1: Reduction rules for forests, series-parallell graphs and partial 3-trees

that a higher priority rule takes precedence. The remaining case of conflict is where adjacent vertices are reducible according to the same rule. To break the ties in this case, we consider a refined list of rules derived from the complete set of safe rules presented in [5].

3 Canonical representation of partial 2-trees

We start our exposition by presenting the algorithm for partial 2-trees and then we generalize it to partial 3-trees.

Our algorithm is based on a construction of (vertex and edge) labels that record the sequence of reductions of the original graph G leading to a set of labeled isolated vertices. From these, we construct the final label that is a canonical representation of G and, as such, would also allow a unique (and efficient) reconstruction of G.

Initially, every node label is (0) and every edge label is (0,0). The second component in an edge label is the orientation information, locally recorded by 'an arrowhead' (say, an ordered pair of the edge's end vertices) or its absence. When recorded permanently, this information is transformed into a 0, a 1, or a -1, (like, for instance, in the label given above).

The algorithm will reduce a given connected graph to the single labeled vertex in phases performed in the following manner. Every phase begins with finding the first applicable reduction rule (closest to the beginning of the list of reductions, to be presented shortly). All instances of this rule in the entire graph are then found and the corresponding reductions are performed, resulting in a modified (reduced) graph, and some new labels. Since each reduction decreases the size of the graph, in $\mathcal{O}(n)$ phases the graph will be reduced to a single vertex.

3.1 Reduction rules

In this subsection we describe the rewriting rules (tailored to the needs of unique labeling) that reduce a given connected partial 2-tree to a single labeled vertex. The rules are given in their scanning order, together with the associated new label(s).

0 multiple vertices and edges. The rules below may create several edges with the same end-points (series rules) or several labels for a vertex (pendant and chain with identical end-vertices). The latter is referred to in the pendant rule 1 as a label being merged into the label of a vertex.

0.1 vertex label merge. The merging of multiple vertex and self-loop labels into a vertex label means that we keep a set of labels with multiplicities to describe the merged labels. Before the vertex itself is reduced, we must construct a proper label for it. This is done simply by lexicographically sorting the labels, and keeping duplicates in the sorted list l. The label will be $(0.1; l)$.

0.2 parallel rule. This rule is applicable when m edges have the same end-points s_1 and s_2 ($s_1 \neq s_2$, arbitrarily ordered). Let the labels of these edges be l_i and let d_i be 1 if edge i has an arrowhead from s_1 to s_2, -1 if an arrowhead is directed from s_2 to s_1, and 0 if it is not present. The parallel edges are replaced by one new edge between s_1 and s_2. This edge has a label $(0.2; l)$, where l is the lexicographically smallest of two sorted lists corresponding to edge orientations from s_1 to s_2 and from s_2 to s_1. The new edge will have arrowhead $(s_1 \rightarrow s_2)$ if the first list is smaller, $(s_2 \rightarrow s_1)$ if the second is smaller, and none if they are equal. The first list has the members $(l_i, d_i)_{i=0}^{m-1}$, the second $(l_i, -d_i)_{i=0}^{m-1}$

1 pendant rule. This rule applies to vertices of degree 1 (pendant vertices) and to edges with both endpoints the same (self-loops).

1.1 dipole. If there is a pair of adjacent pendant vertices, they must form a connected component of the graph consisting of two vertices labeled l_1 and l_2 both incident to an edge labeled l', directed from the first to the second vertex $(d = 1)$ or not at all $(d = 0)$. This graph is reduced to a single vertex with label $(1.1; l_1, (l', d), l_2)$ or $(1.1; l_2, (l', -d), l_1)$, whichever is lexicographically smallest.

1.2 pendant vertices. Assume the set of pendant (degree 1) vertices is independent. One vertex and one edge is removed by a pendant vertex removal. The label l' together with with the orientation d of the edge form a pair (l', d). d is defined as 1 if the arrowhead of the edge is directed towards the pendant vertex, -1 if it is directed the other way, and 0 if it is not present. Combined with the label of the vertex, l, they form the label $(1.2; (l', d), l)$, which is merged subsequently into the label of the separating vertex according to rule 0.1.

1.3 self-loops. The label l' of a self-loop is merged into the label of its end-vertex according to rule 0.1.

2 series rule. This rule applies to vertices of degree 2. A set of such vertices inducing a connected graph will induce a path or a cycle in the graph, as detailed below:

2.1 chain rule. A maximal set of degree 2 vertices inducing a connected graph, and ending in two vertices, s_1 and s_2, of degree higher than 2 consists of a path $(v_1, \ldots, v_m)$ with v_1 adjacent to s_1 and v_m adjacent to s_2. Let l_i be the label of v_i, $i = 1, \ldots, m$ and l'_i of the edge (v_i, v_{i+1}), $i = 0, \ldots, m$, with $v_0 = s_1$ and $v_{m+1} = s_2$. Let d_i be 1, -1 or 0, depending on whether the arrowhead of (v_i, v_{i+1}) is directed from v_i to v_{i+1}, from v_{i+1} to v_i, or not present. The vertices v_i, $i = 1, \ldots, m$ and their incident edges are replaced by an edge connecting s_1 and s_2. The label of this edge is the lexicographically smallest of two labels, $(2.1; (l'_0, d_0), l_1, (l'_1, d_1), \ldots, l_m, (l'_m, d_m))$ and $(2.1; (l'_m, -d_m), l_m, (l'_{m-1}, -d_{m-1}), \ldots, l_1, (l'_0, -d_0))$. The arrowhead will point from s_1 to s_2 if the first alternative is smaller, from s_2 to s_1 if the second is smaller, and is absent if they are equal or if $s_1 = s_2$ (the latter condition describing a self-loop).

2.2 ring rule. A connected two-regular graph is a cycle consisting of vertices $v_0, \ldots, v_{m-1}$, with vertex v_i adjacent to vertex v_{i+1} for $i = 0, \ldots, m - 2$ and with v_0 adjacent to v_{m-1}. It is reduced into a single vertex labeled with the lexicographically smallest of $2m$ labels $(2.2; (l'_{i+jk}, kd_{i+jk}), l_{i+jk})_{j=0}^{m-1}$ where $i = 0, \ldots, m - 1$, $k = 1, -1$, and indices are computed modulo m.

4   Canonical representation of partial 3-trees

The idea behind the algorithm for partial 3-trees is similar to that of the algorithm constructing a canonical representation of partial 2-trees. The reduction of vertices is performed in consecutive stages, where each stage consists either of independent reduction instances or of groups of dependent reduction instances of the same kind. The situation is more complicated for partial 3-trees for two reasons. For one, the reduction information (recorded in labels) often concerns three vertices, and thus the resulting label must be associated with triples of vertices. We will present this as the labeling process for hyperedges of order 3. The second reason for a more complicated algorithm is the larger number of ways in which reduction rules of the same kind can involve adjacent vertices. For instance, more than one vertex of a triangle can have degree 3. Below, we elaborate on these differences leading to an algorithm constructing a canonical label for a given connected partial 3-tree.

4.1 Labeling of hyperedges

Each of the three reduction rules involving a degree three vertex causes the elimination of that vertex and edges incident with it and replaces them by edges between the neighbors of the vertex. (We call this reduction of a single vertex v adjacent to the separator vertices s_1, s_2, s_3 a *basic degree 3 reduction*.) We will separate this *topological* action from the *labeling* action of creating a labeled hyperedge corresponding to the three neighbors of the eliminated vertex.

The *orientation* of a hyperedge is determined with respect to a given permutation of hyperedge vertices and describes vertex symmetries with respect to the hyperedge label. These are represented by a subset of permutations of the vertices which constitute the projection, into the symmetric group of hyperedge vertex permutations, of a coset of the automorphism group for the subgraph separated by the hyperedge from the rest of the graph. This subset of permutations is represented wrt. given vertex permutation (initially, the original ordering of graph vertices) by lexicographically sorted list of the permutations.

When a hyperedge is removed in a reduction operation involving one of its vertices, its label will form a component of the label of the created hyperedge, together with an indication of how the orientation of the removed hyperedge corresponds to the orientation of the created hyperedge. Consider an orientation D of a removed hyperedge and a permutation σ of the union of vertices in the removed hyperedges. D is coded with respect to σ in the following way: Replace each vertex occuring in D by its index in σ. (Recall that D is a set of permutations of a subset of the vertices appearing in σ.) Sort lexicographically the resulting set of integer lists.

As an example, consider reduction of a degree 3 vertex v with neighbors a, b and c. This reduction removes at most one 1-hyperedge, three 2-hyperedges, and three 3-hyperedges. Now, for instance, the permutation $\sigma = (a, b, c, v)$ will cause the orientation $D = \{(a, b, v), (a, v, b)\}$ of a 3-hyperedge $\{a, b, v\}$ to be encoded as $((1, 2, 4), (1, 4, 2))$.

We can now describe the reduction of a set of vertices $R = \{v_i\}_{i=1}^k$ separated from the rest of the graph by vertices $S = \{s_i\}_{i=1}^l$:

Let a *sufficient set of permutations* denote a set of permutations of vertices of a subgraph that is able to represent all symmetries (automorphism group) of the subgraph. (This can be achieved by a smaller than full symmetric group set of permutations when permutations exchanging non-symmetric vertices are omitted.) Produce a sufficient set P of permutations of $R \cup S$. For each permutation p in P, produce a label as follows: consider the set of *(label, orientation)* pairs that contain some vertices of R. Replace each orientation with its encoding wrt p and sort the pairs to get a label wrt p. The subset of permutations that yield lexicographically smallest label l constitute the orientation of the new edge. The minimum value l and the rule number r according to which the reduction is made are used to build the label $(r; l)$ of the new hyperedge.

Observe that the previous label construction for partial 2-trees was only slightly different and can be easily changed to the present one. As an example, in the chain rule, each d_i would be changed from 1 to $((i, i + 1))$, from -1 to $((i + 1, i))$ and from 0 to $((i, i + 1), (i + 1, i))$. An arrowhead on an edge between vertices a and b would be represented as an orientation $((a, b))$ or $((b, a))$ and its absence would be represented by orientation $((a, b), (b, a))$.

4.2 Reduction rules for partial 3-trees

In this subsection we continue our list of refined reduction rules from section 3. As noted before, the main obstacle here will be the presence of conflicting reduction instances, which we consider in the rules 5, 6, and 7 below. We determine the complete set of suh instances and give a reduction rule for each of those. In each rule, we identify the sufficient set of permutations that determines the value of the hyperedge label according to the procedure outlined in the preceding subsection.

3 parallel triangles. This rule is applicable when two or more 3-hyperedges have the same set of vertices. Let the vertices be $S=(s_1, s_2, s_3)$, and the labels of the hyperedges l_i, $i = 1, \ldots, m$. The orientations d_i, $i = 1, \ldots, m$ are sets of permutations of S. Let $d_i^{(p)}$ be the coding of d_i wrt a permutation p of S. For each permutation p of S, form the set $\{(l_i, d_i^{(p)})\}_{i=1}^m$, sort it lexicographically and extract the lexicographically smallest list (over p) as well as those p giving a minimum, and call the smallest list l and the set of permutations P. The new hyperedge with vertices s_1, s_2, s_3 has label $(3; l)$ and orientation P.

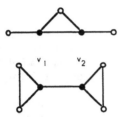

Figure 2: Subgraphs of H with two adjacent degree 1 vertices

4 isolated instances.

4.1 triangle. A vertex that is triangle-reducible and not adjacent to another triangle-reducible vertex can be reduced directly. This reduction of a single vertex adjacent to three separator vertices will be called the *basic degree 3 reduction*. The sufficient set of permutations consists of every permutation of the s_i, each followed by v.

4.2 buddy. Vertices in a buddy configuration not adjacent to another buddy configuration can be reduced simultaneously with the basic degree 3 reduction in all occurences.

4.3 cube. The three reducible vertices in a cube configuration not adjacent to another cube are reduced with the basic degree 3 reduction.

5 conflicting triangles.

5.1 diamonds. We can consider separately the cases when two adjacent triangle-reducible degree 3 vertices v_1, v_2 have two neighbors s_1, s_2 in common.

5.1.1 K_4. More than one occurrence of the diamond rule with adjacent reducible vertices is possible only if a connected component of the graph is the 4-clique K_4. The sufficient set consists of all 24 permutations of the vertices v_i, $i = 1, 2, 3, 4$.

5.1.2 K_4^-. If the vertices s_1 and s_2 are not adjacent, they induce, together with their common neighbors v_1, v_2, the four-clique without an edge, K_4^-. A sufficient set of permutations to consider is each of the two permutations of $\{s_1, s_2\}$ followed by each permutation of $\{v_1, v_2\}$.

5.2 Subgraph H. The remaining configurations of adjacent degree 3 vertices reducible according to the triangle rule and with at most one common neighbor (triangle-reducible for short) are discussed below. For a given partial 3-tree G, let us consider a maximal connected subgraph H consisting of triangle-reducible vertices subject to conflicting reductions according to the *triangle* rule. We will investigate the structure of those subgraphs and will make unique choices of *independently* reducible vertex sets in G. We consider the following cases of reductions in G depending on the degrees of vertices in H:

5.2.1 degree 1 only. Two adjacent vertices of degree 1 in H, v_1 and v_2, represent one of the two subgraphs in G shown in Figure 2. In the case of a four-vertex separator, other reduction rules must be applicable in the rest of the graph ('beyond the four separating vertices'), or else G cannot be a partial 3-tree. In the case of a 3-vertex separator, a separate reduction rule allows us to reduce v_1 and v_2, creating a hyperedge containing the separator vertices, with a unique label describing the reduced subgraph of the original graph (none of the separator vertices is triangle-reducible).

5.2.2 degree 1 and 2 or 3. Nonadjacent vertices of degree 1 in H can be subjects to the basic degree 3 reductions in G, since these do not conflict with each other.

5.2.3 degree 2 only. The vertices of degree 2 form a cycle in H. Let us call an edge t if it is a side of a triangle of G and f otherwise. Notice that end vertices of adjacent t edges have one common neighbor in G. Consider those vertices incident to both a t and an f edge; call this set A. Depending on the relation between A and H, we have three subcases (Figure 3).

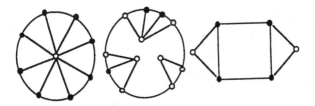

Figure 3: Cycles in H (a) a wheel, (b) a general case, (c) the square

5.2.3.1 wheel. The set A is empty: all edges of H are t edges. This is the wheel configuration, reduced to a single vertex label ("the hub's") according to a separate rule.

5.2.3.2 collection of paths. A does not contain all vertices of H and its vertices partition the set of the remaining vertices of H into connected components. Those can be dealt with in a manner similar to that in rules 4.1, 5.2.1, and 5.2.2.

5.2.3.3 square. If H consists of alternating t and f edges (all the vertices are in A), then we have to consider subcases depending on the number of edges in H (it is trivially greater than 3). When H has 4 edges (only two triangles are involved), the triangles' 'third vertices' form a separator of G. The corresponding configuration (square, Figure 3(c)), is the left hand side of a separate reduction rule. If there are more than two triangles, then there must be another instance of a reduction rule 'beyond the separating vertices' (i.e., in the subgraph of G induced by vertices of G other than in H), or else G is not a partial 3-tree.

5.2.4 degrees 2 and 3. If H consists of both vertices of degree 2 and vertices of degree 3, then the vertices of degree 2 form in H paths that end in degree 3 vertices. When such a path has exactly two degree 2 vertices, these can be reduced according to rule 5.2.1. Otherwise, there are unique and nonadjacent vertices of degree 2 in H and the corresponding vertices in G can be reduced with the basic degree 3 reduction, similarly to the situation in 5.2.2.

5.2.5 degree 3 only (prism). Since a vertex in H is of degree 3 in the original partial 3-tree G, a cubic H is identical with G. To analyse this case, we consider the multigraph $\bar{H}$ obtained from H in the following manner: The set of vertices of $\bar{H}$ is the set of triangles (K_3 subgraphs) of H. The set of edges of $\bar{H}$ is the set of edges of H *not* in the triangles. A vertex of $\bar{H}$ is incident with an edge of $\bar{H}$ if the corresponding triangle contains a vertex of H incident with that edge in H. We will show that $\bar{H}$ is a series-parallel grah, which has important consequences in determining unique triangle reductions in H (or, equivalently, G). A multigraph is *series-parallel* if and only if it does not contain a subgraph homeomorphic to K_4. However, if $\bar{H}$ contains such a subgraph, then H contains as a minor the 4-regular Duffin graph (see Figure 4(b)), and is thus not a partial 3-tree. Since all vertices of $\bar{H}$ have degree 3, there must be instances of independent parallel edge reductions in $\bar{H}$. Unless H has six vertices and nine edges ('the prism', see Figure 4(a)) reduced to a single vertex by a separate rule, an instance of the parallel edge reduction in $\bar{H}$ corresponds to the subgraph of the *5.2.3.3 square* rule in H.

6 conflicting buddies. The 3-leaves v_1, v_2 in an instance of the buddy reduction can be adjacent to 3-leaves u_1, u_2 in another instance of the buddy reduction only if u_1, u_2 are commonly adjacent to a third vertex w. This third vertex may be identical with or different from the third common neighbor of v_1, v_2, leading to two configurations that can be incorporated as the left-hand-sides of new reduction rules. The former is a case of the *5.2.3.1 wheel* rule discussed above, the latter, *cat's cradle*, is shown in Figure 5.

6.1 $K_{3,3}$. Two cat's cradles can overlap only where the graph has a connected component which is $K_{3,3}$.

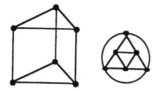

Figure 4: 3-regular subgraphs (a) the prism, (b) a minimal forbidden minor

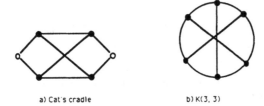

a) Cat's cradle   b) K(3, 3)

Figure 5: Overlapping buddy configurations

6.2 cat's cradle. This is the other case where $w \notin \Gamma(\{v_1, v_2\})$. The separator vertices s_1, s_2 and the cycle (v_1, v_2, v_3, v_4) have edges (s_1, v_1), (s_1, v_3), (s_2, v_2) and (s_2, v_4) between them.

7 conflicting cubes. There are two basic cases of possible conflicts between the reduced vertices in two different instances of the cube reduction.

7.1 qube. In one case, the purported 3-leaf vertices v_1, v_2, v_3 in one instance of the cube reduction are adjacent to 3-leaf vertices of another instance. This occurs only in the 8-vertex, 12-edge three-dimensional cube graph. This is because the vertices u_1, u_2, and u_3 (*cf.* Figure 6) have then degree 3 and are adjacent to a common neighbor (the hub of that instance), also of degree 3.

7.2 hammock. In the second case, the hub x of one instance is a 3-leaf of the other instance. This implies that one of the 3-leaves of the first instance, say v_3, is the hub of the second instance and its other neighbors u_1, u_2 are the remaining 3-leaves of the second instance. The vertices u_1, u_2 must have another common neighbor y. If this vertex is identical with the remaining vertex u_3 of the original cube, u_1, u_2 are triangle-reducible. If y is different than u_3, this leads to a new reduction rule with the left-hand-side configurations given in Figure 6(b).

5 Complexity of the algorithms

Any reasonable implementation of the labeling and reduction process will run in polynomial time for some low-order polynomial. In order to get a *linear time* algorithm we must be more careful. Some apparent obstacles to linear time performance are introduced by the need to perform the following tasks:

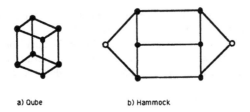

a) Qube   b) Hammock

Figure 6: Overlapping cube configurations

(i) Sorting $\mathcal{O}(n)$ labels, each of length $\mathcal{O}(n)$, when producing labels for applications of parallel rules.

(ii) Finding the lexicographically smallest of m labels each of size k, where mk is $\mathcal{O}(n)$, when deciding label for a 2.2 (ring) reduction.

(iii) Finding instances of applicable reductions.

Actually, there are already methods available to overcome these obstacles. First, the labels to a large extent contain duplicated strings, so by replacing a copied string by a pointer or label number, we essentially get a representation of all labels as vertices in a *dag* of size linear in the graph size, as is done for a tree isomorphism algorithm due to Edmonds and described in, e.g., Aho, Hopcroft and Ullman [1] and Colbourn and Booth [10]. The technique presented there is directly applicable here, and solves problem (i). With this canonical numbering, any of the corresponding algorithms by Sysło [24], Shiloah [23], or Booth [9] solves also problem (ii) above.

The total number of times vertex adjacencies change during the reduction process is proportional to the graph size. Thus, maintaining 'ready lists' for vertices that reach small enough degree to be considered in the reduction rules resolves problem (iii) above, together with a linear time access method for edges in a graph described in detail in [4].

6 Conclusions

The existence of a complete set of safe reduction rules allowing recognition of partial 3-trees led us to a linear time isomorphism algorithm for this class of graphs. We expect that this can be generalized to higher values of k if such rule sets exists.

References

[1] AHO, HOPCROFT AND ULLMAN, Design and Analysis of Computer Algorithms *Addison-Wesley 1972.*

[2] S. ARNBORG, Efficient Algorithms for Combinatorial Problems on Graphs with Bounded Decomposability — A Survey, *BIT 25 (1985), 2-33.*

[3] S. ARNBORG, D.G. CORNEIL AND A. PROSKUROWSKI, Complexity of Finding Embeddings in a k-tree, *SIAM J. Alg. and Discr. Methods 8(1987), 277-287.*

[4] S. ARNBORG, B. COURCELLE, A. PROSKUROWSKI AND D. SEESE, An algebraic theory of graph reduction, *Bordeaux-I University Report I-9002(1990), submitted.*

[5] S. ARNBORG AND A. PROSKUROWSKI, Characterization and Recognition of Partial 3-trees, *SIAM J. Alg. and Discr. Methods 7(1986), 305-314.*

[6] S. ARNBORG AND A. PROSKUROWSKI, Linear Time Algorithms for NP-hard Problems on Graphs Embedded in k-trees, *Discr. Appl. Math. 23(1989) 11-24.*

[7] H.L. BODLAENDER, Polynomial Algorithms for graph isomorphism and chromatic index on partial k-trees. *SWAT, Springer–Verlag LNCS 318 (1988), 227-232.*

[8] J.A. BONDY AND U.S.R. MURTY, *Graph Theory with Applications*, North Holland (1976).

[9] K.S. BOOTH, Finding a lexicographic least shift of a string, *Information Processing Letters 10 (1980), 240-242;*

[10] C.J. COLBOURN AND K.S. BOOTH, Linear time automorphism algorithms for trees, interval graphs, and planar graphs, *SIAM J. Computing 10 (1981), 203-225;*

[11] I.S. FILOTTI AND J.N. MAYER, A Polynomial-time Algorithm for Determining the Isomorphism of Graphs of Bounded Genus, *Proc. 12th ACM Symp. on Theory of Computing (1980), 236-243.*

[12] M. FONTET, A Linear Algorithm for Testing Isomorphism of Planar Graphs, *Proc. 3rd Int. Conf. Automata, Languages, Programming, Springer–Verlag LNCS (1976), 1411-423.*

[13] M.R. GAREY AND D.S. JOHNSON, *Computers and Intractability*, W.H. Freeman and Company, San Francisco (1979).

[14] J.E. HOPCROFT AND J.K. WONG, A Linear Time Algorithm for Isomorphism of Planar Graphs, *Proc. 6th ACM Symp. Theory of Computer Science (1974), 172-184.*

[15] E.M. LUKS, Isomorphism of Graphs of Bounded Valence Can Be Tested in Polynomial Time, *JCSS 25(1982), 42-65*

[16] G.L. MILLER, Isomorphism Testing for Graphs with Bounded Genus, *Proc. 12th ACM Symp. on Theory of Computing (1980), 225-235.*

[17] G.L. MILLER, Isomorphism Tesing and Cannonical Forms for k-contractible Graphs, *Proc. Foundations of Computation Theory, Springer-Verlag LNCS 158 (1983), 310-327.*

[18] A. PROSKUROWSKI, Recursive graphs, recursive labelings and shortest paths, *SIAM J. Computing 10 (1981), 391-397;*

[19] A. PROSKUROWSKI, Separating Subgraphs in k-trees: Cables and Caterpillars, *Discr. Math. 49(1984), 275-285.*

[20] D.J. ROSE, Triangulated Graphs and the Elimination Process, *J. Math. Anal. Appl. 32 (1970) 597-609.*

[21] J.J. ROTMAN, *The theory of Groups (2nd ed.)*, Allyn and Bacon, 1973.

[22] P. SCHEFFLER, Linear time algorithms for NP-complete problems for partial k-trees, *R-MATH-03/87 (1987);*

[23] Y. SHILOAH, A fast equivalence-checking algorithm for circular lists, *Information Processing Letters 8 (1979), 236-238;*

[24] M.M. SYSLO, Linear Time Algorithm for Coding Outerplanar Graphs, in Beitrage zur Graphenteorie und deren Anvendungen, Vorgetragen auf dem internationellen Kolloquium in Oberhof (DDR), H. Sachs (ed.) (1978), 259-269; *TRN-20 1977, Institute of Computer Science, Wrocław University.*

[25] A. WALD AND C.J. COLBOURN, Steiner Trees, Partial 2-trees, and Minimum IFI Networks, *Networks 13 (1983), 159-167.*

[26] T.V. WIMER, Linear algorithms on k-terminal graphs, PhD. Thesis, Clemson University (1988);

On Matroids and Hierarchical Graphs

David Fernández-Baca* and Mark A. Williams[†]

Department of Computer Science, Iowa State University, Ames, Iowa 50011

1   Introduction

Several researchers have proposed and studied models for succinctly representing graphs and other structures [2,7,8,10,12,16,18,21]. In all these models, it is possible to encode large structures using a description whose size is polylogarithmic in the size of the structure, thus allowing considerable savings of storage space. It is natural to ask whether one can speed up the processing of a graph by working with a succinct description of the graph rather than with the graph itself. The answer is often negative in that even polynomially-solvable problems become intractable when inputs are succinctly described. This has been the case, for instance, in the small circuit model [7,18] and the Hierarchical Input Language [2]. Two exceptions have been hierarchical graphs [10,12] and dynamic graphs [8,16]. This paper deals with the former model.

Hierarchical graphs were introduced by Lengauer as a means to succinctly represent graphs having regular structures. Problems such as finding the cost of a minimum spanning forest [10] or determining certain connectivity properties [4,12] can be solved in time polynomial in the size of the hierarchical description of the graph. On the other hand, certain polynomially-solvable problems, such as the maximum-flow problem and the circuit-value problem, become PSPACE-complete in the hierarchical model [11].

There is apparently no correlation between hierarchical and non-hierarchical complexities of graph problems [11]. A more fruitful approach might be to look for families of problems whose hierarchical versions are efficiently solvable. The main result presented in this paper is a characterization of two such families of problems. Both are problems of finding the cost of an optimum base of a matroid defined on the edges of a graph. Their common characteristic is that the circuits of these matroids are the edge sets of graphs homeomorphic from some member of a finite set of graphs. These matroids were studied by Simões Pereira [19], Matthews [14,15], and White and Whiteley [23], and include the cycle matroid, whose bases are the edges of spanning forests of a graph. In addition to matroid theory, our work relies on techniques developed by Lengauer [10,12,13], in particular, his *bottom-up method*, and proof methods introduced in an earlier paper [4]. Our results generalize some of Lengauer's work on minimum spanning forests [10] and also, we believe, puts it in a new perspective by exhibiting to what extent his ideas work for other problems.

This paper is organized as follows. In section 2, we review hierarchical graphs and

---
*Supported in part by NSF Grant CCR-8909626
[†]Supported by an IBM Graduate Fellowship

matroids. In section 3, we describe a hierarchical algorithm for computing optimum bases of certain matroids defined on graphs. In section 4, we introduce the two families of matroids alluded to earlier, and develop efficient hierarchical algorithms for determining costs of their optimum bases. Section 5 contains a summary of our results and a brief discussion of some extensions and open problems.

2 Preliminaries

All graphs are assumed to be undirected. A graph G *simple* if it has no loops or multiple edges. An edge is *incident* on its *endpoints*. A *pendant vertex* is a vertex whose degree is one, and a *pendant edge* is an edge incident on a pendant vertex. A *series vertex* is a vertex with two incident edges, which are said to be in *series*. A loop is *isolated* if its endpoint has degree two. A *minor* of G is obtained from G by a sequence of vertex deletions, edge deletions, and edge contractions. Refer to Wilson [24] for graph terminology not defined here.

2.1 Hierarchical graphs

The following concepts and definitions are adapted from [12,13]. Every graph G has zero or more *pins*, distinguished vertices used to *glue* G onto other graphs as follows:

Definition 2.1 Let H be a graph and $L = [(a_1, b_1), \ldots, (a_r, b_r)]$ a list of pairs where $a_1, \ldots, a_r$ are distinct vertices of H, and $b_1, \ldots, b_r$ are the pins of G. L is called a *gluing list*, and H and L are *compatible* with G. The graph $J = H \circ_L G$ is the result of gluing G onto H using L. If L is empty, J is the disjoint union of G and H. Otherwise, J is obtained by identifying a_i and b_i for $1 \leq i \leq r$. The pins of J are those of H. We will omit gluing lists in situations where no confusion can arise.

For any graph G, let the *environment* of G, denoted $\mathrm{Env}(G)$, be the set of all graph and gluing list pairs (H, L) such that H and L are compatible with G. Clearly, two graphs have the same environment if and only if they have the same pins.

We assume edges have names that are independent of their endpoints, and that graphs which are to be glued together have no common edge names. Thus, $E(H \circ_L G)$ can be viewed as the disjoint union of $E(H)$ and $E(G)$. If, however, G and H have common edge names, we use the following *prefix* operator to make them disjoint: Let s and t be strings, and denote by $s \bullet t$ the concatenation of s and t where the division between s and t is not lost. Extend this notion to an operator on graphs and sets as follows. For a set A, $s \bullet A = \{s \bullet a : a \in A\}$. For a graph G whose pins are $P \subseteq V(G)$, $s \bullet G$ is a graph with edge set $s \bullet E(G)$ and vertex set $P \cup s \bullet (V(G) - P)$.

Definition 2.2 A *hierarchical graph* $\Gamma = (G_1, \ldots, G_n)$ is a finite list of simple graphs called *cells*. For each i, $V(G_i)$ is partitioned into pins, *terminals*, and *nonterminals*. G_i has p_i pins whose names are the integers $1, \ldots, p_i$. N_i denotes the set of nonterminals of G_i. Each nonterminal of G_i has a name of the form $l \bullet t$, where l is an integer, and t, its *type*, is a symbol in the set $\{G_1, \ldots, G_{i-1}\}$. A nonterminal $l \bullet G_j$ has p_j incident edges, each labeled with the name of a unique pin of G_j. Edges between nonterminals are not permitted.

Associated with $\Gamma = (G_1, \ldots, G_n)$ is a *hierarchy tree* T. Each node of T is labeled with the name of a cell of Γ. The root has label G_n. Let $v \in V(T)$ have label G_i, and assume the nonterminals of G_i are $\{1 \bullet G_{i_1}, \ldots, m \bullet G_{i_m}\}$. Then, v has m children $v_1, \ldots, v_m$ where v_j has label G_{i_j}, and edge (v, v_j) has label j. A *pathname* is a sequence $l_1 \bullet l_2 \bullet \ldots \bullet l_r$ consisting of the edge labels on a path from the root to some node of T.

Assume cell G_i contains a nonterminal $v = l \bullet G_j$. Let $v(1), \ldots, v(p_j)$ be the vertices of G_i adjacent to v, and assume for $1 \leq m \leq j$ that edge $(v, v(m))$ has label m We *replace* the nonterminal v with the graph $l \bullet G_j$ by forming the graph $(G_i - v) \circ_L l \bullet G_j$, where $L = [(v(1), 1), \ldots, (v(p_j), p_j)]$.

Definition 2.3 Let $\Gamma = (G_1, \ldots, G_n)$. Then, for $1 \leq i \leq n$, $\Gamma_i = (G_1, \ldots, G_i)$ is a hierarchical graph describing the graph $X(\Gamma_i)$, the *expansion* of Γ_i, which is defined recursively as follows: If G_i has no nonterminals, then $X(\Gamma_i) = G_i$. Otherwise, $X(\Gamma_i)$ is obtained by replacing each nonterminal $l \bullet G_j$ of G_i by the graph $l \bullet X(\Gamma_j)$.

The name of each edge (vertex) of $X(\Gamma_i)$ is a pathname in the hierarchy tree of Γ_i, followed by the name of an edge (vertex) in some cell of Γ_i. We denote $X(\Gamma_n)$ by $X(\Gamma)$. Note that $X(\Gamma)$ may have multiple edges even though its cells are simple. Shown below is a hierarchical graph with three cells, and its expansions. Pins are represented by squares and nonterminals by large circles.

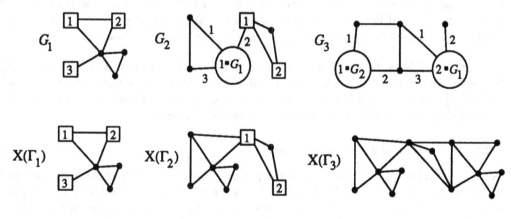

2.2 Matroids

2.2.1 Definitions and properties

A *matroid* $M = (S, I)$ is a finite set of elements S, called the *domain*, and a family I of subsets of S, called *independent sets*, that satisfy the following axioms [22]:

M1: $\emptyset \in I$.

M2: If $A \in I$ and $B \subseteq A$, then $B \in I$.

M3: For any $A \subseteq S$, all maximally independent subsets of A have the same cardinality.

Maximally independent subsets of S are *bases* of M. Subsets of S not in I are *dependent*, and minimally dependent subsets are *circuits* of M. A subset of S is independent if and only if it contains no circuit. Thus, M is uniquely determined by its bases and also by its circuits [22]. The circuits of M satisfy the following axioms [22]:

C1: No circuit is a proper subset of another circuit.

C2: If X and Y are circuits, $X \neq Y$, and $e \in X \cap Y$, then M has a circuit $Z \subseteq (X \cup Y) - e$.

Let W be the set of bases of $M = (S, I)$. Then, the set $W^* = \{S - A : A \in W\}$ is the set of bases of a matroid M^* on S called the *dual* of M. The circuits of M^* are called *cocircuits* of M.

Fact 2.1 [22] *The cocircuits of $M = (S, I)$ are those $C \subseteq S$ such that C, but no proper subset of C, has at least one element in common with every base of M.*

Definition 2.4 Let $M = (S, I)$ be a matroid, and let $A \subseteq S$. $M \uparrow A$, the *restriction* of M to A, is a matroid with domain A whose independent sets are the subsets of A that belong to I. $M - A$, the result of *deleting* A from M, is the matroid $M \uparrow (S - A)$. M/A, the result of *contracting* A from M, is a matroid with domain $S - A$. For $W \subseteq S - A$, W is an independent set of M/A if and only if $W \cup B \in I$ for some base B of $M \uparrow A$. The independence of W in M/A does not depend on any particular base B of $M \uparrow A$ [1]. That is, the phrase "for some base" can be replaced by "for every base".

A *minor* of M is the matroid obtained from M by a sequence of deletions and contractions. The minor obtained is independent of the order in which these operations are applied [1]. Thus, any minor of M can be written as $(M - D)/C$, where D is the set of all deleted elements and C is the set of all contracted elements.

2.2.2 A variant of the greedy algorithm

Let $M = (S, I)$ be a matroid with an integer or real-valued cost function c defined on its domain S. A base B of M is said to be an *optimum* base if $c(B) = \sum_{b \in B} c(b)$ is a minimum. The *minimum-cost base problem* is to find an optimum base of M, and is solved by the *Greedy Algorithm* [22] as follows:

$B := \emptyset$. While B is not a base of M, add to B a lowest cost element $e \in S - B$ such that $B \cup e \in I$.

Maximum cost bases are found by changing the phrase "lowest cost" to "highest cost".

Lawler [9] introduced a variant of the greedy algorithm that relies on deletion and contraction of matroid elements. The algorithm, which we call **L-Greedy**, is nondeterministic in that the deletion and contraction operations can be applied in any order. We shall write L-Greedy using a do-loop of *guarded commands*, as described by Dijkstra [3]. Such a do-loop is of the form

$$\textbf{do } g_1 \to S_1 \; [] \; g_2 \to S_2 \; [] \cdots [] \; g_m \to S_m \; \textbf{od}$$

where each g_i is a boolean expression called a *guard*, and each S_i is a sequence of state-ments. At the start of each loop iteration, all guards are evaluated. If no guards are true, the loop terminates. Otherwise, among the guards that are true, one is chosen and its attached statements are executed. Let M be any matroid.

Algorithm L-Greedy

$F := \emptyset$;
do an element e of M is a highest cost member of a circuit of M
$\quad \rightarrow M := M - e$;
$\quad[]$ an element e of M is a lowest cost member of a cocircuit of M
$\quad\quad \rightarrow M := M/e$; $F := F \cup e$;
od;
return F

Theorem 2.1 [9] *L-Greedy returns an optimum base of any matroid.*

By the preceding theorem, L-Greedy always computes optimum bases. It is also true that any optimum base can be computed by L-Greedy:

Lemma 2.1 *For any optimum base F of any matroid M, there is a valid execution sequence of L-Greedy that computes F.*

The following is a restatement of Theorem 2.1 and Lemma 2.1 in terms of the deletion and contraction operations used by L-Greedy.

Corollary 2.1 *For any matroid $M = (S, I)$, any $e \in S$, and any $W \subseteq S - e$:*

1. *If e is a highest cost member of some circuit of M, then W is an optimum base of $M - e$ if and only if W is an optimum base of M.*

2. *If e is a lowest cost member of some cocircuit of M, then W is an optimum base of M/e if and only if $W \cup e$ is an optimum base of M.*

3   A Greedy Algorithm for Hierarchical Graphs

Let $\mathcal{M}$ be a class of matroids defined on the edges of graphs such that, for any graph G, $\mathcal{M}(G)$ is a matroid on $E(G)$ that satisfies the following property:

D-Closure: For any $e \in E(G)$, $\mathcal{M}(G - e) = \mathcal{M}(G) - e$.

We provide an algorithm **H-Greedy** that, given Γ, returns an optimum base of $\mathcal{M}(X(\Gamma))$. Matroid classes satisfying D-Closure are discussed in Section 4. H-Greedy is based on L-Greedy and the *bottom-up method* of Lengauer [10,12], a dynamic program-ming technique designed to solve problems on hierarchical graphs.

3.1   Algorithm H-Greedy

H-Greedy manipulates *pairs* of the form $\langle G, E \rangle$, where G is a graph and E is a set of edges disjoint from $E(G)$. For a pair $\langle G, E \rangle$, let $s \bullet \langle G, E \rangle = \langle s \bullet G, s \bullet E \rangle$. The graph and edge set of a pair R will be denoted by $R.G$ and $R.E$, respectively. Let $\mathrm{Env}(R)$ be the set of all (S, L) such that $(S.G, L) \in \mathrm{Env}(R.G)$. Pair composition is defined as follows: $R \circ_L S = \langle R.G \circ_L S.G, R.E \cup S.E \rangle$.

The bottom-up method relies on a function called the *burner* [10]. The H-Greedy burner is a function B that takes a pair $\langle G, E \rangle$ and returns a pair $\mathrm{B}(\langle G, E \rangle) = \langle G^b, E \cup C \rangle$, where G^b is a minor of G obtained from some subgraph J of G by contracting $C \subseteq E(J)$. The burner is closely related to algorithm L-Greedy. An edge e of G may be deleted (contracted) provided it satisfies condition HG1 (HG2) given below:

HG1: e is a highest cost element in some circuit of $\mathcal{M}(G)$.

HG2: $\mathrm{Env}(G) = \mathrm{Env}(G/e)$ and, for every $(H, L) \in \mathrm{Env}(G)$,
- $\mathcal{M}(H \circ_L (G/e)) = \mathcal{M}(H \circ_L G)/e$, and
- e is a lowest cost element in some cocircuit $X \subseteq E(G)$ of $\mathcal{M}(H \circ_L G)$.

Algorithm H-Greedy and its burner are given below.

Algorithm H-Greedy

input: $\Gamma = (G_1, \ldots, G_n)$
output: an optimum base
of $\mathcal{M}(X(\Gamma))$

for $i := 1$ to n do
$\quad \tilde{T}_i := \langle G_i - N_i, \emptyset \rangle;$
$\quad$ for each $l \cdot G_j \in N_i$ do
$\quad\quad \tilde{T}_i := \tilde{T}_i \circ l \cdot T_j^b;$
$\quad T_i^b := \mathrm{B}(\tilde{T}_i)$
end;
return $E(T_n^b.G) \cup T_n^b.E$

function $\mathrm{B}(\langle G, E \rangle : \mathrm{pair}):\mathrm{pair}$
begin
$\quad H := G; C := E;$
$\quad$ do $e \in E(H)$ satisfies HG1
$\quad\quad \rightarrow H := H - e;$
$\quad [] \ e \in E(H)$ satisfies HG2
$\quad\quad \rightarrow H := H/e; C := C \cup e;$
$\quad [] \ \mathcal{M}(H)$ has no circuits
$\quad\quad \rightarrow$ exit do-loop
$\quad$ od;
$\quad$ Delete all isolated terminals from H;
$\quad$ return $\langle H, C \rangle$
end

For some matroids it may be difficult to find all edges satisfying HG2. Thus, the last guarded command allows the burner to terminate after no circuits remain, but before all possible contractions have been made. Isolated terminals are deleted because they are not involved in graph gluing and have no bearing on the matroids we consider.

3.2 Correctness of algorithm H-Greedy

We rely on techniques developed in [4] to prove the correctness of algorithm H-Greedy.

Definition 3.1 Let $\langle G, A \rangle$ be any pair, and let G' be obtained from some subgraph J of G by contracting $C \subseteq E(J)$. The pairs $\langle G, A \rangle$ and $\langle G', A \cup C \rangle$ are *similar*, denoted $\langle G, A \rangle \sim \langle G', A \cup C \rangle$, if and only if:

1. $\mathrm{Env}(G) = \mathrm{Env}(G')$, and
2. for every $(H, L) \in \mathrm{Env}(G)$ and every $W \subseteq E(H \circ_L G')$, W is an optimum base of $\mathcal{M}(H \circ_L G')$ if and only if $W \cup C$ is an optimum base of $\mathcal{M}(H \circ_L G)$.

Similarity is an equivalence relation on pairs. The following proves that the burner preserves similarity.

Lemma 3.1 *For any pair R, B(R) ~ R.*

Proof: Let $R = \langle G, E \rangle$, and let $e \in E(G)$ satisfy HG1 or HG2. We show that $R' \sim R$, where R' is the result of applying the appropriate guarded statements of B to R. Then, the lemma follows by the transitivity of similarity. Let $(H, L) \in \text{Env}(G)$.

Case 1: e satisfies HG1.
Then, $R' = \langle G - e, E \rangle$. By HG1, $\mathcal{M}(G)$ contains a circuit X in which e is a highest cost element. G is a subgraph of $H \circ_L G$. Thus, by D-Closure, $\mathcal{M}(H \circ_L G) \uparrow E(G) = \mathcal{M}(G)$. Therefore, every circuit of $\mathcal{M}(G)$ is a circuit of $\mathcal{M}(H \circ_L G)$. The result follows from Corollary 2.1 and since $e \in E(G)$ implies $H \circ_L (G - e) = (H \circ_L G) - e$.

Case 2: e satisfies HG2.
Then, $R' = \langle G/e, E \cup e \rangle$. By HG2, $\mathcal{M}(H \circ_L (G/e)) = \mathcal{M}(H \circ_L G)/e$, and $\mathcal{M}(H \circ_L G)$ has a cocircuit in which e is a lowest cost element. The result now follows immediately from Corollary 2.1. □

We can now prove the correctness of H-Greedy.

Theorem 3.1 *Given any hierarchical graph $\Gamma = (G_1, \ldots, G_n)$, algorithm H-Greedy returns an optimum base of $\mathcal{M}(X(\Gamma))$.*

Proof: A straightforward induction using Lemma 3.1 shows that for each i, $T_i^b \sim T_i$, where $T_i = \langle X(\Gamma_i), \emptyset \rangle$.

Let $T_n^b = \langle G_n^b, C_n \rangle$. Then, $T_n^b \sim T_n = \langle X(\Gamma), \emptyset \rangle$. G_n, G_n^b, and $X(\Gamma)$ each have p_n pins. Let H be a graph with $V(H) = \{1, 2, \ldots, p_n\}$ and $E(H) = \emptyset$. Let $L = [(1, 1), \ldots, (p_n, p_n)]$. Then, $\mathcal{M}(H \circ_L G_n^b) = \mathcal{M}(G_n^b)$ and $\mathcal{M}(H \circ_L X(\Gamma)) = \mathcal{M}(X(\Gamma))$. Thus, by the Definition 3.1, for any $W \subseteq E(G_n^b)$, W is an optimum base of $\mathcal{M}(G_n^b)$ if and only if $W \cup C_n$ is an optimum base of $\mathcal{M}(X(\Gamma))$. But, by construction, $E(G_n^b)$ is independent in $\mathcal{M}(G_n^b)$. Thus, $E(G_n^b) \cup C_n$ is an optimum base of $\mathcal{M}(X(\Gamma))$. This is precisely the set returned by H-Greedy. □

4 Matroidal Families

We shall now describe matroids on graphs whose circuits can be described by subgraphs homeomorphic to a member of a finite set of graphs. Later, we develop efficient hierarchical algorithms that compute the costs of optimum bases of these matroids.

4.1 Definitions and properties

The following definition is given by Matthews in [15].

Definition 4.1 A nonempty family of graphs $\mathcal{F}$ is called a *matroidal family* if, for any graph G, the subgraphs of G isomorphic to members of $\mathcal{F}$ are the circuits of a matroid on $E(G)$. A matroidal family $\mathcal{F}$ is *homeomorphic* if it is closed under homeomorphism: $G \in \mathcal{F}$ and H homeomorphic to G implies $H \in \mathcal{F}$.

Simões Pereira [19] found that the only homeomorphic matroidal families of finite, *connected* graphs are (1) all cycles — graphs homeomorphic from a loop, and (2) all *bicycles* — graphs homeomorphic from one of the graphs given below:

The *cycle matroid* [22] of a graph G is the matroid whose circuits are the edges of subgraphs of G homeomorphic from a loop. The *bicircular matroid* [15] of G is the matroid whose circuits are the edges of subgraphs of G homeomorphic from a bicycle.

Definition 4.2 [15] A graph *contains k independent cycles (k independent bicycles)* if the deletion of k edges is necessary and sufficient to produce a graph containing no cycles (bicycles). A graph is *k-cycle-minimal (k-bicycle-minimal)* if it contains k independent cycles (bicycles), but no proper subgraph of it does. Let C_k (B_k) denote the set of all k-cycle-minimal (k-bicycle-minimal) graphs.

Matthews [15] proved that the only homeomorphic matroidal families of *finite* graphs are the families C_k and B_k, for each positive integer k. It can be shown that for any $k > 0$, C_k and B_k each have a finite number of equivalence classes under homeomorphism. Thus, these families can be described by a finite number of homeomorphism-minimal graphs.

For a graph G and an integer $k > 0$, let $C_k(G)$ ($B_k(G)$) denote the matroid on $E(G)$ whose circuits are the subgraphs of G isomorphic to a member of C_k (B_k). Then, $C_1(G)$ is the cycle matroid of G, and $B_1(G)$ is the bicircular matroid of G.

In what follows, we shall use algorithm H-Greedy to construct efficient hierarchical algorithms for determining costs of optimum bases of $C_k(X(\Gamma))$ or $B_k(X(\Gamma))$. To use H-Greedy, we must show that these matroids classes satisfy D-Closure.

Lemma 4.1 *For any graph G, any $e \in E(G)$, and any $k > 0$, $B_k(G - e) = B_k(G) - e$ and $C_k(G - e) = C_k(G) - e$.*

Proof: We consider only $B_k(G)$. The proof is is essentially identical for $C_k(G)$. Let G be any graph and $e \in E(G)$. The k-bicycle-minimal subgraphs of $G - e$ are the k-bicycle-minimal subgraphs of G that do not contain e. The circuits of $B_k(G) - e$ are the circuits of $B_k(G)$ that do not contain e [22]. Thus, $B_k(G - e)$ and $B_k(G) - e$ have the same circuits, and therefore, are the same matroid. □

Simply deleting edges will not make our algorithms efficient because the size of an optimum base of a matroid on $X(\Gamma)$ can be as large as $X(\Gamma)$ itself. Thus, our burner must also contract edges. The following lemma describes edges e of a graph G for which $B_k(G/e) = B_k(G)/e$ and $C_k(G/e) = C_k(G)/e$.

Lemma 4.2 *Let G be a graph, $e \in E(G)$, and let $k > 0$. If e is a pendant edge or is in series with an edge f, then $C_k(G/e) = C_k(G)/e$ and $B_k(G/e) = B_k(G)/e$. In addition, if e is an isolated loop of G, then $B_k(G/e) = B_k(G)/e$.*

Proof: We consider only $B_k(G)$ — the proof is similar for $C_k(G)$. The circuits of $B_k(G)$ are the edges of k-bicycle-minimal subgraphs of G. We show that $B_k(G/e)$ and $B_k(G)/e$ have the same circuits. If e is a pendant edge or isolated loop, the result follows because e belongs to no circuits of $B_k(G)$.

Suppose e is in series with f. Then, they belong to exactly the same cycles in G, and hence, to the same circuits of $B_k(G)$. Then, for any subgraph X of G containing e, X is k-bicycle-minimal if and only if X/e is k-bicycle-minimal. Thus, the circuits of

$B_k(G/e)$ are $\{E(X/e) : X$ is a k-bicycle-minimal subgraph of $G\}$. This is precisely the set $Q = \{Y - e : Y$ is a circuit of $B_k(G)\}$. Thus, every member of Q is minimal, and, hence, Q is the set of circuits of $B_k(G)/e$ [1]. $\square$

4.2   Cost-finding algorithms

For some $k > 0$, let $\mathcal{M}$ denote the family $\mathcal{C}_k$ or $\mathcal{B}_k$. We present an algorithm **Base-Cost** which, given a hierarchical graph Γ, computes the cost of an optimum base of $\mathcal{M}(X(\Gamma))$.

Base-Cost is a straightforward modification of algorithm H-Greedy, and as such, we present only the burner here. Instead of pairs consisting of a graph and a set of edges, Base-Cost manipulates pairs R containing a graph $R.G$ and a number $R.w$. $R.w$ records the cost of a set of edges, instead of the edges themselves. Pairs are glued as follows: $S \circ_L R = \langle S.G \circ_L R.G, S.w + R.w \rangle$. Base-Cost returns $c(E(T_n^b.G)) + T_n^b.w$, where T_n^b is the last pair it computes on $\Gamma = (G_1, \ldots, G_n)$. The burner is given below.

function B($\langle G, w \rangle$: pair):pair
begin
    STEP 1: Find a subgraph H of G such that $V(H) = V(G)$ and $E(H)$
             is an optimum base of $\mathcal{M}(G)$;
    STEP 2: **for** each pendant edge e with a terminal pendant vertex **do**
             $w := w + c(e);\ H := H/e$
        end;
        for each pair of edges e and f sharing a terminal series vertex **do**
             Assume $c(e) \leq c(f)$;
             $H := H/e;\ w := w + c(e)$
        end;
        if $\mathcal{M}$ is the class $\mathcal{B}_k$ for some $k > 0$ **then**
             for each isolated loop e whose endpoint is a terminal **do**
                $H := H/e;\ w := w + c(e)$
             end;
        Delete all isolated terminal vertices from H;
        return $\langle H, w \rangle$
end

The Base-Cost and H-Greedy burners are closely related. STEP 1 is equivalent to deleting edges satisfying HG1 until no circuits remain. We shall see that STEP 2 is equivalent to repeatedly contracting edges that satisfy HG2.

4.2.1   Correctness of algorithm Base-Cost

Let Γ be a hierarchical graph. In light of Theorem 3.1 and Lemma 4.1, we need only show that the edges contracted by the burner satisfy HG2 to prove that Base-Cost correctly computes the cost of an optimum base of $\mathcal{B}_k(X(\Gamma))$ or $\mathcal{C}_k(X(\Gamma))$. Let $\mathcal{M}$ be one of the classes $\mathcal{B}_k$ or $\mathcal{C}_k$ for some $k > 0$. Let G be any graph.

Lemma 4.3 *Every edge $e \in E(G)$ contracted by the burner satisfies HG2.*

Proof: Given the burner, either (1) e is a pendant edge with a terminal pendant vertex, (2) e shares a terminal series vertex with an edge f for which $c(e) \leq c(f)$, or (3) e is an isolated loop with a terminal endpoint, and $\mathcal{M}$ is the class $\mathcal{B}_k$ for some $k > 0$.

Let $(H, L) \in \text{Env}(G)$, and let $M = \mathcal{M}(H \circ_L G)$. The case that applies to e in G also applies to e in $H \circ_L G$. If (1) or (3) applies, then e belongs to no circuits of M, and hence, by Fact 2.1, $\{e\}$ is a cocircuit of M. Suppose (2) applies. The edges e and f belong to the same cycles in $H \circ_L G$, and hence, to the same circuits of M. In addition, every base of M must contain either e or f, and M has bases containing e and not f, and vice-versa. Thus, by Fact 2.1, $\{e, f\}$ is a cocircuit of M.

In all cases, M contains a cocircuit $X \subseteq E(G)$ in which e is a lowest cost element. By Lemma 4.2, $\mathcal{M}((H \circ_L G)/e) = \mathcal{M}(H \circ_L G)/e$. Since at least one endpoint of e is not a pin of G, $(H \circ_L G)/e = H \circ_L (G/e)$. Therefore, e satisfies HG2. $\square$

We now prove the correctness of algorithm Base-Cost.

Theorem 4.1 *Let $\mathcal{M}$ be the class $\mathcal{C}_k$ or $\mathcal{B}_k$ for some $k > 0$. Then, applied to hierarchical graph Γ, algorithm Base-Cost returns the cost of an optimum base of $\mathcal{M}(X(\Gamma))$.*

Proof: Consider T_n^b, the last pair computed by algorithm Base-Cost. Since H-Greedy is nondeterministic, it follows by Lemmas 4.1 and 4.3 that $T_n^b.G = H$ and $T_n^b.w = c(E)$, where $\langle H, E \rangle$ is the last pair computed by some valid execution of H-Greedy on Γ. Since $E(H)$ and E are disjoint, it follows by Theorem 3.1 that $c(E(T_n^b.G)) + T_n^b.w$ is the cost of an optimum base of $\mathcal{M}(X(\Gamma))$. This is precisely the value returned by Base-Cost. $\square$

4.2.2 Run-time analysis

Assume the burner is applied to the pair $\langle G, w \rangle$. STEP 1 finds a subgraph H of G such that $V(H) = V(G)$ and $E(H)$ is an optimum base of the matroid defined on G. Optimum bases of the matroids we consider can be found in polynomial time using the greedy algorithm. Two special cases for which more efficient algorithms are known are the cycle matroid ($\mathcal{C}_1$) and the bicircular matroid ($\mathcal{B}_1$). A minimum spanning forest can be found in $O(|V| + |E| \log \log^* |E|)$ time [5], while an optimum base for a bicircular matroid, called a *minimum spanning pseudoforest*, can be found in linear time [6]. Next, STEP 2 of the burner contracts edges from H. A linear-time algorithm for STEP 2 can be constructed by using depth-first search [20] to implement each for-loop. Thus, for each $k > 0$, polynomial-time burners exist for $\mathcal{C}_k$ and $\mathcal{B}_k$.

We now show that the burner produces graphs whose sizes are linear in their number of pins. We provide only a brief sketch of the proof here.

Lemma 4.4 *Let G be a graph with p pins, and, for some $k > 0$, let G^b be the result of applying the burner for $\mathcal{B}_k$ or $\mathcal{C}_k$ to G. Then, G^b has $O(p + k)$ vertices and edges.*

Proof: Suppose the burner for $\mathcal{C}_k$ was applied to G. Then, $E(G^b)$ is independent in $\mathcal{C}_k(G^b)$, and thus, G^b contains at most $k - 1$ independent cycles. A graph with $k - 1$ independent cycles is a forest plus an additional $k - 1$ edges. In addition, every pendant and series vertex of G^b is a pin. An analysis of such graphs provides the desired result.

Suppose the burner for $\mathcal{B}_k$ was applied to G. It can be easily verified that a *connected* graph that is not a tree contains r independent bicycles if and only if it contains $r + 1$ independent cycles. Then, an analysis of each component of G^b using the result for $\mathcal{C}_{k+1}$ provides the desired result. $\square$

Let $\Gamma = (G_1, \ldots, G_n)$, and let $\widetilde{G}_i$ (G_i^b) be the graph in the tuple $\widetilde{T}_i$ (T_i^b) created during the ith iteration of Base-Cost for $\mathcal{C}_k$ or $\mathcal{B}_k$. By the preceding theorem, G_i^b has size $O(k + p_i)$. Since k is a constant for a particular matroid class, G_i^b has size $O(p_i)$. Recall that each nonterminal of a cell G_i has as many incident edges as there are pins in the graph that replaces the nonterminal. Thus, it follows that $\widetilde{G}_i$ has size $O(|G_i|)$. Therefore, the time required by the ith iteration of Base-Cost is $O(|G_i|)$. We have the following theorem:

Theorem 4.2 *Applied to hierarchical graph* Γ, *algorithm Base-Cost returns the cost of an optimum base of* $\mathcal{B}_k(X(\Gamma))$ *or* $\mathcal{C}_k(X(\Gamma))$ *in time polynomial in the size of* Γ.

5 Discussion

We developed a hierarchical algorithm H-Greedy for generating optimum bases of matroids defined on the expansions of hierarchical graphs. Using H-Greedy, we constructed polynomial-time hierarchical algorithms for computing costs of optimum bases of matroids from two infinite families of matroids defined on the edge sets of graphs. Straightforward modifications of H-Greedy and Base-Cost also allow for the generation of a succinct representation of an optimum base of a matroid on $X(\Gamma)$ using work-space linear in the size of Γ. One can then list the elements in the optimum base by expanding this representation. Lengauer [10] describes a similar technique for generating minimum-cost spanning trees.

H-Greedy can also be applied to the class of *head partition* matroids [17] and the class of *uniform* matroids [22]. The circuits of these matroid classes can be characterized by finite, non-homeomorphic matroidal families. It would be interesting to know whether there are other classes of problems to which H-Greedy, or a similar method, may be applied.

References

[1] R. E. Bixby, *Matroids and operations research*, Advanced Techniques in the Practice of Operations Research (H. Greenburg, F. Murphy, S. Shaw, eds.), North-Holland, New York, 1982, 333–458.

[2] J. L. Bentley, T. Ottmann and P. Widmayer, The complexity of manipulating hierarchically defined sets of rectangles, *Advances in Computing Research*, 1 (1983), 127–158.

[3] E. W. Dijkstra, *A discipline of programming*, Prentice Hall, Englewood Cliffs, N.J., 1976.

[4] D. Fernandez-Baca and M. Williams, Augmentation Problems on Hierarchically Defined Graphs, TR#89-8, Dept. of Computer Science, Iowa State Univ., August, 1989. (A preliminary version appeared in the Proceedings of WADS 89 (F. Dehne, J.-R. Sack, N. Santoro, eds.), Springer LNCS No. 382 (1989), 563–576.)

[5] H. N. Gabow, Z. Galil, T. Spencer, and R. E. Tarjan, Efficient algorithms for finding minimum spanning trees in undirected and directed graphs, *Combinatorica*, 6(2) (1986), 109-122.

331

[6] H. N. Gabow and R. E. Tarjan, A linear-time algorithm for finding a minimum spanning pseudoforest, *Information Processing Letters*, **27**(5) (1988), 259-263.

[7] H. Galperin and A. Widgerson, Succinct representations of graphs, *Information and Control*, **56** (1983), 183-198.

[8] K. Iwano and K. Steiglitz, Testing for Cycles in Infinite Graphs with Periodic Structure, *Proceedings of 19th ACM STOC*, 1987, 46-55.

[9] E. L. Lawler, *Combinatorial optimization: networks and matroids*, Holt, Rinehart and Winston, New York, 1976.

[10] T. Lengauer, Efficient algorithms for finding minimum spanning forests of hierarchically defined graphs, *J. Algorithms*, **8** (1987), 260-284.

[11] T. Lengauer and K. Wagner, The correlation between the complexities of the non-hierarchical and hierarchical versions of graph properties, Proceedings of STACS 87 (F.J. Brandenburg *et al.* eds.), *Springer LNCS No. 247* (1987), 100-113.

[12] T. Lengauer and E. Wanke, Efficient solution of connectivity problems on hierarchically defined graphs, *SIAM Journal on Computing*, **17** (1989), 1063-1080.

[13] T. Lengauer and E. Wanke, Efficient analysis of graph properties on context-free graph languages, Proceedings of ICALP 88 (T. Lepistö and A. Salomaa, eds.), *Springer LNCS No. 317* (1988), 379-393.

[14] L. R. Matthews, Bicircular Matroids, *Quart. J. Math. Oxford (2)*, **28** (1977), 213-228.

[15] L. R. Matthews, Infinite subgraphs as matroid circuits, *Journal of Combinatorial Theory, Series B*, **27** (1979), 260-273.

[16] J. B. Orlin, *Some problems on dynamic/periodic graphs*, Progress in Combinatorial Optimization, Academic Press Canada, Toronto, 1984.

[17] C. H. Papadimitriou and K. Steiglitz, *Combinatorial optimization – algorithms and complexity*, Prentice Hall, Englewood Cliffs, N.J., 1982.

[18] C. Papadimitriou and M. Yannakakis, A note on succinct representation of graphs, *Information and Control*, **71** (1986), 181-185.

[19] J. M. S. Simões Pereira, On subgraphs as matroid cells, *Math. Z.*, **127** (1972), 315-322.

[20] R. E. Tarjan, Depth-first search and linear graph algorithms, *SIAM Journal on Computing*, **1** (1972), 146-160.

[21] K. W. Wagner, The complexity of combinatorial problems with succinct input representation, *Acta Informatica*, **23** (1986), 325-356.

[22] D. J. A. Welsh, *Matroid Theory*, London Math. Soc. Monographs, Vol. 8, Academic Press, New York/London, 1976.

[23] N. L. White and W. Whiteley, A class of matroids defined on graphs and hypergraphs by counting properties, Manuscript.

[24] R. J. Wilson, *Introduction to Graph Theory*, Academic Press, New York, 1979.

Fast Algorithms for Two Dimensional and Multiple Pattern Matching (*Preliminary version*)

Ricardo Baeza-Yates
Dpto. de Ciencias de la Computación
Universidad de Chile
Casilla 2777, Santiago, Chile

Mireille Régnier
INRIA-78 153 Le Chesnay, France

Abstract

A new algorithm for searching a two dimensional $m \times m$ pattern in a two dimensional $n \times n$ text is presented. It performs on the average less comparisons than the size of the text: $\frac{n^2}{m}$ using m^2 extra space. It improves on previous results in both time: n^2 and space: n or m^2, and is nearly optimal. Basically, it uses multiple string matching on only n/m rows of the text. We also present a new multi-string searching algorithm based in the Boyer-Moore string searching algorithm. A theoretical analysis is performed relying on original methods, word enumeration and probability theory, quite new in this field, and powerful. The average number of comparisons is proved to be asymptotically $\alpha \frac{n^2}{m}$, $\alpha < 1$.

1   Introduction

Text searching is a very important component in many areas, including information retrieval, programming languages, and text processing. We address two problems: the search of two $m \times m$ dimensional patterns in two dimensional $n \times n$ objects (texts), and the multiple string searching problem: search several strings simultaneously in a text. Between the applications of the first problem, we have the detection of edges in digital pictures, local conditions in games, and other image processing and pattern recognition problems. Among the applications of the second, let us cite bibliographic searches, text editing, and other information retrieval problems.

We propose two new algorithms. Our algorithm to search two dimensional (2D) patterns (see Section 3) is based on any multiple string searching algorithm, and we also propose a new such algorithm in Section 4. Then, we extensively discuss performances. We show that using a multi-string searching algorithm yields a 2D algorithm optimal in the worst case (linear). Moreover, it improves drastically on all other algorithms in the average case: $\frac{n^2}{m}$ text-pattern comparisons with $O(m^2)$ extra-space versus $(n^2, O(n))$ or $(Kn^2, O(m^2))$, $(K >> 1)$. Hence, it is *sublinear* on the average, being the first algorithm to achieve this property. Then, we show that optimizing the multi-string searching algorithm may yield an $\alpha \frac{n^2}{m}, \alpha < 1$ cost. This claim relies on a theoretical analysis of our algorithm, in Section 5. Our approach uses original methods, such as combinatorics on words and probability theory.

2 Previous Results

Let $n \times n$ be the 2D text, and $m \times m$ be the 2D pattern. Let $q > 1$ be the size of the alphabet (in two dimensions we may be representing colors or a grey scale). Note that if we say that an algorithm takes linear time, it means n^2, the size of the text. Figure 1 shows a two dimensional pattern and text example. One occurrence of the pattern has the top-left corner at position $(2, 5)$ of the text.

```
                              1 2 3 4 5 6 7 8 9
                           1  a a a a a a b a b
         1 2 3 4 5         2  a b a b a b a b a
       1 a b a b a         3  a a a a a a a a a
       2 a a a a a         4  b b a a b b a a b
       3 b b a a b  pattern 5  a b a b a b a b a  text
       4 a b a b a         6  a a a a a a a a a
       5 a a a a a         7  b b a a b a b a a
                           8  a b a b a b a b b
                           9  a a a a a b a b a
```

Figure 1: Two dimensional searching example

The simplest algorithm for **2D searching** is to try every possible position where the pattern may match. The worst case of this naive algorithm is really bad, namely $O(m^2n^2)$ time. However, on average its behaviour is much better. For random uniform text and pattern, the expected number of comparisons per text character is [BY89b]

$$\frac{C_n}{n^2} = \frac{q}{q-1}\left(1 - \frac{1}{q^{m^2}}\right) \leq 2$$

The first worst case linear time algorithm proposed for 2D searching is due to Bird [Bir77] and, independently, to Baker [Bak78]. It uses Aho-Corasick's multi-string searching algorithm [AC75] as main component, achieving $O(n^2 + m^2)$ searching time (worst and average case). This algorithm needs $O(n + m^2)$ extra space. The basic idea is to run a finite automaton searching for the m strings that form the pattern in every row, while at the same time we run the Knuth-Morris-Pratt (KMP) string searching algorithm on every column searching for the row indices of the pattern. In the example of Figure 1 these indices are $R = (1, 2, 3, 1, 2)$. Thus, they need to memorize the state of the KMP algorithm for the current row.

Probabilistic 2D pattern matching algorithms are proposed by Karp and Rabin [KR87] However, as the constant in the running time ($O(n^2)$) is too expensive, they cannot be used in practice. In this case, the extra storage is either $O(n)$, or $O(m^2)$ by increasing the constant of the linear term in the search time.

Recently, Zhu and Takaoka [ZT89] have presented two algorithms based in Karp-Rabin's algorithm and classical string matching algorithms. Both algorithms preprocess the text computing a hash value for each entry corresponding to a subcolumn of size m, the height of the pattern. This requires $O(n^2)$ time and extra space. The first version searches a hashed version of the columns of the pattern in the preprocessed text using the Knuth-Morris-Pratt algorithm. The searching phase requires $O(n^2 + m^2)$ time. The second version uses a variant of the Boyer-Moore algorithm to search the hashed pattern in average time $O(n^2 \log m/m + m^2)$. They show that both versions are faster than Bird's algorithm **without** including the preprocessing time. Thus, the comparison is not fair, because the preprocessed text can only be reused to search patterns of the same height m; and if we allow the text to be preprocessed, we can search a pattern in $O(m^2)$ time **(the size of the pattern)** using a Patricia tree as index that also uses $O(n^2)$ extra space [Gon88]. In fact, in our experimental results we show that the 8 to 1 ratio presented in [ZT89] goes down to less than 2 to 1, if preprocessing is counted.

Table 1 shows the time and space order of all these algorithms compared with our results.

Algorithm	Worst Case	Average Case	Extra space
Naive	$m^2 n^2$	$\frac{q}{q-1} n^2$	1
Bird, Baker	$n^2 + m^2$	$n^2 + m^2$	$n + m^2$
Karp and Rabin	$K(n^2 + m^2)$	$K(n^2 + m^2) \ (K \gg 1)$	m^2
Zhu and Takaoka 1	$n^2 + m^2$	$n^2 + m^2$	n^2
Zhu and Takaoka 2	$n^2 + m^2$	$n^2 \log m/m + m^2$	n^2
Ours (naive)	mn^2	$n^2/m + m^2$	m^2
Ours (improved)	$n^2 + m^3 + q$	$\alpha n^2/m + m^3 + q \ (\alpha < 1)$	$m^2 + q$

Table 1: Comparison of 2D pattern matching algorithms

The best known algorithm to **search for a set of strings** is due to Aho and Corasick [AC75]. It is based on automata theory and the Knuth-Morris-Pratt [KMP77] string matching algorithm. For L strings of size m and a text of length n, this algorithm runs in time $O(n + m L)$ using $O(m L)$ extra space. Using the key idea of the Boyer-Moore [BM77] string matching algorithm of searching from *right to left* in the string, Commentz-Walter [CW79] proposed a faster algorithm on the average. Variations on this algorithm are proposed by Sridhar [Sri86]. A different algorithm based in boolean matrices is given in [Shy76].

3   A New Two Dimensional Searching Algorithm

We consider the pattern as m superposed strings p_i of length m. Like Bird's algorithm we decompose the search in a one dimensional (say horizontal) multi-string search. However, our **key idea** is to perform the horizontal search only on n/m rows of the text, possibly followed by vertical searches when matches occur. These rows cover all possible positions where the pattern may occur (see example in Figure 3). This allows a drastic improvement of the average case: $\frac{n^2}{m}$ comparisons, and reduces the extra-space to $O(m^2)$. Figure 2 shows the horizontal search in pseudocode, where R is an array of unique indices (see previous example), and A is any multi-string searching automata whose output is a pattern index. For every pattern that appears in one of the n/m rows, the function *Checkmatch* is called to perform the vertical search. This function returns a set with all the occurrences at that point, if any.

```
Search( (text, n × n), (pattern, m × m) )
{
    (S, R) ← Uniquepatterns( pattern )
    (A, q₀,Output) ← Automaton( S )
    for( k ← m; k ≤ n; k ← k + m )
    {
        q ← q₀
        for( j ← 1; j ≤ n; j ← j + 1 )
        {
            q ← A(q, text[k, j])
            if( Output(q) ≠ 0 )
            {
                Rows ← Checkmatch( k, j − m + 1, Output(q)))
                if( |Rows| > 0 )
                    print( match at position (Rowsᵢ₌₁..|Rows|, j − m + 1) )
            }
        }
    }
}
```

Figure 2: New 2D searching algorithm

3.1 Vertical Search and Overlappings

Let us precise the vertical algorithm. For every occurrence of a string p_i in one of the rows, we check if one potential matching, or more (if strings are repeated in the pattern) are true matchings. That is, we check the appearance above or/and below of rows p_{i-1}, p_{i+1}, etc. (see Figure 3 for some examples). If so, the algorithm outputs a match.

A special case occurs when a string is repeated in the pattern ($p_i = p_j$). For example, in the last potential match of row $2m$ in Figure 3. As we said, if we have repeated strings, we search horizontally for the different ones only. If such a string is found, we may store the index of the last complete row found above/below p_i. Thus, before starting comparisons associated to p_j, we check if some complete rows previously matched. This vertical search algorithm is shown in Figure 4.

Experimental results of this algorithm using Aho-Corasick's multi-string searching

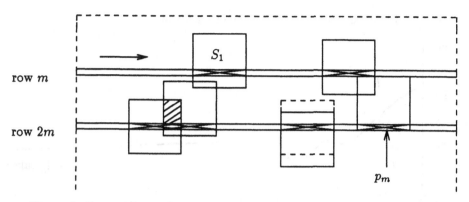

Figure 3: Potential matches for the checking phase of the 2D algorithm

```
Checkmatch( row, column, index ) % text, pattern, A, and R are global
{
      Fᵢ ← 0 for i = 1, ..., 2m − 1
      Fₘ ← index
      Rows ← Ø
      for( i ← 1; i < m; i ← i + 1 )
      {
            if( Rᵢ ≠ index ) continue;
            j ← m − i + 1
            for( r ← row − i + 1; r ≤ row − i + m and r ≤ n; r ← r + 1 )
            {
                  if( Fⱼ = R_{r−row+i} ) continue;
                  q ← q₀
                  for( c ← column; c ≤ column − m + 1; c ← c + 1 )
                        q ← A(q, text[r, c])
                  Fⱼ ← Output(q)
                  if( Fⱼ ≠ R_{r−row+i} ) break;
            }
            if ( r = row − i + m + 1 ) Rows ← Rows ∪ {r − m}
      }
}
```

Figure 4: New 2D searching algorithm

algorithm, and the improvement for vertical overlapping are shown in Figure 5. In this figure, we compare our algorithm with the Naive algorithm, Bird's algorithm, and the two algorithms presented by Zhu and Takaoka. In all cases we have the average time of ten runs to search a random pattern in a random text of size 1000×1000 and a binary alphabet (a typical bit-mapped graphical screen). Our algorithm is the best for $m > 2$.

Interestingly, the brute-force algorithm is the best for $m = 2$.

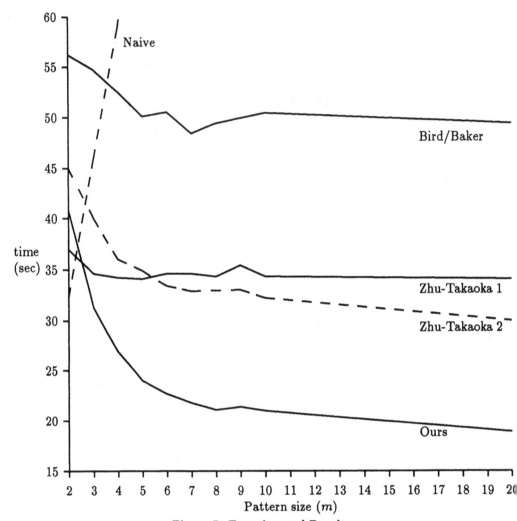

Figure 5: Experimental Results

Let us discuss now the complexity of this version. As discussed below, the horizontal search cost is the dominant term in the average complexity. We will improve it later by using a new automaton for multiple string searching based on Horspool's [Hor80] variant of the Boyer-Moore [BM77] string searching algorithm (see Section 4). We consider here the worst case. The total number of positions where we can find the string p_i is n^2/m ($\lfloor n/m \rfloor$ rows with $n - m + 1$ positions each). Hence, using Aho-Corasick's algorithm, or any automaton, for the horizontal search, we need in the worst case n^2/m comparisons. With a naive vertical search, $m(m-1)$ comparisons are performed for any matching p_i found which yields a $m^2(m-1)$ worst case ($p_1 = \ldots = p_m$), or, for the whole text, a n^2m^2 worst case. However, with our improvement, independently on the pattern, at most $m(2m-1)$ comparisons are performed in a vertical match. This gives a $2mn^2$ worst case.

On average, in every row searched, an occurrence of a pattern row is found in a given position with probability m/q^m. Therefore, a first bound for the average time is $n^2/m + n^2/m \times m/q^m \times (2m - 1)m = O(n^2/m + n^2m^2/q^m)$, or $O(n^2/m)$.

3.2 Horizontal Overlappings

If two strings p_i, p_j overlap on a primary row (for example, in Figure 3, the first and second string matchings in row $2m$), the common region in secondary lines may be searched twice. To improve this and not to repeat comparisons, we delay the vertical comparisons until we know if a horizontal overlapping occurs. If so, we perform first the comparisons in the overlap (see Figure 3).

This improvement guarantees a n^2 worst case. Indeed, no character in secondary lines is read twice when a primary row is searched. Nevertheless, some characters may be read twice when searching two primary rows (1-st matching region S_1 on row m and 2-nd matching region on row $2m$ overlap). But then, there exists one unread (possibly partial) row above S_1, and hence one unread character is associated to any character read twice. (Repercussion on more than one row is possible). To achieve this, we store information about all possible overlappings of the pattern (m^2), such that we know in $O(1)$ time if the common region may belong to two occurrences of the pattern.

Remark that n^2, the size of the text, is a theoretical lower bound on the number of comparisons. That is, if $p_1 = \ldots = p_m$ is a sequence of m a's while text t only contains character a, all possible algorithms have to perform n^2 comparisons. It was achieved by [Bir77] and [Bak78], to the cost of a $O(n)$ extra-space. Note that this bound is attained for all patterns and texts, while our algorithm attains it for only a few configurations. It is **optimal in the worst case**, with a very reduced extra-space, namely the size m^2 of the pattern. No previous algorithm uses less space, except for the naive algorithm, clearly non optimal in time. We delay after the analysis the discussion over the **average optimality**.

We can further improve the previous algorithm on the average, by storing information about partial matchings between two rows. However, this is not worth doing it in practice, as it increases the extra space to $O(n)$.

4   The New Multi-string Searching Algorithm

The main component of the algorithm presented above is a multiple string searching routine. We follow Commentz-Walter's approach [CW79], but using also the Horspool's [Hor80] variant of the Boyer-Moore [BM77] algorithm. The algorithm obtained is simpler, and also faster in practice, using the same extra space as previous solutions. Other variants are presented in Sridhar [Sri86], where $O(n \log M)$ time and $O(\log M)$ space is achieved, with M being the length of the longest pattern.

The algorithm aligns the end of the strings with the first possible matching position of the text (see Figure 6), and starts comparing the text with the set of strings, from right to left. We continue until we find all possible strings matching that position, or realize that none of them are there. The knowlegde of these read characters yields the next possible matching position to the right, as explained below. The set of strings is aligned, or *shifted* with this new position.

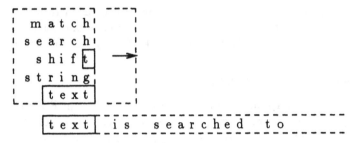

Figure 6: Searching a set of strings from right to left

Let L be the number of strings p_i with sizes m_i. We denote by $m = \min_i(m_i)$. As in Commentz-Walter's algorithm, we first build a trie of the reversed strings, distinguishing final nodes with the index i of the corresponding string p_i. This trie allows us to follow what strings are being matched at any point. Figure 7 shows this trie for the set of strings *match, search, shift, string, text* ($m = 4$). The final nodes are labelled by the corresponding string index.

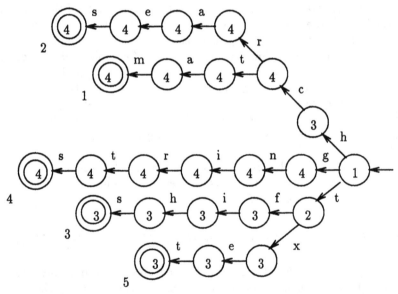

Figure 7: Trie for a set of reversed strings

All nodes are labelled by the *shift* (this is equivalent to the dd table used in the Boyer-Moore algorithm). Theoretically, all information known on the left from previous comparisons could be used. Practically, we consider two heuristics, good when q is small or big.

The *match* heuristic is important when q is small. The set of strings is moved to the first position to the right such that, among the m characters immediately to the left:

1. *all* the characters matched in the last right to left search match at least one string, and

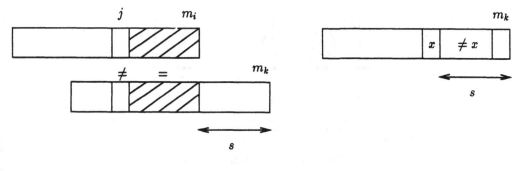

(a) *dd* table   (b) *d* table

Figure 8: Computation of the *dd* and *d* shifts

2. a *different* character is brought to the position in the text that caused the mismatch.

The formal definition, generalizing [BM77] and [KMP77], is:

$$dd(p_i[j]) = \min_{k=1}^{L}\{s|s \geq 1 \text{ and } (s \geq j + m_k - m_i \text{ or } p_k[j - s + m_k - m_i] \neq p_i[j]) \text{ and}$$

$$((s \geq \ell + m_k - m_i \text{ or } p_k[\ell - s + m_k - m_i] = p_i[\ell]) \text{ for } j < \ell \leq m_i)\} ,$$

where $dd(p_i[j])$ denotes the shift associated to the node where we have matched $p_i[j + 1]...p_i[m_i]$ and mismatched $p_i[j]$, measured from $p_i[m_i]$ (root of the trie). See Figure 8(a).

The *occurrence* heuristic, good enough when q is big, is simpler. A single known character in the text is aligned with the first character of the string that matches it. It may be the mismatching character. However, as Horspool noted [Hor80], any of the characters from the text can be used. We make use of a table d, formally defined as

$$d[x] = \min_{k=1}^{L}\{s|s = m_k \text{ or } (1 \leq s < m_k \text{ and } p_k[m_k - s] = x)\} .$$

See Figure 8(b). Table 2 gives the characters with value $d[x] = i$ for $i = 1, ..., 4$ for the same example of Figure 7.

x	c,f,n,x	e,i,r,t	a,h	others
$d[x]$	1	2	3	4.

Table 2: Values for table *d*

The algorithm works as follows:

- We align the root of the trie with position m in the text, and we start matching the text from right to left following the corresponding path in the trie.

- If a match is found (final node), we output the index of the corresponding string.

- After a match or mismatch, we move the trie further in the text using the maximum of the shift associated to the current node, and the value of $d[x]$, where x is the character in the text corresponding to the root of the trie.

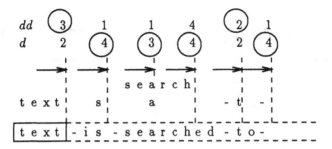

Figure 9: A searching example

- Start matching the trie again from right to left in the new position.

An example is shown in Figure 9. Only one comparison is repeated. This yields 7 less character inspections than Aho-Corasick's algorithm. In practice we can implement the root of the trie with a full table (one entry per alphabet symbol), and all the other nodes using a linked list or a sorted table.

Both shifts can be precomputed based solely on the string and the alphabet. Hence, the space needed is $O(m\,L+q)$. By using the standard preprocessing of the Boyer-Moore algorithm (see [KMP77, Ryt80]) the computation of the dd shifts requires $O(L\sum_i m_i)$, and the computation of the d table requires $O(\sum_i m_i + q)$ time.

5 Multi-string Searching Algorithm Analysis

So far, no average analysis is available for multiple string searching algorithms. Note only that, using a suitable automaton, can always yield n, the size of the text, as an upper bound on the number of comparisons.

The analysis relies first upon algebra and word enumeration or combinatorics. This approach, used in recent works [BYGR90, Reg89] appeared powerful for classical string searching algorithms: Boyer-Moore and Knuth-Morris-Pratt. It allowed to solve the open problems of the average performances of these algorithms. Nevertheless, new theoretical problems arise for multiple string searching.

To simplify the problem, we only analyze the effect of the d table, i.e. the occurrence heuristic (this gives an upper bound, and in practice, strings are not periodic or too similar), and we assume that all the strings have the same length $m_i = m + 1$ ($m + 1$ because the last character is not considered by the d heuristic). We first consider the case of a given set of strings $\{p_i\}$. Then, we generalize part of our results when $\{p_i\}$ ranges over all possible L-sets. Following [BYGR90] we first count the characters in the text that are starting points of a right to left comparison, exhibiting a stationary process, and second, the average number of comparisons performed from these points. Hence, we state:

Definition: A character x in the text is a **head** iff it is read immediately after a shift.

THEOREM 5.1 *[BYGR90] For a given fixed set of L strings $\{p_i\}$ of length $m+1$, let H_n be the probability that the n-th character of the text be a head. Then, H_n converges to a* *stationary probability H_p^∞ defined by:*

$$H_p^\infty = \frac{1}{E_{\{p_i\}}[shift]} \ .$$

Let us now introduce some formalism.

Definition: To any given set of strings is associated in a unique manner some number j and two sequences $(k_i)_{1 \le i \le j}$, $(\sigma_i)_{1 \le i \le j}$ such that:

$$1 \le j \le q, \ 1 \le k_i \ , \ 1 \le \sigma_i$$

and σ_1 (resp. σ_i) characters have their first occurrence at position 1 (resp. $k_1 + \ldots + k_{i-1}$), counted from the right.

Example: $q = 5$, $p_1 = cebccbax$, $p_2 = edcacbbx$, where x is any of the q characters (the last character is not considered for the shift). Here: $(\sigma_3, \sigma_2, \sigma_1) = (2, 1, 2)$, $(k_3, k_2, k_1) = (2, 3, 2)$, as two characters ($a$ and b) occur at position 2 (counting from the right) in some string, one (c in p_1) at position $4 = 2 + 2$, two (d and e) at position $7 = 3 + (2 + 2)$; and $j = 3$. See Figure 10.

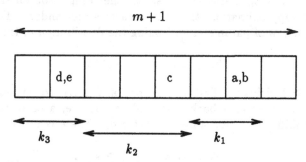

Figure 10: Definition of parameters for the analysis

Lemma 5.1 *For a given set of strings, the expected shift satisfies:*

$$qE_{\{p_i\}}[shift] = \Sigma_j + \sum_{i=1}^{j-1} k_i(\Sigma_j - \Sigma_i) + (q - \Sigma_j)(m + 1) \ ,$$

with $\Sigma_i = \sigma_1 + \ldots + \sigma_i$.

Proof: For any of the σ_i characters occurring at the i-th position, the shift is:

$$s_i = 1 + k_1 + \ldots + k_{i-1} \ .$$

For the $(q - \Sigma_j)$ characters not occurring in the strings, the shift is $m + 1$. Hence:

$$
\begin{aligned}
qE_{\{p_i\}}[shift] &= \sum_{i=1}^{j} s_i \sigma_i + (q - \Sigma_j)(m + 1) \\
&= \Sigma_j + \sum_{i=2}^{j}(k_1 + \ldots + k_{i-1})\sigma_i + (q - \Sigma_j)(m + 1) \\
&= \Sigma_j + \sum_{i=1}^{j-1} k_i(\Sigma_j - \Sigma_i) + (q - \Sigma_j)(m + 1) \ .
\end{aligned}
$$

We are now ready to derive:

THEOREM 5.2 *Let $C_n(\{p_i\})$ be the expected number of text-string comparisons for a given subset of strings $\{p_i\}$ and a random text t.*

$$\frac{C_n(\{p_i\})}{n} = H_p^\infty \left(1 + \phi(q, L) + \frac{\sigma_1}{q}[(1 - 1/q^2)^L - (1 - 1/q)^L] + O\left(\frac{L}{q^3}\right)\right)$$

with

$$\phi(q, L) = \sum_{k \geq 1}(1 - (1 - q^{-k})^L) ,$$

and σ_1 being the number of characters appearing in the rightmost position in at least one pattern.

Proof: The expression of H_p^∞ derives from the preceding lemma. Then, one counts the number of comparisons S_m between two shifts, the comparison on the head excepted. Assume, in a first approximation, that all characters are random. For any string, S_m follows a geometric law of parameter $\frac{1}{q}$. Taking the distribution of the *maximum* yields:

$$P(S_m \leq k) = (1 - q^{-k})^L .$$

Summing over $k \geq 1$ yields the first approximation $\phi(q, L)$. Nevertheless, the last head position may be attained in a backward reading. Hence, a correcting term appears. Assume the last shift was s, for some character a_i. There is no correction if $s_i = m$. Else:

$$P(S_m \geq s) = P(S_m \geq s + 1) ,$$

which yields the correction:

$$(1 - \frac{1}{q^{s+1}})^L - (1 - \frac{1}{q^s})^L .$$

Moreover, $P(S_m \geq s + k), k \geq 2$ is also to be corrected. Assume several chains c_j, each appearing β_j times in the L patterns, have a_i in position s. One, and one only, has been found. Hence:

$$P(S_m \geq s + k) = \frac{1}{q^s}P_{\beta_j}(Comp \geq k - 1) = \frac{1}{q^s}(1 - (1 - q^{-(k-1)})^{\beta_j}) .$$

As a_i appears with probability $\frac{1}{q}$, we get the term:

$$\frac{1}{q} \cdot \frac{1}{q^s} \sum_j \sum_{k \geq 1}(1 - (1 - q^{-k})^{\beta_j}).$$

We get the order of magnitude of the correcting term by considering the shift $s = 1$. One has σ_1 chains, and $\sum \beta_j = L$. As:

$$\sum_{a_1} \sum_j (1 - (1 - 1/q)^{\beta_j}) = O(\sum \sum \beta_j/q) = O(\frac{L}{q}) ,$$

we get the result. ∎

We turn now to the average performance of our multi string searching algorithm. We first study the average shift.

THEOREM 5.3 *The expected value of the shift, when the L strings range over all possible strings, is:*

$$E_{all\ strings}[E_{\{p_i\}}[shift]] = (1-Q)^{-1}\left(1 - Q^{m-1}[m-1+(m-2)Q]\right)$$

with $Q = \left(\frac{q-1}{q}\right)^L$, *which reduces, when* $m- > \infty$ *to:*

$$\frac{1}{1-\left(\frac{q-1}{q}\right)^L}.$$

Remark: When $L = 1$, this reduces to q, as proved for the single string searching algorithm in [BYGR90]. When $L- > \infty$, this reduces to 1: all characters are found as a rightmost letter in at least one string.

Proof: A given character a in Σ appears in a given string p_i, in some position X_i, counted from the right. X_i ranges in $\{1..m\}$ and, if $k < m$,

$$Pr(X_i > k) = Pr(a \text{ does not appear before position } k) = (1 - \frac{1}{q})^k.$$

For a given set of L strings, the rightmost position of a is $\min(X_1, \ldots, X_L)$. Variables X_i are independent. Hence:

$$Pr(shift \le k) = 1 - Pr(shift > k) = 1 - \prod_{i=1}^{L} P(X_i > k)$$

and:

$$Pr(shift = m) = 1 - \sum_{1}^{m-1} Pr(shift = k) = \left(\frac{q-1}{q}\right)^{L(m-1)}.$$

∎

Unfortunately, this is not enough to derive the average number of comparisons. Notably, the average value of the stationary probability H_p^∞ is of interest. In general, $E(\frac{1}{x}) \ge \frac{1}{E(x)}$. We conjecture that the equality holds here. In any case, we can provide a tight upper bound.

THEOREM 5.4 *Let us note* $H_{(m,q,L)}$ *the probability of being a head, when* $\{p_i\}$ *ranges over all possible L-sets of patterns of length* $m+1$ *on a q-ary alphabet. Let* $Ph(q, L)$ *be* $\lim_{m\to\infty} H_{(m,q,L)}$. *Then:*

$$(1 - (1 - \frac{1}{q})^L) \le Ph(q, L) \le \frac{2L}{q+L}, L < q.$$

Remark: $(1 - (1 - \frac{1}{q})^L) \sim \frac{L}{q}$.

Proof: For any set $\{p_i\}$, at most L different letters appear in positions: $1, 2, \ldots, \frac{q}{L}$. Then, we get the lower bound:

$$q E_{\{p_i\}}[shift] \geq L + 2L + \ldots + (q/L)L = \frac{q}{2}(\frac{q}{L} + 1) \ .$$

Hence, the upper bound for H_p^∞ and $Ph(q, L)$. The lower bound is a consequence of Jensen's inequality [Fel68] and the previous theorem. ∎

Practically, computations in [BYGR90], where the case $L = 1$ is considered, show that the lower bound is very tight. Hence, we *conjecture:*

THEOREM 5.5 *(conjectured) When $q- > \infty$, the stationary probability H_p^∞ satisfies:*

$$\lim_{q \to \infty} \frac{H_p^\infty}{Ph(q, L)} = 1 \ .$$

Let us remark that, in any case, our upper bound provides a $\alpha n^2/m, \alpha < 1$ cost, for our 2D searching algorithm. This is better than all previous algorithms, as it appears from the state of the art in Section 2, and the experimental results. We believe that the constant α may be slightly improved, using for instance a suitable automaton. Nevertheless, the main improvement comes from searching only n/m rows which yields the $1/m$ improvement factor. Hence, our algorithm is nearly optimal on the average. Remark that we have neglected the vertical search in this discussion, as it obviously contributes by a $\frac{1}{q} \frac{n}{m}$ factor on average.

6   Conclusions and Extensions

We have proposed a fast algorithm for two dimensional pattern matching, optimal in the worst case, where the main differences with previous solutions are:

- The multi-string searching algorithm runs on n/m rows only. This is an improvement of a $1/m$ factor.

- The multi-string searching algorithm requires less than 1 comparison per character on average.

Both algorithms are nearly optimal on average. If we consider the pattern as a one dimensional string of size m^2 we have a $O(n^2 \log m/m^2)$ lower bound on average [Yao79]. This algorithm is easily extended in the following ways:

- Arbitrary shapes: the pattern is decomposed in strings of different length, and their relative positions are remembered. If a pattern has height longer than width, we should use the automaton through columns instead of rows. A slightly different problem is solved in [AF89] with an $O(\sqrt{mn} \log m)$ algorithm.

- Higher dimensions: we use recursively a $(d - 1)$-dimensional pattern matching algorithm to search in n/m rows of the d-th dimension.

- Multi 2-dimensional pattern matching: The set of patterns is decomposed in strings, and all patterns containing a given string are checked vertically. Comparisons do not need to be repeated if information about pattern overlappings is stored.

Faster searching for 2D and multiple string searching can be done using an index, for example a Patricia tree (see [Gon88, BY89a]). This is worth being done, if we have a text that does not change or if the amount of searching amortizes the preprocessing time needed to construct the index.

We have also derived a theoretical analysis for the average case. The methods are original and combine algebra and probabilities. This leads to an interesting conjecture.

Acknowledgements

We wish to thank Luis Fuentes for the help in obtaining the experimental data.

References

[AC75]   A.V. Aho and M. Corasick. Efficient string matching: An aid to bibliographic search. *C.ACM*, 18(6):333–340, June 1975.

[AF89]   A. Amir and M. Farach. The whole is faster than its parts: Efficient matching of non-rectangular shapes. Technical Report CS-TR-2256, Dept. of Computer Science, Univ. of Maryland, 1989.

[Bak78] T. Baker. A technique for extending rapid exact string matching to arrays of more than one dimension. *SIAM J on Computing*, 7:533–541, 1978.

[Bir77] R. Bird. Two dimensional pattern matching. *Inf. Proc. Letters*, 6:168–170, 1977.

[BM77]   R. Boyer and S. Moore. A fast string searching algorithm. *C.ACM*, 20:762–772, 1977.

[BY89a] R.A. Baeza-Yates. *Efficient Text Searching*. PhD thesis, Dept. of Computer Science, University of Waterloo, May 1989. Also as Research Report CS-89-17.

[BY89b] R.A. Baeza-Yates. String searching algorithms revisited. In F. Dehne, J.-R. Sack, and N. Santoro, editors, *Workshop in Algorithms and Data Structures*, pages 75–96, Ottawa, Canada, August 1989. Springer Verlag Lecture Notes on Computer Science 382.

[BYGR90] R. Baeza-Yates, G. Gonnet, and M. Regnier. Analysis of Boyer-Moore-type string searching algorithms. In *1st ACM-SIAM Symposium on Discrete Algorithms*, pages 328–343, San Francisco, January 1990.

[CW79]   B. Commentz-Walter. A string matching algorithm fast on the average. In *ICALP*, volume 6 of *Lecture Notes in Computer Science*, pages 118–132. Springer-Verlag, 1979.

[Fel68]   W. Feller. *An Introduction to Probability Theory and Its Applications*, volume 1. John Wiley and Sons, New York, 1968.

[Gon88]   G.H. Gonnet. Efficient searching of text and pictures (extended abstract). Technical Report OED-88-02, Centre for the New OED., University of Waterloo, 1988.

[Hor80]   R. N. Horspool. Practical fast searching in strings. *Software - Practice and Experience*, 10:501–506, 1980.

[KMP77]   D.E. Knuth, J. Morris, and V. Pratt. Fast pattern matching in strings. *SIAM J on Computing*, 6:323–350, 1977.

[KR87]   R. Karp and M. Rabin. Efficient randomized pattern-matching algorithms. *IBM J Res. Development*, 31:249–260, 1987.

[Reg89]   M. Regnier. Knuth-Morris-Pratt algorithm: An analysis. In *MFCS'89, Lecture Notes in Computer Science 379*, pages 431–444, Porabka, Poland, August 1989. Springer-Verlag. Also as INRIA Report 966, 1989.

[Ryt80]   W. Rytter. A correct preprocessing algorithm for Boyer-Moore string-searching. *SIAM J on Computing*, 9:509–512, 1980.

[Shy76]   R.K. Shyamasundar. A simple string matching algorithm. Technical report, Nat. Centre for Software Development and Computing Techniques, Tata Institute of Fundamental Research, Bombay, India, 1976.

[Sri86]   M.A. Sridhar. Efficient algorithms for multiple pattern matching. Technical Report Computer Sciences 661, University of Wisconsin-Madison, 1986.

[Yao79]   A.C. Yao. The complexity of pattern matching for a random string. *SIAM J on Computing*, 8:368–387, 1979.

[ZT89]   R.F. Zhu and T. Takaoka. A technique for two-dimensional pattern matching. *Communications of the ACM*, 32(9):1110–1120, September 1989.

BOYER-MOORE APPROACH TO
APPROXIMATE STRING MATCHING

(Extended Abstract)

Jorma Tarhio and Esko Ukkonen

University of Helsinki, Department of Computer Science
Teollisuuskatu 23, SF-00510 Helsinki, Finland

Abstract. The Boyer-Moore idea applied in exact string matching is generalized to approximate string matching. Two versions of the problem are considered. The k mismatches problem is to find all approximate occurrences of a pattern string (length m) in a text string (length n) with at most k mismatches. Our generalized Boyer-Moore algorithm solves the problem in expected time $O(kn(1/(m-k) + k/c))$ where c is the size of the alphabet. A related algorithm is developed for the k differences problem where the task is to find all approximate occurrences of a pattern in a text with $\leq k$ differences (insertions, deletions, changes).

1. Introduction

The fastest known exact string matching algorithms are based on the Boyer-Moore idea [BoM77, KMP77]. Such algorithms are "sublinear" on the average in the sense that it is not necessary to check every symbol in the text. The larger is the alphabet and the longer is the pattern, the faster the algorithm works. In this paper we generalize this idea to approximate string matching. Again the approach leads to algorithms that are significantly faster than the previous solutions of the problem.

We consider two important versions of the approximate string matching problem. In both, we are given two strings, the *text* $T = t_1t_2...t_n$ and the *pattern* $P = p_1p_2...p_m$ in some alphabet Σ, and an integer k. In the first variant, called the *k mismatches problem*, the task is to find all occurrences of P in T with at most k mismatches, that is, all j such that $p_i = t_{j-m+i}$ for $i = 1, ..., m$ except for at most k indexes i.

In the second variant, called the *k differences problem*, the task is to find all substrings P' of T with the edit distance at most k from P. The edit distance means the minimum number of editing operations (the differences) needed to convert P' to P. An editing operation is either an insertion, a deletion or a change of a character. The k mismatches problem is a special case with the change as the only editing operation.

There are several algorithms proposed for these two problems, see e.g. the survey [GaG88]. Both can be solved in time $O(mn)$ by dynamic programming [Sel80, Ukk85b]. A very simple improvement giving $O(kn)$ expected time solution for random strings is described in [Ukk85b]. Later, Landau and Vishkin [LaV88, LaV89], Galil and Park [GaP89], Ukkonen and Wood [UkW89] have given different algorithms that consist of preprocessing the pattern in time $O(m^2)$ (or $O(m)$) and scanning the text in worst-case time $O(kn)$. For the k differences problem, $O(kn)$ is the best bound currently known if the preprocessing is allowed to be at most $O(m^2)$. For the k mismatches problem Kosaraju [Kos88] gives an $O(n\sqrt{m}\,polylog(m))$ algorithm. Also see [GaG86, GrL89].

We develop a new approximate string matching algorithm of Boyer-Moore type for the k mismatches problem and show, under a mild independence assumption, that it scans a random text in expected time $O(kn(\frac{1}{m-k} + \frac{k}{c}))$ where c denotes the size of the alphabet. A related method is (independently) given in

[Bae89a]. We also give an algorithm for the k differences problem and show in a special case that its expected scanning time for a random text is $O(\frac{c}{c-2k} kn (\frac{k}{c+2k^2} + \frac{1}{m}))$. The preprocessing of the pattern needs time $O(m + kc)$ and $O((k + c)m)$, respectively. We have also performed experimental comparison of the new methods with the old ones showing that the Boyer-Moore algorithms are significantly faster, for large m and c in particular, than previous algorithms.

Our algorithms can be considered as generalizations of the Boyer-Moore algorithm for exact string matching, because they are functionally identical with the Horspool version [Hor80] of the Boyer-Moore algorithm when $k = 0$.

The paper is organized as follows. We first consider the k mismatches problem for which we give the Boyer-Moore solution in Section 2. Section 3 develops an extension to the k differences problem. Section 4 reports our experiments. The full version of the paper appears as [TaU90].

2. The k mismatches problem

2.1. Boyer-Moore-Horspool algorithm

The characteristic feature of the Boyer-Moore algorithm [BoM77] for exact matching of string patterns is the right-to-left scan over the pattern. At each alignment of the pattern with the text, characters of the text below the pattern are examined from right to left, starting by comparing the rightmost character of the pattern with the character in the text currently below it. Between alignments, the pattern is shifted from left to right along the text.

In the original algorithm the shift is computed using two heuristics: the match heuristic and the occurrence heuristic. The *match* heuristic implements the requirement that after a shift, the pattern has to match all the text characters that were found to match at the previous alignment. The *occurrence* heuristic implements the requirement that we must align the rightmost character in the text that caused the mismatch with the rightmost character of the pattern that matches it. After each mismatch, the algorithm chooses the larger shift given by the two heuristics.

As the patterns are not periodic on the average, the match heuristic is not very useful. A simplified version of the method can be obtained by using the occurrence heuristic only. Then we may observe that it is not necessary to base the shift on the text symbol that caused the mismatch. Any other text character below the current pattern position will do as well. Then the natural choice is the text character corresponding to the rightmost character of the pattern as it potentially leads to the longest shifts. This simplification was noted by Horspool [Hor80]. We call this method the Boyer-Moore-Horspool or the BMH algorithm.

The BMH algorithm has a simple code and is in practice better than the original Boyer-Moore algorithm. In the preprocessing phase the algorithm computes from the pattern $P = p_1 p_2 ... p_m$ the shift table d, defined for each symbol a in alphabet Σ as

$$d[a] = min\{s \mid s = m \text{ or } (1 \leq s < m \text{ and } p_{m-s} = a)\}.$$

For a text symbol a below p_m, the table d shifts the pattern right until the rightmost a in $p_1...p_{m-1}$ becomes above the a in the text. Table d can be computed in time $O(m + c)$ where $c = |\Sigma|$, by the following algorithm:

Algorithm 1. BMH-preprocessing.
for a in Σ do $d[a] := m$;
for $i := 1, ..., m - 1$ do $d[p_i] := m - i$

The total BMH method [Hor80] including the scanning of the text $T = t_1 t_2 ... t_n$ is given below:

Algorithm 2. The BMH method for exact string matching.
call Algorithm 1;
$j := m$;   (pattern ends at text position j)
while $j \le n$ **do begin**
   $h := j; i := m$;   (h scans the text, i the pattern)
   while $i > 0$ **and** $t_h = p_i$ **do begin**
      $i := i - 1; h := h - 1$ **end**;   (proceed to the left)
   if $i = 0$ **then** report match at position j;
   $j := j + d[t_j]$ **end**   (shift to the right)

2.2. Generalized BMH algorithm

The generalization of the BMH algorithm for the k mismatches problem will be very natural: for $k = 0$ the generalized algorithm is exactly as Algorithm 2. Recall that the k mismatches problem asks for finding all occurrences of P in T such that in at most k positions of P, T and P have different characters.

We have to generalize both the right-to-left scanning of the pattern and the computation of the shift. The former is very simple; we just scan the pattern to the left until we have found $k + 1$ mismatches (unsuccessful search) or the pattern ends (successful search).

To understand the generalized shift it may be helpful to look at the k mismatches problem in a tabular form. Let M be a $m \times n$ table such that for $1 \le i \le m$, $1 \le j \le n$,

$$M[i, j] = \begin{cases} 0, & \text{if } p_i = t_j \\ 1, & \text{if } p_i \ne t_j \end{cases}$$

There is an exact match ending at position r of T if $M[i, r - m + i] = 0$ for $i = 1, \ldots, m$, that is there is a whole diagonal of 0's in M ending at $M[m, r]$. Similarly, there is an approximate match with $\le k$ mismatches if the diagonal contains at most k 1's. This means that any successive $k + 1$ entries of such a diagonal have to contain at least one 0.

Assume then that the pattern is ending at text position j and we have to compute the next shift. We consider the last $k + 1$ text characters below the pattern, the characters $t_{j-k}, t_{j-k+1}, \ldots, t_j$. Then, suggested by the above observation, we glide the pattern to the right until there is at least one match in $t_{j-k}, t_{j-k+1}, \ldots, t_j$. The maximum shift is $m - k$. Clearly this is a correct heuristic: A smaller shift would give an unsuccessful alignment because there are at least $k + 1$ mismatches, and a shift larger than $m - k$ would skip over a potential match.

Let $d(t_{j-k}, t_{j-k+1}, \ldots, t_j)$ denote the length of the shift. The values of $d(t_{j-k}, \ldots, t_j)$ could be precomputed and tabulated. This would lead to quite heavy preprocessing of at least time $O(c^k)$. Instead, we apply a simpler preprocessing that makes it possible to compute the shift on-the-fly with small overhead while scanning.

In terms of M the shifting means finding the first diagonal above the current diagonal such that the new diagonal has at least one 0 for $t_{j-k}, t_{j-k+1}, \ldots, t_j$.

T

	a	b	a	a	c	b	b	a	b	b	b	a
a	0	1	0	0	1	1	1	0	1	1	1	0
b	1	0	1	0	1	0	0	1	0	0	0	1
b	1	0	0	1	0	0	0	1	0	0	0	1
b	1	0	1	0	1	0	0	1	0	0	0	1

P labels the rows.

Figure 1. Determining of shift ($k = 1$)

For example, consider table M in Fig. 1, where we assume that $k = 1$. We may shift from the diagonal of $M[1, 1]$ directly to the diagonal of $M[1, 3]$, as this diagonal contains the first 0 for characters $t_3 = a$, $t_4 = a$. Hence $d(a, a) = 2$ for the pattern $abbb$. Also note that t_4 alone would give a shift 3 and t_3 a shift 2, and $d(t_3, t_4)$ is the minimum over these component shifts.

In general, we compute $d(t_{j-k}, ..., t_j)$ as the minimum of the component shifts for each $t_{j-k}, ..., t_j$. The component shift for t_h depends both on the character t_h itself and on its position below the pattern. Possible positions are $m - k$, $m - k + 1$, ..., m. Hence we need a $(k + 1) \times c$ table d_k defined for each $i = m - k$, ..., m, and for each a in Σ, as

$$d_k[i, a] = min\{s \mid s = m \text{ or } (1 \leq s < m \text{ and } p_{i-s} = a)\}.$$

Here the values greater than $m - k$ are not actually relevant. Table d_k is presented in this form, because the same table is used in the algorithm solving the k differences problem.

Table d_k can be computed in time $O((m + c)k)$ by a straightforward generalization of the BMH-preprocessing which scans $k + 1$ times over P and each scanning creates a new row of d_k.

A more efficient method needs only one scan, from right to left, over P. For each symbol p_i encountered, the corresponding changes are updated to d_k. To keep track of the updates already made, we use a table $ready[a]$, a in Σ, such that $ready[a] = j$ if $d_k[i, a]$ already has its final value for $i = m, m - 1, ..., j$. Initially, $ready[a] = m + 1$ for all a, and $d_k[i, a] = m$ for all i, a. The algorithm is as follows:

Algorithm 3. Computation of table d_k.

```
1.    for a in Σ do ready[a] := m + 1;
2.    for a in Σ do
3.        for i := m downto m - k do
4.            d_k[i, a] := m;
5.    for i := m - 1 downto 1 do begin
6.        for j := ready[p_i] - 1 downto max(i, m - k) do
7.            d_k[j, p_i] := j - i;
8.        ready[p_i] := max(i, m - k) end
```

The initializations in steps 1–4 take time $O(kc)$. Steps 5–8 scan over P in time $O(m)$ plus the time of the updates of d_k in step 7. This takes time $O(kc)$ as each $d_k[j, p_i]$ is updated at most once. Hence Algorithm 3 runs in time $O(m + kc)$.

We have now the following total method for the k mismatches problem:

Algorithm 4. Approximate string matching with k mismatches.

```
1.    compute table d_k from P with Algorithm 3;
2.    j := m;                                          {pattern ends at text position j}
3.    while j ≤ n + k do begin
4.        h := j; i := m; neq := 0;                    {h scans the text, i the pattern}
5.        d := m - k;                                  {initial value of the shift}
6.        while i > 0 and neq ≤ k do begin
7.            if i ≥ m - k then d := min(d, d_k[i, t_h]);   {minimize over the component shifts}
8.            if t_h ≠ p_i then neq := neq + 1;
9.            i := i - 1; h := h - 1 end;              {proceed to the left}
10.       if neq ≤ k then report match at position j;
11.       j := j + d end                              {shift to the right}
```

2.3. Analysis

First recall that the preprocessing of P by Algorithm 3 takes time $O(m + kc)$ and space $O(kc)$. The scanning of T by Algorithm 4 obviously needs $O(mn)$ time in the worst case. The bound is strict for example for $T = a^n$, $P = a^m$.

Next we analyze the scanning time in the average case. The analysis will be done under the random string assumption which says that individual characters in P and T are chosen independently and uniformly from Σ. The time requirement is proportional to the number of the text-pattern comparisons in step 8 of Algorithm 4. Let $C_{loc}(P)$ be a random variable denoting, for some fixed c and k, the number of such comparisons for some alignment of pattern P between two successive shifts, and let $\bar{C}_{loc}(P)$ be its expected value.

Lemma 1. $\bar{C}_{loc}(P) \leq \left(\dfrac{c}{c - 1} + 1\right)(k + 1)$.

Proof. See [TaU90].

Let $S(P)$ be a random variable denoting the length of the shift in Algorithm 4 for pattern P and for some fixed k and c when scanning a random T. Moreover, let P_0 be a pattern that repeatedly contains all characters in Σ in some fixed order until the length of P_0 equals m. Then it is not difficult to see that P_0 gives on the average the minimal shift, that is, the expected values satisfy $\bar{S}(P_0) \leq \bar{S}(P)$ for all P of length m. Hence a lower bound for $\bar{S}(P_0)$ gives a lower bound for the expected shift over all patterns of length m (c.f. [Bae89b]).

Lemma 2. $\bar{S}(P_0) \geq \dfrac{1}{2} min(\dfrac{c}{k + 1}, m - k)$. Moreover, $\bar{S}(P_0) \geq 1$.

Proof. Let $t = min(c - 1, m - k - 1)$. Then the possible lengths of a shift are $1, 2, ..., t + 1$. Therefore

$$\bar{S}(P_0) = \sum_{i=0}^{t} Pr(S(P_0) > i)$$

where $Pr(A)$ denotes the probability of event A. Then

$$Pr(S(P_0) > i) = \left(\frac{c - i}{c}\right)^{k+1}$$

because for each of the $k + 1$ text symbols that are compared with the pattern to determine the shift (step 8 of Algorithm 4), there are i characters not allowed to occur as the text symbols. Otherwise the shift would not be $> i$. Hence

$$\bar{S}(P_0) = \sum_{i=0}^{t} \left(1 - \frac{i}{c}\right)^{k+1}$$

which clearly is ≥ 1, because $t \geq 0$ as we may assume that $c \geq 2$ and that $k < m$.
We divide the rest of the proof into two cases.

Case 1: $m - k < \dfrac{c}{k + 1}$. Then $t = m - k - 1$, and we have

$$
\begin{aligned}
\bar{S}(P_0) \quad &\geq \sum_{i=0}^{m-k-1} \left(1 - \frac{k + 1}{c} \cdot i\right) \\
&= m - k - \frac{k + 1}{c} \cdot \frac{(m - k - 1)(m - k)}{2} \\
&\geq (m - k)\left(1 - \frac{k + 1}{c} \cdot \frac{m - k}{2}\right) \geq \frac{1}{2}(m - k).
\end{aligned}
$$

Case 2: $m - k \geq \frac{c}{k+1}$. Then $t \geq \lceil \frac{c}{k+1} \rceil - 1$, and we have

$$\bar{S}(P_0) \;\geq\; \sum_{i=0}^{\lceil \frac{c}{k+1} \rceil - 1} \left(1 - \frac{i}{c}\right)^{k+1} \;\geq\; \sum_{i=0}^{\lceil \frac{c}{k+1} \rceil - 1} \left(1 - \frac{k+1}{c} \cdot i\right)$$

$$= \lceil \tfrac{c}{k+1} \rceil - \frac{k+1}{c} \cdot \frac{1}{2} \cdot \lceil \tfrac{c}{k+1} \rceil \left(\lceil \tfrac{c}{k+1} \rceil - 1 \right)$$

$$\geq \lceil \tfrac{c}{k+1} \rceil \left(1 - \frac{1}{2} \cdot \frac{k+1}{c} \cdot \frac{c}{k+1}\right) = \frac{1}{2} \lceil \tfrac{c}{k+1} \rceil. \;\square$$

Consider finally the total expected number $\bar{C}(P)$ of character comparisons when a random T is scanned with a pattern P. Let f_i be the number of shifts of length i taken during the scanning, $i = 1, ..., m - k$. Let

$$f = 1 + \sum_{i=1}^{m-k} f_i.$$

Then obviously $\bar{C}(P) = \bar{f} \cdot \bar{C}_{loc}(P)$. Moreover,

$$\sum_{i=1}^{m-k} f_i \cdot i = n - m.$$

Assume that the lengths of different shifts are mutually independent; note that this simplification is not true for two successive shifts such that the first one is shorter than $k + 1$. Then

$$\lim_{n \to \infty} \frac{\sum_{i=1}^{m-k} f_i \cdot i}{f - 1} = \lim_{n \to \infty} \frac{n - m}{f - 1} = \bar{S}(P).$$

Hence

$$\bar{f} = O\left(\frac{n - m}{\bar{S}(P)}\right) = O\left(\frac{n - m}{\bar{S}(P_0)}\right),$$

and by Lemma 2,

$$\bar{f} = O\left(2 \cdot max\left(\frac{k+1}{c}, \frac{1}{m-k}\right) \cdot (n - m)\right).$$

Then, noting Lemma 1, we obtain

$$\bar{C}(P) \leq O\left(2 \cdot max\left(\frac{k+1}{c}, \frac{1}{m-k}\right)(n - m) \left(\frac{c}{c-1} + 1\right)(k + 1)\right)$$

which is $O(\frac{nk^2}{c} + \frac{nk}{m-k})$ as $n \gg m$. Hence we have:

Theorem 1. The expected scanning time of Algorithm 4 is $O(nk(\frac{k}{c} + \frac{1}{m-k}))$, if the lengths of different shifts are mutually independent. The preprocessing time is $O(m + kc)$, and the working space $O(kc)$.

Removing the independence assumption from Theorem 1 remains open.

3. The k differences problem

3.1. Basic solution by dynamic programming

The *edit distance* [WaF75, Ukk85a] between two strings, A and B, can be defined as the minimum number of editing steps needed to convert A to B. Each editing step is a rewriting step of the form $a \to \varepsilon$ (a deletion), $\varepsilon \to b$ (an insertion), or $a \to b$ (a change) where a, b are in Σ and ε is the empty string.

The *k differences problem* is, given pattern $P = p_1 p_2 ... p_m$ and text $T = t_1 t_2 ... t_n$ and an integer k, to find all such j that the edit distance (i.e., the number of differences) between P and some substring of T ending at t_j is at most k. The basic solution of the problem is by the following dynamic programming method [Sel80, Ukk85b]: Let D be a $m + 1$ by $n + 1$ table such that $D(i, j)$ is the minimum edit distance between $p_1 p_2 ... p_i$ and any substring of T ending at t_j. Then

$$D(0, j) = 0, \quad 0 \le j \le n;$$

$$D(i, j) = min \begin{cases} D(i-1, j) + 1 \\ D(i-1, j-1) + \text{if } p_i = t_j \text{ then } 0 \text{ else } 1 \\ D(i, j-1) + 1 \end{cases}$$

Table D can be evaluated column-by-column in time $O(mn)$. Whenever $D(m, j)$ is found to be $\le k$ for some j, there is an approximate occurrence of P ending at t_j with edit distance $D(m, j) \le k$. Hence j is a solution to the k differences problem.

3.2. Boyer-Moore approach

Our algorithm contains two main phases: the scanning and the checking. The scanning phase scans over the text and marks the parts that may contain approximate occurrences of P. This is done by marking some entries $D(0, j)$ on the first row of D. The checking phase then evaluates all diagonals of D whose first entry is marked. This is done by the basic dynamic programming restricted to the marked diagonals. Whenever the dynamic programming refers to an entry outside the diagonals, the entry can be taken to be ∞. Because this is quite straightforward we do not describe it in detail. Rather, we concentrate on the scanning part.

The scanning phase repeatedly applies two operations: mark and shift. The shift operation is based on a Boyer-Moore idea. The mark operation decides whether or not the current alignment of the pattern with the text needs accurate checking by dynamic programming and in the positive case marks certain diagonals. To understand the operations we need the concept of a minimizing path in table D.

For every $D(i, j)$, there is a *minimizing arc* from $D(i-1, j)$ to $D(i, j)$ if $D(i, j) = D(i-1, j) + 1$, from $D(i, j-1)$ to $D(i, j)$ if $D(i, j) = D(i, j-1) + 1$, and from $D(i-1, j-1)$ to $D(i, j)$ if $D(i, j) = D(i-1, j-1)$ when $p_i = t_j$ or if $D(i, j) = D(i-1, j-1) + 1$ when $p_i \ne t_j$. The costs of the arcs are 1, 1, 0 and 1, respectively. The minimizing arcs show the actual dependencies between the values in table D. A *minimizing path* is any path that consists of minimizing arcs and leads from an entry $D(0, j)$ on the first row of D to an entry $D(m, h)$ on the last row of D. Note that $D(m, h)$ equals the sum of the costs of the arcs on the path. A minimizing path is *successful* if it leads to an entry $D(m, h) \le k$.

A *diagonal* h of D for $h = -m, ..., n$, consists of all $D(i, j)$ such that $j - i = h$. As any vertical or horizontal minimizing arc adds 1 to the value of the entry, the next lemma easily follows:

Lemma 3. The entries on a successful minimizing path are contained in $\le k + 1$ successive diagonals of D.

Our marking method is based on the following lemma. For each $i = 1, ..., m$, let the k *environment* of the pattern symbol p_i be the string $C_i = p_{i-k}...p_{i+k}$, where $p_j = \varepsilon$ for $j < 1$ and $j > m$.

Lemma 4. Let a successful minimizing path go through some entry on a diagonal h of D. Then for at most k indexes i, $1 \le i \le m$, character t_{h+i} does not occur in k environment C_i.

Proof. See [TaU90].

Lemma 4 suggests the following marking method. For diagonal h, check for $i = m, m - 1, ..., k + 1$ if t_{h+i} is in C_i until $k + 1$ bad columns are found. Note that to get minimum shift $k + 1$ (see below) we stop already at $i = k + 1$ instead of at $i = 1$. If the number of bad columns is $\leq k$, then mark diagonals $h - k, ..., h + k$, that is, mark entries $D(0, h - k), ..., D(0, h + k)$.

For finding the bad columns fast we need a precomputed table $Bad(i, a), 1 \leq i \leq m, a \in \Sigma$, such that

$Bad(i, a) = $ **true**, if and only if a does not appear in k environment C_i.

Clearly, the table can be computed by a simple scanning of P in time $O((c + k)m)$.

After marking we have to determine the length of shift, that is, what is the next diagonal after h around which the marking should eventually be done. The marking heuristics ensures that all successful minimizing paths that are properly before diagonal $h + k + 1$ are already marked. Hence we can safely make at least a shift of $k + 1$ to diagonal $h + k + 1$.

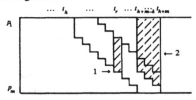

Figure 2. Mark and shift ($k = 2$)

This can be combined with the shift heuristics of Algorithm 4 of Section 2 based on table d_k. So we determine the first diagonal after h, say $h + d$, where at least one of the characters $t_{h+m}, t_{h+m-1}, ..., t_{h+m-k}$ matches with the corresponding character of P. This is correct, because then there can be a successful minimizing path that goes through diagonal $h + d$. The value of d is evaluated as in Algorithm 4, using exactly the same precomputed table d_k. Note that unlike in the case of Algorithm 4, the maximum allowed value of d is now m, not $m - k$, as the marking starts from diagonal $h - k$, not from h. Finally, the maximum of $k + 1$ and d is the length of the shift.

In practice, the marking and the computation of the shift can be merged if we start the searching for the bad columns from the end of the pattern.

Fig. 2 illustrates marking and shifting. For $r = h + m, h + m - 1, ..., h + k + 1$ we check whether or not t_r appears among the pattern symbols corresponding to the shaded block 1 (the k environment). If $k + 1$ symbols t_r that do not appear are found, entries $D(0, h - k), ..., D(0, h + k)$ are marked. Simultaneously we check what is the next diagonal after h containing a match between P and $t_{h+m-k}, ..., t_{h+m}$ (shaded block 2). The next shift is to this diagonal but at least to diagonal $h + k + 1$.

We get the following algorithm for the scanning phase:

Algorithm 5. The scanning phase for the k differences problem.

```
1.    compute table Bad and, by Algorithm 3, table dₖ from P;
2.    j := m;
3.    while j ≤ n do begin
4.        r := j; i := m;
5.        bad := 0;                    {bad counts the bad indexes}
6.        d := m;                      {initial value of shift}
7.        while i > k and bad ≤ k do begin
8.            if i ≥ m − k then d := min(d, dₖ[i, tᵣ]);
9.            if Bad(i, tᵣ) then bad := bad + 1;
10.           i := i − 1; r := r − 1 end;
11.       if bad ≤ k then
12.           mark entries D(0, j − m − k), ..., D(0, j − m + k);
13.       j := j + max(k + 1, d) end
```

The loop in steps 7–9 can be slightly optimized by splitting it into two parts such that the first one handles $k + 1$ text characters and computes the length of shift, and the latter goes on counting bad indexes (a similar optimization also applies to Algorithm 4).

3.3. Analysis

The preprocessing of P requires $O((k + c)m)$ for computing table *Bad* and $O(m + kc)$ for computing table d_k. As $k < m$, the total time is $O((k + c)m)$. The working space is $O(cm)$.

The marking and shifting by Algorithm 5 takes time $O(mn/k)$ in the worst case. The analysis of the average case is similar to the analysis of Algorithm 4 in Section 2. Let $B_{loc}(P)$ be a random variable denoting, for some fixed c and k, the number of the columns examined (step 9 of Algorithm 5) until $k + 1$ bad columns are found and the next shift will be taken. Obviously, $B_{loc}(P)$ corresponds to $C_{loc}(P)$ of Lemma 1. For the expected value $\bar{B}_{loc}(P)$ we show the following rough bound:

Lemma 5. Let $2k + 1 < c$. Then $\bar{B}_{loc}(P) \leq \left(\dfrac{c}{c - 2k - 1} + 1 \right)(k + 1)$.

Proof. See [TaU90].

Let $S'(P)$ be a random variable denoting the length of the shift in Algorithm 5 for pattern P and for some fixed k and c. When scanning a random T, the special pattern P_0 again gives the shortest expected shift, that is, $\bar{S}'(P_0) \leq \bar{S}'(P)$ for all P of length m. Lemma 6 gives a bound for $\bar{S}'(P_0)$.

Lemma 6. $\bar{S}'(P_0) \geq \frac{1}{2} min(\frac{c}{k + 1}, m)$.

Proof. See [TaU90].

As the length of a shift is always $\geq k + 1$, we get from Lemma 6

$$\bar{S}'(P) \geq \bar{S}'(P_0)$$

$$\geq max\left(k + 1, \, min\left(\frac{c}{2(k + 1)}, \frac{m}{2} \right) \right)$$

$$= min\left(max\left(k + 1, \frac{c}{2(k + 1)} \right), \, max\left(k + 1, \frac{m}{2} \right) \right)$$

$$\geq \frac{1}{2} min\left(k + 1 + \frac{c}{2(k + 1)}, \frac{m}{2} \right).$$

The number of text positions at which a right-to-left scanning of P is performed between two shifts is again

$$O\left(\frac{n - m}{\bar{S}'(P)} \right) = O\left(\frac{n - m}{\bar{S}'(P_0)} \right).$$

This can be shown as in the analysis of Algorithm 4. Note that for Algorithm 5 we need not assume explicitly that the lengths of different shifts are independent. They are independent as the length of the minimum shift is $k + 1$.

Hence the expected scanning time of Algorithm 5 for pattern P is

$$O\left(\overline{B}_{loc}(P) \cdot \frac{n-m}{\overline{S}'(P)}\right).$$

When we apply here the upper bound for $\overline{B}_{loc}(P)$ from Lemma 5 and the above lower bound for $\overline{S}'(P)$, and simplify, we obtain our final result.

Theorem 2. Let $2k + 1 < c$. Then the expected scanning time of Algorithm 5 is $O(\frac{c}{c-2k} \cdot kn \cdot (\frac{k}{c+2k^2} + \frac{1}{m}))$. The preprocessing time is $O((k+c)m)$ and the working space $O(cm)$.

The checking of the marked diagonals can be done after Algorithm 5 or in cascade with it in which case a buffer of length $2m$ is enough for saving the relevant part of text T.

For each band of diagonals marked in step 12 of Algorithm 5, the checking can be done in expected time $O(k^2)$. The checking of overlapping bands can be combined, which ensures that the total expected time of the checking phase remains $O(kn)$; here one has to use the $O(kn)$ expected time version of the dynamic programming of [Ukk85b].

Asymptotically, step 12 of Algorithm 5 is executed very seldom. Hence except for small patterns, small alphabets and large k's, the expected time for the checking phase tends to be small in which case the time bound of Theorem 2 is valid for our entire algorithm.

4. Experiments and conclusions

We have tested extensively our algorithms and compared them with other methods. We will present results of a comparison with the $O(kn)$ expected time dynamic programming method [Ukk85b] which we have found to be the best in practice among the old algorithms we have tested [JTU90].

In our tests, we used random patterns of varying lengths and random texts of length 100,000 characters over alphabets of different sizes. The tests were run on a VAX 8800 under VMS. Figures 3–6 show results for Algorithms 4 and 5 and for the corresponding dynamic programming algorithms $DP1$ (for the k mismatches problem) and $DP2$ (for the k differences problem).

Figures 3 and 4 show the total execution times in units of 10 milliseconds when $k = 4$ and m varies for alphabet sizes $c = 2$ and 90. Figures 5 and 6 show the corresponding times when $m = 8$ and k varies for alphabet sizes $c = 4$ and 30. The times of Algorithm 5 include preprocessing, scanning and checking.

Our algorithms, as all algorithms of Boyer-Moore type, work very well for large alphabets, and the execution time decreases when the length of the pattern grows. An increment of the error limit k slows down our algorithms more than the dynamic programming algorithms. Observe also that the Boyer-Moore approach is relatively better in solving the k differences problem than in solving the k mismatches problem.

Our methods turned out to be faster than the previous methods, when the pattern is long enough ($m > 5$), the error limit k is relatively small and the alphabet is not very small ($c > 5$). Results of the practical experiments are consistent with our theoretical analysis. To devise a more accurate and complete theoretical analysis of the algorithms is left as a subject for further study.

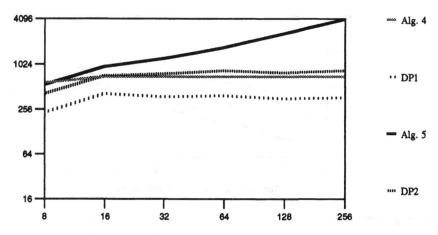

Figure 3. Total times for $k = 4$, $c = 2$ and varying m

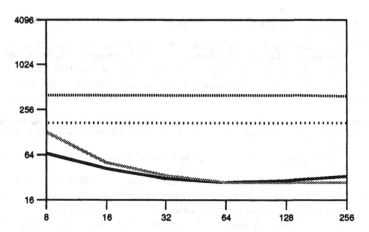

Figure 4. Total times for $k = 4$, $c = 90$ and varying m

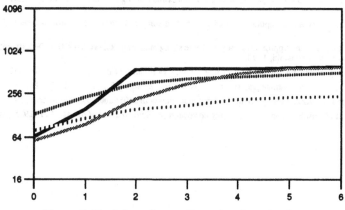

Figure 5. Total times for $m = 8$, $c = 4$ and varying k

359

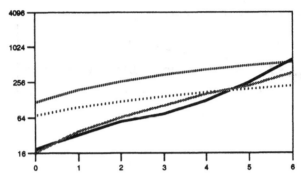

Figure 6. Total times for $m = 8$, $c = 30$ and varying k

Acknowledgement

Petteri Jokinen performed the experiments which is gratefully acknowledged.

References

[Bae89a] R. Baeza-Yates: Efficient Text Searching. Ph.D. Thesis, Report CS-89-17, University of Waterloo, Computer Science Department, 1989.

[Bae89b] R. Baeza-Yates: String searching algorithms revisited. In: *Proceedings of the Workshop on Algorithms and Data Structures* (ed. F. Dehne et al.), Lecture Notes in Computer Science 382, Springer-Verlag, Berlin, 1989, 75–96.

[BoM77] R. Boyer and S. Moore: A fast string searching algorithm. *Communcations of the ACM* 20 (1977), 762–772.

[GaG86] Z. Galil and R. Giancarlo: Improved string matching with k mismatches. *SIGACT News* 17 (1986), 52–54.

[GaG88] Z. Galil and R. Giancarlo: Data structures and algorithms for approximate string matching. *Journal of Complexity* 4 (1988), 33–72.

[GaP89] Z. Galil and K. Park: An improved algorithm for approximate string matching. *Proceedings of the 16th International Colloquium on Automata, Languages and Programming*, Lecture Notes in Computer Science 372, Springer-Verlag, Berlin, 1989, 394–404.

[GrL89] R. Grossi and F. Luccio: Simple and efficient string matching with k mismatches. *Information Processing Letters* 33 (1989), 113–120.

[Hor80] N. Horspool: Practical fast searching in strings. *Software Practice & Experience* 10 (1980), 501–506.

[JTU90] P. Jokinen, J. Tarhio and E. Ukkonen: A comparison of approximate string matching algorithms. In preparation.

[Kos88] S. R. Kosaraju: Efficient string matching. Extended abstract. Johns Hopkins University, 1988.

[KMP77] D. Knuth, J. Morris and V. Pratt: Fast pattern matching in strings. *SIAM Journal on Computing* 6 (1977), 323–350.

[LaV88] G. Landau and U. Vishkin: Fast string matching witk k differences. *Journal of Computer and System Sciences* 37 (1988), 63–78.

[LaV89] G. Landau and U. Vishkin: Fast parallel and serial approximate string matching. *Journal of Algorithms* 10 (1989), 157–169.

[Sel80] P. Sellers: The theory and computation of evolutionary distances: Pattern recognition. *Journal of Algorithms* 1 (1980), 359–372.

[TaU90] J. Tarhio and E. Ukkonen: Approximate Boyer-Moore string matching. Report A-1990-3. Department of Computer Science, University of Helsinki, 1990.

[Ukk85a] E. Ukkonen: Algorithms for approximate string matching. *Information Control* 64 (1985), 100–118.

[Ukk85b] E. Ukkonen: Finding approximate patterns in strings. *Journal of Algorithms* 6 (1985), 132–137.

[UkW89] E. Ukkonen and D. Wood: Fast approximate string matching with suffix automata. Manuscript, 1989.

[WaF75] R. Wagner and M. Fischer: The string-to-string correction problem. *Journal of the ACM* 21 (1975), 168–173.

Complete Problems With *L-samplable* Distributions

Per Grape
Royal Institute of Technology
Stockholm, Sweden

Abstract

The average case complexity classes ⟨P, *L-samplable*⟩ and ⟨NL, *L-samplable*⟩ are defined and we show that Deterministic Bounded Halting is complete for ⟨P, *L-samplable*⟩ and that Graph Reachability is complete for ⟨NL, *L-samplable*⟩.

1 Introduction

Given a problem and an algorithm that solves it, then the expected running time of the algorithm can be regarded as the average case complexity of the problem. The expectation is taken over all instances of the problem. The average case complexity of a problem is, in many cases, a more significant measure than its worst case complexity. Many *NP*-hard problems are practically important and must be solved one way or another in spite of their hardness. There are different approaches in the literature to this problem, like probabilistic and approximate algorithms.

The general question of average case complexity was addressed for the first time by Levin [5], and Johnson [4] provided some intuition behind Levin's definitions and proofs. Levin's work can be viewed as the basis for a theory for average *NP*-compleatness. Since then the theory has developed further and some interesting structural results have been proven, especially by Ben-David, Chor, Goldreich and Luby in [1] and Gurevich in [2].

An average case complexity class consists of pairs, sometimes called distributional problems. Each such pair consists of a decision problem and a probability distribution on problem instances. Of all the definitions in [5] the most controversial one is the association of "simple" distributions with *P-computable*, where *P-computable* is the class of distribution functions that are computable in polynomial time (that is, there exist a polynomial time Turing machine that on input x computes the cumulative probability of all strings $y \leq x$). This might look too restrictive and motivated the definition of *P-samplable* distributions. A distribution is in *P-samplable* if there is a probabilistic polynomial time sampling algorithm that samples it. The *P-samplable* distributions can be regarded as the output distribution of an instance generator to some problem. The *P-computable* distributions are a subset of the *P-samplable* distributions. To sample a *P-computable* distribution μ just choose a random number ρ in $[0,1]$ and by binary search find x such that $\mu(x)$ is the smallest number larger than or equal to ρ. Recently Impagliazzo and

Levin [3] have, very surprisingly, proved that any distributional problem that consists of a *NP-predicate* and a *P-samplable* distribution can be reduced to a distributional problem which consists of another *NP-predicate* and a *P-computable* distribution.

This paper deals with complete problems for $\langle P, L\text{-}samplable \rangle$ and $\langle NL, L\text{-}samplable \rangle$. In complexity theory reductions in *P* or *NL* are required to be done in logarithmic space. Therefore it is natural to restrict distributions in a similar manner in the study of average case complexity of problems in *P* or *NL*. If the space is limited to be logarithmic in the length of x classes of distributions like *L-computable* and *L-samplable* can be defined analogous to *P-computable* and *P-samplable*. There is no obvious connection between *L-computable* and *L-samplable* distributions like there is between *P-computable* and *P-samplable*. This is an indication that there might be some structural difference between classes like $\langle P, L\text{-}computable \rangle$ and $\langle P, L\text{-}samplable \rangle$. A first step to an answer to this question is to find complete problems. In [1] Ben-David et. al. showed that there are complete problems for $\langle P, L\text{-}computable \rangle$ and in this paper complete problems for $\langle P, L\text{-}samplable \rangle$ and $\langle NL, L\text{-}samplable \rangle$ are presented. Yet there is no complete problems for $\langle NL, L\text{-}computable \rangle$.

Section 2 contains most of the definitions needed. And it is shown in section 3 that Deterministic Bounded Halting is complete in $\langle P, L\text{-}samplable \rangle$ and that Graph Reachability is complete in $\langle NL, L\text{-}samplable \rangle$. Both problems with a universal logspace samplable distribution.

2 Definitions

The *standard lexicographic order of strings* order strings first by length and then lexicographically, that is, the first stings are $0, 1, 00, 01, \ldots$ The string preceding x is denoted $x - 1$.

In this paper, probability distribution functions are considered to be from strings to the zero-one interval and letters like μ, α and β are reserved for probability distribution functions. Every probability distribution μ is a nondecreasing function, that is, $\mu(x-1) \leq \mu(x)$ for all x. The density function associated with a probability distribution μ is denoted μ' and satisfies $\mu'(0) = \mu(0)$ and $\mu'(x) = \mu(x) - \mu(x-1)$.

The simple probability distributions are those which can be computed with a reasonable amount of work. Classically they are associated with polynomial time, or more formally:

Definition 2.1 *A probability distribution μ is in P-computable if there is a deterministic polynomial time Turing machine that on input x outputs the binary expansion of $\mu(x)$ (the time is polynomial in $|x|$).*

Note that if μ is in *P-computable* implies that μ' is computable in polynomial time but the converse does not necessarily hold.

As mentioned earlier samplable distributions can be regarded as the output distributions to instance generators. The *P-samplable* distributions are output distributions for simple instance generators.

362

Definition 2.2 *A probability distribution μ is in P-samplable if there is a probabilistic polynomial time algorithm S that outputs x with probability $\mu'(x)$ (the time is polynomial in $|x|$).*

To work nicely with logspace reductions L-computable distributions are needed.

Definition 2.3 *A probability distribution μ is in L-computable if there is a deterministic logarithmic space Turing machine that on input x outputs the binary expansion of $\mu(x)$ (the space is logarithmic in $|x|$).*

And distributions for logspace generators.

Definition 2.4 *A probability distribution μ is in L-samplable if there is a probabilistic polynomial time and logarithmic space algorithm S that outputs x with probability $\mu'(x)$ (the time is polynomial in $|x|$ and the space is logarithmic in $|x|$).*

Now everything are defined needed to make a formal definition of distributional complexity classes.

Definition 2.5 *The distributional complexity class $\langle P, L\text{-samplable}\rangle$ consists of pairs (D,μ) where D is in P and μ is L-samplable. Such a pair is called a distributional problem.*
Analogous for the complexity classes $\langle NL, L\text{-samplable}\rangle$, $\langle NL, L\text{-computable}\rangle$ and $\langle P,L\text{-computable}\rangle$.

When dealing with L-reductions in average case complexity theory it is enough to require the reduction to be logarithmic on average. Therefore there is a need for a formal definition of logarithmic on average.

Definition 2.6 *A function $f : \{0,1\}^* \to \mathbf{N}$ is logarithmic on average if and only if there exists an $\epsilon > 0$ such that*

$$\sum_{x\in\{0,1\}^*} \mu'_n(x)\frac{(2^{f(x)})^\epsilon}{|x|} < \infty.$$

Armed with this definitions logspace reductions between distributional problems turn out to be:

Definition 2.7 *A function f is a logspace reduction from (D_1,μ_1) to (D_2,μ_2) if the following conditions hold:*

1. *The transformation f is computable in space which is logarithmic on average with respect to μ_1.*

2. *For every $x \in \{0,1\}^*$ $D_1(x) = D_2(f(x))$.*

3. *There exists a constant $c > 0$ such that for every $y \in \{0,1\}^*$,*

$$\mu'_2(y) \geq \frac{1}{|y|^c} \sum_{x\in f^{-1}(y)} \mu'_1(x).$$

The third condition is a special case of what is more generally is called domination, and it says that μ_2 dominates the distribution induced by μ_1 on the image of f. A more general definition of domination is.

Definition 2.8 *Let μ and α be probability distributions. Then μ dominates α if there exist a constant $c > 0$ such that for all but finitely many $x \in \{0,1\}^*$*

$$\mu'(x) \geq \frac{1}{|x|^c}\alpha'(x).$$

With these definitions of logarithmic on average and logspace reductions between distributional problems the following theorem can be proved.

Theorem 2.9 *If there is logspace reduction from the distributional problem (D_1, μ_1) to (D_2, μ_2) and D_2 is decidable in logarithmic space on average with respect to μ_2, then so is D_1 with respect to μ_1.*

This theorem follows directly from the following two lemmas.

Lemma 2.10 *Suppose that f logspace reduces (D_1, μ_1) to (D_2, μ_{1f}) where μ_{1f} is the distribution induced by μ_1 on the image of f, that is,*

$$\mu'_{1f}(y) = \sum_{x \in f^{-1}(y)} \mu'_1(x).$$

Then if D_2 is decidable in logarithmic space on average with respect to μ_{1f}, then so is D_1 with respect to μ_1.

PROOF: Let S_f be the space used by f and let S_2 be the space used by the machine that decides D_2 in logarithmic space on average. By assumption the following yields.

$$\sum_{x \in \{0,1\}^*} \mu'_1(x)\frac{(2^{S_f(x)})^\gamma}{|x|} < \infty$$

$$\sum_{x \in \{0,1\}^*} \mu'_{1f}(x)\frac{(2^{S_2(x)})^\eta}{|x|} < \infty.$$

Let $\epsilon = \min(\gamma, \eta)$. Compute $D_1(x)$ by computing $D_2(f(x))$. The space used to decide D_1 is then $S_1(x) = S_f(x) + S_2(f(x))$. Then there is a $\delta \in (0, \epsilon)$, to be determined later such that

$$\sum_{x \in \{0,1\}^*} \mu'_1(x)\frac{2^{(S_1(x))^\delta}}{|x|} = \sum_{x \in \{0,1\}^*} \mu'_1(x)\frac{\left(2^{2\delta S_f(x)}2^{2\delta S_2(f(x))}\right)^{1/2}}{|x|}$$

$$\leq \sum_{x \in \{0,1\}^*} \mu'_1(x)\frac{2^{2\delta S_f(x)}}{2|x|} + \sum_{x \in \{0,1\}^*} \mu'_1(x)\frac{2^{2\delta S_2(f(x))}}{2|x|}. \tag{1}$$

If $2\delta < \epsilon$ then the first term of (1) is

$$\sum_{x \in \{0,1\}^*} \mu'_1(x)\frac{2^{2\delta S_f(x)}}{2|x|} < \infty$$

by assumption. Using that $|f(x)| \leq 2^{S_f(x)}$ and $\sqrt{ab} < (a+b)/2$ when $a > 0$ and $b > 0$, then the second part of (1) yields

$$\frac{1}{2} \sum_{x\in\{0,1\}^*} \mu_1'(x) \frac{2^{2\delta S_2(f(x))}}{|x|} \leq \sum_{x\in\{0,1\}^*} \mu_1'(x) \frac{2^{2\delta S_2(f(x))}}{|f(x)|^{\epsilon/2}} \cdot \frac{\left(2^{S_f(x)}\right)^{\epsilon/2}}{|x|}$$

$$\leq \sum_{x\in\{0,1\}^*} \mu_1'(x) \frac{2^{4\delta S_2(f(x))}}{|f(x)|^{\epsilon}} + \sum_{x\in\{0,1\}^*} \mu_1'(x)\frac{2^{\epsilon S_f(x)}}{|x|^2}$$

$$\leq \sum_{y\in\{0,1\}^*}\sum_{x\in f^{-1}(y)} \mu_1'(x) \frac{2^{4\delta S_2(y)}}{|y|^{\epsilon}} + \sum_{x\in\{0,1\}^*} \mu_1'(x)\frac{2^{\epsilon S_f(x)}}{|x|^2}$$

$$\leq \sum_{y\in\{0,1\}^*} \mu_{1f}'(x) \frac{2^{4\delta S_2(y)}}{|y|^{\epsilon}} + \sum_{x\in\{0,1\}^*} \mu_1'(x)\frac{2^{\epsilon S_f(x)}}{|x|^2}. \tag{2}$$

The second term of 2 is $< \infty$ by assumption. Let $\delta = \epsilon^2/4$ and the first term of 2 is equal to

$$\sum_{y\in\{0,1\}^*} \mu_{1f}'(x) \left(\frac{2^{\epsilon S_2(y)}}{|y|}\right)^{\epsilon} \leq \sum_{y\in\{0,1\}^*} \mu_{1f}'(x) \frac{2^{\epsilon S_2(y)}}{|y|} < \infty.$$

And the lemma follows. $\qquad\square$

Lemma 2.11 Let D be a decision problem, μ_1 and μ_2 two probability distributions such that μ_2 dominates μ_1. Then if D is decidable in logarithmic space on average with respect to μ_2, then so is D with respect to μ_1.

PROOF: Let S be a bound for the space needed to decide D. By assumption there is an $\epsilon > 0$ such that

$$\sum_{x\in\{0,1\}^*} \mu_2'(x) \frac{2^{\epsilon S(x)}}{|x|} < \infty.$$

The average space needed with respect to μ_1 is

$$\sum_{x\in\{0,1\}^*} \mu_1'(x) \frac{2^{\delta S(x)}}{|x|} = \sum_{|x|\geq 2^{\delta S(x)}} \mu_1'(x) \frac{2^{\delta S(x)}}{|x|} + \sum_{|x|< 2^{\delta S(x)}} \mu_1'(x) \frac{2^{\delta S(x)}}{|x|}$$

$$\leq 1 + \sum_{|x|<2^{\delta S(x)}} \mu_1'(x) \left(\frac{2^{\delta S(x)}}{|x|}\right)^{c+1}$$

$$\leq 1 + \sum_{|x|<2^{\delta S(x)}} \mu_2'(x)|x|^c \frac{2^{\delta(c+1)S(x)}}{|x|^{c+1}}$$

$$\leq 1 + \sum_{x\in\{0,1\}^*} \mu_2'(x)\frac{2^{\delta(c+1)S(x)}}{|x|} < \infty$$

if δ is chosen such that $0 < \delta(c+1) < \epsilon$. $\qquad\square$

Following Levin's conventions not all distributions may add up to 1 but to some other constant. A "proper" probability distribution can always be obtained by normalizing these distributions. An example is the uniform distribution on strings $\mu'(x) = 1/(|x|^2 2^{|x|})$ which should be written $\mu'(x) = 1/(|x|^2 2^{|x|})6/\pi^2$.

3 Results

When dealing with logspace reductions the main difficulty lies in handling the distributions. The following lemma is essential.

Lemma 3.1 *There exists an* Universal Logspace Samplable *distribution μ_{ULS} that dominates every distribution in L-samplable.*

PROOF: (Follows ideas in [1, Theorem 9].) The idea is to establish an enumeration of machines sampling *L-samplable* distributions and then define μ_{ULS} as the total sum of all enumerated distributions. To be able to do this enumerate the machines according to the lexicographic order of their binary description.

Assume that all sampling algorithms work in quadratic time in their output. Further more, assume that algorithm i in the enumeration outputs strings of length at least i and that all sampling algorithms uses space $\leq 2\log|x|$. This can be done without loss of generality and the reason is that if the output is too short or too much memory is used it is always possible to pad with some default string so that the running time is quadratic in the output and the output length is at least greater than i for machine i, and last, see to that the space used is less than two times the logarithm of the length of the output. This creates a new instance and can be handled in the following way. Let (D,μ) be a distributional problem and reduce this to (D',α) where the sampling algorithm S_α outputs a pair (x,y) with probability $\mu(x)$ and y is the padding needed for S_α to run in quadratic time in its output. Then $D'(x,y) = D(x)$, that is, D' just throw y away and run D. The length of the padded string y is calculated in the following way. Let b_t be the number of bits needed to fix the running time. If the running time is $c|x|^k$ then obtain b_t by solving $c|x|^k + b_t \leq (|x| + b_t)^2$ which always has one positive root $b_t \geq -|x| + 1/2 + \sqrt{c|x|^k - |x| + 1/4}$ when $k \geq 2$ and $c \geq 1$. Let b_s be the number of bits needed to fix the space. If the space used is $a\log|x|$ then get b_s by solving $a\log|x| \leq 2\log(|x| + b_s)$ which has a positive root $b_s \geq -|x| + \sqrt{|x|^a}$ when $a \geq 2$. The length of y is then $|y| = \max(b_t, b_s, i)$ for machine i in the enumeration.

Let the *obviously logspace and quadratic time algorithms* be those that fulfill the following conditions:

1. First it toss coins to determine the maximum running time t with probability proportional to t^{-3}, and the memory is limited to $2\log t$.

2. It has a "stop watch" that checks the running time and memory usage and stops the sampling if the time is up or the algorithm attempted to use too much memory.

3. Last, if needed, it pads the output with a default string to the proper length, as described above.

The next step is to establish an enumeration of the obviously logspace and quadratic time algorithms. An algorithms number is its binary description interpreted as a number in base 2. If algorithm i is not an obviously logspace and quadratic time algorithm then sample a default algorithm. Otherwise let S_i be the i'th algorithm and it samples according to distribution α_i.

This yields a enumeration of logspace sampling algorithms, and define the universal logspace samplable distribution:

$$\mu'_{ULS}(x) = \sum_{i=1}^{\infty} \frac{\alpha'_i(x)}{i(i+1)}.$$

The next step is to show that for every distribution μ in *L-samplable* there is an algorithm S_i in the enumeration with corresponding distribution α_i that dominates μ. The prime candidate for the sampling algorithm is clearly S_μ extended with coin tossing and a stop watch. Let S_i be this extended algorithm. Note that μ is in *L-samplable* implies that there exist a constant k such that strings of length n are generated in space $\leq k \log n$. The probability for a string x to be generated by S_i is $\alpha'_i(x) = \Pr[x$ is generated $\mid$ time and space are enough $] * \Pr[$time and space are enough$]$. Or with symbols

$$\alpha'_i(x) = \mu'(x) * \Pr[T \geq |x|^2 \text{ and } 2 \log T \geq k \log |x|] = \mu'(x) \sum_{t=t_0}^{\infty} 1/t^3$$

where $t_0 = \max(|x|^2, |x|^{k/2})$. This implies that α_i dominates μ and, of course, μ_{ULS} dominates μ.

To see that μ_{ULS} is in *L-samplable*, first select i with probability proportionally to i^{-2}. The space needed to store i is $|i| = \log i$. Since the output of algorithm α_i in the enumeration is at least i there is space to store i. Now sample algorithm S_i and by definition this is done in space logarithmic in the output. By virtue of that all S_i runs in space $2 \log |x|$ this works fine for all i. $\square$

The following problem is called Deterministic Bounded Halting, DBH.
INPUT: A triple $(M, x, 1^k)$, where M is a deterministic machine, x its binary input and an integer k given in unary.
QUESTION: Is there an accepting computation of M on input x in at most k steps?

In 1988 Ben-David et. al. [1] defined $\langle P, L\text{-computable}\rangle$ and showed that (DBH, μ_{BH}) is complete for it. The distribution is

$$\mu'_{BH}(M, x, 1^k) = \frac{1}{|M|^2 2^{|M|}} \cdot \frac{1}{|x|^2 2^{|x|}} \cdot \frac{1}{k^2}.$$

Theorem 3.2 *The distributional problem* (DBH, μ_{ULS}) *is complete for* $\langle P, L\text{-samplable}\rangle$.

PROOF: Clearly, DBH is in P it is even P-compete. Let (D, μ) be an arbitrary problem in $\langle P, L\text{-samplable}\rangle$ and the aim is to establish the chain $(D, \mu) \leq_L (DBH, \beta) \leq_L (DBH, \mu_{ULS})$.

Let f be the standard Karp reduction of D to DBH. The reduction maps instances x of D to triples $(M, x, 1^{P_M(|x|)})$, where M is a polynomial (in $|x|$) time Turing machine for D and $P_M(|x|)$ is the polynomial bound for the running time of M on input x. Let β be the output distribution of f applied to the output of S_μ. Thus $\beta'(f(x)) = \mu'(x)$. Sample β in the following way. Let S_β first output the description of M. This is done in constant memory and time. Then run S_μ and save the length of x in binary. Last compute $P_M(|x|)$ and write the output in unary. This is all done in space $O(\log |x|)$ and therefore $\beta \in L\text{-samplable}$.

It is clear that $(DBH, \beta) \leq_L (DBH, \mu_{ULS})$ since lemma 3.1 states that μ_{ULS} dominates every distribution in *L-samplable*. $\square$

The following problems called Graph Reachability, GR.

INPUT: A directed graph G in some natural encoding and two vertices v_s, v_e.

QUESTION: Is there a path from v_s to v_e in G?

Armed with the universal distribution it is not hard to show:

Theorem 3.3 *The distributional problem* (GR, μ_{ULS}) *is complete for* $\langle NL,\ L\text{-samplable}\rangle$.

PROOF: Graph Reachability is in NL (actually NL-complete). Let (D, μ) be an arbitrary problem in $\langle NL,\ L\text{-samplable}\rangle$. The aim is to show $(D, \mu) \leq_L (GR, \beta) \leq_L (GR, \mu_{ULS})$.

Let f be the standard Karp reduction from D to GR. Then f maps instances of D to a directed graph G where each vertex represent a configuration of the machine M that solves D. All legal transitions from one configuration to another are represented by edges in G. The graph G is represented by an unordered list of edges.

Let β be the output distribution of f applied to the output of S_μ. Thus $\beta'(f(x)) = \mu'(x)$. Sample β in the following way.

1. Run S_μ for a while to get a bit b. Let the input head look at bit b and for every configuration make an edge where there is a legal transition to another configuration.

2. Repeat step 1 until S_μ stops.

The space needed is $O(\log |x|)$ in each step and the space can be reused. Thus $(D, \mu) \leq_L (GR, \beta)$.

It follows that $(GR, \beta) \leq_L (GR, \mu_{ULS})$ by the same arguments as above. $\square$

4 Acknowledgements

I am very grateful to my advisor Johan Håstad for many helpful discussions and to Mikael Goldmann for helpful comments.

References

[1] S. Ben-David, B. Chor, O. Goldreich and M. Luby, *On the Theory of Average Case Complexity*, Proc. 21st Symp. on Theory of Computing, ACM, 1989, 204–216.

[2] Y. Gurevich, *Complete and Incomplete Randomized NP Problems*, Proc. 28th Annual Symp. on Found. of Computer Science, IEEE, 1987, 111–117.

[3] R. Impagliazzo and L. Levin. Private communication 1989.

[4] D. S. Johnson, *The NP-Completeness Column*, Journal of Algorithms 5 (1984), 284–299.

[5] L. Levin, *Average Case Complete Problems*, SIAM Journal of Computing, 1986.

Upper Envelope Onion Peeling

John Hershberger

DEC Systems Research Center

Abstract

We consider the problem of finding the upper envelope layers of a set of line segments, sequentially and in parallel. The upper envelope of a set of n line segments in the plane can be computed in $O(n \log n)$ time [10]. By repeatedly removing the segments that appear on the envelope and recomputing the envelope, one obtains a natural partition of the set of segments into *layers*. We give an $O(n \log n)$ sequential algorithm to find envelope layers if the segments are disjoint and an $O(n\alpha(n) \log^2 n)$ algorithm if the segments intersect ($\alpha(n)$ is the extremely slowly-growing inverse of Ackermann's function [9]). Finally, we prove that the problem of finding envelope layers in parallel is P-complete, and hence likely to be intractable.

1 Introduction

Finding the convex layers of a point set (the onion-peeling problem) is a classical problem in computational geometry whose parallel complexity is still unresolved. Chazelle's $O(n \log n)$ sequential algorithm [4] does not seem parallelizable, but neither has the problem been proven P-complete. On the other hand, layers of maxima can be computed quickly in parallel. Motivated by these two problems, we consider a related problem, the *envelope layers problem*.

The envelope layers of a set of line segments are analogous to the convex layers of a set of points, with convex hulls replaced by upper envelopes. The upper envelope of a set of (opaque) line segments in the plane is the collection of segment portions visible from the point $(0, +\infty)$. To define the envelope layers, we repeatedly compute the upper envelope of the set and discard the segments that appear on it (if any piece of a segment appears on the envelope, we discard the whole segment). The *envelope layers problem* is to label each segment with the iteration number at which it appears on the envelope.

The complexity of the upper envelope is the number of distinct pieces of segments that appear on it. In the worst case, the upper envelope of a set of n intersecting line segments may have complexity $\Theta(n\alpha(n))$, where $\alpha(n)$ is the functional inverse of Ackermann's function [9, 15]. The envelope can be constructed in optimal $O(n \log n)$ sequential time [10], and in $O(\log n)$ time on an $O(n)$-processor PRAM (Goodrich, personal communication).

If the segments are non-intersecting, then the complexity of their upper envelope is linear in n. Constructing the envelope takes $\Theta(n \log n)$ time. However, if the left-to-right order of the segment endpoints is given, then the upper envelope can be computed in $O(n)$ time [2].

This paper gives two algorithms for the envelope layers problem, a straightforward one for non-intersecting segments and a more complicated one for intersecting segments. The algorithm for non-intersecting segments exploits the fact that disjoint segments can be topologically ordered from top to bottom; it runs in optimal $O(n \log n)$ time and linear space. A similar algorithm was developed simultaneously by Overmars (personal communication). Intersecting segments cannot be ordered, and so the $O(n \log n)$ algorithm is inapplicable to them. The algorithm for this case is based on an output-sensitive method for computing the upper envelope in $O(\log^2 n)$ time per edge of the envelope, with an $O(n \log n)$ overhead; the bound for computing the envelope layers is $O(n\alpha(n) \log^2 n)$.

The second part of the paper addresses the parallel complexity of the envelope layers problem. We show that the problem is complete for P under NC reductions, and hence is probably intractable in parallel. (A P-complete problem is as hard as the hardest problems solvable in deterministic polynomial time; unless $P = NC$, such problems are not solvable in parallel using polylogarithmic time and a polynomial number of processors [7, 12, 13].) Because of space limitations, some of the proofs are omitted.

The P-completeness of the envelope layers problem is important for two reasons. First, it is a step toward resolving the complexity of the convex layers problem. Second, the envelope layers problem is one of only two geometric problems that have been proven P-complete. (At about the same time as this research was performed, Atallah, Callahan, and Goodrich independently proved the P-completeness of the envelope layers problem, using a reduction different from ours, and a lexicographic planar partition problem [3].)

2   A Sequential Algorithm for Disjoint Segments

We first consider the case of a set S of disjoint segments, with $|S| = n$. This case is easier than the general case for two reasons. First, the segments can be ordered topologically from top to bottom. Second, the vertical lines through the segment endpoints partition the plane into *slabs* in which any vertical line intersects the same segments in the same order. Both of these properties contribute to give an $O(n \log n)$ algorithm.

We can define a relation $\succ$, pronounced "above," on the segments of S. We say that $a \succ b$ if segments a and b are intersected by a common vertical line, and the intersection with a is above that with b. Because a and b do not intersect, we cannot have $b \succ a$; that is, the relation is antisymmetric. Guibas and Yao have shown that the "above" relation is a partial order, and hence can be extended to a total order [8]. This total order can be found by computing the vertical visibility graph of the segments, which has linear size, and applying topological sort to it. The key property of the "above" order is that the layer to which a segment belongs is affected only by segments "above" it, as is easy to see by induction.

Partitioning the plane into slabs lets us rephrase the dynamic definition of layer depth in a static form. Consider a vertical ray extending upward from a segment s at x-coordinate $\hat{x}$; let $f(s, \hat{x})$ be the depth of the deepest segment encountered by the ray. Then the depth of the segment s is one greater than the minimum of $f(s, \hat{x})$, taken over all $\hat{x}$ between the endpoints of s. In each slab formed by drawing vertical lines through all the segment endpoints, the ray defining $f(s, \hat{x})$ intersects the same segments in the same order. Thus the problem of minimizing $f(s, \hat{x})$ is discrete rather than continuous: we minimize $f(s, \hat{x})$ over the $O(n)$ slabs that intersect s.

The previous two observations suggest a straightforward $O(n^2)$ algorithm. The algorithm inserts the segments in "above" order from top to bottom; for each slab the algorithm maintains the maximum depth of the segments that span it. The slab depths are initially zero. When the algorithm inserts a new segment, it sets its depth to one greater than the minimum slab depth of the slabs that intersect the segment. It then updates the depth of each of those slabs to the maximum of its old depth and the depth of the new segment. (This raises the depth of some slabs by one and leaves the others unaffected.) It is straightforward to show by induction that this algorithm correctly computes the depth of each segment.

We can implement this algorithm to run in $O(n \log n)$ time using a segment tree. A segment tree is a complete binary tree whose leaves represent the slabs, taken in left-to-right order, and whose internal nodes represent the union of their descendants' slabs [14, pp. 13–15]. Each region associated with a node, whether an original slab or a union of them, is a *canonical slab*. Any segment can be decomposed into $O(\log n)$ subsegments, each with its endpoints on the boundary of some canonical slab. The unique decomposition of a segment into a minimum number of such subsegments is the *canonical decomposition* of the segment. We modify the quadratic algorithm described above to maintain, for each canonical slab, the minimum slab depth of the original slabs contained in it. To find the depth of a new segment, we compute the minimum slab depth over $O(\log n)$ canonical slabs, rather than over $O(n)$ original slabs as above.

We cannot afford to store the minimum depth of each canonical slab at the node representing the slab; to do so would force us to spend $\Theta(n)$ time updating these values when long segments are added. Instead, we store at each node u a value $val(u)$ that satisfies the following invariant:

> For each canonical slab (represented by a node v), the minimum slab depth is given by the maximum of $val(u)$ over all ancestors u of v, including v itself.

We initialize $val(v)$ to zero for all nodes v, establishing the invariant at the beginning of the algorithm.

The canonical decomposition of a segment uses $O(\log n)$ canonical slabs, and hence $O(\log n)$ nodes of the segment tree, which we call *canonical nodes* for the segment. The parents of these nodes lie on two root-to-leaf paths in the tree. By walking down these paths, we can compute the minimum slab depth of all the canonical nodes, and hence the depth of the new segment, in $O(\log n)$ total time. The full paper tells how to update the tree to maintain the invariant, proving the following theorem:

Theorem 2.1 *The segment tree algorithm finds the depth of each segment correctly and runs in $O(n \log n)$ time and linear space.*

3   A Sequential Algorithm for Intersecting Segments

The algorithm for intersecting segments closely follows the definition of envelope layers. It repeatedly computes the upper envelope, then deletes the segments on it. If the algorithm were to build the upper envelope from scratch at each iteration, it could take $\Theta(n^2 \log n)$ time to find the layers. This section shows how to improve on this bound by using an output-sensitive algorithm to compute the upper envelope. The algorithm is too costly to use to compute a single upper envelope—it has an overhead of $\Omega(n \log n)$ time and space, and is more complicated than the known $O(n \log n)$ algorithm [10]—but it is well suited to layer computation. The algorithm produces envelope edges at a cost of $O(\log^2 n)$ apiece, but more importantly, its data structures allow segments to be deleted at an amortized cost of $O(\log^2 n)$ apiece. This means that the cost of finding the layers is $O(n \log^2 n)$ for deletions, plus the total cost of building envelopes. If n_i segments appear in the ith layer, the worst-case size of the ith envelope is $\Theta(n_i \alpha(n_i))$. To find all the envelopes, the algorithm uses at most

$$O\left(\sum_i n_i \alpha(n_i) \log^2 n\right) = O(n\alpha(n) \log^2 n)$$

time and $O(n \log n)$ space.

The output-sensitive algorithm first finds the leftmost segment of the upper envelope, then traces along the envelope. For each envelope segment, it finds where the segment leaves the envelope, computes its successor, and repeats the process until it reaches the right end of the envelope. The key to the algorithm is finding where the current segment leaves the envelope. The data structures used to solve this problem also support finding the leftmost envelope segment to the right of a given x-coordinate, an operation needed, for example, at the beginning of the algorithm.

Finding the end of the current envelope segment is a kind of "shooting" problem [1, 5]. The problem specifies a collection of segments S, a particular segment $e \in S$, and a point $p \in e$ on the envelope of S. Let $\vec{e}$ be the ray with origin p that extends e to the right. The shooting problem asks for the first place where $\vec{e}$ leaves the upper envelope of $\vec{e} \cup S$. If this point is to the right of the right endpoint of e, then e leaves the envelope at its right end; otherwise $\vec{e}$ and e leave the envelope together.

To make the shooting problem easier, we regularize the segments of S. As noted in Section 2, each segment of S can be broken into $O(\log n)$ canonical subsegments based on a segment tree for S. Each canonical subsegment (call it a *fragment*) is associated with some node of the segment tree: the segment's endpoints lie on the boundary lines of the node's slab. The shooting data structure places the canonical fragments into $O(\log n)$ groups, based on the levels of their associated segment tree nodes. Canonical fragments associated with level i of the segment tree belong to group S_i (the root is level 0).

Figure 1: Configurations of $\vec{e}$ and a cup

Each group S_i has a corresponding partition of the plane into at most 2^i vertical slabs. Each segment fragment in S_i belongs to exactly one slab, and extends all the way across it. Thus within a slab, the fragments look like infinite lines. The upper envelope of infinite lines is a single convex face. The upper envelope of a group S_i, therefore, is a sequence of piecewise linear convex functions, one per slab, which we call *cups*.

The shooting algorithm finds the place where $\vec{e}$ leaves the envelope of $\vec{e} \cup S_i$ for each group S_i. Because $S = \cup_i S_i$ and the shooting problem is decomposable, the leftmost such point is the place where $\vec{e}$ leaves the envelope of $\vec{e} \cup S$. Each group S_i is a sequence of cups. Shooting into a cup is simplified by considering the *cup endpoints* (the intersections of the cup with its slab's boundary). For purposes of shooting, a cup's endpoints characterize the cup (see Figure 1).

Lemma 3.1 *The ray $\vec{e}$ leaves the envelope of $\vec{e} \cup S_i$ in the leftmost cup of S_i for which $\vec{e}$ passes strictly below a cup endpoint.*

Once we have found the cup of S_i in which $\vec{e}$ leaves the envelope of $\vec{e} \cup S_i$, completing the shooting query is not difficult, given the right supporting data structures. Binary search on the cup's segments can find the intersection in $O(\log n)$ time.

Before considering the problem of determining which cups to shoot into, let us discuss the representation of the cups. There are $O(n)$ cups, each one associated with a segment tree node, and $O(n \log n)$ canonical segment fragments distributed among them. Each cup is the upper envelope of fragments in a slab, but because the fragments span the slab, it can be regarded as the upper face of an arrangement of lines, clipped to the slab. Because the envelope layers algorithm deletes segments, we need a data structure that represents the upper face of an arrangement of lines, allows deletions from the set of lines, and supports binary search on the boundary of the upper face.

By applying geometric duality [6], we can solve the problem using the *hull tree* data structure of Hershberger and Suri [11]. Duality maps lines to points and maps the upper face of the line arrangement to the convex hull of the points. The problem of intersecting the face with a line maps to the problem of finding the line(s) supporting the convex hull and passing through a query point. The hull tree is ideal for this dual problem: the convex hull of k points can be built in $O(k \log k)$ time and $O(k)$ space; deleting a point takes $O(\log k)$ amortized time; and finding a supporting line through a query point takes $O(\log k)$ time. Because there are $O(n \log n)$ segment fragments in all the cups, it takes

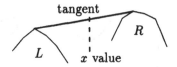

Figure 2: The tangent stored at a hull tree node

$O(n \log^2 n)$ time and $O(n \log n)$ space to build the cup data structures and then to delete the segments. Shooting $\vec{e}$ into a cup takes $O(\log n)$ time.

The shooting algorithm shoots into one cup in each group S_i. It uses the cup endpoints to decide which cups to shoot into. It puts the cup endpoints from all the groups into one left-to-right list (resolving x-coordinate ties according to the order of the cups), then searches rightward in the list from the x-coordinate of p to find a cup endpoint that lies above $\vec{e}$. Let z be the first such endpoint; the shooting algorithm shoots into the cup in each S_i whose slab contains z.

Lemma 3.2 *Let z be the first cup endpoint in L that is right of p and above $\vec{e}$. Then $\vec{e}$ leaves the envelope of $\vec{e} \cup S$ in a cup whose slab contains z.*

The algorithm for finding the point z uses hull trees [11] once again. That data structure does not support insertions, but the updating of cup endpoints requires them. The algorithm resolves this problem by storing the endpoints of all $O(n \log n)$ canonical fragments in the hull tree—these are the present and future cup endpoints. This strategy works because the uppermost fragment endpoint at any slab boundary is also a cup endpoint.

The hull tree is a complete binary tree that stores a set of points in its leaves, sorted in left-to-right order. (In our case the order is not strict.) Each subtree represents the convex hull of the points stored in its leaves; in particular, each node stores the upper common tangent of the convex hulls of its left and right subtrees. Each node also has an associated x-value, either the x-coordinate of its data point, in the case of a leaf, or the x-coordinate of a vertical line separating its left and right subtrees' data points, in the case of an internal node. (See Figure 2.)

The search algorithm begins by searching down the hull tree to find the rightmost leaf to the left of p or with the same x-coordinate. Let T be the set of nodes to the right of the search path whose parents are on the path. The nodes of T are the roots of $O(\log n)$ disjoint subtrees whose union contains all points strictly to the right of p. The algorithm searches these $O(\log n)$ trees in left-to-right order, looking for a point that lies above $\vec{e}$.

To search for the leftmost point that lies above $\vec{e}$ in a subtree, the algorithm compares the position of the ray with that of the tangent edge stored at the root of the subtree, then recursively searches in the children of that node. The search may enter both children, but by distinguishing between successful and unsuccessful searches, we can show that the entire search nonetheless takes only $O(\log^2 n)$ time altogether.

Lemma 3.3 *Using the hull tree data structure for a set of $O(n \log n)$ points, one can find the leftmost point that lies above a rightward-pointing ray $\vec{e}$ in $O(\log^2 n)$ time.*

The discussion so far shows that the shooting algorithm takes $O(\log^2 n)$ time to find the point q where a segment e leaves the upper envelope of S. If e leaves because it intersects some segment, then that segment is the next envelope segment. However, if e leaves because it goes beneath the left endpoint of some other segment (possibly more than one), or because its right endpoint is encountered, then the algorithm must find the uppermost segment immediately to the right of q. This can be done using at most three more shooting queries with vertical and horizontal rays. The same operation is used to find the leftmost segment of the upper envelope at the beginning of each iteration.

The algorithm described in this section finds an upper envelope of complexity k in time $O(k \log^2 n)$. The cost of deleting a segment that appears on the upper envelope is $O(\log^2 n)$, and the overhead of the data structure is $O(n \log^2 n)$ time and $O(n \log n)$ space.[1] This establishes the following theorem:

Theorem 3.4 *The upper envelope layers of a set of n possibly-intersecting line segments can be found in $O(n\alpha(n) \log^2 n)$ time and $O(n \log n)$ space.*

Remark: A simplification of this algorithm can compute the upper envelope layers for disjoint segments in $O(n \log n)$ time and space. The full paper gives more details of this algorithm.

4 Parallel Complexity

This section shows that the envelope layers problem is P-complete even for disjoint segments, and therefore intractable in parallel unless $P = NC$ [7, 13]. The core of the proof is a reduction from the monotone circuit value problem to the envelope layers problem. Given a monotone circuit and an assignment of input values, we produce a segment arrangement that simulates the circuit under the assignment of input values. The outputs of the circuit are encoded by the layer number of certain output segments.

Our proof uses a standard version of the monotone circuit value problem [7]. The circuit is represented by a directed acyclic graph (DAG). The nodes of the graph are *inputs* if they have indegree 0, *outputs* if they have indegree 1 and outdegree 0, and *gates* if they have indegree 2. The gates are of two types, ANDs and ORs. Edges of the graph play the rôle of wires connecting the circuit components. An instance of the problem specifies a circuit and an assignment of the values TRUE and FALSE to the inputs. We assume that the nodes of the DAG are topologically sorted, with all inputs at the beginning of the ordering and all outputs at the end. (Topological sorting is in NC^2.)

Given an instance of the monotone circuit value problem of size n, we construct an arrangement of $O(n^2)$ disjoint horizontal segments and an assignment of circuit outputs

---

[1]Initialization time can be reduced to $O(n \log n)$, but this does not affect the asymptotic complexity of the algorithm.

Figure 3: The two cell configurations

to segments such that the layer depth of an output segment is congruent to zero modulo 3 if and only if the corresponding circuit output is FALSE. We first describe the form of the arrangement, then focus on the details of segment placement.

The segment arrangement we build is laid out on a grid. The grid has one column for each edge of the DAG and one row for each gate, ordered from top to bottom according to the node ordering. (For convenience, we draw input columns to the left of output columns, but this is not essential.) The rows of the grid are numbered from 0 to m, where m is the number of gates; row 0 is used to set depths for the gates in rows 1 through m. All segments are horizontal and have endpoints on the column boundaries. There is no separation between adjacent columns. (The restriction of the endpoints to lie on $O(n)$ vertical lines could be regarded as a degeneracy. The full paper shows how to perturb the segments so that no two endpoints lie on the same vertical line.) Each cell of the grid has one of two structures, as shown in Figure 3.

Each row of the segment arrangement simulates one gate of the circuit. In each column that is not an input or an output of gate k, there is at least one segment above row k with depth greater than $3k$. In the input and output columns, the depths force the output segment of gate k to be at depth $3k$ or $3k + 1$, depending on the depths of the input segments of the gate. The output depth is $3k$ if the output of the gate is FALSE, and $3k + 1$ if the output is TRUE. The segments in the gate gadget (the segments in row k that extend between input and output columns) have depth at most $3k + 1$. Because the depth of coverage in the other columns is at least $3k + 1$, the gate gadget is unaffected by those columns, and vice versa.

Each column of the segment arrangement acts as a wire. Just below the row in which the value of the signal traveling on the wire is set, the arrangement contains a clump of *deepener* segments, which increase the depth of coverage of the column to match the level of the gate or output where the signal is used. Once the signal has been used as a gate input, another clump of deepener segments in the gate row increases the depth of coverage of the column to at least $3m + 1$, preventing interactions between this column and lower gates.

The segment arrangement has three kinds of segments: input segments, deepeners, and gate gadgets. We consider each of these in turn, then prove the correspondence between the segment arrangement and the circuit.

The input slot of row 0 is used only in columns that carry circuit inputs. If the input

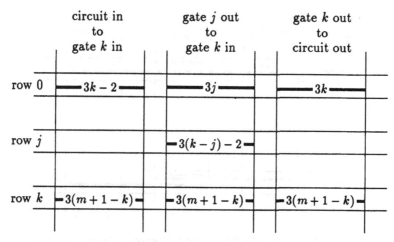

Figure 4: Placing deepeners for two inputs and one output of gate k. The notation "$-\!\!-d-\!\!-$" represents d deepeners.

associated with a particular column is TRUE, then we put a segment spanning the column in the input slot; otherwise, we leave the slot empty.

Deepener segments set the levels of gates and propagate signals from one row of the grid to another. If a column carries the output of gate k, we place $3k$ segments in the deepener slot of row 0 of that column. If a column carries a signal from row j to the gate in row k, with $0 \leq j < k$, we place $3(k-j)-2$ deepeners in row j of that column. Consequently, the deepest segment above row k in this column has depth $3k-2$ or $3k-1$, depending on whether the signal propagating in the column is FALSE or TRUE. If the output of gate k is connected to a circuit output via some column, then we place $3(m+1-k)$ segments in the deepener slot of row k of that column. We also place $3(m+1-k)$ deepeners in row k of the input columns for gate k. Refer to Figure 4 for examples of deepener placement.

The gate gadget for gate k lies entirely in row k. If the inputs of gate k are columns i and j, $i < j$, then column i contains a segment in the "left input" slot of row k; column j contains a segment in the "right input" slot; a segment in the output slot reaches from column i or j to the rightmost column for the output of gate k. The exact configuration of the gadget depends on whether gate k is an AND or an OR, as shown in Figure 5.

The following lemmas show that the segment arrangement solves the given instance of the monotone circuit value problem. We use the notation $depth(i, k)$ to refer to the depth of the maximum depth segment in column i above row k. We adopt the convention that if the depth of a signal-carrying segment may be either d or $d+1$, the lower value means the signal is FALSE and the higher means the signal is TRUE. We call this the *truth convention*.

Lemma 4.1 *Let $\overline{C}_k$ be the set of columns that* do not *carry inputs or outputs of gate k. Suppose that $depth(i, k) > 3k$ for all $i \in \overline{C}_k$, $depth(i, k) \in \{3k-2, 3k-1\}$ for each input i of gate k, and $depth(i, k) = 3k$ for each output i of gate k. Then the output segment of*

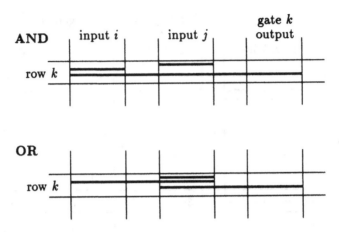

Figure 5: AND and OR gate gadgets

gate k has depth 3k or 3k + 1, satisfying the truth convention as a function of the inputs. The segments in row k do not affect segment depths in columns of $\overline{C}_k$.

Proof: Because $depth(i, k) = 3k$ in each output column i, the output segment of gate k has depth at most $3k + 1$. (Refer to Figure 5.) Since $depth(i, k) > 3k$ for all $i \in \overline{C}_k$, none of the columns in $\overline{C}_k$ affects the depth of the output segment, and vice versa.

We first consider the AND gate gadget. The segments just above the output segment in the two input columns have depths in the range $\{3k - 1, 3k\}$. The depth of the output segment is one more than the minimum of these segments' depths, and so the gate computes the AND function.

In the OR gate gadget, all the computation occurs in the input column on the right. The upper segment of the gadget (the one contained in the right column) has depth $3k - 1$ or $3k$. The lesser of these depths would limit the left segment (the one that spans both input columns) to have depth at most $3k$, but this does not interfere with the value set by the left input column, which is $3k - 1$ or $3k$. The depth of the output segment is one more than the maximum of the upper and left segments' depths, and so the gate computes the OR function. ∎

Lemma 4.2 *Let $\overline{C}_k$ be the set of columns that do not carry inputs or outputs of gate k. Then $depth(i, k) > 3k$ for all $i \in \overline{C}_k$, $depth(i, k) \in \{3k - 2, 3k - 1\}$ for each input i of gate k, and $depth(i, k) = 3k$ for each output i of gate k.*

Proof: The proof is by induction on k. The inputs to gate 1 must be inputs to the circuit, so the input columns have at most one segment in the input slot of row 0 and exactly $(3 - 2) = 1$ segment in the deepener slot. Thus $depth(i, 1) \in \{1, 2\}$ for each input i of gate 1. Clearly, $depth(i, 1) = 3$ for each output i. Each column i in

$\overline{C}_k$ is either an input to some gate $j > 1$ (so $depth(i,1) \geq 3j - 2 > 3$) or the top of an output column for some gate $j > 1$ (so $depth(i,1) = 3j > 3$).

For $k > 1$, each column i that carries an input to the circuit down to gate k has $depth(i,k) = 3k - 2 + \{0$ or $1\}$, which is $3k - 2$ or $3k - 1$. By Lemma 4.1 the output segment of any gate $j < k$ has depth $3j$ or $3j + 1$. Hence if column i carries the output of gate j to an input of gate k, the $3(k-j) - 2$ deepeners below the output segment in row j ensure that $depth(i,k) \in \{3k - 2, 3k - 1\}$. If i is an output column of gate k, the deepeners at the top of the column set the depth to $3k$, and no segment above row k can change this (Lemma 4.1). Each column i in $\overline{C}_k$ is either (a) an input to some gate $j > k$ (so $depth(i,k) \geq 3j - 2 > 3k$), (b) the top of an output column for some gate $j > k$ (so $depth(i,k) = 3j > 3k$), (c) a circuit output from gate $j < k$ (so $depth(i,k) \geq 3m + 3 > 3k$), or (d) the bottom of an input column for gate $j < k$ (so $depth(i,k) \geq (3j - 1) + 3(m + 1 - j) > 3k$). Cases (a), (c), and (d) depend on Lemma 4.1, and case (d) also depends on the gate gadgets of Figure 5. ∎

The depths of certain segments encode the outputs of the circuit. If column i carries a gate output to a circuit output, then $depth(i, m + 1)$ is either $3m + 3$ or $3m + 4$; this is the depth of the deepest segment fully contained in column i, namely a deepener in the gate row. This depth encodes the value of the circuit output according to the truth convention.

Theorem 4.3 *The arrangement of segments described above can be built in NC and simulates the monotone circuit on the given inputs. It follows that the envelope layers problem is P-complete.*

5 Open Problems

Several problems remain unresolved. The first is that of improving the sequential algorithm for computing envelope layers. The lower bound for this problem is just $\Omega(n \log n)$, and an $O(n\alpha(n) \log n)$ algorithm does not seem out of reach.

The convex layers problem, which inspired this research into envelope layers, is still a major open problem in parallel complexity. Does it belong to NC, is it P-complete, or does it lie somewhere in between?

We have shown that the envelope layers problem is P-complete even for disjoint, horizontal segments. To understand the source of this complexity, we can consider a specialization suggested by Goodrich, in which all the segments are horizontal and have constant length. Similarly, we can consider segments with bounded length ratios. Are the resulting problems still P-complete? Our proof does not apply in these cases.

Acknowledgments

I would like to thank Mike Goodrich for proposing the envelope layers problem and for helpful discussions, Richard Anderson for clarifying the subject of P-completeness, and DIMACS, for sponsoring the workshop where this research began.

References

[1] P. K. Agarwal. Ray shooting and other applications of spanning trees with low stabbing number. In *Proceedings of the 5th ACM Symposium on Computational Geometry*, pages 315–325, 1989.

[2] Ta. Asano, Te. Asano, L. Guibas, J. Hershberger, and H. Imai. Visibility of disjoint polygons. *Algorithmica*, 1(1):49–63, 1986.

[3] M. Atallah, P. Callahan, and M. Goodrich. P-complete geometric problems. In *Proceedings of the 2nd ACM Symposium on Parallel Algorithms and Architectures*, 1990.

[4] B. Chazelle. On the convex layers of a planar set. *IEEE Transactions on Information Theory*, IT-31(4):509–517, July 1985.

[5] B. Chazelle and L. Guibas. Visibility and intersection problems in plane geometry. In *Proceedings of the ACM Symposium on Computational Geometry*, pages 135–146, 1985.

[6] H. Edelsbrunner. *Algorithms in Combinatorial Geometry*, volume 10 of *EATCS Monographs on Theoretical Computer Science*. Springer-Verlag, 1987.

[7] R. Greenlaw, H. J. Hoover, and W. L. Ruzzo. A compendium of problems complete for P. Manuscript, 1989.

[8] L. J. Guibas and F. F. Yao. On translating a set of rectangles. In *Proceedings of the 12th ACM Symposium on Theory of Computing*, pages 154–160, 1980.

[9] S. Hart and M. Sharir. Nonlinearity of Davenport-Schinzel sequences and of generalized path compression schemes. *Combinatorica*, 6:151–177, 1986.

[10] J. Hershberger. Finding the upper envelope of n line segments in $O(n \log n)$ time. *Information Processing Letters*, 33:169–174, 1989.

[11] J. Hershberger and S. Suri. Applications of a semi-dynamic convex hull algorithm. In *Proceedings of the 2nd Scandinavian Workshop on Algorithm Theory*. Springer-Verlag, 1990.

[12] R. E. Ladner. The circuit value problem is log space complete for P. *SIGACT News*, 7(1):18–20, 1975.

[13] Ian Parberry. *Parallel Complexity Theory*. Research Notes in Theoretical Computer Science. Pitman Publishing/John Wiley & Sons, 1987.

[14] F. P. Preparata and M. I. Shamos. *Computational Geometry*. Springer Verlag, New York, 1985.

[15] A. Wiernik and M. Sharir. Planar realization of nonlinear Davenport-Schinzel sequences by segments. *Discrete and Computational Geometry*, 3:15–47, 1988.

Applications of a Semi-Dynamic Convex Hull Algorithm

John Hershberger
DEC Systems Research Center

Subhash Suri
Bell Communications Research

Abstract

We obtain new results for manipulating and searching semi-dynamic planar convex hulls (subject to deletions only), and apply them to derive improved bounds for three problems in geometry and scheduling. The new convex hull results are logarithmic time bounds for set splitting and for finding a tangent when the two convex hulls are not linearly separated. We then apply these results to solve the following problems: (1) [**matching**] given n red points and n blue points in the plane, find a matching of red and blue points (by line segments) in which no two edges cross, (2) [**scheduling**] given n jobs with due dates, linear penalties for late completion, and a single machine on which to process them, find a schedule of jobs that minimizes the maximum penalty, and (3) [**covering**] given n points in the plane and two real numbers r_1 and r_2, determine if one can cover all the points with two disks of radii r_1 and r_2. Our time bounds are $O(n \log n)$ for problems (1) and (2) and $O(n^2 \log n)$ for problem (3), an improvement by a factor of $\log n$ over the previous bounds.

1 Introduction

The convex hull is a versatile and well-studied structure in computational geometry. It has numerous applications, including statistical trimming, range searching, and facility-location problems in operations research. Not surprisingly, computing the convex hull is one of the best-investigated problems in computational geometry. In two dimensions, an $O(n \log n)$ time algorithm for computing the convex hull of n points has been known since 1972 [5]—several other algorithms have been discovered since then; see Preparata and Shamos [14] or Edelsbrunner [3] for a survey. A tight bound on the problem was ultimately achieved by Kirkpatrick and Seidel [9], who showed that $\Theta(n \log h)$ is both an upper and lower bound for the planar convex hull problem, where h denotes the number of points on the hull.

In many applications, one needs *dynamic* convex hulls, where points can be inserted or deleted from the set. The best bound currently known for the dynamic convex hull problem is due to Overmars and van Leeuwen [12], who show that a convex hull can be maintained at a worst-case cost of $O(\log^2 n)$ per insertion or deletion. Better results are possible, however, if there are only insertions or only deletions. In the case of insertions, an algorithm due to Preparata [13] can insert a new point and update the convex hull in worst-case time $O(\log n)$. The case of deletions is a little more complicated, but it is possible to process a sequence of n deletions in $O(\log n)$ amortized time per deletion (see Hershberger and Suri [7]. The same result is also implicit, though not fully worked out, in the convex layers algorithm of Chazelle [2].

In all the applications considered in this paper, we work with a deletions-only convex hull. Starting with a set of n points, our algorithms perform an online sequence of deletions intermixed with searches and manipulations on the current convex hull. Our applications require two key operations: (1) finding a common tangent of two convex hulls that are not necessarily separated by a line, and (2) the set-splitting operation, where we split our data structure in two by a vertical line so that the points to the left of the line go into one structure and the points on the right go into the other. We achieve an $O(\log n)$ time bound for both these operations; the bound is amortized for the set-splitting operation. (Overmars and Leeuwen [12] also have a $O(\log n)$ time algorithm for finding a common tangent, but their method only works for linearly-separated convex hulls.) Using these results, we obtain improved algorithms for the following problems.

1. (**Matching.**) Given n red points and n blue points in the plane, find a matching of red and blue points by line segments in which no two segments cross.

2. (**Scheduling.**) Given n jobs with due dates, linear penalties for late completion, and a single machine on which to process them, find a schedule of jobs that minimizes the maximum penalty.

3. (**Covering.**) Given n points in the plane and two real numbers r_1 and r_2, determine if one can cover all the points with two disks of radii r_1 and r_2.

We obtain $O(n \log n)$ time algorithms for problems (1)–(2) and an $O(n^2 \log n)$ time algorithm for problem (3), improving the previous bounds by a factor of $\log n$. The bounds for the first two problems are the best possible in the worst case.

Our data structure and the accompanying analysis can be applied to achieve an $O(\log n)$ time bound for some special cases of intermixed insertions and deletions. These extensions, interesting in their own right, may also be useful for attacking the general problem of maintaining fully dynamic convex hulls.

This paper is organized in six sections. In Section 2, we describe our basic data structure for maintaining and searching a deletions-only convex hull. Sections 3 and 4 show how to strengthen this data structure to allow two additional operations: finding common tangents of two deletions-only hulls, and splitting the point-set at a given x-coordinate. The main applications (matching, scheduling, and covering) are discussed in Section 5. We conclude in Section 6 with some extensions and open problems. Due to space limitations most of the proofs are omitted.

2   The basic data structure

In this section we describe our basic data structure for maintaining and searching the convex hull of a planar set of points during an online sequence of deletions. We are given a fixed set of points S that underlies all the updates, with $|S| = n$. Starting with a subset $P \subseteq S$, we would like to maintain the convex hull of P as points are deleted online from P. For convenience, we consider only the problem of maintaining the "upper hull" of P, denoted $h(P)$.

382

Our underlying data structure is a complete binary tree whose leaves correspond to the points of S in x-order; the same basic structure is also used by Chazelle [2] and Overmars and van Leeuwen [12]. The tree is the *interval tree* for S [14]; we denote it by $I(S)$. With every node $v \in I(S)$, we associate a set $S(v)$ and an x-coordinate $x(v)$, where $S(v)$ is the subset of S corresponding to the leaves below v, and $x(v)$ is the midpoint of the interval bounded by the rightmost point of $S(leftson(v))$ and the leftmost point of $S(rightson(v))$. (If v is a leaf, then $x(v)$ is simply the x-coordinate of the corresponding point.) We represent a subset $P \subseteq S$ by a subgraph of $I(S)$. It is this subgraph that changes as points are deleted from P, and not the interval tree $I(S)$, which remains fixed throughout. The subgraph representing P is denoted $T(P)$. The rule that defines $T(P)$ is quite simple: a node $v \in I(S)$ is copied into $T(P)$ if and only if $P(v) \equiv P \cap S(v)$ is nonempty. Refer to Figure 1 for an example. We endow this basic subgraph with additional node-fields to represent the upper hull $h(P)$. We call the full data structure a *hull tree*.

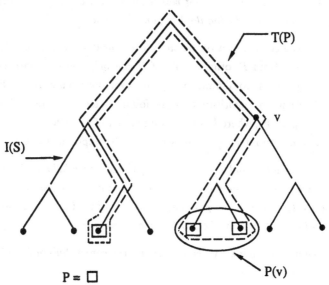

Figure 1: An example

A hull tree node v represents $h(P(v))$, the upper hull of the points that correspond to leaves below v in $T(P)$. Storing the entire hull $h(P(v))$ at each node v would be convenient for searching and manipulating $h(P)$, but unfortunately would require a nonlinear amount of storage. In order to reduce the space complexity to linear, we store only a portion of $h(P(v))$ at the node v. To explain what points are stored where, we need the notion of the *level* of a point, which is also crucial for our amortization arguments.

Consider a point $p \in P$, and let u be the corresponding leaf of $T(P)$. Now, p obviously is a vertex of the convex hull $h(P(u))$, but it may also belong to hulls represented by ancestors of u. Let v be the highest ancestor of u for which p is a vertex of $h(P(v))$; node v may be the root of $T(P)$. We define the *level of p*, written $level(p)$, to be the depth of v in $T(P)$. The point $p \in P$ is stored at node $v \in T(P)$—that is, the node that gives p its level.

It might appear that this storage scheme fragments the hulls rather arbitrarily. However, this is not the case: hull partitioning has a particularly simple form, due to the fact that the line $x = x(v)$ separates the points associated with the left and right children of v. To illustrate, let l and r denote the left and right children of v. We determine the hull $h(P(v))$ by finding the common tangent of $h(P(l))$ and $h(P(r))$. Then the portions of $h(P(l))$ and $h(P(r))$ hidden by the common tangent are precisely the fragments stored at l and r. The root of $T(P)$ stores the current hull $h(P)$ in its entirety.

Returning to our discussion of the data structure, we store at v the common tangent of $h(P(l))$ and $h(P(r))$ and its two endpoints, which gives us the capability to form $h(P(l))$ and $h(P(r))$ from $h(P(v))$ and vice versa in constant time. With this high-level description of our data structure at hand, we now state several technical lemmas, which together establish our main result in this section (Theorem 2.5). Due to space limitations, we omit the proofs.

Lemma 2.1 *Let $P \subseteq S$ be a given set of points on which we perform a sequence of deletions. Then level(p) is nonincreasing during the lifespan of any point $p \in P$.*

This lemma forms a critical part of the analysis of our data structure. The lemma says that when we delete a point from P, other points of P can only move upward in $T(P)$. Since the height of $T(P)$ is $\lceil \log_2 n \rceil$, the total number of point-level movements (*percolations*) is $O(n \log n)$. To process each deletion, our algorithm uses the following lemma, which allows us to perform a deletion in time $O(\log n + k)$, where k is the total number of percolations caused by the deletion. We use the notation $a..b$ to refer to a convex chain with endpoints a and b.

Lemma 2.2 *Suppose that L and R are two x-separated upper hulls and that lr is their upper common tangent, where $l \in L$ and $r \in R$. Let $p \in L$ and $q \in R$ be two points that lie to the left of l and r, respectively. Then, given p and q, we can compute l and r in time $O(k)$, where k is the total number of hull vertices in the subchains $p..l$ and $q..r$.*

Using the previous two lemmas, we can construct the convex hull of P and then delete points online in $O(\log n)$ amortized time each.

Lemma 2.3 *Given a set of points $P \subseteq S$, $|P| = m$, we can initially construct the hull tree $T(P)$ in $O(m \log n)$ time. Then we can delete points from P online and update $T(P)$ in $O(\log n)$ amortized time per deletion.*

Proof Sketch: To construct $T(P)$ initially, we perform a bottom-up merging, based on Lemma 2.2. To update the tree after a deletion, we use a similar merging of hulls along the path from the leaf that contains the deleted point up to the root of $T(P)$. We apply Lemma 2.2 to bound the running time by $O(\log n + \langle\# \text{ of percolations}\rangle)$. ∎

Finally, we show that our representation of the convex hull supports the standard repertoire of binary searches. A typical binary search proceeds by performing tests with the tangent edges stored at the nodes of $T(P)$. In an earlier paper [7], we represented $h(P)$ by a search tree that emulated the hull-chain stored at the root of $T(P)$. The following lemma removes this complication.

Lemma 2.4 *Using the bridge edges stored at nodes of $T(P)$, we can perform a binary search on the node-list of $h(P)$. In particular, the following primitives can be implemented in worst-case time $O(\log n)$: (1) find a tangent to $h(P)$ from a point outside the hull, (2) find an extreme vertex of $h(P)$ in a query direction, (3) find the intersections of $h(P)$ with a query line, and (4) determine whether or not a point lies inside $h(P)$. In addition, if L and R are two x-separated subsets of S stored in the hull trees $T(L)$ and $T(R)$, respectively, then we can find the outer common tangents of $h(L)$ and $h(R)$ in time $O(\log n)$.*

The main result of this section is summarized in the following theorem.

Theorem 2.5 *Let S be a fixed set of points, $|S| = n$, and let $P \subseteq S$ be a subset, $|P| = m$. In $O(m \log n)$ preprocessing time, we can build a data structure representing the convex hull of P such that the data structure (1) uses $O(\min(n, m \log n))$ space, (2) supports the search queries of Lemma 2.4, and (3) allows online deletion of points from P in $O(\log n)$ amortized time per deletion.*

Our data structure as described so far is quite similar to, although a bit simpler than, the one we use for circular hulls in [7]. We now extend this basic structure to allow more general types of search and manipulation, in particular, the two-hull tangent finding and set-splitting operations. These results are discussed in the next two sections.

3 Searching on Two Convex Hulls

Let R (red) and B (blue) be two disjoint sets of points in the plane, where $|R|$ and $|B|$ are $O(n)$. We wish to maintain the (upper) convex hulls of R and B, allowing deletions, such that at any time a common tangent of $h(R)$ and $h(B)$ can be computed in logarithmic time. Recall that Theorem 2.5 allows us to maintain the two hulls, $h(R)$ and $h(B)$, at an amortized cost of $O(\log n)$ per deletion. The tangent-finding problem, however, can be tricky: given two arbitrary convex polygons of n vertices each, we need $\Omega(n)$ time in the worst case to find a common tangent. In general two convex hulls may have a linear number of common tangents between them, but we are content with finding just one.

In view of the lower bound mentioned above, we clearly need some additional information about the two hulls to speed up the tangent-finding operation. Overmars and van Leeuwen have shown that if we know a line that is guaranteed to cut the tangent, then the tangent can be found in $O(\log n)$ time [12]. Our method also uses this fact, but the difficulty in our case stems from not knowing such a line *a priori*. (In Overmars and van Leeuwen's case, the existence of such a line is trivially guaranteed, since their two hulls are always separated by a vertical line.) Indeed, most of our effort is spent in determining a line that is guaranteed to cut the tangent. Once such a line is found, we invoke the tangent-finding method of [12].

Our main result in this section is this: given two sets of points R and B, we can maintain their upper convex hulls with deletions so that at any instant if the x-intervals of the two sets do not nest (neither contains the other), then we can find their upper common tangent in logarithmic time. It is crucial for the logarithmic bound that we know $R \cup B$ in advance; if R and B are not preprocessed together, then $\Omega(\log^2 n)$ time is a lower bound for finding a tangent [6].

Our method for finding a common tangent of two hulls is based on the hull tree data structure of Section 2. Rather than solving the problem of finding an upper common tangent of $h(R)$ and $h(B)$ directly, we work on the "dual" problem, which is to find an intersection point of the two hulls. That is, we maintain the upper convex hulls of R and B, allowing deletions, so that at any time if the two hulls intersect then we can find an intersection point in logarithmic time. Observe that given an intersection point, we can use Overmars and van Leeuwen's algorithm to find a tangent—a vertical line through the intersection point cuts the tangent edge. In the following discussion, it is convenient to assume that an upper convex hull begins and ends with a semi-infinite vertical ray to $-\infty$. We denote these extended hulls also by $h(R)$ and $h(B)$, respectively. Observe that if the x-intervals of R and B overlap but do not nest, then their vertical rays alternate along the x axis.[1] If the x-intervals are disjoint, we can apply Lemma 2.4 to find a tangent directly.

Lemma 3.1 *Let R and B be two disjoint subsets of S that are stored in the hull trees $T(R)$ and $T(B)$, respectively. If the x-intervals of R and B overlap but do not nest, then we can find a point of intersection between $h(R)$ and $h(B)$ in worst-case time $O(\log n)$.*

> **Proof Sketch:** The algorithm traces a root-to-leaf path in the interval tree $I(S)$. At each path node v, we evaluate a function $f(x(v))$, defined to be the difference in the heights of $h(R)$ and $h(B)$ at $x(v)$; recall that $x(v)$ is the x-coordinate associated with v. We branch to the left or the right child of v depending on the sign of $f(x(v))$. The non-nesting assumption on R and B guarantees that the function $f(x)$ has opposite signs at the two extreme x-values. Therefore, by the mean-value theorem, we can find a zero of $f(x)$, which obviously indicates an intersection of $h(R)$ and $h(B)$. To prove the time complexity, we show that f can be computed along a root-to-leaf path in $O(1)$ time per node; this suffices since the height of $I(S)$ is $O(\log n)$. ∎

In view of the earlier discussion, we have established the following theorem.

Theorem 3.2 *Let R and B be two disjoint subsets of S that are stored in the hull trees $T(R)$ and $T(B)$, respectively. If the x-intervals of R and B do not nest, then we can find an upper common tangent of $h(R)$ and $h(B)$ in worst-case time $O(\log n)$.*

4 Set Splitting

In this section we show how to split our convex hull data structure in two along a vertical line. We get an $O(\log n)$ amortized splitting bound by charging the splitting cost to percolations, as in the deletion algorithm of Section 2. Suppose that we want to split P at $x = x_{split}$ to get two sets L and R. From the point of view of L, the result of splitting is the same as if

---

[1] This also ensures that the intersection $h(R) \cap h(B)$ is nonempty.

all the points of R had been deleted individually. Changes in the structure of $T(L)$ are due to percolations of points of L. Instead of deleting the points of R piecemeal, our algorithm batches all the (apparent) deletions into one operation with $O(\log n)$ overhead. This produces $T(L)$ in $O(\log n + k)$ time, where k is the number of percolations in L. The algorithm simultaneously performs a symmetric operation to produce $T(R)$.

Theorem 4.1 *Given $T(P)$ and a vertical line that separates P into L and R, we can build $T(L)$ and $T(R)$ in $O(\log n)$ amortized time and $O(\log n)$ additional storage.*

> **Proof Sketch:** Without loss of generality assume that the vertical separating line has x-coordinate equal to $x(z)$ for some internal node $z \in T(P)$. The splitting algorithm copies the nodes on the path from the root of $T(P)$ to z, putting one copy of each in $T(L)$ and one in $T(R)$. All other nodes belong to exactly one of $T(L)$ and $T(R)$. Tangent edges along the path are recomputed in each tree by bottom-up merging; no other tangent edges are affected. By using the old tangent endpoints as hints for finding the new tangent edges, the algorithm bounds the splitting time by $O(\log n + \langle\# \text{ of percolations}\rangle)$. ∎

5 Applications

This section applies the convex hull data structure to solve three problems. We derive optimal algorithms for two problems, namely, finding a non-intersecting matching of n red and n blue points in the plane, and minimizing the maximum weighted tardiness in a single-machine scheduling problem. The third problem is a disk-covering problem: given n points and two radii r_1 and r_2, determine whether the points can be covered by two disks of the specified radii. We use a variant of our data structure to solve the problem in $O(n^2 \log n)$ time, improving the previous $O(n^3)$ bound.

5.1 A Matching Problem

Let R and B be two planar sets of points in general position, where $|R| = |B| = n$. We refer to the members of R and B, respectively, as *red* and *blue* points. A *matching* of R and B is a collection of n straight line segments linking red and blue points, no two of which share an endpoint. We want to find a matching of R and B in which no two edges intersect. That there always exists a non-intersecting matching of R and B is a celebrated Putnam Competition problem (see [10]).

Our interest in the matching problem is purely algorithmic. The best algorithm previously known for this problem runs in $O(n \log^2 n)$ time and is due to Atallah [1]. Our result is an optimal $O(n \log n)$ time algorithm; a simple reduction from sorting provides a matching lower bound.

Our algorithm, described below in pseudocode, consists of a procedure Match that takes two parameters R and B, which are the input sets of points, and returns a non-intersecting matching of R and B. We use $H(\hat{x})$ to denote the closed left halfplane defined by the vertical line $x = \hat{x}$.

Procedure Match (R, B)
    while R and B are non-empty **do**
      if the x-intervals of R and B are non-nesting **then**
        Compute an upper common tangent rb of $h(R)$ and $h(B)$;
        Delete r from R and b from B
      else
        Find an x-value $\hat{x}$ such that the sets $R_1 = R \cap H(\hat{x})$ and $B_1 = B \cap H(\hat{x})$
          have the same number of points and $R_1 \neq R$, $B_1 \neq B$;
        Call **Match**(R_1, B_1) and **Match**$(R \setminus R_1, B \setminus B_1)$;
      endif
    end while
end Procedure

The correctness of this algorithm is straightforward. To bound the running time of the algorithm, we use results from the previous sections on maintaining and searching convex hulls. We claim that the running time of the algorithm is dominated by n tangent-finding steps, $2n$ deletions, and at most $n - 1$ set-splitting operations. This follows since each tangent-finding operation results in the deletion of two points, one each from R and B, and each set-splitting step partitions R and B into strict subsets. We implement the algorithm by organizing our points into the hull trees $T(R)$ and $T(B)$. The initial construction of these trees takes $O(n \log n)$ time, and by Theorems 2.5 and 3.2, they allow the tangent-finding operation as well as the deletion of a point to be performed in logarithmic amortized time. We show that the remaining set-splitting operation also can be implemented in $O(\log n)$ amortized time. We find a splitting x-coordinate $\hat{x}$ and then apply Theorem 4.1 to partition the data structures $T(R)$ and $T(B)$ into $T(R_1)$, $T(R \setminus R_1)$ and $T(B_1)$, $T(B \setminus B_1)$. The amortized cost of splitting the data structures is $O(\log n)$; it only remains to show how to find $\hat{x}$ in the same time bound. Our method is similar to the one used in Lemma 3.1. We search down a root-to-leaf path in the interval tree $I(R \cup B)$. To determine branching at a node v, we evaluate a function $g(x(v))$, defined to be the number of red points minus the number of blue points lying to the left of the line $x = x(v)$. We set $\hat{x} = x(\hat{v})$, where $\hat{v}$ is the node for which $g(x(\hat{v})) = 0$. That such a node exists is guaranteed by the boundary conditions of g, which state that g has opposite signs at the second leftmost and the second rightmost x-coordinates. Details of this search procedure and its time complexity will be provided in the full paper. The main result of this subsection is stated in the following theorem.

Theorem 5.1 *Let R and B be two planar sets of points in general position, with $|R| = |B| = n$. We can find a matching of R and B in which no two edges intersect in $O(n \log n)$ time and linear space. These bounds are asymptotically optimal.*

5.2 A Scheduling Problem

We present an optimal $O(n \log n)$ time algorithm for the problem of minimizing the maximum weighted tardiness for a set of n jobs. The problem belongs in the area of machine scheduling and is formulated as follows.

Suppose that n distinct jobs need to be scheduled on a single machine. Job i has length p_i, due date d_i, and a weight w_i, which are all assumed to be nonnegative. Under a given schedule (ordering) of the jobs, let f_i be the finish time of job i. The *weighted tardiness* of job i is $\max\{0, w_i(f_i - d_i)\}$. The *maximum weighted tardiness* problem is to find a schedule that minimizes the maximum weighted tardiness of any job.

An $O(n^2)$ time algorithm for the maximum weight tardiness problem was proposed by Lawler [11]. His algorithm allows precedence constraints as well as arbitrary weight functions. Quite recently, Hochbaum and Shamir [8] showed that if the weight functions are linear, then Lawler's algorithm can be implemented in $O(n \log^2 n)$ time.

Our result in this section is an $O(n \log n)$ time algorithm for the maximum weighted tardiness problem with linear weight function. This time bound is easily shown to be optimal by a reduction from sorting. Recently, and independently of us, Fields and Frederickson [4] also have obtained the same result; their algorithm uses Chazelle's [2] deletion method.

Our method is similar to that of Hochbaum and Shamir [8], which in turn is based on Lawler's original algorithm. Lawler's algorithm uses the following "greedy backwards" method: find the job with least cost at the current endpoint of the schedule, place it last in the schedule, and remove it from the list of jobs. Geometrically, we want to maintain the "lower envelope" of n lines so that at any time we can determine which line is lowest at, say, $x = \hat{x}$. This can be viewed, via a familiar point-line duality, as a problem of maintaining convex hulls.

We use the duality transform T, which maps a point $p : (a,b)$ to a line $T_p : y = ax + b$ and a line $l : y = cx + d$ to a point $T_l : (-c, d)$. The transform T has the important property of preserving vertical distances—the vertical distance between a point p and line l in the primal plane is the same as the vertical distance between the point T_l and the line T_p in the dual plane; one can easily verify this by simple algebra. This transformation maps the set of linear constraints $(y = w_i x - w_i d_i)$ to a set of points S, and maps the current schedule endpoint $(\hat{x}, 0)$ to a line $\hat{l} : y = x\hat{x}$. Because T preserves vertical distances, the constraint whose cost is minimized at $\hat{x}$ in primal maps to the point whose vertical distance to the line $\hat{l} : y = x\hat{x}$ is minimum in the dual plane. Finding this point is easy: it is one of the extreme vertices of the convex hull of S in the direction normal to $\hat{l}$.

In short, the following three operations suffice to implement Lawler's "greedy backwards" method:

1. Compute the convex hull of a set of points S, where $|S| = n$.

2. Find an extreme point of the convex hull of S in a given direction.

3. Delete a point from S, and update the convex hull.

By Theorem 2.5, after spending $O(n \log n)$ time to construct the convex hull of S, we can implement the other two operations in amortized time $O(\log n)$ each. Thus, we can solve the maximum weighted tardiness problem for n jobs in time $O(n \log n)$ and space $O(n)$.

This method works even with precedence constraints, as follows. We maintain a *precedence graph*, where jobs are vertices and there is a directed edge from i to j if i must precede j. The *precedence count* for a job i at any time is the number of jobs that must precede i and that are unfinished. Whenever our algorithm selects a job i for scheduling, we first check if the precedence count of i is zero. If so, we schedule it, and subtract one from the precedence count of all j that have an incoming edge from i. Otherwise we place i in a queue and schedule it as soon as its precedence count becomes zero. If there are m precedence constraints, then our algorithm requires $O(m + n \log n)$ time.

Finally, we note that our algorithm also works for concave piecewise linear weight functions. Our system of constraints consists of the linear extensions of all the line segments of all the weight functions. If the total number of line segments in all the weight functions is s, then our algorithm runs in $O(s \log s)$ time. We have established the following theorem.

Theorem 5.2 *Given n jobs with linear weight functions, we can find an optimal schedule for the maximum weighted tardiness problem in $O(n \log n)$ time. If m precedence constraints are additionally specified, our algorithm requires $O(m + n \log n)$ time. If the weight functions are concave piecewise linear, consisting of s line segments altogether, then our algorithm works in time $O(s \log s)$. The space requirement is linear in all cases.*

5.3 A Covering Problem

In a recent paper on tailored partitions, we asked whether there is an algorithm that takes $o(n^3)$ time to place two disks of specified radii so as to cover n data points [7]. In the full paper we give an $O(n^2 \log n)$ algorithm based on circular hulls: a *circular hull* (or α-hull [3]) of a set of points is the intersection of all unit-radius disks that contain the set. The circular hull data structure [7] has essentially the same properties as the convex hull structure of this paper. (If we implemented the algorithm using Overmars and Van Leeuwen's dynamic convex hull structure [12], it would run in $O(n^2 \log^2 n)$ time.)

The covering problem specifies a set S of n points and two radii; without loss of generality assume the radii are 1 and r, with $r > 1$. If we draw the arrangement of n circles of radius r centered on the points, each of the $O(n^2)$ faces of the arrangement corresponds to a subset of S that can be covered by an r-disk. We want to determine, for each subset, whether its complement with respect to S can be covered by a 1-disk.

To save space, we do not build the whole arrangement, but instead explore it piecemeal. Because each region of interest is bounded by the inside of at least one r-disk, we explore the regions along the boundary of each r-disk in turn. Let C be the current r-disk. We intersect all the other r-disks with C and sort the intersections along C. Now we can walk along C, passing from face to face of the arrangement, maintaining the complement of the points corresponding to the current face. Each step from one face to the next adds or deletes a point of the set. Altogether, the operations of intersecting, sorting, and walking take $O(n \log n)$ time for each r-disk.

Walking through the regions along C corresponds to dragging an r-disk along C and maintaining the set of points outside the moving r-disk. At each step during the walk, we want to know whether the points outside the moving disk can be covered by a unit disk. A covering disk exists if and only if the intersection of unit disks centered on the points is nonempty.

It is easy to see that we can drag the r-disk halfway around C (180°) without re-covering any data point that was uncovered during the motion. We compute the desired intersection by maintaining intersections of two disjoint sets A and D; A is subject only to insertions and D is subject only to deletions. We initialize A to be empty and D to contain all the data points outside the initial placement of the moving r-disk. As the disk moves, it covers new data points, which we delete from D, and uncovers others, which we add to A. For each of the sets A and D, we maintain the intersection of the unit circles centered at its points. Any time either A or D changes, we intersect these two regions of the plane. If the result is nonempty, then the problem has a solution: any unit circle whose center lies in the intersection will cover the points outside the moving d-circle.

We cover the full 360° of possible placements for the moving r-disk by repeating this process a second time, starting the moving disk from where it stopped the first time.

By a point-circle duality, intersecting unit disks is equivalent to finding the circular hull of their centers. We use the deletions-only data structure to maintain the circular hull of D. The same data structure, with the deletion operation replaced by insertion, works for A, or we can use a variation on Preparata's technique [13]. Because either A or D may be too large to fit in a unit disk, we enhance the data structure to support this case. Whenever A or D changes during the walk around C, we can determine in $O(\log n)$ time whether $A \cup D$ will fit in a unit disk. (Details omitted.) There are $O(n)$ changes to A and D, and so circular hull operations during the walk along C take $O(n \log n)$ time. Summed over all r-disks, the algorithm takes $O(n^2 \log n)$ time altogether.

Theorem 5.3 *Given n points in the plane and two radii r_1 and r_2, we can determine in $O(n^2 \log n)$ time and linear space whether two disks of radii r_1 and r_2 can be placed to cover all the points.*

6   Extensions and Open Problems

Our data structure achieves the $O(\log n)$ bound per deletion by amortizing the total number of percolations (i.e., point-level movements) over a sequence of deletions. We showed that, during an online sequence of n deletions, there are $O(n \log n)$ percolations. In the deletions-only case, the bound on the total number of percolations was derived by using the fact the a point always moves up in the hull tree, never down. Since the tree has height $O(\log n)$, each point can contribute just that many percolations.

This analysis works for any set of operations that causes $O(n \log n)$ percolations altogether. To what extent can we push this analysis to accommodate both insertions and deletions? The difficulty with arbitrary insertions is that a single new point can cause almost all other points of P to move down in the tree, thus increasing their level. The repeated insertion and deletion of a "bad" point can therefore kill the amortization argument. But what if the insertions have some special form? We describe a few instances where our data structure can process insertions and deletions in $O(\log n)$ amortized time per operation.

Suppose we are given a set S of n points in the plane. We want to maintain the convex hull of those points of S that fall inside a rectangular window whose sides are parallel to the coordinate axes and that is moving from left to right. (Think of a camera scanning a scene.) In this case, a point is inserted (resp. deleted) when it hits the right (resp. left) wall of the window. Our hull tree data structure can maintain the convex hull inside the window in $O(\log n)$ amortized time per insertion/deletion. The proof shows that the level of a point moves up and down only a constant number of times.

Next, consider a wedge having a fixed angle $\theta < \pi$, rotating around a fixed center O. We want to maintain the convex hull of points of S that are in the wedge. A transformation to the (r, θ) coordinate system reduces this problem to the one of a moving window in the Cartesian coordinate system, and hence we obtain the same result.

Finally, we can extend the moving slab (window) idea to allow some insertions in the middle of the slab, as follows. (The deletions still occur when a point hits the left boundary of the slab.) We require that all new insertions are at least a constant fraction of the points away from the left boundary of the slab; that is, if a point p is inserted when the number of points in the slab is k, then there are at least αk points in the slab to the left of p (for some $\alpha > 0$). Then our data structure achieves $O(\log n)$ amortized cost per insertion and deletion.

In terms of the open problems, of course, the most outstanding one is to achieve an $O(\log n)$ bound, amortized or worst-case, for maintaining the convex hulls during an arbitrary sequence of insertions and deletions.

There are other, possibly easier, instances of intermixed insertions and deletions where we don't know how to beat the $O(\log^2 n)$ bound. For example, in the machine-scheduling problem, if the weight functions are *convex piecewise linear*, then implementing Lawler's algorithm requires both the insertion and the deletion of linear functions. When phrased as a convex hull maintenance problem, the deletions and insertions have a very special form, as follows. Each convex piecewise linear function dualizes to a convex chain. At each step, we only need to maintain the convex hull of the leftmost points of all the chains. Whenever we delete the leftmost point of a chain, its successor is inserted in our convex hull. Is it possible to achieve an $O(\log n)$ bound per insertion and deletion in this case?

References

[1] M. Atallah. A matching problem in the plane. *Journal of Computer and System Sciences*, 31:63–70, 1985.

[2] B. Chazelle. On the convex layers of a planar set. *IEEE Transactions on Information Theory*, IT-31(4):509–517, July 1985.

[3] H. Edelsbrunner. *Algorithms in Combinatorial Geometry*, volume 10 of *EATCS Monographs on Theoretical Computer Science*. Springer-Verlag, 1987.

[4] M. Fields and G. Frederickson. A faster algorithm for the maximum weighted tardiness problem. Manuscript, 1989.

[5] R. Graham. An efficient algorithm for determining the convex hull of a finite planar set. *Information Processing Letters*, 1:132–133, 1972.

[6] L. Guibas, J. Hershberger, and J. Snoeyink. Compact interval trees: A data structure for convex hulls. In *Proceedings of the First Annual ACT-SIAM Symposium on Discrete Algorithms*, January 1990.

[7] J. Hershberger and S. Suri. Finding tailored partitions. In *Proceedings of the 5th ACM Symposium on Computational Geometry*, pages 255–265. ACM, 1989.

[8] D. Hochbaum and R. Shamir. An $O(n \log^2 n)$ algorithm for the maximum weighted tardiness problem. *Information Processing Letters*, 31:215–219, 1989.

[9] D. Kirkpatrick and R. Seidel. The ultimate planar convex hull algorithm? *SIAM J. Comput.*, 15:287–299, 1986.

[10] L. C. Larson. *Problem-Solving Through Problems*. Springer Verlag, New York, 1983.

[11] E. L. Lawler. Optimal sequencing of a single machine subject to precedence constraints. *Management Science*, 19:544–546, 1973.

[12] M. Overmars and H. van Leeuwen. Maintenance of configurations in the plane. *Journal of Computer and System Sciences*, 23:166–204, 1981.

[13] F. P. Preparata. An optimal real time algorithm for planar convex hulls. *Communications of the ACM*, 22:402–405, 1979.

[14] F. P. Preparata and M. I. Shamos. *Computational Geometry*. Springer Verlag, New York, 1985.

Intersection Queries in Sets of Disks*

Marc van Kreveld[1]   Mark Overmars[1]   Pankaj K. Agarwal[2]

[1] Dept. of Computer Science, Utrecht University
P.O. Box 80.089, 3508 TB Utrecht, the Netherlands
[2] DIMACS Center, Rutgers University
P.O. Box 1179, Piscataway, NJ 08855-1179, USA
and
Computer Science Department, Duke University
Durham, NC 27706, USA

Abstract

In this paper we develop some new data structures for storing sets of disks such that different types of queries can be answered efficiently. In particular we study intersection queries with lines and line segments and shooting queries. In the case of non-intersecting disks we obtain structures that use linear storage and have a query time of $O(n^{\beta+\epsilon} + k)$ for intersection queries, where n is the number of disks, $\beta = \log_2(1 + \sqrt{5}) - 1 \approx 0.695$, ϵ is an arbitrarily small positive constant, and k is the number of disks reported. For sets of intersecting disks we obtain a structure that uses $O(n \log n)$ storage and answers a query in time $O(n^{2/3} \log^2 n)$. For ray shooting we obtain a structure that uses linear storage and has $O(n^{\beta+\epsilon})$ query time, for any $\epsilon > 0$.

1   Introduction

Data structures for storing geometric objects, allowing for a variety of different types of queries, play an important role in computational geometry. Many different data structures have been designed for different types of objects and queries. One important type of query is the intersection query. Here we ask which objects in the data structure intersect a given query object. This problem has particularly been studied for the case in which the structure stores a set of points and the query object is an axis-parallel rectangle, the so-called range searching problem. Also data structures exist that store axis-parallel objects like horizontal line segments or rectangles.

Only recently structures have been designed that store sets of arbitrarily oriented line segments and allow for intersection queries with arbitrary line segments (see [1, 6, 10, 13]).

---
*The research of the first two authors was supported by the ESPRIT II Basic Research Actions Program of the EC under contract No. 3075 (project ALCOM). The work of the third author was supported by DIMACS (Center for Discrete Mathematics and Theoretical Computer Science), a National Science Foundation Science and Technology Center - NSF-STC88-09648.

Almost no work has been done on intersection searching in curved objects. The only result we are aware of is a forthcoming paper by Sharir for searching with a point in a set of disks [14]. In this paper we will give some first data structures for storing sets of disks such that intersection queries with lines and line segments can be performed efficiently.

The results in this paper use as underlying data structure the so-called conjugation tree designed by Edelsbrunner and Welzl [9]. This structure is based on an elegant way to partition the plane recursively. (The structure is also called a ham-sandwich tree in [7].) We augment this structure by associating data structures to internal nodes that store subsets of the disks. These associated structures are based on hierarchical convex hulls of disks.

The rest of this paper is organized as follows.

In Section 2 we briefly recall the main properties of the conjugation tree we use. Moreover we indicate the basic method of storing disks in the conjugation tree.

In Section 3 we restrict ourselves to sets of non-intersecting disks. First we describe how intersection queries with a line can be performed in time $O(n^\beta + k)$ using only linear storage, where $\beta = \log_2(1 + \sqrt{5}) - 1 < 0.695$ and k is the number of reported answers. Next, we extend this result to queries with line segments instead of lines. The query time becomes $O(n^{\beta+\epsilon} + k)$ for any $\epsilon > 0$, while still using only $O(n)$ storage.

In Section 4 we handle intersection queries in arbitrary (i.e., intersecting) disks. In this case we use a quite different approach. We transform the problem to a three-dimensional problem, mapping the disks to points in $\mathbb{R}^3$. The structure we obtain uses $O(n \log n)$ storage and has a query time of $O(n^{2/3} \log^2 n + k)$.

In Section 5, we look at the related problem of ray shooting. Here we are given a query ray emanating from a point q in direction d, and we ask for the first disk we hit as we move from q along the ray. We give a data structure for this problem that uses linear storage (for arbitrary disks) and has a query time of $O(n^{\beta+\epsilon})$, for any $\epsilon > 0$.

Finally, in Section 6, we conclude with some extensions and open problems. In particular, we briefly indicate how the query time of the algorithms, described in Section 3 and Section 5, can be improved to $O(n^{2/3+\epsilon})$, for any $\epsilon > 0$, without increasing the space requirement.

2 The conjugation tree

The basic data structure for most solutions presented in this paper is the conjugation tree (also called the ham-sandwich tree), introduced by Edelsbrunner and Welzl in [9] (see also [7]). The tree was originally used for storing a set of points in the plane, and for solving half planar range queries: how many (or which) points lie in a query halfplane. The structure can also be used to determine how many (or which) points lie in a triangle, or on a query line. The conjugation tree is a binary tree in which each node corresponds to a convex region of the plane. The root corresponds to the whole plane. The region corresponding to a node δ is divided by a straight line l_δ, called *bisecter*, into two subregions, each containing about the same number of points. These subregions correspond to the sons of δ. The important property of conjugation trees is that any line intersects with the region of at most $O(n^\beta)$ nodes, where $\beta = \log_2(1 + \sqrt{5}) - 1 < 0.695$. As a result one can efficiently determine regions that are completely contained inside the query halfplane or lie completely outside the halfplane. In this way only $O(n^\beta)$ nodes have to be visited when performing a query. A conjugation tree can be constructed in $O(n \log n)$ time. (For details see [9].)

In this paper, we will use conjugation trees to store disks rather than points. Let $\mathcal{D} = \{D_1, \ldots, D_n\}$ be a set of disks, and let $\mathcal{P} = \{p_1, \ldots, p_n\}$ be the set of their centers. We construct a conjugation tree T on $\mathcal{P}$. For each disk D, we search with its center p in the tree T and locate the highest node δ on the path, such that the bisecter l_δ intersects D. We store D in some type of associated structure S_δ at this node. This associated structure S_δ depends on the type of query we want to answer. Each disk is stored in the structure S_δ for exactly one node δ in the tree.

Storing the disks in the way described above has an important property: Whenever a line l intersects a disk D stored in S_δ, then l intersects the region associated with δ. This easily follows from the fact that D is stored at the highest possible node. As a result D will lie completely inside the region associated with δ. Hence, l must intersect the region if D intersects l. Thus, all disks intersecting a line l can be determined by querying the S_δ structures at only $O(n^\beta)$ nodes. Let $R(n)$ denote the query time of the associated structure storing n disks. It easily follows from [9] that the total query time of the structure is given by the recurrence $Q(n) \le Q(n/2) + Q(n/4) + R(n)$. Also, if the storage requirement of an associated structure is $M(n)$, then the whole structure needs $O(n) + O(M(n))$ storage.

The following lemma will be useful in estimating time bounds for the data structures we obtain:

Lemma 1 *Let T be a data structure storing n objects, having a query time of $Q(n)$. If $Q(n)$ satisfies $Q(n) \le Q(n/2) + Q(n/4) + c \cdot n^\eta$ for constants $c > 0$ and $\eta < \beta$, then $Q(n) = O(n^\beta)$ (where $\beta = \log_2(1 + \sqrt{5}) - 1$).*

Proof. Claim: $Q(n) \le c'(n^\beta - n^\mu)$, where $\eta < \mu < \beta = \log_2(1 + \sqrt{5}) - 1$, and c' is some fixed constant.

$$
\begin{aligned}
Q(n) &\le Q(n/2) + Q(n/4) + c \cdot n^\eta \\
&\le c'((n/2)^\beta - (n/2)^\mu) + c'((n/4)^\beta - (n/4)^\mu) + c \cdot n^\eta \\
&\le c' \cdot n^\beta((1/2)^\beta + (1/4)^\beta) - c' \cdot n^\mu((1/2)^\mu + (1/4)^\mu) + c \cdot n^\eta \\
&\le c' \cdot n^\beta - c' \cdot n^\mu - c' \cdot \zeta \cdot n^\mu + c \cdot n^\eta \quad \text{(for a positive constant } \zeta) \\
&\le c'(n^\beta - n^\mu) \quad \text{(for } c' > c/\zeta).
\end{aligned}
$$

$\square$

3 Intersection queries in disjoint disks

We will first consider intersection queries with a line in a set of disjoint disks. We are given a set $\mathcal{D} = \{D_1, \ldots, D_n\}$ of n non-intersecting disks of arbitrary size. We wish to preprocess $\mathcal{D}$ for the following type of query: Given a query line l, report all disks in $\mathcal{D}$ that intersect l.

To solve the problem, we use the conjugation tree for disks, as described in the previous section. So we only have to describe the associated structure S_δ. Note that all disks stored in S_δ intersect the bisecter l_δ. Let $p = l_\delta \cap l$ be the intersection point of the query line and l_δ. We first describe a solution to a special case in which all disks intersect l_δ on the same side of p, and then extend it to the general case.

Suppose a set $\mathcal{D}'$ of disks $D_1, \ldots, D_m$ is given, all of which intersect the fixed line l_δ (which we assume to be horizontal), and we perform an intersection query with a line l which intersects l_δ in a point p, left of all intersections of disks with l_δ (see Figure 1). Let l_δ^+ (resp. l_δ^-) denote the halfplane lying above (resp. below) l_δ. Define the sets $\mathcal{D}'^+ = \{D_i^+ \mid D_i^+ = D_i \cap l_\delta^+, 1 \le i \le m\}$ and $\mathcal{D}'^- = \{D_i^- \mid D_i^- = D_i \cap l_\delta^-, 1 \le i \le m\}$. $D_i^+ \ne D_i$ (or D_i^-) is the intersection of a disk with a halfplane, which we will call a *disk part*. We describe how to preprocess $\mathcal{D}'^+$; $\mathcal{D}'^-$ can be preprocessed similarly.

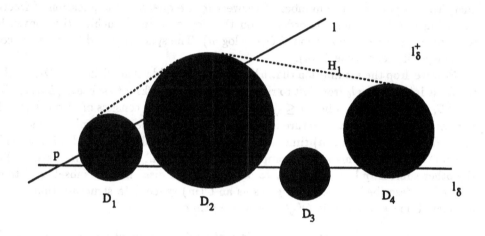

Figure 1: Disjoint disks intersected by a line

Compute the convex hull H_1 of $\mathcal{D}'^+$. The boundary of H_1 consists of a sequence of circular arcs and connecting line segments. For all disk parts of $\mathcal{D}'^+$ that do not contribute to H_1, again compute the convex hull H_2, and repeat it until all disk parts contribute to some convex hull. The convex hulls $H_1, \ldots, H_j$ are called *convex layers* of $D_1^+, \ldots, D_m^+$.

Lemma 2 *The convex layers $H_1, \ldots, H_j$ have the following properties with respect to a line l:*

- *if $l \cap H_i = \emptyset$, then $l \cap H_j = \emptyset$ for all $j \ge i$,*

- *the disk parts of H_i that intersect l are consecutive in the ordering along the boundary of H_i,*

- *if $l \cap H_i \ne \emptyset$, then there is a disk part contributing to H_i which intersects l.*

Proof. The first part follows because H_j, for all $j > i$, lies in the interior of H_i.

To prove the second part, suppose that there are three disk parts D_1^+, D_2^+, D_3^+ in some layer H_i, which intersect l_δ in this order from left to right, and that l intersects only D_1^+

and D_3^+. Then D_2^+ lies inside the convex hull of D_1^+ and D_3^+, and therefore does not contribute to H_i, a contradiction.

For the third part, assume that l intersects two edges h_1 and h_2 of H_i. If h_1 or h_2 is the boundary of some disk part, we are done, so assume that both h_1 and h_2 are line segments connecting circular arcs of H_i. Obviously, $h_1, h_2 \subset l_\delta^+$, $h_1 \neq h_2$, and there must be a circular arc γ between h_1 and h_2 on H_i. Since the disk part D^+, whose boundary contains γ, is also incident upon l_δ, l has to intersect D^+. □

In view of the above lemma, the disk parts of $\mathcal{D}'^+$ intersecting l can be reported as follows. Let $i = 1$. If l does not intersect H_i, we are done, otherwise report the answers of H_i, and continue with $i = i+1$. Similarly, we can search in $\mathcal{D}'^-$. Observe that no disk is reported twice.

A straightforward implementation of the search algorithm leads to $O((k+1)\log m)$ query time, where k is the number of answers to the query. An application of fractional cascading (see [3]), using the ordering on the slopes of lines touching the convex hulls, leads to an improved query time of $O(k + \log m)$. The space required to store the convex layers is easily seen to be $O(m)$.

Now we drop the assumption that p lies on the same side of all disks. Let $D_1, \ldots, D_m$ be the disks intersecting l_δ from left to right. Partition the disks into subsets $\{D_1, \ldots, D_{\sqrt{m}}\}$, $\ldots, \{D_{i\sqrt{m}+1}, \ldots, D_m\}$, where $i \leq \sqrt{m}$. Hence, each subset consists of at most $\sqrt{m}$ disks. For each subset, build a structure as described above. When performing a query with a line l, determine in $O(\log m)$ time, the subset in which disks intersect l_δ on both sides of $p = l_\delta \cap l$. Traverse this subset completely and report all disks that intersect l. For all the other subsets, p lies completely to one side of the disks in that subset, and they are queried as described above. This gives us an $O(m)$ space data structure that solves the associated query problem in $O(\sqrt{m}\log m + k)$ time.

Theorem 1 *Let $\mathcal{D}$ be a set of n non-intersecting disks in the plane. There exists a linear space data structure that stores $\mathcal{D}$, such that all k disks in $\mathcal{D}$ intersecting a query line can be reported in time $O(n^\beta + k)$.*

Proof. Construct the conjugation tree for disks and the associated structures, as described above. It uses linear space, as each disk is stored in the associated structure S, of only one node, and S_δ requires only linear space.

All associated structures that may return non-empty sets are visited, because a query line that avoids some region of the conjugation tree will never intersect a disk lying completely in that region. Correctness in the associated structures follows from the first and the second part of Lemma 2, and the above discussion.

The bound on query time follows from the third part of Lemma 2, the above discussion, and Lemma 1. □

Let us now consider queries with a line segment rather than with a line. The key to our solution is the following lemma.

Lemma 3 *Let s be an arbitrary line segment in the plane, and let R be the infinite strip containing s, whose bounding lines are perpendicular to s, and contain the endpoints of s. Then any disk D that intersects s, either contains one of the endpoints of s, or has its center in R, or both.*

Proof. Let s and R be as in the lemma. Suppose there is a disk D that intersects s, but does not have its center m in R. Let q_1 and q_2 be the endpoints of s, and let q be an intersection point of s and D. Since m is not in R, it is easy to see that the point of s closest to m is one of its endpoints, say say q_1. Then the distance from m to q_1 is not greater than the distance from m to any point on s, in particular, to q. Thus q_1 is contained in D. □

The lemma shows that if we want to report all disks in a set that intersect a line segment s, it suffices to report the disks that: (i) contain one of the endpoints of s, and (ii) intersect the line containing the segment s and have their centers in the strip R. Notice that these two sets of disks are not disjoint, but it is easy to circumvent reporting a disk more than once, and we leave the details to the reader.

The set $\mathcal{D}$ of non-intersecting disks $D_1, \ldots, D_n$ partitions the plane into $n + 1$ regions. Add to this partition n horizontal line segments, starting at the leftmost point of every disk, and extending leftward until it hits another disk. Similarly, add horizontal segments at the rightmost points of the disks. This results in a subdivision of the plane into $O(n)$ regions, which is monotone in the horizontal direction. Preprocess this subdivision for efficient point location, choosing one of the methods that works for circular arcs too, such as Cole's [5], or the structure of Edelsbrunner et al. [8]. This solves the problem of finding the disks that contain the endpoints of the query segment s in $O(n)$ space and $O(\log n)$ query time (there are at most two answers).

To find the other disks that intersect s, the following structure is used. Let T be a standard conjugation tree on the centers of the disks to be stored, such that every center of a disk is stored with one leaf. Every internal node corresponds to a set of centers that lie in the subtree rooted at this node. Choose a small constant $\epsilon > 0$. For every internal node δ on the level $i \cdot \lfloor \frac{1}{2} \epsilon \log n \rfloor$, $1 \leq i \leq \lfloor 2/\epsilon \rfloor$, associate with δ a conjugation tree for the disjoint disks to which δ corresponds, as described above. This main tree is used to extract the disks with their center in the strip R. The other internal nodes have no associated structure. (This technique of building hierarchical structures is based on ideas of Dobkin and Edelsbrunner [6].)

A query with the line segment $s = \overline{q_1 q_2}$ is performed as follows. Let l_1 and l_2 be the lines perpendicular to s, containing q_1 and q_2 respectively. Search with l_1 and l_2, by recursively continuing the search in the children of a node δ if the corresponding region is intersected by l_1 or l_2. If the region associated with δ lies completely on one side of both l_1 and l_2, then return from recursion. If the region lies completely between l_1 and l_2, then descend from δ to the first (highest) level in the tree where associated structures are stored. Search in these associated structures with the line containing s as the query line.

Theorem 2 *Let $\mathcal{D}$ be a set of n non-intersecting disks in the plane. $\mathcal{D}$ can be stored in a linear size data structure, such that all k disks in $\mathcal{D}$ intersecting a query line segment can be reported in $O(n^{\beta+\epsilon} + k)$ time, for any $\epsilon > 0$.*

Proof. The tree described above, together with the point location structure, yields the bounds. The point location structure takes care of the disks containing an endpoint of the query segment, and the tree T is used to report the other disks. In the main tree all disks are selected for which the center lies between l_1 and l_2. For these disks, searching with the line segment or the line containing this segment gives the same answers (cf. Lemma 3).

To prove the space bound, observe that every $\lfloor \frac{1}{2}\epsilon \log n \rfloor^{th}$ level of the tree stores each disk in exactly one associated structure. Consequently, the total space requirement is $O(2n/\epsilon) = O(n)$.

The query with the lines l_1 and l_2 on the main tree select $O(n^\beta) \cdot O(2^{\lfloor \frac{1}{2}\epsilon \log n \rfloor}) = O(n^{\beta+\epsilon/2})$ nodes of the main tree in which the associated structure is searched. The total query time (excluding the number of answers) $Q(n)$ satisfies $Q(n) \leq Q(n/2) + Q(n/4) + O(n^{\beta+\epsilon/2}) = O(n^{\beta+\epsilon})$ (cf. Lemma 1), for any $\epsilon > 0$. □

4 Intersection queries in intersecting disks

We will now generalize our algorithms to sets of arbitrary (possibly intersecting) disks. Again we will first consider the case of queries with a line and then extend it to line segments. The solutions we obtain are based on a quite different approach. We will use transformations that map disks to points in three-dimensional space, and then apply known techniques to answer the query. We are given a set $\mathcal{D}$ of n possibly intersecting disks in the plane, and we want to report all intersections between $\mathcal{D}$ and a query line l.

Let φ be a function that maps a disk D in the plane with center (a, b) and radius r to a point in $\mathbb{R}^3$ such that

$$\varphi(D) = (a, b, r).$$

Let $\mathcal{D}^* = \{\varphi(D) \,|\, D \in \mathcal{D}\}$ be the set of n points in $\mathbb{R}^3$. For a given line l define the wedge W_l as follows. If l is the vertical line $x = c$, then

$$W_l = \{(x, y, z) \,|\, z \geq |x - c|\}.$$

On the other hand, if l is the line $y = mx + c$, then

$$W_l = \left\{(x, y, z) \,|\, z \geq \left| \frac{mx - y + c}{\sqrt{m^2 + 1}} \right| \right\}.$$

Lemma 4 *A disk D intersects a line l if and only if $D^* \in W_l$.*

Proof. Follows from elementary geometry. □

Thus the problem reduces to reporting the points of $\mathcal{D}^*$ lying inside the wedge W_l. Chazelle and Welzl [4] have shown that a given set of n points in $\mathbb{R}^3$ can be preprocessed into a data structure of size $O(n \log n)$ so that all k points lying inside a query tetrahedron can be reported in time $O(n^{2/3} \log^2 n + k)$. Clearly, the wegde is a special case of a tetrahedron. Therefore we have

Theorem 3 *Given a set $\mathcal{D}$ of n disks in the plane, one can preprocess $\mathcal{D}$ into a data structure of size $O(n \log n)$, so that given a query line l one can report all k disks intersecting l in time $O(n^{2/3} \log^2 n + k)$.*

Let us now consider queries with a line segment s. Using Lemma 3 we can again split the problem into two subproblems: (i) find the disks that contain an endpoint of s, and (ii) find the disks with center in the strip R perpendicular to s, which intersect the line l containing s. To solve the second subproblem, we transform the problem to $\mathbb{R}^3$, and let the line segment s lie in the xy–plane. Let H_s be the vertical slab in $\mathbb{R}^3$ that contains the segment s, and is bounded by vertical planes normal to s and passing through the endpoints of s. Define $\triangle_s = W_l \cap H_s$. It is easy to see that a disk D whose center lies in R intersects s if and only if $\varphi(D) \in \triangle_s$. Since $\triangle_s$ is bounded by four planes, we can use Chazelle and Welzl's result [4] to report all k_1 points of $\mathcal{D}^* \cap \triangle_s$ in time $O(n^{2/3} \log^2 n + k_1)$ using $O(n \log n)$ space.

Next we describe how to solve the first subproblem: given a set $\mathcal{D}$ of disks, store them so that for a query point p, the disks of $\mathcal{D}$ containing p can be reported efficiently. Let φ' be a function that maps a disk D with center (a, b) and radius r to the point $(a, b, a^2 + b^2 - r^2)$ in $\mathbb{R}^3$. Let $\hat{\mathcal{D}} = \{\varphi'(D) | D \in \mathcal{D}\}$. Also, map a point $p = (c, d)$ to the plane $\hat{p} : z = 2cx + 2dy - (c^2 + d^2)$. Now the following lemma holds:

Lemma 5 *A point $p \in D$ if and only if the point $\varphi'(D)$ lies below the plane $\hat{p}$.*

Using the results of Chazelle and Welzl [4], we obtain a data structure of size $O(n \log n)$ that reports, in time $O(n^{2/3} \log^2 n + k_2)$, all k_2 points of $\hat{\mathcal{D}}$ lying below a query plane. Combining this with the result for subproblem (ii) we obtain

Theorem 4 *Given a set $\mathcal{D}$ of n (possibly intersecting) disks in the plane, we can construct a data structure of $O(n \log n)$ size, such that for a query segment s, all k disks of $\mathcal{D}$ intersecting s can be reported in time $O(n^{2/3} \log^2 n + k)$.*

5 Ray shooting in a set of disks

A ray shooting query in a set $\mathcal{D}$ of disks asks for the first disk hit by a ray in a given direction d, emanating from a given point q. Such a query is well-defined only when the starting point q does not lie in any disk of $\mathcal{D}$. So in the sequel we assume that the starting point q for the shooting query lies outside the union of the disks. This can easily be tested in $O(\log n)$ time by building a point location structure on the union of the disks, which has linear size (see [8], [12]). We use the following lemma to simplify the problem:

Lemma 6 *Given a set D of (possibly intersecting) disks, and a ray shooting query with point q and direction d, then the first disk that is hit (i.e., the answer to the query) has its center in the halfplane H, which is bounded by the line through q and perpendicular to d, and contains the query ray.*

Proof. Immediate by Lemma 3. □

To solve the ray shooting problem, we construct a two-level data structure. The first level is used to extract the disks that have their centers in the correct halfplane. The second level is used to determine the first disk hit when we shoot in these disks from 'minus infinity' along the line containing the query ray.

As in the previous sections, we build a conjugation tree T on the set of centers of the disks, and every node δ on a level $i \cdot \lfloor \frac{1}{2}\epsilon \log n \rfloor$ has an associated structure, to be described below, for shooting from infinity in the disks whose centers lie in the subtree rooted at δ. To perform a shooting query we first determine the correct nodes whose associated structures contain the disks with center in the halfplane H and perform a shooting query from infinity in the associated structures.

Shooting from infinity asks for the first disk intersected by a directed line l. To solve this problem we store all disks in a conjugation tree as described in Section 2, that is, we construct a conjugation tree on the set of centers and a disk D is associated with the highest node δ where l_δ intersects D. Let $\mathcal{D}_\delta$ be the set of disks associated with the node δ. We now visit each node δ whose region intersects l and, using the associated structure S_δ, determine the first disk of $\mathcal{D}_\delta$ hit by l.

It remains to describe the structures S_δ. We will show that S_δ requires linear storage and answers a query in time $O(\log^2 n)$. A similar analysis as in Section 3 will show that this leads to a linear size data structure with query time of $O(n^{\beta+\epsilon})$, for any $\epsilon > 0$.

All disks in $\mathcal{D}_\delta$ intersect l_δ. Without loss of generality we assume that l_δ is horizontal. Let l be the query line. Assume that l intersects l_δ in a point p and that the starting point q of the ray lies above l_δ. (The special case that l does not intersect l_δ can easily be dealt with.) Now there are two cases: either l intersects a disk above l_δ or not. To determine the first case, let C be the outer-contour of the set of disks plus the halfplane below l_δ. C is a (infinite) polygon with circular arcs and straight line segments (also called a splinegon) of linear complexity. See Figure 2 for an example. Shooting inside a splinegon can be done relatively easy in time $O(\log^2 n)$ using $O(n \log n)$ storage by adapting the shooting algorithm for simple polygons by Chazelle and Guibas [2]. Using the special nature of the splinegon involved, the storage can be reduced to linear. Details are left to the reader. If we hit a circle arc, the corresponding disk is the required answer. On the other hand, if we hit l_δ (in point p), then l does not intersect any disk of $\mathcal{D}_\delta$ above l_δ.

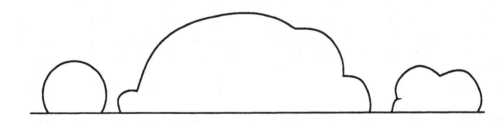

Figure 2: The top splinegon

Note that the contour of the disks below l_δ can be viewed as a number of 'pockets'. The pockets are the connected components of the boundary of the outside of the union of the disks, intersected by the halfplane l_δ^-. See Figure 3. Each of these pockets is a splinegon itself. Each of these splinegons we preprocess for shooting. We first determine the pocket p belongs to. This can easily be done in time $O(\log n)$ by binary search on the line l_δ. Next we shoot from p in the corresponding splinegon. In this way we find in time

$O(\log^2 n)$ either an answer or know that no answer exists (when we don't hit anything). Because the total size of all the splinegons is linear (because the contour of a set of circles has linear size) S_δ uses linear storage.

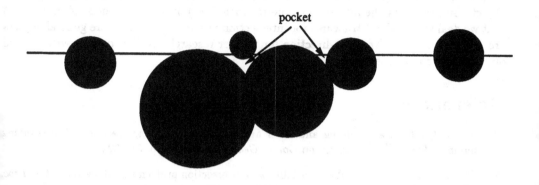

Figure 3: The pockets

Using the special properties of the splinegons obtained the shooting can in fact be done in time $O(\log n)$. Because this does not influence the total time bound we did not describe this technique. Putting all of it together, using a similar analysis as in Section 3 we conclude:

Theorem 5 *Given a set $\mathcal{D}$ of n (possibly intersecting) disks in the plane, it can be stored in a data structure of linear size, such that the first disk hit by a query ray can be determined in $O(n^{\beta+\epsilon})$ time, for any $\epsilon > 0$.*

6   Conclusions

In this paper we have presented some new data structures for storing sets of disks, which allow for efficient intersection queries. For sets of disjoint disks we obtained structures that use only linear storage and answer a query in time $O(n^{\beta+\epsilon}+k)$, where $\beta = \log_2(1+\sqrt{5})-1$, k is the number of answers, and $\epsilon > 0$ an arbitrarily small positive constant. For arbitrary disks, we constructed data structures of size $O(n\log n)$ with query time $O(n^{2/3}\log^2 n+k)$. We also presented efficient algorithms for shooting queries.

The query time of algorithms described in Section 3 and Section 5 can in fact be improved. To this end we replace the conjugation tree by a structure based on ϵ-nets as described by Haussler and Welzl [11]. This structure is also based on convex decomposition of the plane. Structures can be associated to internal nodes in a similar way as described above, although the details are a bit more involved. This improves all time bounds to $O(n^{2/3+\epsilon})$ rather than $O(n^{\beta+\epsilon})$ without increasing the space requirement.

In the preceding sections we have never mentioned preprocessing time bounds for the data structures and only talked about query time and amount of storage required. All structures however can be constructed deterministically in polynomial time.

Many directions for further research remain open. At the moment we are working on improved data structures based on spanning trees with low stabbing number (see e.g. [1]). Query times in such structures lie close to $O(\sqrt{n})$ but the storage consumption is normally a bit higher. An important open problem concerns the extension to more general objects than disks. Although some of the results presented can be extended to more general objects (in particular the intersection queries with lines), most results depend on the fact that we deal with disks. Structures for intersection queries that store more general objects are still unavailable. Another direction for further research involves queries with curved objects, e.g. circle arcs.

References

[1] Agarwal, P.K., Ray shooting and other applications of spanning trees with low stabbing number, *Proc. 5th ann. symp. on Comp. Geometry* (1989), pp. 315-325.

[2] Chazelle, B., and L.J. Guibas, Visibility and intersection problems in plane geometry, *Proc. 1st ann. symp. on Comp. Geometry* (1985), pp. 135-146.

[3] Chazelle, B., and L.J. Guibas, Fractional cascading: I. A data structuring technique, *Algorithmica 1* (1986), pp. 133-162.

[4] Chazelle, B., and E. Welzl, Quasi-optimal range searching in spaces of finite VC-dimension, *Discr. & Comp. Geometry 4* (1989), pp. 467-489.

[5] Cole, R., Searching and storing similar lists, *J. of Algorithms 7* (1986), pp. 202-220.

[6] Dobkin, D.P., and H. Edelsbrunner, Space searching for intersecting objects, *J. of Algorithms 8* (1987), pp. 348-361.

[7] Edelsbrunner, H., *Algorithms in Combinatorial Geometry*, Springer-Verlag, Berlin, 1987.

[8] Edelsbrunner, H., L.J. Guibas, and J. Stolfi, Optimal point location in a monotone subdivision, *SIAM J. Comput. 15* (1986), pp. 317-340.

[9] Edelsbrunner, H., and E. Welzl, Halfplanar range search in linear space and $O(n^{0.695})$ query time, *Inf. Proc. Lett. 23* (1986), pp. 289-293.

[10] Guibas, L., M. Overmars, and M. Sharir, *Ray shooting, implicit point location, and related queries in arrangements of segments*, Techn. Rep. No. 433, New York University, 1989.

[11] Haussler, D., and E. Welzl, ε-nets and simplex range queries, *Discr. & Comp. Geometry* (1987), pp. 127-151.

[12] Kedem, K., R. Livne, J. Pach, and M. Sharir, On the union of Jordan regions and collision free translational motion amidst polygonal obstacles, *Discr. & Comp. Geometry 1* (1986) 59-71.

[13] Overmars, M.H., H. Schipper, and M. Sharir, Storing line segments in partition trees, *BIT* to appear.

[14] Sharir, M., *Efficient algorithm for reporting disc stabbing by points in the plane*, preliminary note, 1989.

Dynamic Partition Trees*

Haijo Schipper[1]   Mark H. Overmars[2]

[1] Department of Computer Science, University of Groningen
P.O. Box 800, 9700 AV Groningen, The Netherlands
[2] Department of Computer Science, Utrecht University
P.O. Box 80.089, 3508 TB Utrecht, The Netherlands

Abstract

In this paper we study dynamic variants of conjugation trees and related structures that have recently been introduced for performing various types of queries on sets of points and line segments, like half-planar range searching, shooting, intersection queries, etc. For most of these types of queries dynamic structures are obtained with an amortized update time of $O(\log^2 n)$ (or less) with only minor increases in the query times. As an application of the method we obtain an output-sensitive method for hidden surface removal in a set of n triangles that runs in time $O(n \log n + n \cdot k^{\log_2(1+\sqrt{5})-1})$ where k is the size of the visibility map obtained.

1 Introduction

In recent years a class of new data structures, called *partition trees*, has been designed for storing sets of points to answer particular types of queries, in particular half-planar range queries in which one wants to determine which points (or their number) lie on one side of a given line (or inside a given triangle). Partition trees recursively divide the plane into smaller parts in such a way that any line intersects a relative small number of parts and that every part contains a progressively decreasing number of points. In this way one can efficiently detect large parts of the plane which lie totally inside or totally outside the query range.

The first type of partition tree was introduced by Willard [21]. His *near-ideal J-way polygon trees* have a query time for half-planar range searching of $O(n^{\log_6 4}) = O(n^{.774})$ and a worst-case preprocessing time of $O(n^2)$ (although the expected preprocessing time is $O(n \log^2 n)$). Edelsbrunner and Welzl [7, 8] improved this by introducing their *conjugation trees*. This tree has $O(n \log n)$ preprocessing time and a query time of $O(n^{\log_2(1+\sqrt{5})-1}) = O(n^{.695})$. Both structures use linear storage.

Later Haussler and Welzl [9] presented *partition trees based on ϵ-nets*. The query time for this type of tree is $O(n^{\frac{2}{3}+\gamma})$, for any $\gamma > 0$. Recently Matoušek [10] showed how to construct these trees in $O(n \log n)$ time. (Before only a randomized construction

---

*Research of the second author was partially supported by the ESPRIT II Basic Research Actions Program of the EC, under contract No. 3075 (project ALCOM).

algorithm was known.) The most recent and best results are the *optimal spanning trees, with low stabbing number*, presented by Welzl [20]. As shown by Agarwal [1] these can be constructed in time $O(n\sqrt{n}\log^{1+\omega} n)$, for some $\omega < 3.3$. The query time is $O(\sqrt{n})$ (which is optimal due to the lower bound results of Chazelle [3]).

All the structures mentioned above are static structures. However it is sometimes desirable to have a structure which allows for insertions and deletions in a reasonable efficient way. In this paper we will study techniques for obtaining dynamic partition trees. We will restrict ourselves to the dynamic version of the conjugation trees. This is done because the conjugation tree can be constructed efficiently, which is a necessity to get efficient update times using the presented techniques, and conjugation trees are simple to adapt and use for many other types of queries. We will show how to adapt the conjugation tree to a so-called δ-*conjugation tree* such that the partial rebuilding technique as introduced by Overmars [12] can be used to dynamize them. We also investigate how to apply dynamization techniques for decomposable searching problems (as introduced by Bentley [2] and Overmars [12]) to conjugation trees.

In a recent paper [14] it is shown how to store line segments in conjugation trees to perform different types of queries, like e.g. intersection queries and shooting queries. This is accomplished by associating extra data structures to nodes of the conjugation tree, storing subsets of the line segments. We will show that the techniques for dynamizing conjugation trees also apply to these augmented structures. As one of the applications we obtain an output-sensitive solution to hidden surface removal in a set of triangles.

The paper is organized as follows: In section 2 the δ-conjugation tree is introduced. It is shown that queries on such a tree can be performed in almost the same time bounds as for the original conjugation tree. The advantage of δ-conjugation trees is that they have more flexibility. In section 3 we use this extra flexibility to perform updates in the structure using the technique of partial rebuilding. The resulting structure still uses linear storage, has an amortized update time of $O(\log^2 n)$ and a query time of $O(n^{\log_2(1+\sqrt{5})-1+\gamma})$ for arbitrary small $\gamma > 0$.

Next, in section 4 we shown how the techniques for dynamizing decomposable searching problems can be used with conjugation trees. To this end we show how "weak deletions" (see [12]) can be performed on conjugation trees. The resulting structure has an insertion time of $O(\log^2 n)$ and a deletion time of $O(\log n)$ (worst-case). The query time remains of the same order of magnitude. Of course, this technique can only be applied if the searching problem we want to solve with the structure is decomposable.

As stated above conjugation trees can also be used to answer queries on line segments. To this end two augmented versions of conjugation trees are introduced in [14]: segment conjugation trees and interval conjugation trees. Section 5 shows how to obtain dynamic equivalents of these structures.

One of the applications of these dynamic structures is an output-sensitive algorithm for hidden surface removal for a set of triangles in space which we present in Section 6. Although the bound obtained is not very good, the method is interesting and might have other applications as well. Finally, in Section 7 we give some concluding remarks and state some open problems.

2   The δ-conjugation tree

Conjugation trees are based on a partition of the set of points by a line in two equal halves. To introduce more freedom in our structure we will define a δ-conjugation tree in which lines split the set of points in about equal halves. Such a splitting line is called a δ-bisector.

Definition 1 *Let S be a set of n points in the plane. Let δ be a constant with $\frac{1}{2} \leq \delta < 1$. A line l is called a δ-bisector of S if and only if the number of points to the left of l is at most δn and the number of points to the right of l is also at most δn.*

We also say that l *δ-bisects* S. Obviously, such a δ-bisector always exists. (From now on we assume that $\frac{1}{2} \leq \delta < 1$ without explicitly stating so.) A normal bisector is a $\frac{1}{2}$-bisector.

l splits S into two subsets. $Left(S, l)$ is the set of points to the left of l and $Right(S, l)$ is the set of points to the right of l (if l is horizontal replace left by above and right by below or vice-versa). When l contains points of S we denote them by $On(S, l)$. The points in $On(S, l)$ we distribute among $Left(S, l)$ and $Right(S, l)$ in the following way. Let n_l, n_r and n_o denote the number of points to the left, to the right and on l. If $n_l + n_o \leq \delta n$ then we put all point on l in $Left(S, l)$. Otherwise we put in $Left(S, l)$ the top $\delta n - n_l$ points on l and in $Right(S, l)$ the remaining points. Let s_l denote a position on l than separates the points that go to $Left(S, l)$ from those that go to $Right(S, l)$. In this way S is indeed partitioned into two subsets, each with size at most δn.

Definition 2 *Given a set S of points in the plane and a line l, another line m is called a δ-conjugate of l for S if m is a δ-bisector of both $Right(S, l)$ and $Left(S, l)$.*

A normal conjugate is a $\frac{1}{2}$-conjugate. A useful simple lemma on δ-conjugates is the following:

Lemma 2.1 *Given a λ-conjugate m of l for S, then for every $\delta \geq \lambda$, m is also a δ-conjugate of l for S.*

In particular, a $\frac{1}{2}$-conjugate is a δ-conjugate for any δ. Willard [21] has shown that a $\frac{1}{2}$-conjugate always exists, and Megiddo [11] gives an algorithm for finding such a $\frac{1}{2}$-conjugate in $O(n)$ time. Hence, we obtain the following result:

Lemma 2.2 *Given a set S of n points and a line l, for each $\frac{1}{2} \leq \delta < 1$ a δ-conjugate always exists and can be found in $O(n)$ time.*

Using the concept of conjugates we can now define our data structure which will be called a *δ-conjugation tree*.

Definition 3 *Let S be a set of n_v points in the plane. Let l_v be a δ-bisector for S. A δ-conjugation tree T for S and l_v is a binary tree, defined as follows: If $n_v = 0$ then T is an empty leaf. If $n_v = 1$ then T is a leaf containing the point. Else, let l' be a δ-conjugate of l_v for S. T consists of a root v storing l_v, s_l, n_v, $On(S, l_v)$ and n_o (the number of points in $On(S, l_v)$), a left subtree that is a δ-conjugation tree for $Left(S, l_v)$ and l', and a right subtree that is a δ-conjugation tree for $Right(S, l_v)$ and l'.*

Note that if $\delta = \frac{1}{2}$ the structure described above is the normal conjugation tree given by Edelsbrunner and Welzl [7, 8]. (In Edelsbrunner [6, pp. 345 - 348] the structure is called a ham-sandwich tree.) If $\delta > \frac{1}{2}$ the tree need not be perfectly balanced. This gives us the freedom to perform updates on it. The depth of a δ-conjugation tree however is still bounded by $O(\log n)$ as is stated the following lemma (the trivial proof of which has been omitted).

Lemma 2.3 *The depth of a δ-conjugation tree is bounded by $\lceil \log_{\frac{1}{\delta}} n \rceil = O(\log n)$.*

With each node v of a δ-conjugation tree T we can associate a part of the plane, called its *cell* or *range*, denoted as ran_v. If v is the root of T ran_v is the whole plane, else ran_v is the intersection between $ran_{father(v)}$ and the open half-plane to left or right of $l_{father(v)}$. The *boundary* of v, denoted as $bound_v$ is the closure of ran_v.

The following observations immediately follow from the definition of δ-conjugation trees.

Observation 1 *Let v be a node of a δ-conjugation tree. The following properties hold: (i) All points stored in a subtree rooted at v lie in ran_v or on $bound_v$. (ii) ran_v is a convex area. (iii) $bound_v$ is a, not necessarily closed, connected convex polygon. (iv) $bound_v$ consists of (parts of) bisectors at ancestors of v. (v) No bisector at an ancestor of v properly intersects ran_v or $bound_v$.*

Note that points can be stored more than once in a δ-conjugation tree. For some nodes v they can be part of both $On(S, l_v)$ and $Left(S, l_v)$ (or $Right(S, l_v)$). This is necessary to keep the tree balanced when updates occur. Still the tree uses only linear storage and can be constructed efficiently:

Theorem 2.4 *Let $\frac{1}{2} \leq \delta < 1$. Let S be a set of n points in the plane. A δ-conjugation tree T for S can be constructed in $O(n \log n)$ time and uses $O(n)$ storage.*

Proof. Edelsbrunner and Welzl [8] have shown how to construct a $\frac{1}{2}$-conjugation tree T in $O(n \log n)$ time. Using lemma 2.1 T is also a δ-conjugation tree.

It is obvious that T has $O(n)$ nodes. As noted above, some points are duplicated and stored more than once. It is even possible that some points are stored up to $\Omega(\log n)$ times. Note however that a point that is stored both in the subtree rooted at v and at an ancestor of v must lie on $bound_v$. Also note that a point that is stored both at v and in the subtree rooted at v must lie on l_v. But l_v can intersect $bound_v$ at most twice, since $bound_v$ is connected and convex (see the above observation). Thus at v there are stored at most two points which are also stored above v and below v in the tree. The storage bound follows. $\square$

The important property of normal conjugation trees is that any line intersects the ranges ran_v for only a small number of nodes v. This is based on the trivial observation that, whenever a line l intersects ran_v, for at least one of the four grandsons of v l does not intersect the range. The same property holds for δ-conjugation trees. We obtain the following result:

Theorem 2.5 *For every $\gamma > 0$ a δ exists such that any line crosses no more than $O(n^{\log_2(1+\sqrt{5})-1+\gamma})$ ranges of a δ-conjugation tree containing n points.*

Proof. Let $C(n)$ be the maximum number of nodes which a line l can cross in a δ-conjugation tree containing n points. Then we have

$$C(n) \leq C(\delta n) + C(\delta(1-\delta)n) + O(1).$$

If $\delta = \frac{1}{2}$ the solution to this recurrence is (see [7, 8]) $C(n) = O(n^{\log_2(1+\sqrt{5})-1})$. It seems hard to give an exact solution to the recurrence (in both n and δ) but for our purposes an estimation is good enough. Let α be such that $C(n) = O(n^{\alpha+\epsilon})$ for every $\epsilon > 0$. Such an α clearly exist. It easily follows that any α such that $\delta^\alpha + (\delta(1-\delta))^\alpha \leq 1$ will do.

Define a function $\alpha(\delta)$, such that $\delta^{\alpha(\delta)} + (\delta(1-\delta))^{\alpha(\delta)} = 1$. Because for $\frac{1}{2} \leq \delta < 1$, $\alpha(\delta)$ is continuous and monotone increasing, and $\alpha(\frac{1}{2}) = \log_2(1+\sqrt{5}) - 1$, it is clear that for any $\gamma' > 0$ there exists a δ such that $\alpha(\delta) < \log_2(1+\sqrt{5}) - 1 + \gamma'$. Now choose $\gamma' = \gamma - \epsilon$ and the lemma follows. $\quad\square$

One of the basic types of queries for which conjugation trees were designed are half-planar range queries: Given a set of points S in the plane and a line h, report those points (or their number) that lie to the left (or to the right) of h. We can store the points in a δ-conjugation tree T and answer the query in the following way (see [8]): We search with h in T. At a node v we have three possible cases. If h does not intersect ran_v and ran_v lies to the left of h we report all points in the subtree rooted at v (or we add n_v to a global counter). If h does not intersect ran_v and ran_v lies to the right of h then nothing needs to be done. Otherwise (h intersects ran_v) we continue the search at both sons of v. (Some care has to be taken when h lies on l_v. In this case we use the set $On(S, l_v)$ to find the correct answers. See [8] for details.)

Theorem 2.6 *Let S be a set of n points. For each $\gamma > 0$ there exists a δ such that in a δ-conjugation tree for S half planar range queries can be performed in $O(n^{\log_2(1+\sqrt{5})-1+\gamma} + k)$ time using linear storage, where k is the number of reported answers. Counting queries can be performed in time $O(n^{\log_2(1+\sqrt{5})-1+\gamma})$.*

Proof. Whenever we visit a node v, v is either the root or the range of its father was intersected by h. From theorem 2.5 it follows that we visit at most $O(n^{\log_2(1+\sqrt{5})-1+\gamma})$ nodes. $\quad\square$

3 Performing updates using partial rebuilding

We will now show how the δ-conjugation tree can be updated when insertions or deletions occur. The method applies the technique of *partial rebuilding* introduced by Overmars [12, Chapter 4].

Assume T is a δ-conjugation tree. We perform updates on T by simply adding or deleting the point p without rebalancing the structure. Such updates can easily be performed on δ-conjugation trees. We search with p in the tree and add it to (or delete it from) the appropriate $On(S, l_v)$ sets and insert (or delete) a leaf for the point. To be precise, an insertion proceeds as follows:

Algorithm: Insert

Suppose a point p has to be inserted in a tree $\mathcal{T}$. (p not already present in $\mathcal{T}$.) Let v be the current node, with subtree $\mathcal{T}_v$. Then there are three cases.

1. $\mathcal{T}_v$ is empty. Then a new tree is created, containing p only.

2. $\mathcal{T}_v$ is not empty, and p lies on l_v. Then if p lies above (or below) s_l, p is inserted in $On(\mathcal{S}, l_v)$ and in left subtree of v (or the right subtree of v). Also the values n_v and n_o are adjusted.

3. $\mathcal{T}_v$ is not empty, and p does not lie on l_v. Then if p lies right (left) of l_v, p is inserted in the right (left) subtree of v. Also n_v is adjusted.

The deletion method works in a similar way. Such an operation clearly takes time the depth of the tree, i.e., $O(\log n)$.

Obviously, after such an update the tree might no longer be balanced, i.e., for some nodes v (which must lie on the search path of p) either $Left(\mathcal{S}, l_v)$ or $Right(\mathcal{S}, l_v)$ might have become too large ($> \delta n_v$). We rebalance the tree by rebuilding part of it. The naïve approach would be to rebuild only the subtree below the highest node that became out of balance. Looking at the definition of the δ-conjugation tree it can be concluded that this approach is not correct. When a subtree $\mathcal{T}_v$ rooted at a node v is rebuilt its new bisector l_v^{new} will normally be different from the old bisector l_v^{old}. But because the brother of v must have the same bisector as v, the structure does no longer satisfy the definition of a δ-conjugation tree. Therefore, (if v is not the root) we rebuild both subtrees $\mathcal{T}_v$ and $\mathcal{T}_w$ for the brother w of v. To avoid that we have to rebuild them again soon we rebuild them as $\frac{1}{2}$-conjugation trees.

Such a rebuilding takes time $O(n_v \log n_v)$ because the subtree rooted at w contains about the same number of points as the subtree rooted at v. This can sometimes be very large, for example when v is the root of the tree. Partial rebuilding though is based on the fact that, once a subtree is rebuilt it takes many updates before it has to be rebuilt again (as we will show below). Hence, the amortized update time will be small.

Lemma 3.1 Let $\delta > \frac{1}{2}$. Let T be a newly built $\frac{1}{2}$-conjugation tree of n points. It takes $\Omega(n)$ updates on T before the root v will be out of balance, i.e., before one of the subtrees of v contains $> \delta m$ points, where m is the current number of points in the set.

Proof. In the worst case insertions only take place in one son of v and deletions in the other (in this way, the root goes out of balance as quickly as possible). Before any updating, both sons contain at most $\lceil \frac{n}{2} \rceil$ points. Suppose v gets out of balance after i insertions and d deletions. Define $u = i + d$ to be the number of updates. Then, because v is not longer in balance $\lceil \frac{n}{2} \rceil + i > \delta(n + i - d)$. Thus

$$\frac{n}{2} > \delta(n + i - d) - i - 1 \Rightarrow (2 - 2\delta)i + 2\delta d > (2\delta - 1)n - 1$$

As $2\delta > 2 - 2\delta$ (because $\delta > \frac{1}{2}$) we obtain

$$2\delta u > (2\delta - 1)n - 1 \Rightarrow u > \frac{2\delta - 1}{2\delta}n - 1$$

which is $\Omega(n)$ because $\delta > \frac{1}{2}$. $\qquad\qquad\qquad\qquad\qquad\qquad\qquad\qquad\qquad\qquad$ □

So whenever we rebuild the subtree of a node v (together with the subtree at the brother w of v) it take $\Omega(v_n)$ updates in $\mathcal{T}_v$ before we again have to rebuild the subtree a v. Stated differently, whenever we rebuild a subtree at v there have been $\Omega(n_v)$ updates in $\mathcal{T}_v$ at which no work needed to be done on rebuilding at node v. Hence, we can charge the $O(n_v \log n_v)$ work for rebuilding to these updates, making for $O(\log n_v) = O(\log n)$ per update.

The same update will be charged at most once from node v but it might get charged from other nodes as well. All these nodes must lie on the search path of the update. Hence, there can be at most $O(\log n)$ nodes that can charge costs to the update. So each update is charged at most $O(\log n)$ time $O(\log n)$ work. (See [12] for a more extensive description of this technique of charging costs.) We conclude:

Theorem 3.2 *For each $\delta > \frac{1}{2}$ the amortized update time in a δ-conjugation tree is bounded by $O(\log^2 n)$.*

As a result we obtain e.g. the following theorem:

Theorem 3.3 *Let S be a set of n points. For each $\gamma > 0$ there exists a data structure for storing S which uses linear storage and can be built in time $O(n \log n)$ such that half planar range queries can be performed in $O(n^{\log_2(1+\sqrt{5})-1+\gamma}+k)$ time, where k is the number of reported answers. Counting queries can be performed in time $O(n^{\log_2(1+\sqrt{5})-1+\gamma})$. Insertions and deletions can be performed in $O(\log^2 n)$ amortized time.*

4 Using decomposability

We will now apply a different dynamization technique that provides even better bounds. The draw-back of the method is that it can only be applied for a certain subclass of searching problems.

Bentley [2] was the first to define the class of *decomposable searching problems* (see also Overmars [12]). Let $PR(x, V)$ denote the answer to the searching problem PR for a query object x and a set V.

Definition 4 *[2] A searching problem PR is called* decomposable *if and only if for any partition $A \cup B$ of V and any query object x, $PR(x, V) = \Box(PR(x, A), PR(x, B))$ for some constant time computable operator $\Box$.*

For decomposable searching problems it is not necessary to keep all points in one large structure. They can be divided over a number of smaller structures and the answers can be composed using the operator $\Box$. Many searching problems are decomposable. For example range searching is decomposable, using the operator $\Box = \cup$, which can be performed in constant time because the sets are disjoint.

Many general techniques are known for dynamizing structures for decomposable searching problems (see [12]). Unfortunately, these techniques only yield good insertion time bounds. To be able to perform deletions efficiently as well, Overmars [12] introduces the concept of *weak deletions*.

Definition 5 *[12] Deletions are called* weak *if and only if there exists a constant α, with $0 < \alpha < 1$ such that the routines to carry them out on a newly built (i.e. perfectly balanced) structure of n points guarantee that after $\leq \alpha n$ deletions, the query time, the amount of storage required and the weak deletion time on the resulting structure of m points are no more than a factor k_α worse than of a newly built structure of m points, for some constant k_α depending solely on α and the type of structure.*

Weak deletion can easily be carried out on conjugation trees:

Lemma 4.1 *Deletions in a conjugation tree can be performed weakly in time $O(\log n)$.*

Proof. We perform the weak deletion by simply removing the point from the conjugation tree without any rebalancing. Such a deletion takes time the depth of the tree, which is $O(\log n)$ when the conjugation tree is balanced. Take $\alpha = \frac{1}{2}$. After $\leq \alpha n$ deletions we have $m \geq \frac{1}{2}n$. Each line will still cross at most $O(n^{\log_2(1+\sqrt{5})-1})$ ranges in the tree. This should have been $O(m^{\log_2(1+\sqrt{5})-1})$. Because $\log_2(1 + \sqrt{5}) - 1 < 1$ the structure is less than twice as bad. Hence, for most types of queries, the query time becomes less than a factor 2 larger. The amount of storage required will become $O(m)$ as required and the weak deletion time remains $O(\log n)$ which is at most one larger than $O(\log m)$. Hence, the deletions are indeed weak. $\square$

Overmars [12] shows that, when a structure allows for weak deletions, it can be turned into a structure that allows for real deletions in almost the same time bounds. The technique is called *global rebuilding* and works by rebuilding the complete structure after αn deletions. For the conjugation tree this results in a deletion time of $O(\log n)$. (See [12] for a description of the technique.) The resulting structure has still $O(n \log n)$ preprocessing time and requires $O(n)$ storage.

We can now use the decomposability of the query problem PR to obtain a fully dynamic data structure. Let $Q(n)$ be the query time for PR on a conjugation tree. Assuming that $Q(n) = \Omega(n^\epsilon)$ for some $\epsilon > 0$ (as is the case for all types of queries that use conjugation trees) we can apply Theorem 7.3.3.2 of [12] to obtain

Theorem 4.2 *Let S denote a set of n points in the plane. Then there exists a data structure for S, which requires $O(n)$ storage and $O(n \log n)$ time for construction, such that PR can be solved in time $O(Q(n))$, insertions take time $O(\log^2 n)$ and deletions $O(\log n)$.*

Note that these are worst-case time bounds. Using the results in [12] also other trade-offs between query and insertion time can be obtained. Applying the above theorem to half planar range searching we obtain the following result:

Theorem 4.3 *Let S be a set of n points. There exists a data structure for storing S which uses linear storage and can be built in time $O(n \log n)$ such that half planar range queries can be performed in $O(n^{\log_2(1+\sqrt{5})-1} + k)$ time, where k is the number of reported answers. Counting queries can be performed in time $O(n^{\log_2(1+\sqrt{5})-1})$. Insertions can be performed in $O(\log^2 n)$ time and deletions in $O(\log n)$ time.*

5 Segment partition trees

Partition trees were introduced to store sets of points and answer particular queries. Recently, it was shown by Overmars, Schipper and Sharir [14] that one can adapt partition trees, by associating particular structures to the internal nodes, such that they can be used to store line segments and answer e.g. intersection queries. In this section we will show that these augmented structures can (often) be dynamized as well.

The first type of structure introduced by [14] is the *segment partition tree*. Let us briefly describe the structure (for details see [14]). Let S be a set of line segments. The segment partition tree has a conjugation tree T on the set of endpoints of line segments in S as underlying structure. At each internal node v a data structure B_v is associated, depending on the query which is to be performed, storing a subset of the line segments. Moreover we store at each node an ordinary segment tree S_v (see e.g. [16] for a description of the segment tree). Line segments are stored in the following way: For each segment s we search in the tree. At each node v one of the following cases can occur:

1. s intersects the cell ran_v and has no endpoints in the interior of ran_v. Then the part of s which intersects ran_v is stored in the associated structure B_v.

2. s lies on the splitting line l_v. In this case s is stored in the segment tree S_v.

3. Otherwise, the search is continued at those sons of v where s (partially) intersects the range.

It is easy to see that each segment will be stored in at most $O(\log n)$ B-structures and in at most 2 S-structures. Now assume the data structure B has a preprocessing time of $P_B(n)$, an (amortized) insertion time of $I_B(n)$, an (amortized) deletion time of $D_B(n)$, a query time of $Q_B(n)$ and it uses $M_B(n)$ storage. The following lemma follows easily (see [14]):

Lemma 5.1 *A segment partition tree T uses $O(M_B(n)\log n)$ storage and can be constructed in time $O(P_B(n)\log n)$.*

To dynamize segment partition trees we can adapt both of the techniques described in sections 3 and 4. Let us first look at the technique using partial rebuilding.

We replace the underlying conjugation tree T by a δ-conjugation tree. To perform an insertion of a segment s we first insert the two endpoints of s in T in the way described in section 3. This might cause part of the tree to be rebuilt as a perfectly balanced segment partition tree. Next we search with s in the tree in the way described above and insert s into the at most $O(\log n)$ B-structures and the at most 2 S-structures. This takes total time $O(I_B(n)\log n)$. (Segment trees allow for insertions and deletions, using partial rebuilding, in $O(\log n)$ amortized time. These bounds can also be obtained as worst-case time bounds.)

If a line segment s has to be deleted we first delete it from all the B- and S-structures it appears in. Next we remove the two endpoints from T, possibly rebuilding some subtrees.

Lemma 5.2 *Insertions in a segment partition tree can be performed in amortized time $O(I_B(n)\log n + \frac{P_B(n)}{n}\log^2 n)$. Deletions take $O(D_B(n)\log n + \frac{P_B(n)}{n}\log^2 n)$ amortized time.*

Proof. Following the same arguments as in section 3 the insertion of the endpoints in the underlaying structure and the possible rebuilding of subtrees take amortized time $O(\frac{P_B(n)}{n}\log^2 n)$ per insertion. Inserting the segment in the at most $O(\log n)$ B-structures takes time $O(I_B(n)\log n)$. Inserting the segment in the at most 2 segment trees takes time $O(\log n)$. The bound for deletions follows in the same way. □

Most queries on segment partition trees take time $O(n^{\log_2(1+\sqrt{5})-1})$ (assuming the query time $Q_B(n)$ is bounded by $O(n^\alpha)$ for some $\alpha < \log_2(1+\sqrt{5})-1$). (See [14].) Replacing the conjugation tree by a δ-conjugation tree, such queries will take time $O(n^{\log_2(1+\sqrt{5})-1+\gamma})$ for arbitrary small $\gamma > 0$.

Let us demonstrate this with one type of queries treated in [14]: Let S be a set of non-intersecting line segments, determine those segments that intersect a particular query line segment s. In this case all segments that have to be stored at a particular node v of the segment partition tree fully intersect ran_v and do not intersect each other. Note that they all intersect the edge of $bound_v$ belonging to the splitting line $l_{father(v)}$ at the father of v. We simply store the segments ordered by intersection with this edge in a tree. So B is a simple binary tree. Those segments at node v intersecting the query segment s are located by binary search on B (plus a stabbing query on S, see [14]) in time $O(\log n)$ plus the number of answers found. So $Q_B(n) = O(\log n)$, $I_B(n) = D_B(n) = O(\log n)$, $P_B(n) = O(n\log n)$ and $M_B(n) = O(n)$. Applying the dynamization method we obtain the following result:

Theorem 5.3 *Let S be a set of n non-intersecting line segments in the plane. Then for every $\gamma > 0$ there exists a data structure for storing S which uses $O(n\log n)$ storage and which can be constructed in $O(n\log^2 n)$ time such that line segment intersection queries can be performed in time $O(n^{\log_2(1+\sqrt{5})-1+\gamma}+k)$, where k is the number of reported answers, and updates can be performed in $O(\log^3 n)$ amortized time.*

Proof. The preprocessing time, amount of storage required and update time follow from lemma 5.1 and 5.2. The query time bound follows in the following way: For each γ there exists a δ such that in a δ-conjugation tree each line l passes through the ranges for at most $O(n^{\log_2(1+\sqrt{5})-1+\gamma})$ nodes. Clearly, only associated structures in nodes for which the query segment s passes through the range have to be checked. Carefully adding up the query times in the associated structures for these nodes yields the desired result (see [14]). □

By constructing the tree in a slightly more careful way, the preprocessing time can be reduced to $O(n\log n)$ (see [14]). In this way the update time is reduced to $O(\log^2 n)$. Similar results can be obtained for the other types of queries treated in [14], like e.g. triangle stabbing queries.

A second possibility is to apply the techniques for decomposable searching problems to segment partition trees. To this end we have to show how weak deletions can be performed on segment partition trees. This is rather easy. We simply remove the segment s to be deleted from all associated B- and S-structures it belongs to. These structures can be determined in time $O(\log n)$ and the actual deletions will take time $O(D_B(n)\log n)$. Now applying Theorem 7.3.3.2 of [12] we obtain

Theorem 5.4 *Let PR be a decomposable searching problem that can be solved using a segment partition tree with query time $Q(n)$. Then there exist a dynamic structure for solving PR with the following performances:*

storage	$O(M_B(n)\log n)$
preprocessing	$O(P_B(n)\log n)$
query time	$O(Q(n))$
insertion time	$O(\frac{P_B(n)}{n}\log^2 n)$
deletion time	$O(\frac{P_B(n)}{n}\log n + D_B(n)\log n)$

These are worst-case update time bounds. We can apply this to the segment intersection query problem which is obviously decomposable.

Theorem 5.5 *Let S be a set of n non-intersecting line segments in the plane. Then there exists a data structure for storing S which uses $O(n\log n)$ storage and which can be constructed in $O(n\log^2 n)$ time such that line segment intersection queries can be performed in time $O(n^{\log_2(1+\sqrt{5})-1} + k)$, where k is the number of reported answers, and insertions can be performed in time $O(\log^3 n)$ and deletions in time $O(\log^2 n)$.*

Again we can reduce the preprocessing time to $O(n\log n)$ which improves the insertion time to $O(\log^2 n)$.

The second type of structure for storing line segments defined by Overmars, Schipper and Sharir [14] is the *interval partition tree*. Interval partition trees have also a conjugation tree T on the set of endpoints of the segments as main structure and nodes contain associated structures, storing subsets of the segments. But this time segments are stored in a different way. Each segment s is stored at most twice in the tree. We determine the highest node v such that s intersects l_v. If s lies on l_v we store s in an ordinary interval tree S_v associated with v. (See e.g. [5, 16] for a description of interval trees.) Otherwise we store at both sons the part of s that lies inside its range in a B-structure, depending on the type of problem. The advantage over segment partition trees is that the amount of storage used will be reduced. Unfortunately, the structure can not be used for all searching problems.

Again one can apply both partial rebuilding (using a δ-conjugation tree as underlying structure) or decomposability (when the searching problem is decomposable) to dynamize these structures. Details are left to the reader.

6 Hidden surface removal

In this section we will apply the dynamic segment partition tree to solve the hidden-surface removal problem: Given a set V of triangles in space, compute which parts of the triangles are visible when looking from a particular viewpoint. Assuming that there is no cyclic overlap among the triangles, a simple transformation turns the problem into an instance where the triangles lie parallel to the xy-plane and the viewpoint lies at $z = -\infty$.

The basic idea behind the method is the following: We add the triangles one by one, starting at the bottommost triangle. At each step we keep track of the contour C of the

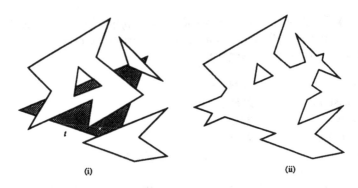

Figure 1: (i) determine the parts of t outside the contour and (ii) update the contour

triangles treated so far. For the next triangle t we compute the parts outside $\mathcal{C}$, which are visible and can be reported. Next we update the contour. See figure 1 for an illustration.

So we need a dynamic structure for maintaining the contour. The contour $\mathcal{C}$ consists of a number of polygonal regions, possibly with holes. The boundary of the contour consists of a set of non-intersecting line segments. These line segments we store in the dynamic segment partition tree $\mathcal{T}$ used in theorem 5.5 such that we can efficient determine the segments intersected by the boundary of the triangle t. We also store the vertices of the contour in a dynamic structure $\mathcal{T}'$ for half planar range searching as given in theorem 4.3. This structure can also efficiently determine those points that lie inside a given triangle. Finally we store $\mathcal{C}$ as a dynamic structure $\mathcal{P}$ for planar point location as described in Overmars [13] (or Preparata and Tamassia [17]) that allows for updates in time $O(\log^2 n)$ while we can determine in time $O(\log^2 n)$ whether a given point lies inside the contour or not.

In the beginning all these structures are empty. Let t be the next triangle to treat. For each edge s of the triangle we use the structure $\mathcal{T}$ to determine the intersections of s with the contour. From the list of intersections we can easily decide which parts of s are visible. If s does not intersect any edge of the contour, we search with one of its endpoints in $\mathcal{P}$ to see whether s lies completely inside the contour (in which case it is not visible) or completely outside. In this way we find all parts of the boundary of t that are visible.

To update the structures we proceed as follows. All parts of the edges that are visible must be inserted in $\mathcal{T}$ and $\mathcal{P}$ and their endpoints in $\mathcal{T}'$. From the query we also find those edges of the contour that are intersected by the triangle boundary. These must be shortened to the piece(s) outside t. This can be done by deleting them and inserting the reduced pieces. Finally we have to remove those edges of the contour that lie completely inside t. To this end we perform a query with t on $\mathcal{T}'$ to find the vertices inside t and, hence, the edges inside t. We simply delete these edges from all the structures. It can easily be seen that in this way the contour is correctly updated.

Theorem 6.1 *The hidden surface removal problem for a set of n triangles (without cyclic overlap) can be solved in time $O(n \log n + n \cdot c^{\log_2(1+\sqrt{5})-1} + k \log^2 c)$ where k is the output size and c the maximal contour size during the construction.*

Proof. Sorting the triangles by depth will take time $O(n \log n)$. The size of all the data structures is bounded by c. On each structure we perform $O(n)$ queries. The most expensive structure is the segment partition tree which has a query time of $O(c^{\log_2(1+\sqrt{5})-1} + k')$. Each reported intersection (in T) or point (in T') can be charged to part of the output visibility map. Hence, the total amount of time spend for queries is $O(n \cdot c^{\log_2(1+\sqrt{5})-1} + k)$ where k is the output size.

The total number of updates on the structures is $O(k)$. The update time on each of the structures is bounded by $O(\log^2 c)$. Hence, the total time spend on updates in $O(k \log^2 c)$.

□

The time bound can be restated as $O(n \log n + n \cdot k^{\log_2(1+\sqrt{5})-1})$ because $c \le k$ and $k \log^2 c = O(n \cdot k^{\log_2(1+\sqrt{5})-1})$ (because $k = O(n^2)$).

7 Concluding remarks

In this paper two dynamic versions of conjugation trees have been presented. The first version slightly relaxes the balance condition of the structure and uses partial rebuilding to perform updates in $O(\log^2 n)$ time. The second technique uses the decomposability of most of the searching problems for which partition trees are used and applies standard techniques from [12]. Both methods also apply to two related data structures, segment partition trees and interval partition trees, that store line segments and answer various types of queries on them. As an application we obtained an output-sensitive method for hidden surface removal. (Very recently some new, more efficient methods have been designed, see [15, 18, 19].)

Some open problems do remain. In particular, it would interesting to see in which way the techniques presented here can be applied to other partition trees. Using decomposability will in general be possible (although it is not always clear how to perform weak deletions). Partial rebuilding on the other hand requires a special type of balance criteria and a method for efficiently rebuilding subtrees that do not affect the total tree. Especially the partition trees of Haussler and Welzl [9] form an interesting candidate for this research.

References

[1] Agarwal, P.K., A deterministic algorithm for partitioning arrangements of lines and its applications, *Proc. 5th ACM Symp. on Computational Geometry*, 1989, pp. 11–22.

[2] Bentley, J.L., Decomposable searching problems, *Inform. Proc. Letters* **8** (1979), 244–251.

[3] Chazelle, B., Polytope range searching and integral geometry, *Proc. 28th Symp. on Foundations of Computer Science*, 1987, pp. 1–10.

[4] McCreight, E.M., Priority search trees, *SIAM J. Computing* **14** (1985), 257–276.

[5] Edelsbrunner, H., Intersection problems in computational geometry, Forschungsberichte F93, Inst. f. Informationsverarbeitung, TU Graz, 1982.

[6] Edelsbrunner, H., *Algorithms in combinatorial geometry*, Springer-Verlag, Berlin, 1987.

[7] Edelsbrunner, H. and E. Welzl, Halfplanar range search in linear space and $O(n^{0.695})$ query time, Forschungsberichte F111, Inst. f. Informationsverarbeitung, TU Graz, 1983.

[8] Edelsbrunner, H. and E. Welzl, Halfplanar range search in linear space and $O(n^{0.695})$ query time, *Inform. Proc. Letters* **23** (1986), 289–293.

[9] Haussler, D. and E. Welzl, Epsilon nets and simplex range queries, *Discrete Computational Geometry* **2** (1987), 127–151.

[10] Matoušek, J., Construction of ϵ-nets, *Proc. 5th ACM Symp. on Computational Geometry*, 1989, pp. 1–10.

[11] Megiddo, N., Partitioning with two lines in the plane, *J. Algorithms* **6** (1985), 430–433.

[12] Overmars, M.H., *The design of dynamic data structures*, Springer-Verlag, Berlin, 1983.

[13] Overmars, M.H., Range searching in a set of line segments, *Proc. 1st ACM Symp. on Computational Geometry*, 1985, pp. 177–185.

[14] Overmars, M.H., H. Schipper and M. Sharir, Storing line segments in partition trees, *BIT*, to appear.

[15] Overmars, M.H., and M. Sharir, Output-sensitive hidden surface removal, *Proc. 30th IEEE Symp. on Foundations of Computer Science*, 1989, pp. 598–603.

[16] Preparata, F.P., and M.I. Shamos, *Computational geometry: An introduction*, Springer-Verlag, Berlin, 1985.

[17] Preparata, F.P., and R. Tamassia, Dynamic techniques for point location and transitive closure in planar structures, *Proc. 29th IEEE Symp. on Foundations of Computer Science*, 1988, pp. 558–567.

[18] Sharir, M., and M.H. Overmars, A simple output-sensitive algorithm for hidden surface removal, *ACM Trans. on Graphics*, to appear.

[19] Sharir, M., and M.H. Overmars, An improved technique for output-sensitive hidden surface removal, Techn. Rep. RUU-CS-89-32, Dept. of Computer Science, Utrecht University, 1989.

[20] Welzl, E., Partition trees for counting and other range searching problems, *Proc. 4th ACM Symp. on Computational Geometry*, 1988, pp. 23–33.

[21] Willard, D.E., Polygon retrieval, *SIAM J. Computing* **11** (1982), 149–165.

Vol. 408: M. Leeser, G. Brown (Eds.), Hardware Specification, Verification and Synthesis: Mathematical Aspects. Proceedings, 1989. VI, 402 pages. 1990.

Vol. 409: A. Buchmann, O. Günther, T. R. Smith, Y.-F. Wang (Eds.), Design and Implementation of Large Spatial Databases. Proceedings, 1989. IX, 364 pages. 1990.

Vol. 410: F. Pichler, R. Moreno-Diaz (Eds.), Computer Aided Systems Theory – EUROCAST '89. Proceedings, 1989. VII, 427 pages. 1990.

Vol. 411: M. Nagl (Ed.), Graph-Theoretic Concepts in Computer Science. Proceedings, 1989. VII, 374 pages. 1990.

Vol. 412: L. B. Almeida, C. J. Wellekens (Eds.), Neural Networks. Proceedings, 1990. IX, 276 pages. 1990.

Vol. 413: R. Lenz, Group Theoretical Methods in Image Processing. VIII, 139 pages. 1990.

Vol. 414: A. Kreczmar, A. Salwicki, M. Warpechowski, LOGLAN '88 – Report on the Programming Language. X, 133 pages. 1990.

Vol. 415: C. Choffrut, T. Lengauer (Eds.), STACS 90. Proceedings, 1990. VI, 312 pages. 1990.

Vol. 416: F. Bancilhon, C. Thanos, D. Tsichritzis (Eds.), Advances in Database Technology – EDBT '90. Proceedings, 1990. IX, 452 pages. 1990.

Vol. 417: P. Martin-Löf, G. Mints (Eds.), COLOG-88. International Conference on Computer Logic. Proceedings, 1988. VI, 338 pages. 1990.

Vol. 418: K. H. Bläsius, U. Hedtstück, C.-R. Rollinger (Eds.), Sorts and Types in Artificial Intelligence. Proceedings, 1989. VIII, 307 pages. 1990. (Subseries LNAI).

Vol. 419: K. Weichselberger, S. Pöhlmann, A Methodology for Uncertainty in Knowledge-Based Systems. VIII, 136 pages. 1990 (Subseries LNAI).

Vol. 420: Z. Michalewicz (Ed.), Statistical and Scientific Database Management, V SSDBM. Proceedings, 1990. V, 256 pages. 1990.

Vol. 421: T. Onodera, S. Kawai, A Formal Model of Visualization in Computer Graphics Systems. X, 100 pages. 1990.

Vol. 422: B. Nebel, Reasoning and Revision in Hybrid Representation Systems. XII, 270 pages. 1990 (Subseries LNAI).

Vol. 423: L. E. Deimel (Ed.), Software Engineering Education. Proceedings, 1990. VI, 164 pages. 1990.

Vol. 424: G. Rozenberg (Ed.), Advances in Petri Nets 1989. VI, 524 pages. 1990.

Vol. 425: C. H. Bergman, R. D. Maddux, D. L. Pigozzi (Eds.), Algebraic Logic and Universal Algebra in Computer Science. Proceedings, 1988. XI, 292 pages. 1990.

Vol. 426: N. Houbak, SIL – a Simulation Language. VII, 192 pages. 1990.

Vol. 427: O. Faugeras (Ed.), Computer Vision – ECCV 90. Proceedings, 1990. XII, 619 pages. 1990.

Vol. 428: D. Bjørner, C. A. R. Hoare, H. Langmaack (Eds.), VDM '90. VDM and Z – Formal Methods in Software Development. Proceedings, 1990. XVII, 580 pages. 1990.

Vol. 429: A. Miola (Ed.), Design and Implementation of Symbolic Computation Systems. Proceedings, 1990. XII, 284 pages. 1990.

Vol. 430: J. W. de Bakker, W.-P. de Roever, G. Rozenberg (Eds.), Stepwise Refinement of Distributed Systems. Models, Formalisms, Correctness. Proceedings, 1989. X, 808 pages. 1990.

Vol. 431: A. Arnold (Ed.), CAAP '90. Proceedings, 1990. VI, 285 pages. 1990.

Vol. 432: N. Jones (Ed.), ESOP '90. Proceedings, 1990. IX, 436 pages. 1990.

Vol. 433: W. Schröder-Preikschat, W. Zimmer (Eds.), Progress in Distributed Operating Systems and Distributed Systems Management. Proceedings, 1989. V, 206 pages. 1990.

Vol. 435: G. Brassard (Ed.), Advances in Cryptology – CRYPTO '89. Proceedings, 1990. XIII, 634 pages. 1990.

Vol. 436: B. Steinholtz, A. Sølvberg, L. Bergman (Eds.), Advanced Information Systems Engineering. Proceedings, 1990. X, 392 pages. 1990.

Vol. 437: D. Kumar (Ed.), Current Trends in SNePS – Semantic Network Processing System. Proceedings, 1989. VII, 162 pages. 1990. (Subseries LNAI).

Vol. 438: D. H. Norrie, H.-W. Six (Eds.), Computer Assisted Learning – ICCAL '90. Proceedings, 1990. VII, 467 pages. 1990.

Vol. 439: P. Gorny, M. Tauber (Eds.), Visualization in Human-Computer Interaction. Proceedings, 1988. VI, 274 pages. 1990.

Vol. 440: E. Börger, H. Kleine Büning, M. M. Richter (Eds.), CSL '89. Proceedings, 1989. VI, 437 pages. 1990.

Vol. 441: T. Ito, R. H. Halstead, Jr. (Eds.), Parallel Lisp: Languages and Systems. Proceedings, 1989. XII, 364 pages. 1990.

Vol. 442: M. Main, A. Melton, M. Mislove, D. Schmidt (Eds.), Mathematical Foundations of Programming Semantics. Proceedings, 1989. VI, 439 pages. 1990.

Vol. 443: M. S. Paterson (Ed.), Automata, Languages and Programming. Proceedings, 1990. IX, 781 pages. 1990.

Vol. 445: A. J. M. van Gasteren (Ed.), On the Shape of Mathematical Arguments. VIII, 181 pages. 1990.

Vol. 447: J. R. Gilbert, R. Karlsson (Eds.), SWAT 90. Algorithm Theory. Proceedings, 1990. VI, 417 pages. 1990.